Unless Recalled Earlier

DATE DUE

JUL 1 2 1996		
OCT 1 3 1996		
NOV 1 1 1996		
AUG 1 1997		
NOV 1 7 1997 JUN 2 1998		
MAY 1 7 1999		
OCT -3 2000		
OCT 2 2 2000		
NOV 2 9 2000		
JAN -2 2000		

Demco, Inc. 38-293

Principles and Practice of Modern Chromatographic Methods

Principles and Practice of Modern Chromatographic Methods

K. Robards
School of Science and Technology
Charles Sturt University, Wagga Wagga
Australia

P. R. Haddad
Department of Chemistry
University of Tasmania, Hobart
Australia

P. E. Jackson
Waters Chromatography
Lane Cove
Australia

ACADEMIC PRESS
Harcourt Brace & Company, Publishers
London   Boston   San Diego   New York
Toronto   Sydney   Tokyo

ACADEMIC PRESS LIMITED
24–28 Oval Road
LONDON NW1 7DX

U.S. Edition Published by
ACADEMIC PRESS INC.
San Diego, CA 92101

This book is printed on acid-free paper

A catalogue record for this book is available from the British Library

ISBN 0-12-589570-4

Typeset by Keyset Composition, Colchester, Essex
Printed in Great Britain by Hartnolls Ltd, Bodmin, Cornwall

Contents

Preface

Chromatography is an established analytical procedure with a history of use spanning at least seven decades. It is, nevertheless, a continuously evolving technique with new variants and modified procedures. In many ways this has led to a plethora of terms that are confusing to the specialist and beginner alike. Most texts currently available are written for the specialist and concentrate on one particular form of chromatography. This is understandable, given the volume of information available on each of them but is not of much help to the user requiring an overview of developments across all areas of chromatography. Certainly, a unified approach is essential for the novice but should also be of assistance to the specialist suddenly faced with the need to switch from one technique to another.

The authors have taught and researched extensively in both academic and industrial areas. The intention in writing this text was to appeal to as wide an audience as possible. To the non-chemist it is hoped that this material will provide an easy-to-read overview in an area that has had a profound effect in fields as diverse as clinical chemistry, geology and food science. For many scientists engaged in these areas, their first real contact with chromatography comes when faced with an analytical problem requiring the separating power that only chromatography can provide. To the practising chromatographer involved in research or routine analyses it will provide an update in those techniques with which they are less familiar. Students will find the material suitable as an undergraduate text.

Although the book is introductory, it is comprehensive in topic coverage. The bibliography is based on on-line data base searching and provides detailed sources of information to facilitate further in-depth study and application.

The authors are indebted to Dr E. Patsalides, for contributions to Sections 7.1, 7.2.1 and 7.2.2.

Introduction and Overview 1

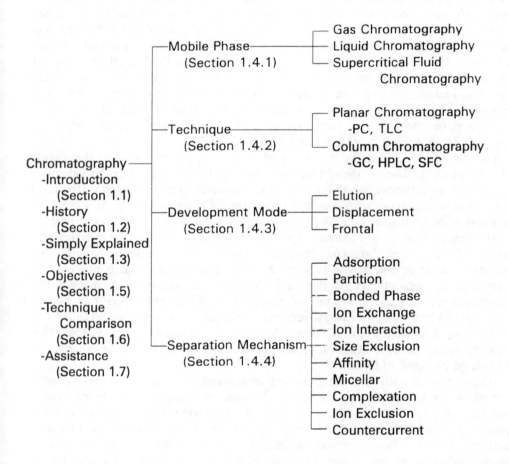

Mobile Phase
(Section 1.4.1)
— Gas Chromatography
— Liquid Chromatography
— Supercritical Fluid
    Chromatography

Technique
(Section 1.4.2)
— Planar Chromatography
    -PC, TLC
— Column Chromatography
    -GC, HPLC, SFC

Chromatography
-Introduction
    (Section 1.1)
-History
    (Section 1.2)
-Simply Explained
    (Section 1.3)
-Objectives
    (Section 1.5)
-Technique
    Comparison
    (Section 1.6)
-Assistance
    (Section 1.7)

Development Mode
(Section 1.4.3)
— Elution
— Displacement
— Frontal

Separation Mechanism
(Section 1.4.4)
— Adsorption
— Partition
— Bonded Phase
— Ion Exchange
— Ion Interaction
— Size Exclusion
— Affinity
— Micellar
— Complexation
— Ion Exclusion
— Countercurrent

1.1   Introduction

'What is chromatography?' This is a logical and seemingly simple question. Chromatography was originally developed by the Russian botanist M. S. Tswett (1872–1919) as a technique for the separation of coloured plant pigments (see Fig. 1.1). Tswett gave a very pragmatic definition [1]. The first detailed definition appears to be by Zechmeister [2] and various subsequent definitions have since been formulated. These definitions are, of themselves, unimportant. What is of interest is that with each successive definition the criteria for a process to be called chromatography have generally been liberalized. A generalized definition was provided by a special committee of the International Union of Pure and Applied Chemistry [3] which regards chromatography as '. . . a method, used primarily for separation of the components of a sample, in which the components are distributed between two phases, one of which is stationary while the other moves. The stationary phase may be a solid, liquid supported on a solid, or a gel. The stationary phase may be packed in a column, spread as a layer, or distributed as a film. . . . The mobile phase may be gaseous or liquid.'

This definition neglects the possibility of using a supercritical fluid as the mobile phase, which highlights the difficulties associated with providing an adequate definition. Nevertheless, we should not allow ourselves to be distracted by the need for a clear and concise definition, but rather regard chromatography as a group of separation methods that are undergoing continuous development and refinement.

The origins of the word 'chromatography' are no less obscure [4, 5]. In Tswett's papers, it was coined by combining two Greek words, *chroma*, 'colour' and *graphein*, 'to write' selected to indicate the individual coloured bands observed by Tswett in his separations. At the same time Tswett emphasized that colourless substances can be separated in the same way. However, it may well be that Tswett, who was involved in a bitter controversy with his peers, gave reference to the Greek words only as an excuse, for as Purnell [6] states '. . . it would be nice to think that Tswett, whose name, in Russian, means colour, took advantage of the opportunity to indulge his sense of humour.'

Irrespective of such considerations chromatography is a universal and versatile technique. It is equally applicable in all areas of chemistry and biochemistry, biology, quality control, research, analysis, preparative-scale separations and physicochemical measurements. It can be applied with equal success on the macro and micro scale. Chromatography is used industrially in the purification of such diverse materials as cane sugar, pharmaceuticals and rare earths. It is also widely used in the laboratory for the separation of minute quantities of substance, as in the initial chromatographic experiments leading to the discovery of element number 100, which involved only about 200 atoms; this surely represents one of the most remarkable achievements of modern science [7].

The importance of chromatography in science can be illustrated in a number of ways. One of these is the number of publications involving chromatography relative to the total number of science-based publications. Such a comparison shows that 3.3% or a total of approximately 8100 of citations appearing in *Chemical Abstracts* for 1989 involve chromatography, a significant proportion for any one technique. However, this is an underestimate as a significant number of citations will have

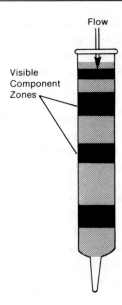

Fig. 1.1. System as used by Tswett in his original experiments. Prior to 1935 the column packing was removed from the column after use and the separated zones were extracted in order to recover the 'pure' components.

exploited chromatography as a technique but will not have referred to it in either the title or abstract. A further illustration is the fact that two Nobel Prizes in Chemistry (to A. Tiselius of Sweden in 1948 and to A. J. P. Martin and R. L. M. Synge of Great Britain in 1952) were awarded for work directly in the field of chromatography. In addition, chromatography played a vital role in work leading to the award of a further twelve Nobel Prizes between 1937 and 1972 [8].

Sales of chromatographic and related equipment have led the field in analytical instrument sales for some years and still represent an expanding market. This is illustrated by the following extract from *Trends in Analytical Chemistry* [1991, **10**: V]: 'For the 1990s shipments of Chromatographic instruments—the leading product category—are projected to increase in dollar terms from $725 million in 1990 to $990 million in 1994. Included are analytical gas, liquid, ion and supercritical fluid chromatographs as well as detectors employed in these instruments. Columns, supplies and accessories or preparatory chromatographic systems are not included.' However, the last of these is a rapidly expanding market. Thus, there is a vast amount of chromatography being performed both in academia and industry.

1.2 Historical Aspects

There is a temptation to ignore the work of the past. However, in order to exploit fully current developments, an awareness of past advances is desirable. A brief historical excursion at this point should place in perspective the development of thought and activity in a technique that has had an important fundamental and applied influence in chemistry and in other areas. This discussion represents an overview rather than a detailed and comprehensive account of the evolution of thought and practice in each chromatographic technique. The reader interested in a more detailed account of the history of chromatography is referred to the various articles published since 1970 [7–12] and to the text by Zechmeister and Cholnoky [13], although this may be difficult to obtain.

Credit for the introduction of chromatography is difficult to assign and is perhaps of academic interest only. Many natural processes, such as the underground migration of fluids through clays and sediments, can be regarded as chromatographic processes. Various ancient texts describe procedures that undoubtedly involve the unconscious use of chromatography. An instance is the conversion, described by Aristotle, of salty and bitter water into potable water with clay. However, credit for the invention of chromatography is not attributed to the observation of these natural processes. Certain investigations in the second half of the last century may be regarded as the precursors of chromatography. Prominent among these is the work of the American petroleum chemist D. T. Day (1859–1925). Zechmeister and Cholnoky [13] first drew attention to Day's activities. This was followed by a heated exchange of ideas [14–17] culminating in claims that Day was the inventor of chromatography. Significant though Day's contributions were, his role in the development of chromatography has probably been exaggerated. Indeed, the Editorial Board of the journal *Biokhimiya* [1951, **16**: 478] regarded these claims as a Western plot to 'reduce the merit of the Russian scientist M. S. Tswett.' The principles of chromatography as practised nowadays were first outlined by Tswett (for details of his life see, for example, [19–21]) on March 21, 1903 (March 8 according to the old Russian calendar in use at that time) at a meeting of the Biological Section of the Warsaw Society of Natural Sciences. This represents the first report of Tswett's systematic investigation of the chromatographic separation of plant pigments. Tswett soon became embroiled in a bitter controversy regarding his procedure, which was rejected by his contemporaries [22] as being of little merit. Kohl, an authority on carotene pigments, cited Tswett's failure to reference Kohl's own book as proof of the inaccuracy of Tswett's results [23]. However, Tswett was so aware of the importance and scope of his discovery that he insisted, although his experiments did not result in isolation of pure substances, against all opposition that chlorophyll was a mixture of two components.

Tswett's involvement with chromatography had almost ceased by 1912 and there followed a 20-year period of dormancy in which few researchers used the technique [10]. By 1930, the interest in natural substances and the need to separate and purify these provided a fertile environment for the acceptance of chromatography. The rebirth occurred in the laboratories of the Kaiser Wilhelm Institute for Medical Research at Heidelberg. Edgar Lederer [24], in a careful study of the literature, found a reference to Tswett's work and decided to apply chromatography in his own research with Richard Kuhn on carotenoids. The work was published in a series of papers [25–27] in 1931 and the 'new' technique soon spread and became accepted as a standard laboratory procedure. An important development occurred in 1937 when Schwab and Jockers [28], at the University of Munich, adapted the technique to the analysis of inorganic ions.

Following the rediscovery, uses of chromatography continued to expand and modifications and variants were introduced. In the procedure as practised before 1935, a column was packed with a suitable adsorbent (e.g. calcium carbonate, alumina) and the sample added to the top of the column. Individual components were separated by allowing a solvent (termed the mobile phase or eluent) to pass through the column. The process was stopped before the first component emerged from the bottom of the column and the column packing was slowly removed in sections and the 'pure' compounds recovered by extraction. In the second half of the 1930s it became the accepted practice to 'wash' or elute the components out of the column by continued addition of mobile phase. An important achievement was the development of a procedure [29] for continuously monitoring column effluent, by measuring its refractive index. In response to the need for faster separations (and easier detection and recovery of sample), 'open column' chromatography developed [30] in the late 1930s. In this variant, solutes were separated on a thin layer of adsorbent coated on a flat, rigid support (e.g. glass). Meinhard and Hall [31] in 1949 were the first to use starch binder to hold the adsorbent to the rigid support. Twenty years elapsed before the technique became widely accepted as thin-layer chromatography, and then only following the systematic investigations of Stahl [32] and the commercial availability of standardized adsorbents.

Ion-exchange chromatography also developed in the late 1930s when Taylor and Urey [33] separated lithium and potassium isotopes on zeolites. The real advance in this variant came with the application of synthetic ion-exchange resins [34] which became commercially available in the early 1940s. Their value was demonstrated [35] in the separation of rare earth and transuranium elements in connection with the Manhattan Project of World War II.

The decade beginning in 1941 has been termed 'The golden decade of Chromatography' by Ettre [11]. Three milestones in chromatography occurred during this period and all are associated with one person: Archer John Porter Martin. In 1941, Martin and Synge [36] developed partition chromatography in which the stationary phase was a liquid (e.g. water) retained on a solid support (e.g. silica gel). This work was carried out at the Wool Industries Research Association Laboratories in England. Partition chromatography (in columns) had a tremendous impact but was further strengthened with the development of paper chromatography [37], which was originally developed for the analysis of organic compounds but was soon extended to inorganic applications [38, 39]. The thoroughness which characterizes Martin's work is illustrated by a problem encountered in their initial work on separating amino acids by paper chromatography using a ninhydrin spray for detection. Following spraying, the purple amino acid spots were accompanied by 'pink fronts' which were traced to copper salts of amino acids formed from traces of copper dust originating from an unshielded d.c. generator [12]. Considering the advantages of simplicity and the ability to analyse several samples simultaneously, it is not surprising that paper chromatography soon became universally accepted. Sanger's use of paper chromatography [40] in 1955 to separate amino acids from insulin—the first protein to be sequenced—demonstrated clearly the immense separating power of paper chromatography.

The development of reversed-phase chromatography [41] and gradient elution [42] paved the way for the introduction of modern column chromatography. These procedures represent variants where the mobile phase is more polar than the stationary phase (reversed-phase chromatography), or the polarity of the mobile phase is continuously varied throughout the analysis (gradient elution).

The next major step was the development of gas–liquid chromatography in 1952 by James and Martin while at the National Institute for Medical Research at Mill Hill in London [43]. They demonstrated the separation of amines and carboxylic acids using a gaseous mobile phase. Their system was very simple by modern standards but achieved a dramatic improvement in separating ability relative to other techniques then in use such as fractional distillation. The new technique of gas–liquid chromatography found immediate important applications that eclipsed all other chromatographic techniques. The titrimetric detector used in the initial work suffered serious limitations, partly overcome by adoption of the thermal conductivity detector which was already known in the relatively inefficient technique of gas–solid chromatography. The most significant step in the acceptance of gas–liquid chromatography was the invention of the flame ionization detector in 1958 by McWilliam and Dewar in Australia [44] and Harley et al. in South Africa [45]. This detector provided an almost universal system for detection of organic solutes and increased the sensitivity of gas–liquid chromatography by several orders of magnitude which, in turn, enabled the use of smaller samples and more efficient columns. It was followed by introduction of the argon ionization and electron affinity detectors [46], the latter being the forerunner of the selective electron capture detector.

Principal developments thereafter were the introduction of open tubular or capillary columns by Golay [47], size exclusion on cross-linked dextran gels as a result of the work of Porath and Flodin [48] and affinity chromatography [49]. The 'open' tubular columns dramatically increased the separating power of gas–liquid chromatography and were of immediate interest to the petroleum industry [50]. However, their true potential has only recently been realized with the commercial availability of initially glass and now fused silica open tubular columns. The widespread interest in the technique of gas–liquid chromatography initiated basic research on the theory of chromatography which cross-fertilized liquid chromatography. The rekindled interest in liquid chromatography resulted in a new explosion — the development of modern liquid column chromatography, which was termed high-performance liquid chromatography [51–56]. Unlike gas–liquid chromatography, there was no single event here that heralded the introduction of the modern technique but rather there were a series of stages, each representing a small advance. The major problem with the introduction of a truly high-performance system in liquid chromatography was that theoretically desirable small size (3–10 μm) column packings [57, 58] were not available. In the early stages this was overcome by the use of pellicular packings (37–44 μm) with a thin porous surface layer (2 μm deep) of stationary phase over an inert core. These materials gave efficient separations but had limited sample capacity and their major contribution was that they led to development of pumping systems and detectors [59] essential to the next stage of development, which was the introduction of truly microparticulate packings (10 μm, 5 μm or 3 μm). In early work, problems of high back-pressure (up to 40 MPa)[1] resulted from use of long columns (50–100 cm × 1 mm i.d.) and led to

---

[1]Various pressure units are used in chromatography. Those commonly encountered, and relevant conversions are: 1 MPa = 10 bar = 9.869 atm = 145.038 psi = 10.197 kg cm^{-2}

the technique being called high-pressure liquid chromatography. Subsequently, shorter columns (25 cm, 10 cm and more recently, 3 cm) have been the norm and the preferred name for this technique is now high-performance liquid chromatography (HPLC). HPLC has now expanded enormously to the point where it has provided for some years the largest sales area for scientific equipment. Interest has recently been rekindled in the use of long, narrow-bore columns comparable with those currently in vogue in gas–liquid chromatography.

Historical developments in the theory of chromatography

Chromatography has frequently been regarded as more art than science. However, this view is far from the truth and contributions to chromatographic theory have paralleled the development of each technique. The theoretical foundations of chromatography were recognized by its discoverers. Martin and Synge, in their classic paper of 1941, presented the results of their work together with the theoretical considerations and laid the 'foundations' for the later development of gas–liquid chromatography. Furthermore, they discussed the physical factors affecting the separation and indicated that further improvements could be achieved by using very small particles as the stationary phase and a high pressure difference along the column. This advice was not acted upon until 25 years later when HPLC was developed. The 1944 paper by Martin and his group on paper chromatography also developed the theoretical framework for this new technique. In 1956 van Deemter *et al.* [60] presented their rate theory which expressed column efficiency as a function of the mobile-phase flow velocity and the characteristics of the chromatographic system such as solute diffusivity and particle diameter of the column packing. In a number of instances, theoretical considerations have been used to predict practical developments. Such was the case when Giddings [61] pointed out that small particle size packings would be essential in order to achieve efficiencies in liquid chromatography that were comparable with those in gas chromatography. This would require high column inlet pressures to achieve reasonable mobile phase flow. This led to the development of high-performance liquid chromatography.

The final paragraph in this short history cannot be written as developments in chromatography are continuing. The development of chromatography to the present time has been characterized by exponential growth, with one development fertilizing another in the areas of new equipment, column technologies, procedures and applications. As pointed out by Lochmuller [62] in an article titled 'The Future in Chromatography' the only thing certain about the future of chromatography is whatever is said will undoubtedly prove incorrect. It is equally certain that any predictions will grossly underestimate future developments and applications in this still expanding field. As an illustration, in 1948 the resolution of amino acids from a protein hydrolysate required 8 days [63]. Ten years later the separation of the 19 common amino acids could be achieved in 22 h. By 1982 the separation required less than 30 minutes with a detection sensitivity increased by several orders of magnitude. In 2002 who knows.

1.3 Chromatographic Separation Simply Explained

In chromatography, the components of a sample are separated by distribution between two phases, one of which is stationary (a solid or liquid) and the other moving or mobile (a liquid, gas or supercritical fluid). Consider a two-component mixture which is introduced at time, t_0, into a moving phase that is in contact with a second phase, the stationary phase. A continuous supply of fresh mobile phase is then provided to transport the sample components through the stationary phase. As the analytes come into contact with the stationary phase, they distribute or partition between the two phases depending on their relative affinities for the phases, as determined by molecular structures and intermolecular forces. This process is depicted in Fig. 1.2 where analyte A has a higher affinity than analyte B for the stationary phase and thus spends a greater proportion of the available time in the stationary phase. When an analyte is present in the mobile phase, it will pass through the system with the same velocity as the mobile phase, but when it is in the stationary phase its velocity will be zero. Hence, analytes with a high affinity for the stationary phase will move through the system very slowly, whereas analytes with a lower affinity will migrate more rapidly. This differential migration rate of analytes results in separation of the components as they move through the system, as shown in Fig. 1.2 at time t_1 and t_2, under ideal and real conditions.

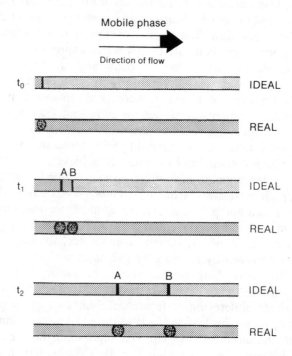

Fig. 1.2. Schematic representation of the chromatographic process showing the separation of two analytes under ideal conditions and in the situation pertaining in real systems, where both separation and spreading of analyte bands occurs during the separation process.

Even though the system is dynamic, it must be operated as close to equilibrium conditions as possible by optimizing the mobile phase velocity and designing the stationary phase to allow rapid equilibration to be achieved; i.e., the time-scale for distribution of solute molecules between phases must be rapid compared with the velocity of the mobile phase. Under these conditions the system can be characterized by a thermodynamic partition or distribution coefficient, K, which is usually expressed as the ratio of analyte concentration in the stationary phase, C_s to that in the mobile phase, C_m:

$$K = \frac{C_s}{C_m} \tag{1.1}$$

The distribution coefficient is a characteristic physical property of an analyte which depends only on the structure of the analyte, the nature of the two phases and the temperature. Phenolic solutes, for example, would be expected to form intermolecular attractions with phenolic stationary phases to a much higher degree than would hydrocarbon solutes exposed to the same stationary phase. Thus, the K value of a phenol is higher than that of a hydrocarbon of corresponding chain length in a phenolic phase. The separation of two compounds on a particular chromatographic system requires that they have different distribution coefficients. Conversely, two compounds with the same distribution coefficient will not be separated. In this case, the separation can be improved by varying the mobile phase, the stationary phase or the temperature of the system. In practice it is often difficult to predict the effects of changing the mobile phase or stationary phase and the only method is to try the change experimentally. In gas chromatography, the partition properties of the gases used as mobile phase are similar and the mobile phase is described as non-interactive, so that only the stationary phase and temperature can be varied to improve separation. The greater versatility of liquid and supercritical fluid chromatography is possible because the mobile phase is interactive and all three variables can be altered, although temperature changes are very restricted.

Equation 1.1 is an oversimplification, since K, like any thermodynamic equilibrium constant is really a quotient of analyte activities. However, in chromatographic systems we are normally dealing with solutions that tend towards infinite dilution and therefore the activity coefficient is one. This equation also assumes that the analyte is present as only one molecular structure or ion and that the analyte does not interact with other analyte molecules at infinite dilution. Considering the low levels of analytes involved, this is a reasonable assumption. The concentration profiles depicted in Fig. 1.2 as ideal are never achieved in practice. At the molecular level, various solute diffusional effects and random statistical motion of molecules causes spreading of the analyte bands, which assume the normal distribution (provided adsorptive effects are absent; discussed in later chapters) also depicted in Fig. 1.2.

1.4   Classification of Chromatography

Classification simplifies and aids the study of chromatography. Any one of several factors (stationary or mobile phase, separation process or even the type of solute, e.g. ion chromatography, protein chromatography) can serve as a basis for classification. Thus, chromatographic separations can be classified in a number of ways, depending on interests. Unfortunately, this multiplicity of overlapping classification systems, together with the diversity of chromatography as now practised, has led to a proliferation of labels which can be confusing for both the novice and specialist chromatographer alike.

1.4.1 Mobile phase

One system of classification recognizes the importance of the mobile phase and divides chromatography into three broad areas of liquid chromatography (LC), gas chromatography (GC) and supercritical fluid chromatography (SFC) (Fig. 1.3), depending on whether the mobile phase is a liquid, gas or supercritical fluid, respectively. Further classification is possible by specifying both the mobile and stationary phases leading, for example, to gas–solid and gas–liquid chromatography in which the mobile phase is a gas and the stationary phase is a solid or a liquid. More recently, supercritical fluids have been employed as mobile phases and these techniques are, at present, termed supercritical fluid chromatography irrespective of the state of the stationary phase.

Reversed-phase and normal-phase chromatography

In liquid chromatography, systems involving a polar stationary phase and a non-polar mobile phase are termed normal-phase systems. With this combination of phases, solute retention generally increases with solute polarity. Conversely, if the stationary phase is less polar than the mobile phase, the system is described as reversed-phase and polar molecules have a lower affinity for the stationary phase and elute faster. The choice of these terms is purely historical and no special significance (beyond that indicated) is attached to the use of the term 'normal'. Indeed, reversed-phase systems are far more common in liquid chromatography than normal-phase. With normal-phase systems, increasing the mobile phase polarity makes it more like the stationary phase so that the mobile phase competes more effectively with the stationary phase for solute molecules. The solute molecules therefore spend less time in the stationary phase and elute faster. Using a similar argument we predict slower elution as mobile phase polarity is increased in reversed-phase chromatography.

1.4.2 Technique

The technique refers to the equipment and operational procedures or manner in which the chromatographic process is performed. There are two broad categories, namely column and planar chromatography, but there are a number of techniques encompassed by these.

Planar and column chromatography

There are two techniques in which the stationary phase is supported on a planar surface: paper chromatography (PC) and thin-layer chromatography (TLC), which

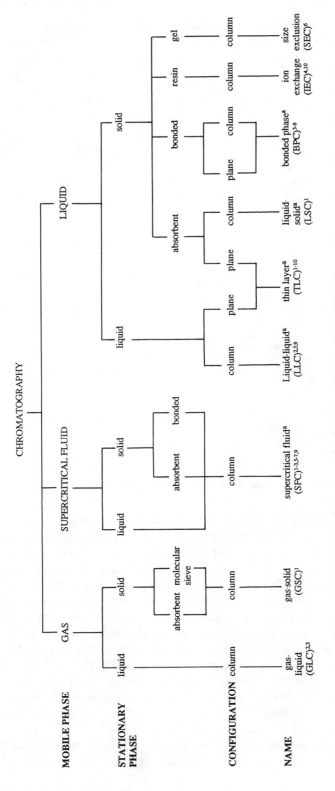

Fig. 1.3. Classification of chromatographic systems.

[a]For these techniques the combination of mobile and stationary phase can be varied to generate either a normal phase or reversed phase system. Mechanisms which have been exploited in the various techniques are identified as: [1]adsorption, [2]partition, [3]bonded phase, [4]ion exchange, [5]ion interaction, [6]size exclusion, [7]affinity, [8]micellar, [9]chelation, [10]ion exclusion.

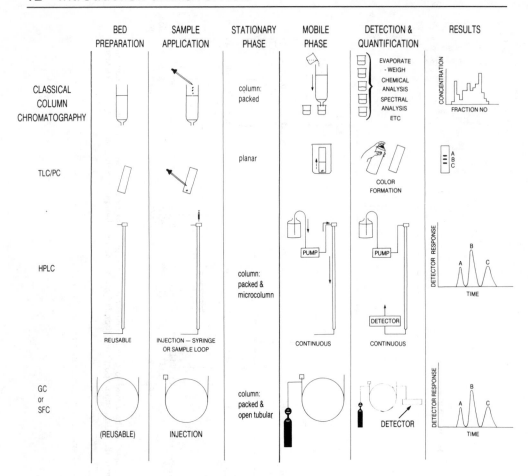

Fig. 1.4.   Comparison of chromatographic techniques.

collectively are termed planar chromatography. With PC, a sheet of paper comprises the stationary phase whereas in TLC, the stationary phase consists of a thin layer of solid spread uniformly over a flat sheet of glass, plastic or aluminium. Alternatively, the stationary phase may be packed in a closed column and the technique is referred to as column chromatography. If the stationary phase is a liquid, it must be immobilized on the thin layer or in the column and this is conveniently achieved by coating or chemically bonding the liquid stationary phase to an inert solid support, which is then packed in the column or spread in a thin layer over a flat plate. Planar procedures have been restricted to liquid chromatography because of the technical difficulties associated with confining a gas or supercritical fluid to a planar surface. In contrast to planar procedures, column chromatography is used in LC, GC and SFC. In the case of liquid chromatography, there are two variants which may be termed, because of their chronological development, classical column and modern (or high-performance liquid) column chromatography (HPLC).

Methods involving gaseous, liquid and supercritical fluid mobile phases will be treated individually in later chapters. Although there are few fundamental reasons

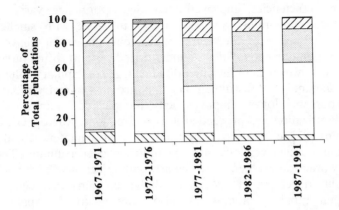

Fig. 1.5. Relative share of techniques in the total of chromatographic publications as determined by an on-line CAS data search of index terms. The total number of publications involving chromatography increased from 13 365 in 1967–1971 to 35 129 in 1987–1991. ▨, Paper chromatography; ▨, thin layer chromatography; ▢, gas chromatography; □, column chromatography; ▨, gel chromatography.

for separate treatment, it is nonetheless warranted by differences in equipment and operational procedures. For now it is sufficient to make a comparison of the different techniques on the basis of operational procedures and results, as shown in Fig. 1.4.

A comparison of the usage of the various techniques (see Fig. 1.5) is informative. Such a comparison for the years 1967–1991 shows a significant decline in the number of publications involving paper chromatography which has gradually been replaced by thin layer chromatography. The period 1967–1971 is clearly significant and it was in 1968 that modern liquid column chromatography (or HPLC) commenced its rapid growth. SFC is now in a growth phase but still remains a minor contributor to the overall number of chromatographic publications. In 1987, for example, 115 papers (approximately 0.4% of the total number of papers involving chromatography) were published on SFC [64]. It must also be remembered that publications reflect research activity rather than the importance of a technique as a routine procedure.

Column types

Column procedures may be further classified according to the nature and dimensions of the column. Conventional column procedures both in GC and LC and, more recently, SFC, exploit 'wide'-bore packed columns with internal diameters exceeding 1.0 mm. In GC the internal diameters of such columns are typically 2–4 mm and 4.6 mm in HPLC. These packed columns contain a stationary phase consisting of either a solid or a liquid coated or bonded to an inert solid support.

There are many benefits associated with column miniaturization and the first step in this direction was made in 1957 with the development of capillary columns for GC [47]. It is now realized that miniaturized separation columns, whether used in various forms of GC, LC or SFC share similar technologies and instrumental requirements [65]. In 1981, Novotny [66] identified the advantages of microcolumns as higher column efficiencies, improved detection performance, various benefits of drastically reduced flow-rates, and the ability to work with smaller samples. Priorities have changed over the years as different applications have varied the emphasis of these unique capabilities of miniaturized systems. Miniaturization is not without problems however, and the chief disadvantages of capillary columns are that they are more demanding of instrument performance, less forgiving of poor operator technique and possess a lower sample capacity than packed columns.

In GC, miniaturization has proceeded in two directions. Micropacked or packed capillary columns [67], characterized by small internal diameters, usually less than 1.0 mm, are miniaturized versions of conventional packed columns. Their use has been limited by practical problems, particularly with injection at high backpressures. In contrast, the development of open tubular columns has been immensely successful. Open tubular columns are also referred to as capillary columns. However, the characteristic feature of these columns is their openness, which provides an unrestricted gas path through the column. Hence, open tubular column is a more apt description, although both terms will undoubtedly continue to be used and can be considered interchangeable. If the stationary phase is coated or bonded directly to the internal wall of the column, then it is known as a wall-coated (WCOT) or bonded-phase open tubular (BPOT) column, respectively. The capacity of WCOT and BPOT columns can be adjusted by varying the column diameter and the film thickness of the stationary phase. Alternatives to the WCOT and BPOT column are the porous-layer open tubular column (PLOT) and surface-coated open tubular column (SCOT). The inner wall of the column is extended in PLOT columns by addition of a porous layer such as fused silica. In SCOT columns, the stationary phase is applied to a solid support, which is coated on the internal wall of the column. SCOT columns were popular because of higher sample capacity. However, the popularity of both SCOT and PLOT columns has declined in recent years because wide-bore or mega-bore (0.53–1.00 mm internal diameter) WCOT and BPOT columns are easier to use and are more stable. The various column types have been compared by Duffy [68].

Three types of microcolumn are currently in use in liquid chromatography. Microbore columns are similar in construction to conventional packed columns except that the column diameter is reduced to 1 mm. Packed microcapillaries have a column diameter of 70 μm or less and are loosely packed with particles having diameters of from 5 to 30 μm. Open microtubular columns are the equivalent of the capillary or open tubular column in gas chromatography. Ideally, they have diameters of 10–30 μm and contain a stationary phase or an adsorbent either coated on, or chemically bonded to, the column wall [69].

Virtually all separations by GC are now performed with BPOT columns whereas conventional packed columns still dominate routine separations by HPLC. The situation is less clearcut with SFC, where neither column type predominates. The development of SFC, has occurred in the last decade at a time when column technology in both GC and HPLC has been well developed. Thus, column technology for SFC has been largely borrowed from HPLC for the packed column format or from GC for the open tubular format. Generally, the diameter of open tubular columns for SFC must be smaller than 100 μm to maintain both reasonable analysis times and high resolution [70].

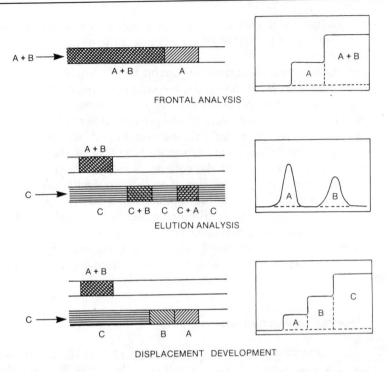

Fig. 1.6. Schematic representation of the different development modes showing the effect on migration of sample components and the resulting zone profile. A and B represent sample components and C the displacer.

1.4.3 Development mode

The various chromatographic techniques such as GC, SFC, TLC, classical column chromatography and HPLC can be performed in three different development modes; namely, elution, displacement or frontal analysis. The term 'development mode' refers to the manner in which the sample and mobile phase are applied to the stationary phase bed (column or plane) and, as shown in Fig. 1.6, the nature of the resulting peak profile, termed the chromatogram, differs between the three modes. Of the three modes elution is the most common in analysis. In this mode, the sample is applied as a compact band to the mobile phase (or eluent) followed by a continuous flow of fresh mobile phase. The individual components move through the column in the form of separate zones mixed with the mobile phase that carries them.

A less popular form of development is frontal analysis, which is useful for obtaining thermodynamic data from chromatographic measurements. In this mode the sample is swept continuously onto the column by the mobile phase during the entire course of the process. When the column becomes saturated with respect to a particular component, that component is then eluted from the column. When the zone of pure component has completely eluted, it is followed by a mixture with the next component, and so on. A complete separation cannot be achieved and the method has limited application for quantitative measurements. A typical application would be in the estimation of a trace impurity in a high purity substance, where the impurity can be concentrated in front of the main constituent, provided that it was the less strongly retained.

Displacement development is particularly useful in preparative-scale separations in column chromatography. With this mode, the sample is applied to the system as a discrete plug as in elution, but unlike elution, the mobile phase has a higher affinity for the stationary phase than any sample component. Alternatively, a substance more strongly retained than any of the components of the sample is added continuously to the mobile phase. This substance is known as the displacer and it pushes the sample components down the column. The mixture resolves itself into zones of pure components in order of the strength of retention on the stationary phase. Each pure component displaces the component ahead of it, with the last and most strongly retained component being forced along by the displacer. The record depicting the concentration of component coming from the column (Fig. 1.6) is seen to resemble the record obtained with frontal analysis but with an important difference: in displacement, the steps in the chromatogram represent pure components.

1.4.4 Separation mechanism

The nature of the interaction between sample components and the two phases forms a further basis for classification. This is perhaps the most common basis for classification. These interactions involve various physicochemical processes occurring in the system, reflecting the relative attraction and repulsion that the particles of the competing phases show for the solute and for each other. The forces involved in these interactions are usually weak intermolecular forces such as van der Waals forces or hydrogen bonding. In some instances, ionic interactions are exploited [71] and in rare cases specific interactions such as charge-transfer forces [72, 73]. This is therefore the most fundamental of all classifications but in many ways it is the most difficult since, in a number of instances, it is not clear exactly what mechanism is involved. Nevertheless, a knowledge of the mechanism is crucial to our understanding of the chromatographic process, to enable predictions about the expected behaviour of a system and in choosing a stationary phase/mobile phase combination to obtain a desired separation. In most instances it is possible to specify the predominant mechanism operating in a particular situation even though the nominated mechanism is rarely, if ever, the sole mechanism.

The mechanisms can be classified into a number of types as:

- Adsorption.
- Partition.
- Bonded phase.
- Ion-exchange.
- Ion-interaction.
- Size exclusion.
- Affinity.
- Micellar.
- Complexation.

- Ion-exclusion.
- Countercurrent.

Separations exploiting each of these mechanisms have been developed in LC, whereas GC is restricted to separations involving one or more of the first three named mechanisms. The flexibility of SFC is intermediate between the two, although it appears likely that systems will ultimately be developed that enable most of these mechanisms to be exploited.

In this chapter only sufficient detail as is necessary for an understanding of the general principles involved in each mechanism will be given. This will be supplemented in the relevant sections on GC, LC and SFC.

Adsorption chromatography

Adsorption was exploited by Tswett in the form of liquid–solid chromatography in columns and thus represents the oldest of the chromatographic techniques. In separations involving adsorption [74–77], solute and mobile-phase molecules compete for active sites on the surface of the solid stationary phase, which is called the adsorbent. A number of mechanisms [78] have been proposed to account for the adsorption process. The competition model developed by Snyder and others [75–77] assumes, in the case of mobile phases which interact with the adsorbent surface largely by dispersive and weak dipole interactions (i.e. non-polar and moderately polar mobile phases), that the entire adsorbent surface is covered by a monolayer of mobile phase molecules. Solute retention then occurs by a competitive displacement of a mobile-phase molecule from the adsorbent surface. The solvent interaction model [74] proposes the formation of mobile phase bilayers adsorbed onto the adsorbent surface. The composition and extent of bilayer formation depends on the concentration of polar solvent in the mobile phase. Solute retention occurs by interaction (association or displacement) of the solute with the second layer of adsorbed mobile-phase molecules. Adsorption onto the surface of the adsorbent is distinguished from partition processes in which the solute also diffuses into the interior of the stationary phase. A more appropriate term for the latter would probably be absorption, in which case the general process could be termed sorption. Nevertheless, the terminology used in this monograph conforms to usual practice and refers to adsorption and partition.

Separation in adsorption chromatography results from the interaction of polar functional groups on the solute with discrete adsorption sites on the adsorbent surface. The selectivity of the separation is dependent on the relative strength of these polar interactions. The extent to which a solute can be accommodated on an adsorbent surface depends on its spatial configuration and its ability to hydrogen bond with the adsorbent surface. Adsorption processes are therefore sensitive to spatial differences in solutes and are ideally suited to separations of molecules that have slight differences in shape (i.e. geometric isomers). Adsorbents also demonstrate a unique ability to differentiate solutes possessing different numbers of electronegative atoms, such as oxygen or nitrogen, or for molecules with different functional groups. This leads to the use of adsorption for class separations. Conversely, partition processes depend on a competitive solubility between two liquid phases and are quite sensitive to small differences in molecular mass. For this reason, members of a homologous series are generally best separated by a partition system. Moreover, partition is usually more suitable than adsorption for highly polar substances such as amino acids and carbohydrates. In this case, the need to use highly polar mobile phases in adsorption would negate any small differences between adsorptive properties of the solutes and produce no separation.

Adsorption still finds use in liquid column chromatography (both classical and high-performance) and is widely exploited in TLC, whereas applications of adsorption in GC are limited mainly to separations where the analytes are permanent gases. Because the stationary phase in such separations is a solid, the systems are referred to as liquid–solid chromatography and gas–solid chromatography. The practical application of gas–solid chromatography (based on adsorption) antedated the now more popular form of gas–liquid chromatography (based on partition). One of the first important accounts of gas–solid chromatography was published in 1946 by Claesson [79], who used displacement development for the separation of hydrocarbons on columns packed with activated carbon. Phillips and coworkers used the same method [80] but this approach was abandoned after 1952 in favour of partition systems such as those developed by James and Martin [43]. Nevertheless, for certain separations, namely that of permanent gases, gas–solid chromatography has come back into favour. Typical adsorbents for gas–solid chromatography are zeolites (aluminium silicates), Porapaks (cross-linked poly-styrene) and molecular sieves, in addition to the more common adsorbents such as silica gel, charcoal and alumina encountered in liquid chromatography.

The particle size is an important characteristic of an adsorbent. To a first approximation sample retention is proportional to surface area which, in turn, depends on the particle size and on the internal structure of the adsorbent particles. The smaller the average particle size, the greater the surface area of the adsorbent and hence, the number of active sites available for adsorption. Most adsorbents are available in a range of particle sizes to suit the needs of the various chromatographic techniques. For TLC, particle sizes of 20 to 40 μm have been most common, whereas for classical liquid chromatography in columns the particles are larger (100–30 μm). For the modern counterpart (i.e. HPLC) of this particular technique they are smaller (10 μm, 5 μm or 3 μm) and for high-performance TLC, particle sizes of 5 μm are used. Adsorbents for gas chromatography are typically in the size range of 125–150 μm up to 177–250 μm.

Partition chromatography

Partition chromatography originated with the Nobel Prize winning work of Martin and Synge in 1941 and has, as its basis, the partitioning of a solute between two immiscible liquids, as in solvent extraction, except that one of the liquids is held stationary on a solid support such as silica gel, diatomaceous earth, cellulose, polytetrafluoroethylene (PTFE) or polystyrene. The solid support is, in principle, inert and solely provides a large surface area on which the stationary phase is retained. Partition chromatography exploits the fact that a solute in contact with two immiscible liquids (or phases) will distribute itself between them according to its distribution coefficient, K (see equation 1.1). The principal intermolecular forces involved are dispersion, induction, orientation and donor–acceptor interactions, including hydrogen bonding. These forces provide the framework for a qualitative understanding of the separation process.

The importance of partition systems has declined in all areas of chromatography with the development of bonded phases. Nevertheless, in GC with packed columns, partition systems are still used. Conversely, use of partition systems in LC and SFC is restricted by instability of coated liquid stationary phases. This is caused by the small but finite solubility of the liquid stationary phase in solvents used as mobile phases, leading to stripping of the stationary phase from the column. One application area in LC where the separation mechanism is predominantly partition is paper chromatography. Here, the water sorbed on cellulose functions as the stationary phase.

Bonded phase chromatography

Bonded phases, in which the stationary phase is chemically bonded to either a solid support or cross-linked and bonded to the internal wall of the column, are popular in all forms of chromatography. The popularity of bonded phases is testimony to their many advantages. Compared with adsorption systems, they equilibrate faster, do not exhibit irreversible sorption and are available with a wide range of functionalities. Bonded phases were originally developed for GC in an attempt to stabilize the stationary phase at elevated temperatures. These early phases were susceptible to decomposition by traces of water or oxygen in the system and have been replaced by newer phases in current open tubular columns [81]. Bonded phases are also available in LC and SFC, where they overcome the problems of column bleed associated with physically bonded phases.

The mechanism of bonded phase chromatography is complex but appears to involve a combination of partition and adsorption. In a number of instances bonded phase chromatography has been referred to as partition chromatography because the organic surface layer is regarded as a 'bound liquid film'. However, Locke [82] concluded that bonded phases acted more like modified solids than thin liquid films. Nevertheless, the mechanism involved with bonded phases is sufficiently different from both adsorption and partition to warrant separate treatment.

Ion-exchange chromatography

Ion-exchange entails a reversible, stoichiometric exchange between sample ions in the mobile phase and ions of like charge associated with the ion-exchange surface. The stationary phase is a rigid matrix, the surface of which carries fixed positively or negatively charged functional groups (A). Counter ions (Y) of opposite charge are associated with each site in the matrix and these can exchange with similarly charged ions in the mobile phase. If the matrix contains negatively charged acidic functional groups then it is capable of exchanging cations and is called a cation-exchanger; if it bears positively charged basic groups it is an anion-exchanger capable of exchanging anions. If the sample ions are depicted as M^+ or X^- the process can be represented as:

cation exchange

$$\text{Matrix–}A^-Y^+ + M^+ \quad \rightarrow \quad \text{Matrix–}A^-M^+ + Y^+ \qquad K_{MY}$$

anion exchange

$$\text{Matrix--A}^+\text{Y}^- + \text{X}^- \rightarrow \text{Matrix--A}^+\text{X}^- + \text{Y}^- \qquad K_{XY}$$

In order to achieve a separation the matrix must exhibit some affinity for the sample ions. The counter ion already on the resin must also not be too strongly held that it cannot be displaced by sample ions. Secondary effects can arise owing to adsorption or hydrophobic interaction with the matrix itself. Because of the ionic nature of the interactions, ion-exchange is restricted to aqueous liquid chromatography.

Ion-interaction chromatography

Ion-pair extraction is a valuable liquid–liquid separation technique for isolating water-soluble ionic compounds by partitioning them between water and an immiscible liquid. The ionic solutes partition formally as ion pairs according to the equilibrium:

$$n\text{A}^{m+}_{(aq)} + m\text{B}^{n-}_{(aq)} \longleftrightarrow (n\text{A}^{m+} \cdot m\text{B}^{n-})_{(organic)}$$

in which A^{m+} represents the solute and B^{n-} the pairing ion or vice versa. This principle was extended to chromatography during the 1970s by Eksborg and Schill [83], and quickly gained acceptance as a versatile technique for the separation of ionized and weakly ionized solutes. The early successes promoted general interest in ion-interaction chromatography which uses conventional high-efficiency microparticulate, normal-phase or reversed-phase packings. However, the value of ion-interaction chromatography lies in its ability to separate simultaneously ionic and molecular species. In reversed-phase chromatography, ionic species generally show little, if any, retention and are eluted as an unresolved mixture. Ion-interaction chromatography does have some disadvantages; the ionic solutions can result in short column life or affect metal components of the system.

The technique as described by Schill and his group became known as extraction chromatography but was dubbed soap chromatography by Knox and coworkers [84] because of their use of the detergent cetyltrimethylammonium bromide as the pairing ion in the mobile phase. Subsequent terms used to describe the procedure have included ion pair, paired ion and mobile phase ion chromatography (proposed by the Dionex Corporation), solvent generated ion-exchange, ion association chromatography and dynamic ion-exchange, although the authors prefer the use of ion-interaction chromatography. It is unfortunate that the diversity of terms and debate over the mechanism of retention have caused considerable confusion.

Size exclusion chromatography

Separations in size exclusion chromatography are based on a physical sieving process and thus differ from all other mechanisms in the respect that neither specific nor nonspecific interactions between analyte molecules and the stationary phase are involved. In fact, every effort is made to eliminate such interactions because they impair column efficiency. Various names have been used to describe this form of chromatography, including gel permeation, gel filtration and steric exclusion. Historically, gel filtration referred to separations of biopolymers, such as proteins, on dextran or agarose gels using aqueous mobile phases, whereas separations of organic polymers in organic mobile phases on a polystyrene phase were termed gel permeation.

Internal surface reversed-phase supports, or Pinkerton columns, have been developed in the last 5 years [85] and involve a dual mechanism: size exclusion and reversed-phase bonded supports. These materials contain stationary phase on the internal surface of the pores of the support with an external surface which is nonadsorptive. Thus, large biomolecules are eluted unretained whereas smaller molecules penetrate the pores and are separated by a conventional reversed-phase mechanism.

Affinity chromatography

Affinity chromatography is at the opposite extreme to size exclusion in that very specific analyte–stationary phase interactions are exploited to achieve separation. The stationary phase consists of a bioactive ligand bonded to a solid support (e.g. cross-linked agarose or polyacrylamide). Since the latter may sterically hinder the ligand's accessibility, the concept of a spacer arm was introduced. This consists of a short alkyl chain inserted between the ligand and solid support to reduce or eliminate the steric influence of the matrix. Separation relies on biospecific interactions, such as antibody–antigen interactions, chemical interactions, such as the binding of *cis*-diol groups to boronate, or other interactions whose nature is not fully understood, such as the attraction of albumin to Cibacron Blue F3G-A dye. The specificity of the ligand sets these bonded phases apart from all others. Ligands may show absolute specificity for a single substance or may be group specific [86]. The interaction between ligand and analyte must be specific but reversible.

On adding the sample in a suitable mobile phase, the 'active' components with an affinity for the ligand are bound and retained while the unbound material is eluted in the mobile phase. The composition or pH of the mobile phase is then altered to weaken the specific interaction of ligand and active analyte, which is released and eluted.

Micellar or pseudophase liquid chromatography

The popularity of modern liquid chromatography relates partly to the unique selectivities that can be generated in the mobile phase by the addition of modifiers. In ion-interaction chromatography, this is achieved by adding a low concentration of modifier to the mobile phase. Here, the concentration of ion-interaction reagent was intentionally maintained below the critical micellar concentration. However, Armstrong and Henry [87] demonstrated the use of reversed-phase mobile phases containing higher concentrations of surfactant, exceeding the critical micellar concentration [88]. There are a number of reports concerning the theory [89] and unique chromatographic advantages of micellar chromatography [90]. One major advantage of micellar systems is their selectivity. Retention of solutes generally decreases with increasing micelle concentration but the rate of decrease varies considerably, producing inversions in retention order.

Complexation chromatography

Complexation or chelation chromatography can be considered as a generic term to encompass all chromatographic separations dependent on the rapid and reversible formation of a complex between a Lewis acid (metal ion) and a Lewis base. The versatility of complexation chromatography is due, in part, to its suitability in all areas of chromatography. Other reasons are the vast range of Lewis acids, Lewis bases and complexes that can be utilized, the different ways that they can be incorporated in the chromatographic column and the fact, that in many cases, conventional chromatographic columns or packings (e.g. silica adsorption, reversed-phase and ion-exchange) may be utilized. Potentially important applications involve the incorporation of Lewis acid, Lewis base or complex in the mobile or stationary phase to effect the separation of a wide range of inorganic, organic and biochemical species. Ligand exchange chromatography, chelate affinity chromatography and other terms have been used to describe various aspects of this application.

Ion-exclusion chromatography

Ion-exclusion chromatography is defined [91] as a technique used to separate weak acids, amino acids, sugars, alcohols and other substances on an ion-exchange column. Because of Donnan exclusion, ionic material is excluded from the ion-exchange resin and passes quickly through the column. Non-ionic substances are not excluded and partition between the aqueous mobile phase and occluded water within the resin beads. Because of differing partitioning effects and van der Waals forces, non-ionic solutes are retarded by the column and separated. As an illustration of an ion-exclusion separation, inorganic and organic acids may be chromatographed on a cation-exchange resin. The strongest acids elute at the column void volume because they are highly ionized and are repelled by the immobilized negative charge of the resin. Weaker acids exist largely in the unionized molecular form and are separated by partitioning between the mobile phase and the occluded solvent [92]. Other terms used to describe the process include ion-moderated partition chromatography [93] and ion-exclusion partition chromatography [94].

Countercurrent chromatography

Countercurrent chromatography is a recent development [95–97] which is similar to conventional liquid–liquid partition chromatography, with the distinction that the stationary liquid phase is retained in the apparatus without use of an adsorptive or porous support. In one variant, called droplet countercurrent chromatography, the stationary phase is retained by gravitational force in a narrow vertical tube (c. 2 mm i.d.) while droplets of an immiscible mobile phase are passed through it. The mobile phase is either added at the top or base of the tube, depending on whether it has a higher or lower density than the stationary phase. Typically, 300 tubes are connected in series using capillary-bore polytetrafluoroethylene (PTFE) tubing to provide a column of high efficiency. With the more common systems, the stationary phase is retained in more or less segmented compartments within a coiled tubing while the mobile phase is passed through it. Coils (2–3 mm i.d.) are usually made of PTFE and range in length from a few metres to more than 100 m. In contrast to conventional liquid–liquid chromatography where the volume of stationary phase in the column is relatively small, the stationary phase in countercurrent chromatography occupies from 40–90% of the total column volume. One of the problems of countercurrent chromatography is the relatively long analysis times. For example, a relatively simple separation of 10 compounds may require anything from a few to several hours using centrifugally operated units, and up to 1–3 days for units operated in a unit gravitational field. The main role for countercurrent chromatography is for preparative-scale separations in the milligram to gram range. Newer devices also function as efficient extractors to concentrate trace components from environmental samples, such as riverwater and biological fluids, such as urine by replacing the relatively long separation column with a short column.

1.4.5   Other systems of classification

Other methods of classification are in current use and no doubt as new developments occur further systems will be introduced. One method that should be mentioned is classification according to the nature of the analyte. This system of classification gives rise to terms such as ion chromatography and fast-protein liquid chromatography. Ion chromatography originally described the particular system of ion-exchange chromatography in which a low capacity ion-exchange column, a suppressor column and a conductivity detector were used to measure inorganic ions. The definition has now been broadened to include all modern chromatographic separations of ionic species. The word 'modern' implies that the separation is high-performance and that automatic, on-line detection is employed. The bulk of ion chromatography is concerned with ion-exchange separations using low capacity ion-exchange resins; however, ionic species are also separated by ion-interaction chromatography and ion-exclusion chromatography.

1.5   Information and Objectives of Chromatography

Chromatography is used to solve a diverse range of problems in numerous application areas as indicated by the following titles of articles taken from the journal *Analytical Chemistry*.

- Effects of raw material change in manufacturing process resolved. Mitchell J. (1974). *Anal. Chem.*, **46**: 804A.
- Industrial analytical chemists and OSHA regulations for vinyl chloride. Levine S. P., Hebel K. G., Bolton J. and Kugel R. F. (1975). *Anal. Chem.*, **47**: 1075A.
- Drug monitoring in the news. Focus (1983). *Anal. Chem.*, **55**: 1185A.
- Ion chromatography in bombing investigations. Reutter D. J., Buechele R. C. and Randolph T. L. (1983). *Anal. Chem.*, **55**: 1468A.
- Anatomy of an off-flavor investigation: The medicinal cake mix. Sevenants M. R. and Sanders R. A. (1984). *Anal. Chem.*, **56**: 293A.

- Identification of the musty component from an off-odor packaging film. McGorrin R. J., Pofahl T. R. and Croasmun W. R. (1987). *Anal. Chem.*, **59**: 1109A.
- Solving mysteries using infrared spectrometry and chromatography. Brown D. J., Schneider L. F. and Howell J. A. (1988). *Anal. Chem.*, **60**: 1005A.

Despite the diversity, the chromatographer is usually seeking the answer to one of three questions in performing a chromatographic separation: what is present, how much is present or, how can a pure substance be isolated from a mixture? Obviously, these various aspects of chromatography are related. For instance, measuring the amount of substance present clearly entails identification of the component to be quantified as a first step. For convenience, however, we will look at the answer to these questions separately and consider them as qualitative, quantitative and preparative chromatography. The intent of qualitative and quantitative analysis varies. It may be performed to provide information on the purity of a sample, to collect physicochemical data (e.g. reaction rates), to quantify the amount of analyte in a clinical sample to enable diagnosis and treatment, etc.

1.5.1 Qualitative applications

Chromatography is frequently used to confirm either the presence or absence of a compound in a sample. This is done by comparing a chromatogram of the pure substance with that of the unknown, performed under identical conditions. The chromatographer must be confident that a substance identified in a sample is actually present (e.g. a prohibited substance in the urine of an athlete) and, equally as important, that a substance not found is indeed not present (e.g. pesticide in a foodstuff). One of the difficulties in the comparison is that the chromatogram is not unique; many substances will exhibit the same chromatographic behaviour under identical conditions. The magnitude of the problem is compounded by the fact that there are approximately 10 million compounds known and over 60 000 of these are in commercial use. In short, chromatography can be used for qualitative analysis in a limited set of circumstances, but its main use is not for screening unknowns. The best methods of qualitative analysis are the 'coupled' or 'hyphenated' techniques which combine the excellent separating ability of chromatography with the capabilities of spectrometry for identification. Such techniques include GC—mass spectrometry, GC—infrared spectrometry, and the corresponding HPLC techniques.

Chromatography provides information on the complexity of a sample. The number of peaks (GC, HPLC or SFC) or spots (TLC or PC) indicates the minimum number of components; conversely, the purity of a compound can be checked and the presence of a single spot or peak is taken as an indication of purity. In other instances, all that is required for quality control purposes is a fingerprint chromatogram showing the pattern of peaks which can be compared with the 'normal' pattern. Patterns have been used to look for the presence or absence of metabolites in certain diseases, for example in phenylketonuria. One situation in which qualitative analysis is widely used is the examination of reaction products from organic and inorganic syntheses to determine which conditions give the cleanest products and which reactions do not occur at all (giving only reactants).

1.5.2   Quantitative applications

Chromatography is used to establish the amount of individual components in a sample by comparison with suitable standards and calibrations. Quantitative data are widely used in industry for quality control, in clinical chemistry for the assay of body fluids, and in environmental science for monitoring air, water and soil samples. Chromatography has also enabled the measurement of product distributions in reaction mixtures. Such data are useful in several areas, including physicochemical measurements (e.g. kinetic studies).

1.5.3   Preparative applications

The ability of chromatography to separate the components of a sample can be used preparatively to produce pure constituents on either a laboratory or industrial scale. A single thick layer (up to 1 mm) TLC plate can be used for up to 10 mg of a sample. Multiple runs can be used to separate larger samples. Preparative GC and HPLC is also possible by scaling up the column dimensions to provide up to gram quantities of pure components. In industry, much larger quantities ranging up to several tonnes are processed and purified by chromatography. An example is provided by the refining of cane sugar. Operating chromatographic equipment at this opposite extreme to that involved in analysis presents its own intriguing problems of process engineering and is discussed in detail elsewhere [98].

1.6   Comparison of Chromatographic Techniques

The choice between different chromatographic techniques is a daunting prospect for the novice chromatographer. However, the answers to a few simple questions will often indicate the most suitable technique. The nature of the analyte and the sample matrix are the first considerations. The intent of the analysis is also important, as are practical considerations, such as the availability of a particular piece of equipment. For example, GC is the obvious choice for the analysis of atmospheric gases. In other instances the choice is less clear-cut and equipment availability and the operator's experience and personal bias will probably determine the final choice. Nevertheless, different chromatographic techniques should not be considered to be competing with each other for the solution of every problem. Each technique should be considered on its merits and used where it is most appropriate. It is only by recognizing the complementary nature of the various techniques that maximum benefit will be achieved. Indeed, this concept extends beyond the traditional boundaries of chromatography, as elucidated by Ruzicka and Christian [99] who compared the synergism of flow injection and chromatographic techniques.

The comparisons presented in this section should provide a guide to a sensible decision-making process. It is difficult to reduce the selection of a method to a simple flow chart (Fig. 1.7); either the chart becomes so complex or contains so little information that it becomes misleading. Normally, non-volatile samples are analysed by LC, as indicated in Fig. 1.7, but consideration should be given to the possibility of derivatizing the sample to obtain adequate volatility for GC, particularly if

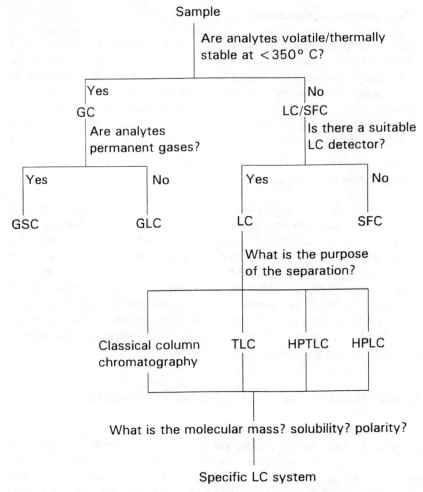

Fig. 1.7. Selection tree for choosing a chromatographic technique.

detection presents a problem in LC. An understanding of the capabilities and limitations of the various techniques is a good starting point in method selection. Experience is also a very good guide but the versatility of chromatography must always be remembered, e.g. for many years ion-exchange was limited to the separation of ionic substances largely because contemporary thought restricted it to such separations; now many excellent separations of non-ionic solutes are achieved by ion-exchange-type processes.

Analyte characteristics

In principle, the decision between GC, LC and SFC is relatively simple. For gas chromatographic analysis, the analyte must be volatile (in general this requires a boiling point below 500°C or a vapour pressure below this temperature of several kPa) and thermally stable. These limitations are not imposed on either LC or SFC. Conversely, the greater versatility of LC for analyte type imposes detector restrictions. As a result of volatility considerations, GC is limited to relatively low molecular mass organic solutes for which a universal detector (the flame ionization detector, FID; see Chapter 3) exists. Strictly, the FID is not universal but it comes very close. The versatility of LC means that it is amenable to all analyte types (providing a suitable solvent exists) including inorganic and organic species, low and high molecular mass substances including polymers and ionized and non-ionized materials alike. A detector capable of sensing this broad range of analytes has not been developed and so the strength of LC also becomes its weakness. Derivatization of the analyte can be used to improve the volatility/thermal stability in GC or the detectability in LC. The ease of detection is not a consideration in selecting SFC, as detectors developed for both GC and LC can be used.

Is either GC, LC or SFC inherently more efficient or faster? Giddings [100, 101] attempted to answer this question in relation to GC and LC. He derived an expression relating column pressure and efficiency which predicts that LC would be more efficient than GC by a factor of 10^3 if run at the same pressure. Looking at the problem the other way, GC would be as efficient as LC if the GC were performed at a pressure 10^3 times as high as LC. The problem with the first conclusion is that the time required to achieve the efficiencies in LC would be unacceptably long. The problem with the second conclusion is that it is very difficult to perform GC at the high pressures required. In conclusion, theory has not provided much assistance and the initial choice between GC, LC and SFC is seldom on the basis of differences in efficiency or time. Indeed, it is a reasonable maxim in chromatography that a technique should not be claimed to give a faster separation than is possible with other techniques; the number of chromatographic variables is sufficiently large that any such example could be demonstrated as incorrect.

Characteristics of the sample matrix

In many instances, the chromatographic technique can be used to separate the analyte from its matrix and to determine its identity and concentration. In other cases, prefractionation is necessary to avoid accumulation of sample residues in the chromatographic system. This is probably a more common occurrence in GC where accumulation and charring of non-volatile residues can lead to analyte decomposition. Nevertheless, accumulation of sample components also occurs in LC and SFC, where it can lead to various anomalies such as unstable baselines. Prefractionation may involve simple filtration, solvent extraction or some form of chromatography such as planar or liquid column chromatography (see Chapter 8). In general, classical column chromatography is ideal for this purpose. It is relatively simple, inexpensive and provides a crude separation which is frequently all that is required at this stage.

With samples containing both volatile and non-volatile components, less sample preparation will probably be required for LC than for GC. Conversely, TLC is ideally suited to separations where it is desirable to observe all the components of a sample, as any residues left at the sampling point are easily observed. For very complex matrices, a high-resolution technique is clearly desirable. Although HPLC and high-performance TLC (HPTLC) provide high resolving power, the performance of open tubular column GC is unsurpassed in this respect. For components present at the trace level, a sensitive and/or selective detector is necessary.

Purpose of the analysis

For analyses where accuracy and sensitivity are not prime requirements, TLC is ideal. An example of this would be the qualitative examination of the reaction products from an inorganic or organic synthesis. Simultaneous analysis of several samples is also possible with TLC, which is therefore well suited to routine screening of large numbers of samples, as in monitoring biological fluids for the presence/absence of drugs where speed and economy are needed. Analyses requiring greater selectivity and quantitative accuracy can normally only be achieved with more sophisticated instrumental techniques such as GC, HPLC and HPTLC. Such methods are also more easily automated for process control or analysis of large numbers of similar samples.

The following series of Tables (1.1–1.3) present an overview of various chroma-

Table 1.1 Comparison of column techniques: gas, supercritical fluid and high-performance liquid chromatography.

	Gas chromatography	Supercritical fluid chromatography	High-performance liquid chromatography
Sample	Limited to thermally stable, volatile species although this can be extended by derivatization	Well suited to samples not easily handled by GC or HPLC	Few, if any sample restrictions
Mobile phase	Non-interactive	Interactive	Interactive
Resolving power	In routine practise, open tubular column GC provides unparalleled separations	Excellent	Excellent
Detectors	Very well developed and, in general, more sensitive than HPLC detectors. Flame ionization detector gives almost universal detection for organics	In theory, both GC and HPLC detectors can be used. Limitation is on mobile phase/detector compatibility	No universal, sensitive detector as a result of the broad applicability of HPLC
Sample recovery	Simple	Simple	Less convenient
Coupled techniques	GC/MS and GC/FTIR are routine procedures	Well suited to coupling with supercritical fluid extraction for on-line sample preparation	LC–MS is emerging as a routine procedure
Capital and running costs	Capital costs vary greatly depending on the level of sophistication of the equipment. Running costs depend very much on local circumstances		

Table 1.2   Comparison of liquid column chromatographic techniques.

	Classical column chromatography	High-performance liquid chromatography
Convenience	Tedious, time consuming	Rapid, easily automated
Columns	Columns usually discarded after a single use	Columns reusable
Procedure	Detection and quantification achieved by fraction collection and further off-line processing	Separation, detection and quantification achieved on-line
Application	Sample preparation, preparative scale separations	Sample preparation and preparative scale possible but also applicable to high resolution separations
Capital costs	Low	High

Table 1.3   Comparison of liquid chromatographic techniques.

	Classical column chromatography	Thin layer chromatography	High-performance thin layer chromatography
Sample introduction	Single, sequential	Multiple, concurrent	Multiple, concurrent
System	Closed column	Open bed	Open bed
Separation time	30 min to several hours	30–200 min	3–20 min
Resolving power	Low	Medium	High
Detection	Fraction collection	Mostly static, *in situ*, permanent record. Can be adapted to quantitative analysis	Versatile including quantitative analysis

tographic techniques. However, it is worth stating, once again, that the different techniques are complementary.

1.7   Obtaining Assistance

Many distributors of chromatographic equipment have specialist applications departments that are able to provide assistance in choosing or developing a new method. A number publish regular newsletters and product bulletins. There are also numerous literature sources that may assist in method selection. *Analytical Chemistry* publishes biennial reviews with a coverage that includes various chromatographic techniques and, in alternate years, chromatographic applications. Data bases with a chromatography coverage include general sources such as *Chemical Abstracts* and *Analytical Abstracts* which can be searched by computer. More specialist data bases are provided by Elsevier Applied Science Publishers (*Chromatography Abstracts*, Elsevier Applied Science Publishers, Barking, UK, or Elsevier Science Publishing, New York, USA) and Preston Publications (GC and LC Literature: Abstracts and Index, Preston Publications, Niles, IL, USA).

References

1. Tswett M.S. (1906). *Ber. Dtsch. Bot. Ges.*, 316 and 384.
2. Zechmeister L. (1950). *Progress in Chromatography 1938–1947*. Chapman and Hall, London.
3. IUPAC Analytical Chemistry Division, Commission on Analytical Nomenclature (1974). Recommendations on Nomenclature for Chromatography. *Pure Appl. Chem.*, **37**: 447.
4. Williams T.I. and Weil H. (1952). *Nature*, **170**: 503.
5. Williams T.I. and Weil H. (1953). *Ark. Kemi*, **5**: 283.
6. Purnell H. (1962). *Gas Chromatography*. Wiley, New York, p. 1.
7. Seaborg G.T. and Higgins G. (1979). In Ettre L. S. and Zlatkis A., Editors, *75 Years of Chromatography—A Historical Dialogue*. Elsevier, Amsterdam.
8. Ettre L.S. (1980). In Horvath C., Editor, *High-Performance Liquid Chromatography*: *Advances and Perspectives*. Academic Press, New York.
9. Ettre L.S. (1971). *Anal. Chem.*, **43**: 20A.
10. Ettre L.S. and Horvath C. (1975). *Anal. Chem.*, **47**: 422A.
11. Ettre L.S. (1991). *Analyst*, **116**: 1231.
12. Gordon H. (1991). *Analyst*, **116**: 1245.
13. Zechmeister L. and Cholnoky L. (1941). *Principles and Practice of Chromatography* (English Edition). Chapman and Hall, London.
14. Weil H. and Williams T.I. (1950). *Nature*, **166**: 1000.
15. Farradane J. (1951). *Nature*, **167**: 120.
16. Zechmeister L. (1951). *Nature*, **167**: 405.
17. Weil H. and Williams T.I. (1951). *Nature*, **167**: 906.
18. Weil H. (1951). *Petroleum*, **14**: 5.
19. Sakodynskii K.I. (1970). *Chromatographia*, **3**: 92.
20. Sakodynski K.I. (1970). *J. Chromatogr.*, **49**: 2.
21. Sakodynski K.I. (1972). *J. Chromatogr.*, **73**: 303.
22. Meinhard J.E. (1949). *Science*, **110**: 387.
23. Kohl F.G. (1906). *Ber. Dtsch. Bot. Ges.*, **24**: 124.
24. Lederer E. (1972). *J. Chromatogr.*, **73**: 261.
25. Kuhn R. and Lederer E. (1931). *Naturwiss.*, **19**: 306.
26. Kuhn R., Winterstein A. and Lederer E. (1931). *Z.Physiol.Chem.*, **197**: 141.
27. Kuhn R. and Lederer E. (1931). *Chem. Ber.*, **64**: 1349.
28. Schwab G.-M. and Jockers K. (1937). *Angew. Chem.*, **50**: 546.
29. Tiselius A. (1940). *Ark. Kem. Mineral. Geol.*, **14B**(22): 1.
30. Shraiber M.S. (1972). *J. Chromatogr.*, **73**: 367.
31. Meinhard J.E. and Hall N.F. (1949). *Anal. Chem.*, **21**: 185.
32. Stahl E. (1969). *Thin Layer Chromatography* (English translation), 2nd edn. Academic Press, New York.
33. Taylor T.I. and Urey H.C. (1938). *J. Chem. Phys.*, **6**: 429.
34. Samuelson O. (1939). *Fresenius Z. Anal. Chem.*, **116**: 328.
35. Spedding F.H., Fulmer E.I., Butler T.A., Gladrow E.M., Gobush M., Porter P.E., Powell J.E. and Wright J.M. (1947). *J. Am. Chem. Soc.*, **69**: 2812.
36. Martin A.J.P. and Synge R.L.M. (1941). *Biochem. J.*, **35**: 1358.
37. Consden R., Gordon A.H. and Martin A.J.P. (1944). *Biochem. J.*, **38**: 224.
38. Pollard F.H., McOmie J.F.W. and Elbeih I.I.M. (1949). *Nature*, **163**: 292.
39. Lederer M. (1948). *Anal. Chim. Acta*, **2**: 261.
40. Dixon B. (1989). *Curr. Contents*, **32** (16): 3.

41. Howard G.A. and Martin A.J.P. (1950). *Biochem. J.*, **46**: 532.
42. Alm R.S., Williams R.J.P. and Tiselius A. (1952). *Acta Chem. Scand.*, **6**: 826.
43. James A.T. and Martin A.J.P. (1952). *Biochem. J.*, **50**: 679.
44. McWilliam I.G. (1983). *Chromatographia*, **17**: 241.
45. Harley J., Nel W. and Pretorius V. (1958). *Nature*, **181**: 177.
46. Lovelock J.E. and Lipsky S.R. (1960). *J. Am. Chem. Soc.*, **82**: 431.
47. Golay M.J.E. (1957). *Anal. Chem.*, **29**: 928.
48. Porath J. and Flodin P. (1959). *Nature*, **183**: 1657.
49. Axen R., Porath J. and Ernback S. (1967). *Nature*, **214**: 1302.
50. Desty D.H., Goldup A. and Whyman B.H.F. (1959). *J. Inst. Petroleum*, **45**: 287.
51. Halasz I. and Walking P. (1969). *J. Chromatogr. Sci.*, **7**: 129.
52. Horvath C. and Lipsky R.S. (1969). *J. Chromatogr. Sci.*, **7**: 109.
53. Huber J.F.K. (1969). *J. Chromatogr. Sci.*, **7**: 172.
54. Kirkland J.J. (1969). *Anal. Chem.*, **41**: 218.
55. Scott R.P.W. and Lawrence J.G. (1969). *J. Chromatogr. Sci.*, **7**: 65.
56. Snyder L.R. (1967). *Anal. Chem.*, **39**: 698.
57. Snyder L.R. (1969). *J. Chromatogr. Sci.*, **7**: 352
58. Knox J.H. and Saleen M. (1969). *J. Chromatogr. Sci.*, **7**: 352.
59. Felton H. (1969). *J. Chromatogr. Sci.*, **7**: 13.
60. van Deemter J.J., Zuiderweg F.J. and Klinkenberg A. (1956). *Chem. Eng. Sci.*, **5**: 271.
61. Giddings J.C. (1963). *Anal. Chem.*, **35**: 2215
62. Lochmuller C.H. (1987). *J. Chromatogr. Sci.*, **25**: 583.
63. Snyder L.R. (1991). *Analyst*, **116**: 1237.
64. Palmieri M.D. (1988). *J. Chem. Ed.*, **65**: A254.
65. Novotny M. (1988). *Anal. Chem.*, **60**: 500A.
66. Novotny M. (1981). *Anal. Chem.*, **53**: 1294A.
67. Cramers C.A. and Rijks J.A. (1979). In Giddings J.C., Editor, *Advances in Chromatography*, Vol. 17. Marcel Dekker, New York, p. 101.
68. Duffy M.L. (1985). *Am. Lab.*, **17**(10): 94.
69. Ishii D. and Takeuchi T. (1983). In Giddings J.C., Editor, *Advances in Chromatography*, Vol. 21. Marcel Dekker, New York, p. 131.
70. Smith R.D., Wright B.W. and Yonker C.R. (1988). *Anal. Chem.*, **60**: 1323A.
71. Pacholec F. and Poole C.F. (1983). *Chromatographia*, **17**: 370.
72. Karger B.L. (1967). *Anal. Chem.*, **39**: 24A.
73. Laub R.J. and Pecsok R.L. (1975). *J. Chromatogr.*, **113**: 47.
74. Scott R.P.W. (1982). In Giddings J.C., Editor, *Advances in Chromatography*, Vol. 20. Marcel Dekker, New York, p. 167.
75. Snyder L.R. (1968). *Principles of Adsorption Chromatography*. Marcel Dekker, New York.
76. Snyder L.R. (1983). *J. Chromatogr.*, **255**: 3.
77. Snyder L.R. and Poppe H. (1980). *J. Chromatogr.*, **184**: 363.
78. Yashin Y.I. (1982). *J. Chromatogr.*, **251**: 269.
79. Claesson S. (1946). *Arkiv Kemi., Min. Geol.*, **A23**: 1.
80. Phillips C.S. (1949). *Discuss. Faraday Soc.*, (**7**): 241.
81. Gilpin R.K. (1984). *J. Chromatogr. Sci.*, **22**: 371.
82. Locke D.C. (1974). *J. Chromatogr. Sci.*, **12**: 433.
83. Eksborg S. and Schill G. (1973). *Anal. Chem.*, **45**: 2092.
84. Knox J.H. and Laird G.R. (1976). *J. Chromatogr.*, **122**: 17.
85. Hagestam I.H. and Pinkerton T.C. (1985). *Anal. Chem.*, **57**: 1757.
86. Walters R.R. (1985). *Anal. Chem.*, **57**: 1099A.
87. Armstrong D.W. and Henry S. (1980). *J. Liq. Chromatogr.*, **3**: 657.
88. Cline Love L.J., Habarta J.G. and Dorsey J.G. *Anal. Chem.*, **56**: 1132A.
89. Arunyanart M. and Cline Love L. (1984). *Anal. Chem.*, **56**: 1557.
90. Landy J. and Dorsey J. (1984). *J. Chromatogr. Sci.*, **22**: 68.
91. Gjerde D.T. and Fritz J.S. (1987). *Ion Chromatography*, 2nd edn. Huethig, Heidelberg.
92. Harlow G.A. and Morman D.H. (1964). *Anal. Chem.*, **36**: 2438.
93. Jupille T., Gray M., Black B. and Gould M. (1981). *Am. Lab.*, **13**: 80.

94. Harlow G.A. and Morman D.H. (1964). *Anal. Chem.*, **36**: 2438.
95. Sutherland I.A., Heywood-Waddington D. and Peters T.J. (1984). *J. Liquid Chromatogr.*, **7**: 363.
96. Sutherland I.A., Heywood-Waddington D. and Peters T.J. (1985). *J. Liquid Chromatogr.*, **8**: 2315.
97. Foucault A.P. (1991). *Anal. Chem.*, **63**: 569A.
98. Poole C.F. and Schuette S.A. (1984). *Contemporary Practice of Chromatography*. Elsevier, Amsterdam, Ch. 6.
99. Ruzicka J. and Christian G.D. (1990). *Analyst*, **115**: 475.
100. Giddings J.C. (1965). *Dynamics of Chromatography, Chromatographic Science Series, Vol. 1.* Marcel Dekker, New York, pp. 293–301.
101. Giddings J.C. (1965). *Anal. Chem.*, **37**: 60.

Bibliography

History

Done J.N., Kennedy G.J. and Knox J.H. (1972). Revolution in liquid chromatography. *Nature*, **237**: 77.
Ettre L.S. (1972). The development of gas adsorption chromatography. *Am. Lab.* **4**, October 10.
Ettre L.S. (1980). Evolution of liquid chromatography: a historical overview. *High Perform. Liquid Chromatogr.*, **1**: 1.
Ettre L.S. and Horvath C. (1975). Foundations of modern liquid chromatography. *Anal. Chem.*, **47**: 422A.
McWilliam I.G. (1983). The origin of the flame ionization detector. *Chromatographia*, **17**: 241.
Sakodynski K. (1970). M.S. Tswett — his life. *J. Chromatogr.*, **49**: 2.
Sakodynski K. (1972). The life and scientific works of Michael Tswett. *J. Chromatogr.*, **73**: 303.
Strain H.H. and Sherma J. (1967). Micheal Tswett's contribution to sixty years of chromatography. *J. Chem. Educ.*, **44**: 235.
Weil H. and Williams T.I. (1950). History of chromatography. *Nature*, **166**: 1000.
Wintermeyer U. (1990). Historical review of chromatography. *Chromatogr. Sci.*, **47** (*Pack. Stationary Phases Chromatogr. Tech.*), pp. 1–42.

Nomenclature

Ettre L.S. (1981). The nomenclature of chromatography. III. General rules for future revisions. *J. Chromatogr., Chromatogr. Rev.*, **220**: 65.

Reversed-phase chromatography

Colin H. and Guiochon G. (1980). Selectivity for homologous series in RP-LC. I. Theory. *J. Chromatogr. Sci.*, **18**: 54.
Horvath C. (1981). Reversed-phase chromatography. *Trends Anal. Chem.*, **1**: 6.
Horvath C. (1983). Contributions of physicochemical phenomena to retention in RPC. *LC Magazine*, **1**: 552.

Normal-phase chromatography

Antle P.E. (1982). Solvent optimization in normal-phase LC of selected steroids. *Chromatographia*, **15**: 277.
Snyder L.R. and Schunk T.C. (1982). Retention mechanism and the role of the mobile phase in normal-phase separation on amino-bonded-phase columns. *Anal. Chem.*, **54**: 1764.

Adsorption chromatography

Glajch J.L. and Snyder L.R. (1981). Solvent strength of multicomponent mobile phases in LSC. Mixtures of three and more solvents. *J. Chromatogr.*, **214**: 21.

Glajch J.L., Kirkland J.J. and Snyder L.R. (1982). Practical optimization of solvent selectivity in LSC using a mixture-design statistical technique. *J. Chromatogr.*, **238** 269.

Saunders D.L. (1974). Solvent selection in adsorption liquid chromatography. *Anal. Chem.*, **46**: 470.

Snyder L.R. (1971). Solvent selectivity in adsorption chromatography on alumina. *J. Chromatogr.*, **63**: 15.

Snyder L.R. and Glajch J.L. (1981). Solvent strength of multicomponent mobile phases in LSC. Binary-solvent mixtures and solvent localization. *J. Chromatogr.*, **214**: 1.

Snyder L.R., Glajch J.L. and Kirkland J.J. (1981). Theoretical basis for systematic optimization of mobile phase selectivity in LSC. Solvent-solute localization effects. *J. Chromatogr.*, **218**: 299.

Soczewinski E. and Jusiak J. (1981). A simple molecular model of adsorption chromatography. XIV. R_f or R_m? Secondary retention effects in TLC. *Chromatographia*, **14**: 23.

Bonded-phase chromatography

Snyder L.R. (1983). LSC on bonded phases. *LC Magazine*, 478.

Size-exclusion chromatography

Aubert J.H. and Tirrell M. (1983). Flow rate dependence of elution volumes in size exclusion chromatography: a review. *J. Liq. Chromatogr.*, **6**(Suppl. 2): 219.

Dubin P.L. (1992). Problems in aqueous size exclusion chromatography. In Giddings J.C. and Brown P.R., Editors, *Advances in Chromatography, Vol. 31*. Marcel Dekker, New York, pp. 119–151.

Podzimek S. (1992). A review of the application of HPLC and GPC (gel permeation chromatography) to the analysis of synthetic resins. *Chromatographia*, **33**: 377.

Affinity chromatography

Chaiken I.M. (1986). Analytical affinity chromatography in studies of molecular recognition in biology: a review. *J. Chromatogr.*, **376**: 11.

Jones K. (1991). A review of biotechnology and large scale affinity chromatography. *Chromatographia*, **32**: 469.

Micellar chromatography

Hernandez M., Jose M., Alvarez-Coque G. and Celia M. (1992). Solute–mobile phase and solute–stationary phase interactions in micellar liquid chromatography. A review. *Analyst*, **117**: 831.

Countercurrent chromatography

Conway W.D. (1990). *Countercurrent Chromatography—Apparatus, Theory and Applications*. VCH, New York.

Ion chromatography

Colenutt B.A. and Trenchard P.J. (1985). Ion chromatography and its application to environmental analysis: a review. *Environ. Pollut., Ser. B*, **10**: 77.

Tarter J.B. (1987). A review of ion chromatography: a bibliography. *Chromatogr. Sci.*, **37** (*Ion Chromatogr.*), pp. 369–413.

Weiss J. (1987). Ion chromatography—a review of recent developments. *Fresenius Z. Anal. Chem.*, **327**: 451.

Preparative chromatography

Jones K. (1988). A review of very large scale chromatography. *Chromatographia*, **25**, 547.

Verzele M. and Dewaele C. (1985). Preparative liquid chromatography. A critical review: 1980–1984, *LC Magazine*, **3**: 22.

Theory of Chromatography 2

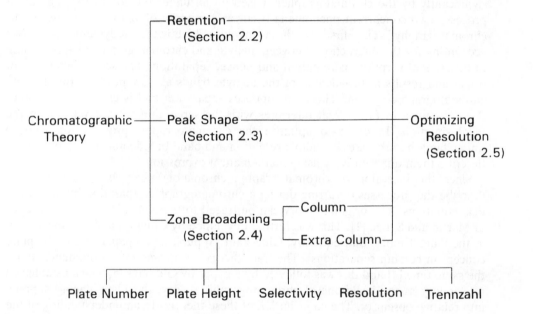

2.1   Introduction

The chromatographer must consider a number of factors in establishing and optimizing a chromatographic system. First, the resolution of the compound(s) of interest from each other and from other sample components must be considered. Second, an important practical consideration is the time taken for the analysis. Third, a means must be available to detect the analyte(s). The latter is often a function of the detector selectivity and the presence of interferences. The remaining factors of resolution and analysis time are interrelated and can often be improved significantly by the chromatographer, based on an understanding of the separation process. Two concurrent phenomena occur as a mixture of analytes is subjected to chromatography. The first of these involves kinetic/thermodynamic processes accounting for the interactions between analyte and chromatographic phases leading to differential sorption or retention and hence, separation. The second is kinetic in origin and results in broadening of the analyte bands as they progress through the chromatographic system. The latter opposes separation and is clearly undesirable. As was shown in Fig. 1.2, it interferes with the separation process relative to the ideal situation. In order to optimize the chromatographic process, we need to understand the processes of sample retention and band broadening and to be able to describe them quantitatively using mathematical expressions.

Since the inception of chromatography, chromatographers have attempted to describe the processes occurring during a chromatographic separation by mathematical equations. The original theory can be traced to the Nobel Prize winning paper of Martin and Synge [1]. This was further developed by Craig [2] and became known as the Plate Theory. Subsequently, Glueckauf [3] published a paper titled 'The plate concept in column separations'. The 'rate theory' also came into prominence about the same time [4] and this was followed by a paper by Giddings [5] which established rate theory as the backbone of chromatographic theory; the Plate Theory slipping into relative obscurity. The contribution of these theories to an understanding of the practice of chromatography is examined in this chapter. Discussion is concentrated on those aspects that can be manipulated by the operator to improve the separation process. The basic concepts and theories underlying all chromatographic techniques are effectively the same regardless of the nature of the stationary or mobile phases. A more detailed discussion of the theoretical basis and thermodynamic principles of the chromatographic process can be found in standard texts quoted in the bibliography.

2.2   Chromatographic Retention

The retention of an analyte by a chromatographic system can be characterized by its rate of migration relative to that of the mobile phase. There are important differences between planar and column techniques which require separate treatment. However, before proceeding it is necessary to define the terms elution, development, band, zone and peak. As commonly used in chromatography, the last three terms describe the concentration or mass profile of an analyte in space. Band represents this distribution while the analyte is still in or on the system, whereas peak describes the distribution of analyte once it has left the system and been monitored by some sort of detector. Zone is a generic term covering both bands and peaks used when further distinction is unnecessary. The process by which the mobile phase is caused to flow over the stationary phase to effect the chromatographic separation is known as development. When the analytes are actually 'washed out' of the chromatographic system, they are said to have been eluted or elution has taken place. In elution chromatography, the mobile phase is frequently referred to as the eluent (also spelt eluant) and the eluate is the combination of eluent and analyte exiting the chromatographic system.

A number of other definitions, symbols and terms will be introduced in this chapter. Unfortunately, universal agreement has not been reached on these and, in some cases, the terminology varies between the different chromatographic techniques as a consequence of the historical development of certain of the techniques in isolation. Nonetheless, recommendations have been made by a number of groups including ASTM [6, 7] and IUPAC [8]. Because chromatography is still evolving, the recommended terminology is often contrary to current usage and the style adopted in the principal chromatography journals. Ettre [9] has summarized the situation applying to liquid [10] and gas chromatography [11].

2.2.1   Parameters for column techniques

In column procedures, it is possible to control the flow of mobile phase and, in these circumstances, it is now accepted practice to continue addition of mobile phase until all analytes are recovered from the column. The data from such a separation are presented in the form of a chromatogram, which is a plot of the detector signal representing analyte concentration or mass versus volume of mobile phase or time. The simplified chromatogram in Fig. 2.1, shows a peak corresponding to a retained component plus a second smaller peak from an unretained solute which, in many instances, may not appear in the chromatogram [12–14].

In Fig. 2.1 the symbols t_r, t_m or the corresponding V_r, V_m refer to the retention time or retention volume of the analyte and unretained solute, respectively. The retention time is the time needed for a compound to move from the point of introduction into

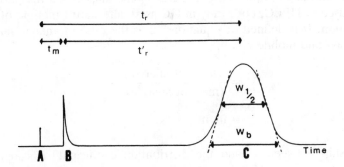

Fig. 2.1. Schematic chromatogram showing peaks resulting from a retained component (C) and an unretained component (B). The point where the chromatogram commences is denoted by A.

the system to the point of detection, and the retention volume is the volume of mobile phase corresponding to this time. In theory, it is the latter that is more significant but for purely practical reasons it is retention time that is more commonly measured. Since the chromatographic peak in Fig. 2.1 represents analyte concentration (or mass) as it emerges from the column, it is necessary to measure time or volume to an arbitrary but common point. By convention, the point selected to characterize an analyte is the peak maximum since this represents, for symmetrical peaks, an average molecule.

The time an analyte spends in the chromatographic system, its retention time, is composed of two components. The first contribution is due to the time it takes a solute to pass through the space occupied by the mobile phase (i.e. holdup volume, dead space or column void volume). This time is called the dead time, t_m and it is the same for all analytes in a given chromatographic system. During this time all solutes migrate with the same velocity—that of the mobile phase. Therefore, regardless of their retention times, all solutes spend the same length of time, equal to t_m, in the mobile phase. Hence, the dead time is characteristic of a system and it is obtained by measuring the time required for an unretained solute to traverse the system. Since an unretained solute spends all of its time in the mobile phase, it is also the time required for the mobile phase front to pass through the column. In this time there is no contribution to the separation process. The second contribution to the retention time of a solute is the time the solute spends retained by the stationary phase. It is this time that is more characteristic of a solute and it is usual, therefore, to define the adjusted retention time, t_r', which represents the actual time during which true retention occurs, i.e. the average time spent by analyte molecules in the stationary phase. Expressed another way, the adjusted retention time is the time an analyte is retained compared with an unretained compound (equation 2.1)

$$t_r' = t_r - t_m \qquad 2.1$$

and is therefore directly related to the interaction of the analyte with the stationary phase.

Capacity factor, k'

The partition ratio is an important parameter (sometimes called the partition ratio, k) routinely used in HPLC, but rarely in GC as an alternative method of expressing analyte retention. It is defined by equation 2.2 as the ratio of analyte masses in the stationary phase and mobile phase:

$$k' = \frac{(\text{mass of analyte})_s}{(\text{mass of analyte})_m} \qquad 2.2$$

$$= m_s/m_m \qquad 2.3$$

The relationship between k' and the distribution coefficient, K can be seen by expanding equation 1.1 as follows:

$$K = C_s/C_m$$

$$= \frac{m_s/V_s}{m_m/V_m} \qquad \text{2.4}$$

$$= m_s/m_m \cdot V_m/V_s \qquad \text{2.5}$$

$$= k' \cdot \frac{V_m}{V_s} \qquad \text{2.6}$$

where V_m and V_s are the volume of mobile and stationary phase, respectively. The ratio V_m/V_s is called the phase ratio, β. It is one of several parameters used to characterize a column particularly in gas chromatography (see Chapter 3). In effect, β is a measure of the 'openness of the column'. Typical values for open tubular columns (50–1500) are much higher than values for packed columns (5–35), where the support both limits the volume available for mobile phase and increases the area over which the stationary phase is distributed.

Substituting for the phase ratio in equation 2.6 gives

$$K = k'\beta \qquad \text{2.7}$$

Equation 2.7 is significant since it relates the equilibrium distribution of analyte within the column to the thermodynamic distribution coefficient, K. The latter is independent of the particular column (depending only on temperature and the nature of the stationary and mobile phases) although both the capacity factor and phase ratio are characteristic for a particular column. Thus, if the value of the capacity factor is higher, then the phase ratio must be smaller and vice versa.

In many situations it is difficult to measure the phase ratio and hence to obtain k'. A more practical method relating k' directly to easily measurable quantities is required. Since equation 2.2 represents the ratio of the probability that a solute molecule will be in the stationary phase versus the mobile phase it can be rewritten as

$$k' = P_s/P_m \qquad \text{2.8}$$

where P_s and P_m represent the probabilities of a single solute molecule being in the stationary phase or mobile phase. But this probability ratio is also equal to the ratio of the average time the molecule spends in the stationary and mobile phases, and therefore to the ratio of the total time spent in the stationary and mobile phase, respectively; i.e.,

$$k' = \frac{t_r - t_m}{t_m} \qquad \text{2.9}$$

$$= \frac{t'_r}{t_m} \qquad \text{2.10}$$

It is apparent from equation 2.10 that the capacity factor compares the adjusted retention time of an analyte to the retention time of the mobile phase, i.e., the ratio of the additional time required for an analyte to traverse the system relative to an unretained solute. This is effectively the ratio of the time the analyte molecules spend in the stationary phase (where they are stationary) to their time in the mobile phase (where they are transported down the column)

Equation 2.10 enables us to calculate k' directly from parameters in the chromatogram. The capacity factor is important because it is independent of the mobile phase flow-rate and the physical dimensions of the column and can therefore be used for comparing retention on different instruments (see Section 9.2). In practice the accurate measurement of column void volume in HPLC is difficult and different 'unretained' marker compounds give different results [13, 14]. Moreover, no international agreement has been reached on the measurement of column void volume. Its measurement in gas–liquid chromatography is somewhat simplified and more accurate results are obtainable [12] which suffice for all but the most exacting work.

Retention time

Rearrangement of equation 2.9 gives

$$t_r = t_m(1 + k')$$ 2.11

and by substituting for k' from equation 2.6

$$t_r = t_m(1 + KV_s/V_m)$$ 2.12

Equation 2.12 relates the experimental variables t_r and t_m to the thermodynamic distribution coefficient, K.

If the average linear mobile phase velocity is denoted by u, then for a column of length, L

$$u = L/t_m$$ 2.13

By substituting for t_m from equation 2.13 in equation 2.12, a useful expression for the retention time is obtained

$$t_r = L/u(1 + KV_s/V_m)$$ 2.14

From this expression it is possible to predict the effect of changes in column length, linear velocity of mobile phase, phase ratio and distribution coefficient on retention time.

The expression for retention time in equation 2.14 is a special case which is valid when the mobile phase velocity is constant throughout the column. This applies when the mobile phase is noncompressible, which is the situation with a liquid mobile phase. However, in GC and SFC, the mobile phase is a compressible fluid and linear velocity (cm s^{-1}) varies along the column length in such a way that it is higher at the column outlet than at the inlet even though the flow-rate (ml min^{-1}) is constant.

The variations in mobile phase linear velocity and column pressure along the column are shown in Fig. 2.2 as a function of the column inlet and column outlet pressures, P_i and P_o, respectively. For large values of the ratio of column inlet/outlet pressure, the linear velocity changes more rapidly and the region of most rapid

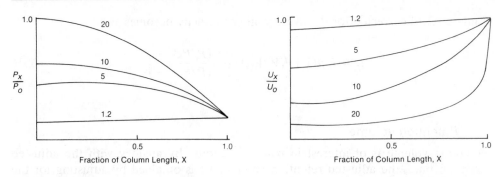

Fig. 2.2. Variation of pressure and velocity along a column for P_i/P_o values of 1.2, 5, 10 and 20 where P_i and P_o are the column inlet and outlet pressures. P_x and u_x are the column pressure and mobile phase velocity at distance x down the column.

change is near the column exit. In Section 2.4, we shall see that optimum column efficiency can only be achieved at a single value of linear velocity and that at higher and lower values of this linear velocity the efficiency decreases. Clearly, it is desirable that as much of the column as possible is operated at the optimum velocity. Figure 2.2 shows that the least change in linear velocity throughout the length of the column is encountered when the ratio, P_i/P_o, is close to unity; that is, the column pressure drop is least, as is the case with open tubular columns. In practice, this is achieved by raising the column inlet pressure above atmospheric to produce the required linear velocity, rather than by lowering the outlet pressure below atmospheric.

In cases of a compressible mobile phase, the average linear mobile phase velocity, $\bar{u} = j.u$, is obtained by multiplying the velocity measured at the column outlet by a correction factor called the compressibility, j, given in equation 2.15.

$$j = \frac{3}{2} \cdot \frac{(P_i/P_o)^2 - 1}{(P_i/P_o)^3 - 1} \qquad\qquad 2.15$$

It can be seen from the values for j in Table 2.1, that this correction is particularly significant at high P_i/P_o ratios. In common practice, GC instrument gauges measure the pressure drop across the column; thus, the inlet pressure used for calculating j in equation 2.15 is commonly the value read from the gauge plus the value for P_o.

Table 2.1 Typical values of the compressibility, j.

P_i/P_o	j	P_i/P_o	j
1.1	0.95	2.0	0.64
1.2	0.91	2.5	0.54
1.4	0.83	3.0	0.46
1.6	0.76	4.0	0.36
1.8	0.70	10.0	0.15

Equation 2.14 corrected for the average linear velocity becomes

$$t_r = L/u(1 + KV_s/V_m) \cdot \frac{2}{3} \cdot \frac{(P_i/P_o)^3 - 1}{(P_i/P_o)^2 - 1} \qquad 2.16$$

Retention volume

Retention volume is of interest in many situations. By analogy with the adjusted retention time, the adjusted retention volume, V_r' is obtained by adjusting for the holdup volume (dead volume) of the system:

$$V_r' = V_r - V_m \qquad 2.17$$

If the volume flow-rate, F, of mobile phase is constant, then t_r and V_r are related by the relationship

$$V_r = Ft_r \qquad 2.18$$

Similarly, the total volume (or holdup volume), V_m, of mobile phase in the column is given by

$$V_m = Ft_m \qquad 2.19$$

Substituting for t_r and t_m in equation 2.12 gives

$$V_r = V_m(1 + KV_s/V_m) \qquad 2.20$$

or

$$V_r = V_m + KV_s \qquad 2.21$$

Equation 2.21 is a fundamental equation in chromatography. However, it must be modified in adsorption and bonded phase chromatography which involve adsorption rather than partition. With bonded phases, the stationary phase volume can alter with mobile phase composition because of changes in orientation of the surface-bonded groups. This makes it difficult to assign a value to V_s for such phases. In adsorption chromatography equation 2.21 should be written as

$$V_r = V_m + K_A A_s \qquad 2.22$$

where K_A and A_s are the adsorption coefficient and adsorbent surface area, respectively.

Strictly speaking, equation 2.21 (and equation 2.22) assumes that K is constant and independent of the solute concentration in the mobile phase. Furthermore, kinetic processes that broaden the solute zone as it passes through the column have

been ignored. The outcome when these conditions do not apply is e:
Section 2.4.

Net and specific retention volume

It is necessary in GC, because of the compressibility of the mobile phase, to correct the retention volume to a mean column pressure using the compressibility factor, j. This corrected retention volume V_r^o is given by

$$V_r^o = jV_r \qquad 2.23$$

The same convention applies to all other volumes in GC. Thus, V_m^o is the corrected holdup volume.

It is also customary in GC to measure the mobile phase flow-rate with a soap bubble flowmeter at the column outlet at ambient temperature. For accurate measurements it is necessary to correct the measured flow, F_o, for the difference in temperature between the column and flowmeter and for the vapour pressure of the soap film (assumed to equal that of pure water). With these corrections, the corrected mobile phase flow-rate is given by equation 2.24.

$$F_c = F_o \cdot \frac{T_c}{T_a} \cdot \frac{P - p_w}{P} \qquad 2.24$$

where T_a and T_c are the ambient and column temperatures, respectively in Kelvin; P is the atmospheric pressure and p_w is the equilibrium water vapour pressure at T_a.

The corrected retention volume may be calculated from retention times using

$$V_r^o = jF_c t_r \qquad 2.25$$

In order to standardize retention volumes in GC, they are frequently corrected (for average flow) and adjusted (by subtraction of dead volume). The resulting retention volume, the net retention volume, V_N is calculated from equation 2.26.

$$V_N = jV_r' = V_r^o - V_m^o \qquad 2.26$$

Finally, the specific retention volume, V_g is the net retention volume of analyte per gram of stationary phase at 0°C, thus

$$V_g = V_N \cdot \frac{273}{T \cdot m} \qquad 2.27$$

where T is the column temperature in Kelvin and m is the mass (in grams) of stationary phase in the column. Initial hopes that V_g would provide a standardized method of reporting retention data for GC have not been fulfilled and the term is not commonly used.

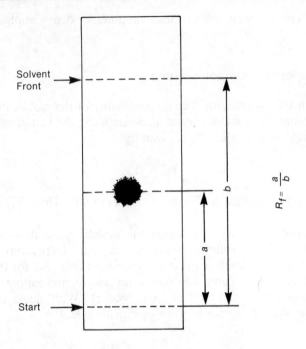

Fig. 2.3.   The calculation of retardation factor.

2.2.2   *Parameters for planar techniques*

In paper chromatography and TLC, development (i.e. addition of mobile phase to the system) is usually stopped once the desired separation of analytes has been achieved, in which case the analytes will have migrated only part way along the paper or thin layer. In this situation, the position of an analyte is characterized by the retardation factor or R_f value. It is obtained as the ratio (see Fig. 2.3) of the distance moved by the analyte to the distance moved by the mobile phase:

$$R_f = \frac{\text{distance moved by analyte band}}{\text{distance moved by mobile phase}} \qquad 2.28$$

The distance is measured to the centre of the spot. If the spot is irregular in shape an attempt should be made to locate a centre of mass for measurement, in this way reasonable accuracy can be achieved with irregular but compact spots. No useful result can be obtained in cases of bad streaking; the separation should be repeated with a different mobile phase. R_f values are usually quoted to two significant figures and by definition are between 0 and 1.

The retardation factor as defined above simply describes numerically the position of a band in a chromatogram relative to the mobile phase front. It is not an absolute physical quantity being instrument- and procedure dependent, in contrast to the thermodynamic R_f value, which can be defined in a number of ways, for example, as the fraction of analyte molecules in the mobile phase. It is this latter quantity that is related to the chromatographic equilibrium process involving analyte distribution between mobile and stationary phases. As discussed by Geiss [15] the measured R_f can only equal the thermodynamic value if: (i) the phase ratio is constant along the layer; (ii) the physicochemical composition of the layer is constant; (iii) the velocity of the mobile phase front equals the velocity at the location of the analyte band; and (iv) there are no mobile phase molecules or any other mobile liquid molecules on the layer before development is started.

These conditions are seldom all fulfilled and the observed R_f value is smaller than the true thermodynamic value, which can be obtained by multiplication of the observed value by a factor that varies between 1.0 and 1.6. For a more detailed discussion, the reader is referred to the book by Geiss [15].

2.3 Peak Shape

In section 2.2, it was assumed that the distribution coefficient, K, was a constant independent of sample concentration, in which case a plot of C_s against C_m is linear with a slope equal to the distribution coefficient. Such a plot is called an isotherm because it is obtained at a single temperature. The peak resulting for an analyte obeying a linear relationship is symmetrical and Gaussian and can be described by the equation for the normal distribution. The distribution of a peak about its mean position (i.e. retention time or retention volume) is normally measured in terms of the peak variance, σ^2, or standard deviation, σ, as plotted in Fig. 2.4. Also shown on the plot are tangents to the points of inflection. The latter describe a region at the baseline which is called the peak width at baseline, w_b, and which is equal to 4σ. The peak width at half height is of interest in a number of situations and has a value of 2.354σ. When the distribution is Gaussian, all analyte molecules migrate at the same rate (ignoring diffusion effects causing zone broadening) and the retention time is independent of sample size (Fig. 2.5).

A number of situations arise in which nonlinear isotherms are observed and the peaks are not symmetrical. For example, Langmuir isotherms are commonly observed in adsorption systems owing to sorption at heterogeneous energy sites, as caused by free silanol groups or Lewis acid sites, on the surface of silica supports. This behaviour arises when solute–stationary phase interactions are strong and solute–solute interactions are relatively weak, as occurs with a stationary phase containing sites capable of hydrogen bonding with analyte molecules. The process can be viewed as involving sorption of analyte molecules on the surface of the stationary phase until a monolayer is formed. Because the interactions between solute molecules are weak, the extent of sorption decreases following monolayer formation even though the concentration in the mobile phase is increased. Hence, the distribution coefficient is large at low C_s values but decreases as C_m increases. As the total quantity of analyte increases, the fraction in the mobile phase increases and, accordingly, the areas of highest concentration (i.e. centre of the band) migrate with the greatest velocity. The analyte molecules in the centre of the band 'catch up' with the molecules at the band front, producing a peak with a sharp front but a diffuse tail. The resulting analyte peak is described as a tailing peak.

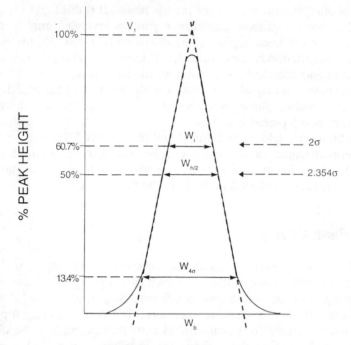

Fig. 2.4. A Gaussian peak.

Anti-Langmuir type isotherms are more common in partition systems where solute–stationary phase interactions are relatively weak compared with solute–solute interactions or where column overload occurs as a result of application of an excessive amount of solute to the system. In this case, sorption of additional analyte molecules is facilitated by other sorbed analyte molecules. Thus, at low concentrations, K has a low value that increases rapidly as analyte molecules 'attract' additional analyte molecules to the stationary phase. The resulting peak has a diffuse front and a sharp tail, and is described as a fronting peak.

A fourth type of isotherm arises from chemisorption rather than physical adsorption of analyte molecules. The resulting peak is characterized by a very slow return to the baseline because the desorption process is so slow, as often seen in complexation chromatography. In severe cases, total retention of the solute by the stationary phase occurs and no peak is observed.

A further consequence of nonlinear isotherms is that solute retention time varies significantly with sample size. This is illustrated in Fig. 2.6 for a solute exhibiting a Langmuir isotherm. The dependence of retention on sample size is a distinct disadvantage in both qualitative and quantitative analysis. It is frequently desirable to reduce such effects by suitable derivatization of the solute or by modification of the stationary or mobile phase.

Peak asymmetry can arise from a variety of other sources [16]. Unresolved solute components are sometimes the cause of distorted peak shapes. In HPLC, column void volumes result from shrinkage of the stationary phase bed with column use and the outcome is progressive peak broadening or distortion. A void near the column

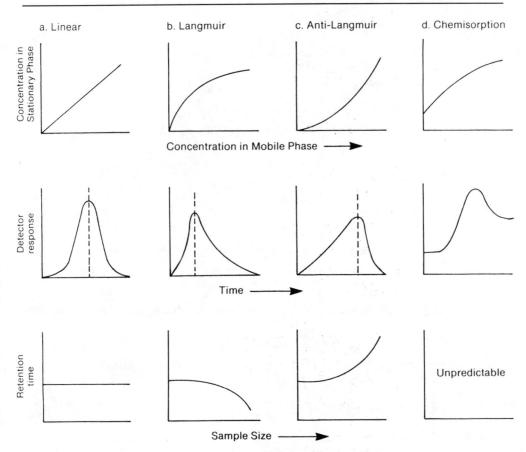

a. Linear   b. Langmuir   c. Anti-Langmuir   d. Chemisorption

Concentration in Stationary Phase

Concentration in Mobile Phase ──────▶

Detector response

Time ──────▶

Retention time

Unpredictable

Sample Size ──────▶

Fig. 2.5.   The effects of different distribution isotherms on peak shape and retention time.

inlet usually produces peak broadening but voids occupying only part of the cross-section of the bed can produce pronounced tailing or fronting. In the extreme case, zones may be split into resolved or unresolved doublets. The phenomenon is far less significant in GC because diffusion is much faster in gases.

Several measures of peak asymmetry have been devised. The simplest is the tailing factor or peak asymmetry factor, A_s.

$$A_s = \frac{b}{a}$$

2.29

where a and b are measured at a specified fraction of total peak height, the most common being 10%, as shown in Fig. 2.7. A value of 1.0 indicates a symmetrical peak whereas values exceeding 1 are tailing peaks and values less than 1 are fronting peaks. Most skewed peaks have a tailing factor of 1.10–2.50. Values within the range 0.8 to 1.2 can be tolerated without great loss in either separation efficiency or quantitative accuracy in measuring peak area or retention time. However, the cause

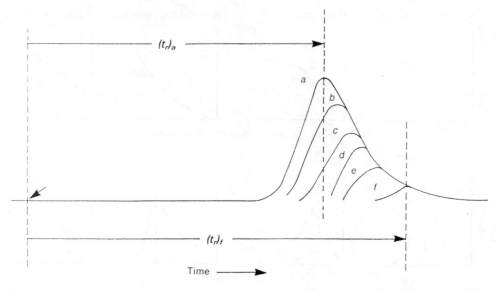

Fig. 2.6.   Effect of sample size on retention time for a hypothetical solute obeying a Langmuir isotherm.

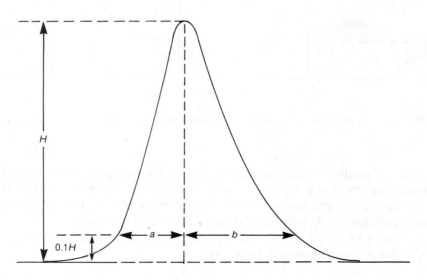

Fig. 2.7.   Procedure for measurement of the asymmetry factor.

of the peak asymmetry should be investigated for peaks with higher or lower values, as this can indicate on-column degradation of analyte, poor column packing, unfavourable interaction between analyte and stationary phase or poor injector design or setup.

   More accurate data are obtained by measuring the statistical moments of the peak (see Section 2.4), although these are not commonly used in routine work [17].

2.4   Zone Broadening and Measures of Efficiency

It has already been noted that chromatographic separation is accompanied by a concomitant broadening of the analyte zone, which is detrimental to the separation. An important goal in chromatography is to achieve sharp, symmetrical peaks as this optimizes the chance of analyte separation and improves detection. The sharpness of the peaks represents the efficiency of the chromatographic column or, more correctly, the entire system [18]. Considerable effort has been expended in devising methods of measuring the column efficiency, explaining how zone broadening processes occur and how columns are to be designed to give sharp peaks and rapid separation times. It is important to realize that efficiency is used in this context as a measure of the broadening of the sample zone as it passes through the system.

The first theoretical treatment of chromatography was developed for LC. Subsequently, various modifications have been presented which are applicable to particular situations such as GC or open tubular systems. The first two-dimensional dynamic theory of TLC was presented in 1967 [19] and the understanding of the laws governing spot spreading and layer efficiency has been extended by Guiochon *et al.* [20–25]. The theory of band broadening in TLC is the most difficult and least understood of all. The complication stems from the fact that mobile phase velocity in TLC is not constant over a run. Nevertheless, in principle, the same theory can be applied equally to all forms of chromatography [26] although such a simplified approach does have limitations and the meaning of various terms must be modified in planar chromatography. The treatment in this text relates specifically to column chromatography but it is easily extended to planar chromatography [15].

Two general approaches, the Plate theory and rate theory have emerged and their historical development prior to 1965 has been reviewed by Giddings [27]. Other general reviews [28–36] of the band broadening process are available which expand on the treatment presented below. There have been several criticisms of the Plate theory. However, apart from any other considerations it is useful to consider the Plate theory briefly in order to indicate the genesis of two terms employed in rate theory.

2.4.1   Plate theory

This is attributed to Martin and Synge [1] and describes the chromatographic process in terms similar to distillation and countercurrent extraction theory. Plate theory models a chromatographic column or layer as a series of narrow discrete sections called theoretical plates. At each plate, equilibration of analyte between the two phases is assumed to occur. Movement of analyte and mobile phase is viewed as a series of transfers from one plate to the next. The efficiency of a column increases as the number of equilibrations, i.e. theoretical plates, increases. Thus, the number of theoretical plates or plate number, n is used as a measure of column efficiency. It is defined as the square of the ratio of the retention of the analyte divided by the peak broadening (equation 2.30).

$$n = (t_r/\sigma)^2 \qquad\qquad 2.30$$

Since the variance, σ^2, of a chromatogram is difficult to measure directly, several approximate methods have been suggested. The most common approach assumes a Gaussian peak for which the relationship between peak width and standard deviation is known (see Fig. 2.4). Hence equation 2.30 can be written as

$$n = 16(t_r/w_b)^2 = 5.54(t_r/w_{h/2})^2 \qquad 2.31$$

Thus, n is a dimensionless quantity as retention time and peak width have the same units. A second plate number, called the effective plate number or number of effective theoretical plates, N or n_{eff} was introduced to allow for the inflated value of n obtained for early eluting peaks, where the void volume makes a significant contribution to the overall retention. The effective plate number is calculated according to equation 2.32 using the adjusted retention time instead of the retention time.

$$N = 16[t_r - t_m)/w_b]^2 = 5.54[(t_r - t_m)/w_{h/2}]^2 \qquad 2.32$$

Both plate number, n, and effective plate number, N, are not constant but vary with increasing retention time (Fig. 2.8). Nevertheless, N is a better parameter than n for

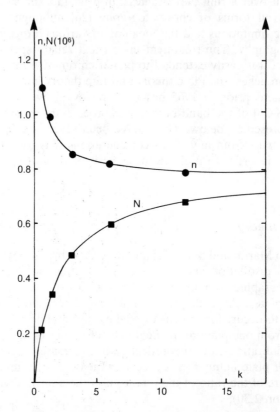

Fig. 2.8.   Variation of N and n with partition ratio. Modified from Krupcik J., Garaj J., Guiochon G. and Schmitter J.M. (1981). *Chromatographia*, **14**: 501. Reprinted with permission.

comparing column efficiencies because of the greater theoretical significance of adjusted retention time compared with retention time. As t_r approaches t_m we predict that N approaches zero and this is consistent with the poorer separation observed for early eluting peaks in all forms of chromatography.

The calculation of efficiency is more complex if the peak is asymmetrical or non-Gaussian, which is the usual situation. Here, approximations using equation 2.31 are widely used. Several other methods for calculating efficiency, but which make no assumptions about the zone distribution or profile, have been proposed. For example, a graphical method [37] and a computational method [38] have been suggested which do not require the use of a computer. Using the computational approach, n is calculated from

$$n = \frac{41.7(t_r/w_{0.1})^2}{A_s + 1.25}$$ 2.33

where A_s and $w_{0.1}$ are the asymmetry factor and peak width at 10% of peak height.

The most versatile and accurate method of calculating efficiency uses statistical moments [39–42]. The approach adopted by one instrument manufacturer [43] to obtain statistical moment data involves data logging from a fast-response detector to a computer volatile memory at a rate of 15 measurements per second. A peak-sensing algorithm then defines the analyte peak in the stored memory thereby eliminating error-prone real-time techniques for sensing peak thresholds. The various statistical moments are calculated by numerical integration and may be summarized as:

- The zero moment measures peak area.
- The first moment measures the peak mean and hence retention time.
- The second moment measures peak variance (σ^2), which is used to calculate the plate number using equation 2.30.
- The third moment measures the peak asymmetry or skew. A positive value indicates a tailing peak and a negative number a fronting peak. Statistical moments higher than the second have a value of zero for Gaussian peaks.
- The fourth moment measures the extent of vertical flattening of the peak. A negative value indicates a flattening of the peak profile relative to a Gaussian peak, while a positive value indicates a sharpening of the upper portions of the peak profile.

The only practical way to make statistical moment measurements is with a microcomputer and appropriate software for data collection and manipulation. Direct numerical integration [44–46] may lead to errors arising from the limits used in the integration, extra-column contributions, baseline drift and noise [47, 48]. More accurate moment data are obtained by curve fitting to an appropriate model [37, 38, 40], such as a Gaussian peak profile onto which an exponential decay function is superimposed. The peak components are now described by a symmetrical component due to the original Gaussian distribution and a nonsymmetrical contribution due to the exponential decay. The mathematics involved in this process

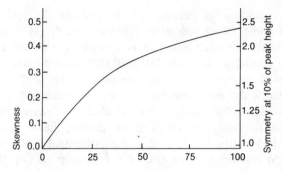

Fig. 2.9. Effect of peak shape on the accuracy of plate number calculations. Brownlee Labs Technical Notes 912. Reprinted with permission from Perkin Elmer.

are fairly complex and the interested reader is referred to the reviews by Foley and Dorsey [38, 40] for further details.

The results of moment skewness calculations are compared with the commonly used b/a ratio at 10% peak height in Fig. 2.9. Plate numbers calculated by statistical moments are consistently lower than values obtained with equation 2.31. The error introduced into plate number calculations by using equation 2.31 increases with peak asymmetry, as shown in Fig. 2.9 [17].

Plate height

The plate number and effective plate number depend on the length of the column, making intercolumn comparisons difficult. To overcome this difficulty a new parameter, the height equivalent to a theoretical plate or plate height, h or h_{etp} or hetp, has been defined according to equation 2.34.

$$h = L/n \qquad\qquad 2.34$$

The corresponding effective plate height H or height equivalent to an effective theoretical plate $HETP$ is

$$H = L/N \qquad\qquad 2.35$$

Both plate heights have units of length, usually centimetres or millimetres. Typical values of N and H are given in Table 2.2 for various column types.

Plate theory has been criticized on several occasions because of what are regarded as serious defects. The assumption that mobile phase flow occurs in a discontinuous fashion is obviously incorrect. Axial diffusion, which is assumed to be negligible, contributes significantly to band broadening. However, the largest shortcoming of Plate theory is that it fails to relate the band broadening process, as measured by N to the experimental variables such as stationary phase film thickness and particle size. This view of Plate theory is not shared by Said [26] who views the Plate theory and rate theory as complementary. Regardless of such considerations, plate number (either n or N) and plate height (h or H) are useful parameters for characterizing column efficiency.

Table 2.2   Typical values of effective plate number and height.

Column type	N (per column)	H (mm)
Gas chromatography		
Packed, 1–3 m	500–2000	1–6
Open tubular, 25 m		
0.1 mm i.d.	30 000–100 000	0.2–0.6
0.5 mm i.d.	20 000–50 000	0.5–1.3
High-performance liquid chromatography		
25 cm C18 reversed-phase		
10 μm particles	2500–5000	0.05–0.1
3 μm particles	8000–18 000	0.02–0.05
25 cm silica (10 μm)	2500–5000	0.05–0.1
ion exchange (10 μm)	1000–3000	0.08–0.25

2.4.2   Rate theory

The origin of this model is attributed to a group of Dutch chemical engineers J. J. van Deemter, F. J. Zuiderweg and A. Klinkenberg [4]. It was developed initially for packed GC columns but, because of its generality, it has readily been extended to all other types of chromatography, including packed and open tubular columns and, with minor modifications, to planar chromatography. This model focuses on the contribution of various kinetic factors to zone or band broadening as represented by H. In essence, H is assumed to be the sum of the individual contributions of the kinetic factors. The resulting summation is called the van Deemter equation and in its most general form can be written as shown in equation 2.36.

$$H = A + B/u + Cu \qquad 2.36$$

In this general form of the equation, u is the average linear mobile phase velocity, the A term represents the contribution to zone broadening by eddy diffusion, B represents the contribution of longitudinal diffusion (molecular diffusion in the axial direction) and C the contribution of resistance to mass transfer in both the stationary and mobile phases. Although this classical model is partially obsolete, it is retained in many texts because of its simplicity and didactic value.

Several modifications and refinements have been proposed [27, 49, 50] to improve the validity and utility of equation 2.36. More recently, it has been suggested [51] that the rate equation should take the form

$$H = B/u + Cu \qquad 2.37$$

$$= B/u + (C_s + C_m)\, u \qquad 2.38$$

$$= \frac{2\gamma D_m}{u} + q \cdot \left(\frac{k'}{(1+k^1)^2} \cdot \frac{d_f^2}{D_s} + \frac{\mathrm{fn}(d_p^2, d_c^2, v)}{D_m} \right) u \qquad 2.39$$

Here, the mass transfer or C term has been split into its component parts, mass transfer in the stationary phase and mass transfer in the mobile phase. The term for stationary phase mass transfer relates strictly to the situation of a liquid stationary phase. In the case of an adsorbent this term is more appropriately written as

$$\frac{k'}{(1 + k')^2} \cdot 2t_d \qquad\qquad 2.40$$

where t_d is the mean desorption time, i.e. the mean time that an analyte molecule remains attached to the adsorbent surface. Equation 2.39 resembles superficially the Golay equation [52] which refers specifically to efficiency in open tubular columns.

The symbols used in the above equations may be defined thus:

H	= column dispersivity
A	= eddy diffusion
B	= coefficient of longitudinal diffusion
C	= coefficient of mass transfer in the stationary (C_s) or mobile phase (C_m)
d_c	= column diameter
d_f	= film thickness of stationary phase
d_p	= particle diameter
D_m, D_s	= coefficient of molecular diffusion in the mobile or stationary phase
fn	= function of
k'	= capacity factor
q	= factor describing shape of stationary phase. Values generally vary from 2/15 for spherical particles to 2/3 for a uniform film
u	= average linear velocity, cm of column traversed per second
v	= volume
γ	= obstruction factor; assumes the value of unity with open tubular columns

Comparisons between columns packed with different size particles are facilitated by using reduced parameters [33, 35, 53–55]. The reduced plate height, $h_r = H/d_p$ is the number of particles to the theoretical plate. The reduced velocity, $v = u.d_p/D_m$, measures the rate of flow of the mobile phase relative to the rate of diffusion of the analyte over one particle diameter. These reduced parameters have been invaluable in developing the theory of HPLC in terms of the relationship between particle size, analysis time and column pressure drop.

Factors contributing to zone broadening: the A, B and C terms in the van Deemter equation

The individual contributions from the A, B and C terms (equation 2.36) to the overall zone spreading can be considered as independent variables, except in certain circumstances where the eddy diffusion (or A) term is coupled to the mobile phase mass transfer (C_m) term [56]. The relative contribution of the three terms is not the same in GC, LC and SFC because of the large differences in physical properties (see Table 2.3) between gases, liquids and supercritical fluids. In GC, longitudinal molecular diffusion in the mobile phase and mass transfer effects in the stationary

Table 2.3 Approximate values of parameters affecting zone broadening in gas, liquid and supercritical fluid chromatography [57].

Parameter	Gas	Supercritical fluid	Liquid
Diffusion coefficients ($cm^2 s^{-1}$)	10^{-1}	10^{-3}–10^{-4}	$<10^{-5}$
Density ($g\,cm^{-3}$)*	10^{-3}	0.3	1
Viscosity ($\mu Pa. s$)	10	10–100	1000

*$1\,g\,cm^{-3} = 10^3\,kg\,m^{-3}$.

phase are important, while in LC molecular diffusion can often be ignored but mass transfer in both phases is of importance. In SFC, the situation is intermediate between these two. The precise value of the terms will not concern us, but rather trends and the effect on each of the terms of altering practical aspects of the system, such as mobile phase velocity and stationary phase viscosity.

The A term — eddy diffusion

The contribution of eddy diffusion to the overall spreading is represented by the A term in equation 2.36. This is the most controversial term in the van Deemter equation and recently Hawkes [51] has suggested that it could be omitted altogether. Its overall significance has been reviewed [58]. When a solute zone migrates through a packed bed, the individual flow paths around the particles of packing are of different lengths. These variations, or multipaths, lead to zone broadening, the extent of which depends on the particle size of the column packing and a geometrical packing factor (related to the density and homogeneity of the column packing) whose value increases with decreasing particle size. Thus, small uniformly packed particles produce the most efficient columns. This term is zero for open tubular columns because the stationary phase in such columns is deposited directly on the walls of the column.

The B term — molecular diffusion

The B/u term arises from random molecular motion of analyte molecules in the mobile phase. As the mobile phase moves through the column analyte molecules diffuse in all directions. The longitudinal component of this diffusion, which is along the axis of the column, results in axial spreading of the zone. Its value is proportional to the time the sample spends in the column and to the diffusion coefficient of the analyte in the mobile phase. The diffusion rate is dependent on temperature and pressure of the mobile phase and spreading is reduced by low temperatures and high column pressures. The contribution of molecular diffusion to the overall plate height can usually be disregarded in LC because of the low diffusion coefficients in liquids. However, diffusion rates are much larger in gases and this term is quite significant in GC particularly at low mobile phase velocities. In order to minimize the contribution from this term, diffusion in the mobile phase must be kept small. Thus, a gas of high molecular mass, where diffusion rates are lower, is favoured as the mobile phase in GC. However, this term becomes less important as mobile phase velocity increases because of the inverse dependence on the velocity and other factors favour gases of high diffusivity (as discussed for the C term).

The C term — resistance to mass transfer

The Cu term is the most important factor in GC, LC and SFC, especially at normal operating mobile phase velocities. It describes the contribution from resistance to mass transfer in both the mobile and stationary phases and arises because the flow of mobile phase is such that equilibrium distribution of analyte between the two phases cannot be established. For example, at the front of the analyte zone, where the mobile phase encounters fresh stationary phase, equilibrium is not instantaneous and some analyte molecules are therefore carried further than would be expected under true equilibrium conditions. The effects of nonequilibrium mass transfer become greater as the mobile phase velocity is increased because less time is available for equilibrium to be approached.

The effects of resistance to mass transfer in the stationary phase can be largely ignored for solid phases because the transfer of analyte on and off the surface of an adsorbent is very rapid. With a liquid stationary phase the mass transfer term depends on the thickness of the film, the diffusion coefficient of the analyte in the stationary phase and a geometrical factor dependent on the nature of the packing. In practice, liquid stationary phases of low viscosity, present as a thin film are favoured. Sample capacity and analyte retention are frequently increased by increasing the amount of stationary phase in a column. This results in a thicker film of the phase regardless of whether an open tubular or packed column is involved and hence, entails some loss of efficiency.

The second part of the Cu term represents the effects of resistance to mass transfer in the mobile phase. Its magnitude depends on the capacity factor of the analyte and the particle diameter of the stationary phase or the internal diameter of the column in the case of an open tubular column. In either case, efficiency increases as the particle size or column diameter is decreased [59–62]. However, the effects of column pressure drop must also be considered. For example, in routine practice, open tubular columns are restricted to internal diameters above 0.15 mm to maintain an acceptable pressure drop across the column. The most important factor limiting C_m is the dependence on the diffusion coefficient of the analyte in the mobile phase. It is for this reason that gases of high diffusivity, hydrogen and helium, are favoured as mobile phases in GC. Although the effects of longitudinal diffusion (B term) can often be ignored in LC because of the low diffusion coefficients in liquids, the importance of resistance to mass transfer in the mobile phase becomes quite significant owing to the inverse dependence of C_m on the diffusion coefficient. The diffusion coefficient in a liquid is directly proportional to temperature and molecular mass of the mobile phase and indirectly related to mobile phase viscosity. Thus, increasing the temperature of a liquid chromatographic separation usually improves efficiency significantly (particularly for high molecular mass analytes) because of the effect on diffusion rates. Moreover, the reduced efficiency caused by slow solute diffusion in LC is partly overcome by operating at much lower mobile phase velocities than in gas chromatography, although this is achieved at the expense of an increased analysis time.

Concluding remarks

In addition to its theoretical significance in understanding the processes occurring during chromatographic separation, the van Deemter equation has been very useful in improving column and instrument design [34, 35, 63] and in day to day setting of operating conditions (i.e. by the use of van Deemter plots) to minimize H. In general, we can say that a minimum value of H is obtained, i.e. efficiency is maximized if stationary phase thickness is minimized, columns are packed with small diameter particles or, alternatively, open tubular columns with small internal diameters are used and mobile phases and stationary phases of low viscosity and high diffusivity are utilized.

2.4.3 Van Deemter plot

A graph of plate height against average linear mobile phase velocity is termed a van Deemter plot. Such plots are of considerable practical use even though the values of parameters in the van Deemter equation are seldom known accurately. In practice, data for the van Deemter plot are determined experimentally using measured values of retention time, dead time and peak width to obtain N and hence H at varying mobile phase velocities. The plot is then used to either minimize H; that is, to minimize zone dispersion and maximize efficiency or to minimize separation time by using the fastest mobile phase velocity (see Table 2.4) consistent with an acceptable value of H.

Van Deemter curves for GC, LC and SFC exhibit a similar form, showing a minimum value of H at some optimum mobile phase velocity. A hypothetical curve is presented in Fig. 2.10 showing the individual contributions of the three terms, A, B and C terms, to the overall efficiency, H. In this plot, the A term is a constant independent of mobile phase velocity. At mobile phase velocities below the optimum, the overall efficiency is strongly dependent on diffusion effects (the B

Table 2.4 Range of linear mobile phase velocities and flow-rates for gas and high-performance liquid chromatography.

Technique	Mobile phase velocity (cm s^{-1})		Flow-rate (ml min^{-1})	
	Optimum	Operating	Optimum	Operating
Gas chromatography 4 mm i.d. packed column	2–4	4–8	40–50	40–80
0.25 mm i.d. open tubular column, hydrogen carrier gas*	30–40	60–80	1–2	1–4
High-performance liquid chromatography 4.6 mm i.d. packed column	0.05–0.1	0.1–0.2	0.2–0.5	1–2

*For helium, multiply the values by about 0.55; for nitrogen multiply by about 0.3.

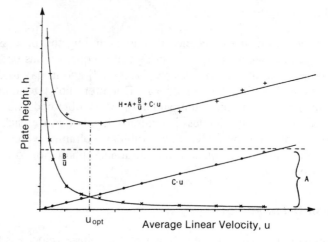

Fig. 2.10.   Hypothetical curve showing the relationship between efficiency and average linear mobile phase velocity.

term) and increases rapidly. At higher velocities, the efficiency decreases because mass transfer terms become more important. However, despite reduced efficiency it is common to operate at mobile phase velocities above the optimum because of the reduced analysis time and the savings this represents in instrument and operator time.

2.4.4 Extra-column effects in zone broadening

The discussion has been limited, so far, to zone broadening in the chromatographic column only. However, zone broadening can arise from phenomena occurring outside the stationary phase bed. Hence, measured values of H correspond to broadening in the column plus extra-column effects [64] since the column cannot be used in isolation from the rest of the chromatographic system. A thorough treatment of these 'end-effects' is given by Sternberg [28] and Gill and Sankasubramanian [65]. There is also an inherent zone broadening in GC because of the pressure drop across the column, which results in an expansion of the gaseous analyte. Expressing the total zone broadening as a sum of the variance of the column and extra-column effects we can write

$$\sigma_{T(otal)}^2 = \sigma_{C(olumn)}^2 + \sigma_{E(xtra-column)}^2 \qquad 2.41$$

Extra-column effects include zone broadening contributions from the method of sample introduction, the time constants associated with the detector amplifier and recorder and particularly dead volumes in the injector, the detector, and connector tubing. Hence, the final term in equation 2.41 may be expanded and the total zone broadening may be expressed as a sum of the variance of the individual contributions:

$$\sigma_T^2 = \sigma_C^2 + \sigma_{I(njection)}^2 + \sigma_{In(jector)}^2 + \sigma_{Tu(bing)}^2 + \sigma_{D(etector)}^2 \qquad 2.42$$

This equation emphasizes the importance of the entire chromatographic system in contributing to efficiency. It is futile purchasing an expensive column in an effort to minimize σ_C^2 if extra-column contributions to zone broadening, as measured by the variances σ_I^2, σ_{In}^2, σ_{Tu}^2 and σ_D^2 have not also been minimized.

For example, columns are sold with a stated efficiency, frequently expressed as a plate number. It is all too common to blame column performance for lower efficiencies than stated when the fault may lie elsewhere. This can be easily checked by measuring performance with the column connected and repeating the measurement with a low dead volume union connecting the injector and detector. This is illustrated by the following example, as detailed below.

A column was purchased with a stated plate number of 30 400 using aqueous methanol as mobile phase and aniline ($t_r = 6.54$ min; $w_b = 9$ s) as a test solute. The column was checked in the purchaser's laboratory and gave a plate number of only 6047, as calculated from the retention time of 6.48 min and baseline peak width of 20 s. These data reflect the performance of the entire chromatographic system and do not allow any decisions as to the efficiency of the column. However, with the column replaced by a low dead volume connector, the peak width was 17.5 s, corresponding to the extra-column zone broadening.

The column contribution to zone broadening, $w_{b,column}$ cannot be obtained directly but can be calculated from the following relationships.

$$\sigma_T^2 = \sigma_C^2 + \sigma_E^2 \qquad 2.43$$

and

$$(w_{b,total})^2 = (w_{b,column})^2 + (w_{b,extra\text{-}column})^2 \qquad 2.44$$

In this equation, both $w_{b,total}$ and $w_{b,extra\text{-}column}$ are known. Hence,

$$(20)^2 = (w_{b,column})^2 + (17.5)^2$$

and

$$(w_{b,column})^2 = 93.75$$

$$w_{b,column} = 9.5 \text{ s}$$

In a properly designed system, the contribution of non-column zone broadening will be small and hence the total variance will equal the column variance. Thus, we can write

$$\sigma_T^2 \approx \sigma_C^2$$

and

$$t_{r,total} \approx t_{r,column}$$

Substituting for $t_{r,column}$ in the plate equation gives

$$n_{column} = 16(t_{r,total}/w_{b,column})^2$$

and a plate number for the column of

$$n_{\text{column}} = 16(6.48 \times 60 \text{ s}/9.5 \text{ s})^2$$
$$= 26\,800$$

The column efficiency is still excellent and the problem lies somewhere in the non-column sections of the chromatographic system. In this instance, the chromatographer would be well advised to check for dead volume in connector tubing.

Practical considerations

The discussion thus far has assumed an infinitely thin input volume of sample. This is not practicably attainable because of the design of sample inlet systems and the finite volume of sample introduced. The latter is particularly true in GC, where the sample is introduced by syringe through a septum and often vaporized in the injection system. For example, one microlitre of a volatile solvent (a typical injection volume in packed column GC) when vaporized at 250°C and 200 kPa pressure, will occupy approximately 0.5 ml—a considerable volume, representing several centimetres of column length. Moreover, the finite rate of vaporization spreads the analyte zone into an exponential distribution. Injection with a valve employing a fixed volume sample loop at ambient temperature is most common in liquid chromatography. In this situation slow removal of traces of sample from the walls of the valve may contribute to zone broadening.

A major concern in considering extra-column effects is dead volume, a term used loosely over many years, but now taken to be the volume between the injection point and detection point after deducting the stationary phase volume. Thus, any section of the system in which solute and mobile phase are mixed, but not exposed to stationary phase, constitutes dead volume in which diffusion of solutes can occur. Dead volume effects become more severe as the scale of the instrument decreases and thus are a major concern with open tubular and microbore column systems. A number of the factors contributing to dead volume are largely beyond the control of the chromatographer and their limitation and reduction is inherent in good mechanical design and manufacture of chromatographic instruments. Nonetheless, there are some steps that can be taken by the operator to minimize dead volume. In GC, dead volume is only a problem with open tubular column systems because of the high diffusion rates in gases. Steps that can be taken to minimize the effects of dead volume are the correct installation of the open tubular column in both the injector and detector. In HPLC, the lower diffusion rates in a liquid reduce mixing and dead volume can be very significant. The column packing is susceptible to compression owing to the high operating pressures and this is a serious source of dead volume, particularly where a column has been subjected to sudden and rapid changes in pressure. Moreover, the connector tubing between the column and the detector, or the injector, is particularly sensitive to zone broadening being dependent on the radius (r) and the length (L) of the tubing. The contribution to zone broadening in an open tube is given by

$$\sigma_{\text{Tu}}^2 = \pi r^4 \, FL/24D_{\text{m}} \qquad\qquad 2.45$$

A practical consequence is that short, narrow-bore tubing should be used to link the column to the injector and detector in conjunction with zero dead volume couplings.

Chromatography detectors have a finite cell volume in which analyte zones are detected. If this cell volume greatly exceeds the volume of mobile phase in which a zone is eluted, the detector volume will act as a mixing chamber, resulting in exponential dilution of the analyte zone. This situation is common in low-pressure column chromatography where large dead volume outlets are common. In a well-designed detector, laminar flow will be preserved and the zone broadening in the detector can be calculated as a variance

$$\sigma_D^2 = \frac{(\Delta t)^2}{12} \qquad\qquad 2.46$$

where Δt is given by the ratio of detector volume, V, to mobile phase flow-rate, F.

For example, let $\sigma_C = 2$ s for a particular separation involving HPLC using a mobile phase of methanol + water $(40 + 60)$ at a flow-rate of 1 ml min^{-1}. What is the maximum detector cell volume permissible if zone broadening in the detector is to contribute less than 1% to the total? (Ignore other non-column contributions to zone broadening.)

We require that σ_T be no more than 2.02 seconds (i.e. $100 + 1\%$). Thus,

$$\sigma_T^2 = \sigma_C^2 + \sigma_D^2$$

or,

$$\sigma_D^2 = (2.02)^2 - (2.00)^2$$

$$= 0.0804 \text{ s}^2$$

and

$$\sigma_D = 0.283 \text{ s}$$

Substituting into equation 2.46

$$\sigma_D^2 = \frac{(\Delta t)^2}{12} = \frac{V^2}{12F^2}$$

Rearranging

$$V = \sqrt{12}\,\sigma_D F$$

$$= (3.46)\,(0.283 \text{ s})\,(16.7 \;\mu l.s^{-1})$$

$$= 16.4 \;\mu l$$

Detectors meeting this requirement, or better, are readily available. Typical detector volumes in a modern HPLC instrument, for example, are 10, 5 or 3 μl.

2.4.5 Other measures of separation efficiency

Column efficiency, as measured by plate height (and number), is a measure of the zone broadening that an analyte experiences during its passage through a chromatographic system. However, this provides, at best, an indirect indication of how good a separation will be. The separation of solutes by a chromatographic system depends on selective retention by the stationary phase. This selectivity can be expressed as a relative adjusted retention. This is referred to as the selectivity or separation factor, α, defined as follows:

$$\alpha = \frac{t_{r2}}{t_{r1}} = \frac{k_2}{k_1} = \frac{K_2}{K_1} = \frac{V_{r2}}{V_{r1}} \qquad 2.47$$

The relationship of the selectivity factor to the distribution coefficients emphasizes the thermodynamic basis of chromatographic separation. The selectivity factor may be used as the basis for standardizing retention times (see Chapter 9) by quoting values relative to a 'standard solute'.

Resolution

Although the selectivity factor describes the separation of zone centres it takes no account of peak widths. A better measure of separation is provided by resolution, R_s, which takes into account both retention difference and column efficiency; and for symmetrical peaks of Gaussian shape is given by

$$R_s = \frac{2(t_{r2} - t_{r1})}{w_{b1} + w_{b2}} \qquad 2.48$$

Equation 2.48 can be written in terms of retention volumes by replacing t_{r1} and t_{r2} with V_{r1} and V_{r2}, respectively and using volume units for the baseline widths such that R_s has the same dimensionless value. Baseline resolution of Gaussian peaks is considered to be achieved with a resolution of 1.5. For symmetrical peaks, a resolution of 1.0 corresponds to approximately 2% peak overlap, which is adequate for most quantitative work. The ability to recognize two peaks becomes progressively more difficult as resolution decreases, particularly when the analyte peaks are of different intensities. Snyder [66] has suggested a practical approach to estimating resolution by comparison with computer-simulated chromatograms (Fig. 2.11). Such a method provides a quick, convenient method of estimating resolution, but it is at best only an approximation. In Fig. 2.11(a), two peaks are clearly identified at a resolution of 0.6 whereas in Fig. 2.11(c), the second peak is only apparent as an asymmetry on the tail of the first peak. The situation will be exacerbated with nonsymmetrical peaks. In the case of overlapping peaks, the position of the true and apparent zone centres of the two components do not coincide, as shown in Fig. 2.11, where the true zone centres are indicated by black dots. This situation is deleterious for quantitative analysis (see Chapter 9).

Peak capacity and Trennzahl

The peak capacity is the maximum number of peaks that can be resolved ($R_s = 1.0$) by a given system in a specified time. It was devised by Giddings [67] to facilitate comparison of the separating capabilities of the various chromatographic techniques.

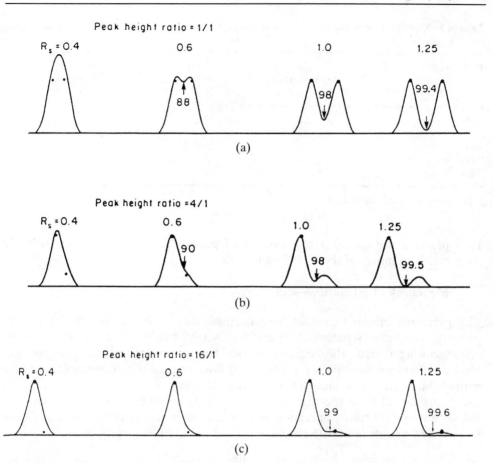

Fig. 2.11.   Comparison of resolution values for peaks of equal and nonequal heights. Reproduced from Snyder L.R. (1972), *J. Chromatogr. Sci.*, **10**: 200, by permission of Preston Publications, A division of Preston Industries, Inc.

As we have seen, there is a lower limit to the retention time (or volume), this being the dead time. With the exception of size exclusion chromatography, theoretically there is no upper limit. Nevertheless, there is a practical limit and, because the peaks have a finite width as defined by the plate number, only a limited number of peaks can be accommodated in the range accessible to them. In many gas and most liquid chromatographic separations, the practical limit of t_r/t_m is 10 (see Section 2.5), although this can be extended in GC under favourable conditions to a t_r/t_m ratio of 50. Table 2.5 shows theoretical values of the peak capacity obtained by Giddings for a given number of theoretical plates.

   Kaiser [68] introduced a similar concept to the peak capacity called the separation number or Trennzahl, TZ, which is defined as the resolution between two consecutive members of an homologous series.

$$TZ = \frac{t_{r(x+1)} - t_{r(x)}}{w_{b(x+1)} + w_{b(x)}} - 1 \qquad\qquad 2.49$$

Table 2.5　Calculated peak capacities of chromatographic systems for a given number of theoretical plates.

Plate number	Gas chromatography $(t_r/t_m = 50)$	Liquid* chromatography $(t_r/t_m = 10)$	Size exclusion chromatography $(t_r/t_m = 2.3)$
100	11	7	3
400	21	13	5
1000	33	20	7
2500	51	31	11
5000	69	41	14

*Adsorption or partition systems.

In a practical sense, TZ is the number of peaks that can be resolved between consecutive members of the homologous series.

Measures of efficiency—summary

The practising chromatographer is concerned only with the system's ability to perform a desired separation. Since the system hardware is largely fixed by manufacturing design, the concern focuses largely on the column performance. Various arguments have been presented [69] favouring the use of one or other of the various parameters as a measure of column efficiency. For example, TZ, n (or N) and h (or H) all have specific advantages. TZ is useful since it can be used for temperature-programmed operation as well as isothermal operation. In addition, it is related to Kovats indices (Sections 3.5.2 and 9.2.1) by the equation

$$TZ = \frac{100}{\Delta I} - 1$$

2.50

Plate numbers emphasize peak width, assuming that a system eluting a sharper peak at a given retention time is more efficient. As we have seen, the exact meaning is that it causes less zone broadening. However, 'more efficient' is often interpreted [70] as meaning separating more substances within a given time. In many instances, the chromatographer is just as concerned with analysis time as with absolute efficiency and a better indication of column performance may be provided by the number of theoretical plates generated per second.

It should be appreciated that none of the efficiency parameters can provide an infallible guide to performance in all situations. Nevertheless, such parameters do provide an indication of likely performance and can be used to monitor column and system performance for quality control purposes.

2.5　Optimizing Resolution

The ultimate goal of the chromatographer is the attainment of sharp well-resolved peaks in the minimum analysis time. If components 1 and 2 have capacity factor k'_1

and k_2', respectively, then their retention times, t_{r1} and t_{r2}, from equations 2.11 and 2.13 are equal to

$$t_{r1} = L/u(1 + k_1') \qquad\qquad 2.51$$

and

$$t_{r2} = L/u(1 + k_2') \qquad\qquad 2.52$$

The peak separation is obtained as the difference of equations 2.52 and 2.51:

$$t_{r2} - t_{r1} = L/u(k_2' - k_1') \qquad\qquad 2.53$$

indicating that the separation is proportional to the column length. An obvious solution to improved resolution is therefore to increase column length. However, zone broadening, which opposes separation, increases as the analyte traverses the column. Zone broadening or peak width is proportional to the square root of the column length. Thus, attempts to improve resolution on long columns will be difficult because of the increased peak width of the analytes. Nevertheless, the dominant effect, because of the direct proportionality, is the increase in peak separation as column length is increased. If, for example, column length is doubled, peak width will increase by a factor of $\sqrt{2}$ whereas separation distance will double. Therefore, any two solutes with at least some difference in their respective distribution coefficients can be separated by using a column of sufficient length.

In practice, altering column length is not usually the easiest way to improve resolution and achieve separation. With a longer column, more pressure is required to move the mobile phase through the column, so that column length is ultimately governed by the available pressure. Furthermore, the problems of packing and housing very long columns limit their length in practice, although this is not a consideration in the case of flexible open tubular columns.

Equation 2.48 does not provide any useful information about the kinetic or thermodynamic properties of the column, or as to how resolution could be improved. However, by combining the equations for efficiency (equation 2.31) and resolution (equation 2.48) with equations 2.9 and 2.47 for two peaks with similar retentions, a more useful form of the resolution equation for this purpose is obtained (equation 2.54).

$$R_s = \frac{1}{2} \cdot \frac{\alpha - 1}{\alpha + 1} \cdot \frac{k_2'}{1 + k_2'} \cdot \sqrt{N} \qquad\qquad 2.54$$

or,

$$R_s = \frac{1}{2} \cdot \frac{\alpha - 1}{\alpha + 1} \cdot \frac{k_2'}{1 + k_2'} \cdot \sqrt{(L/h)} \qquad\qquad 2.55$$

This is a popular form of the resolution equation, of which there are now various forms but all have in common [71, 72] a selectivity term dependent on α, a rate of migration term dependent on k' and an efficiency term dependent on L and h. The first two terms are essentially thermodynamic, whereas the L/h term is mainly associated with the kinetic features of the separation process. Since a large value of R_s indicates good resolution, the three terms should be maximized. It is instructive to consider the three parts of the equation in more detail, considering each as independent functions in order to further investigate their effect on resolution. Strictly, the three terms are not independent but can be treated as such to a first approximation.

2.5.1 Maximizing resolution by optimizing partition ratio

When using equation 2.55 to optimize a given separation, the term involving k' should be considered first. Equation 2.55 can be written in the form

$$R_s/Q = \frac{k_2'}{1 + k_2'} \qquad\qquad 2.56$$

which is plotted in Fig. 2.12. When k_2' for the initial separation is less than 0.5, a rapid enhancement in resolution is achieved by increasing k_2' into the optimum range $1 < k_2' < 10$. No other change in separation conditions will give as large an increase in resolution for as little effort. However, analysis time is significantly increased [59]. It is clear from Fig. 2.12 that values of k_2' exceeding 10 are to be avoided as this provides little increase in resolution but markedly increases the analysis time.

Since k' is proportional to the distribution coefficient, it depends on the intermolecular interactions between analyte molecules and each of the two phases. Hence, k' can be varied by altering the temperature or stationary phase in GC. Alternatively, k' can be altered by changing the relative volume of stationary and mobile phases ($k' = KV_s/V_m$). To increase k' at constant temperature, the phase ratio must be decreased. In LC, the stationary phase or mobile phase can be varied although, in practice, the latter is easier and is the usual approach except in size exclusion where mobile phase composition usually has little effect on zone migration. In SFC, temperature, pressure and mobile or stationary phase composition can be varied.

The optimum value of k_2' should consider both resolution and time. Indeed, the chromatographer's interest is generally in the ability of a system to deliver a certain efficiency per unit time [73, 74]. A time-dependent equation for resolution can be obtained by combining equations 2.11, 2.13 and 2.35 to give the time of analysis as:

$$t_r = \frac{N(1 + k')H}{u} \qquad\qquad 2.57$$

Substituting this expression in equation 2.54 and eliminating n gives:

$$t_r = 4 R_s^2 \cdot \left[\frac{\alpha + 1}{\alpha - 1}\right]^2 \cdot \frac{(1 + k_2')^3}{k_2'^2} \cdot \frac{H}{u} \qquad\qquad 2.58$$

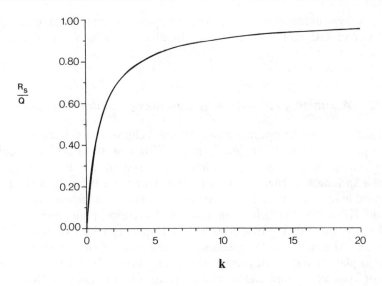

Fig. 2.12. Effect of capacity factor on resolution.

Hence, the time of analysis is a function of the required resolution and the column operating conditions in terms of selectivity, capacity factor, plate height and mobile phase linear velocity. Thus, for example, analysis time can be halved by reducing resolution by 0.71 (i.e., $\sqrt{1/2}$) or by doubling mobile phase velocity. However, u is related to H (see rate theory) and any increase in mobile phase linear velocity will also increase H and so resolution would decrease. Similarly, halving the plate height will halve the analysis time but this would also degrade resolution. The optimum value of k' for minimum analysis time can be obtained by differentiating equation 2.58 with respect to k'. This yields an optimum k' value of 2, but in practice it depends on circumstances and varies between about 2 and 5 [75].

General elution problem

When a sample containing components with a large spread of k' values is chromatographed under isocratic conditions (i.e. conditions of constant mobile phase composition) the resulting chromatogram commonly exhibits three distinct regions; the early eluting components (with $k' < 2$) bunched together and showing poor resolution, a middle region with well-resolved symmetrical peaks and a final region of broad diffuse peaks. This behaviour is typical of all forms of chromatography and has been termed the general elution problem by Snyder [76, 77]. Attempts to improve separation under isothermal or isocratic conditions will serve only to rearrange the three regions. A solution to this problem requires that column conditions be changed during the separation so as to optimize k' for individual zones as they move through the column. This has been achieved by gradient elution (changing mobile phase composition) in LC (Chapters 5 and 6), by temperature programming in GC (Chapter 3) and by temperature, pressure and mobile phase gradients in SFC (Chapter 7). The use of coupled columns, a technique now encompassed by multidimensional chromatography, has been used as an alternative in both LC and GC and is particularly effective in the analysis of very complex mixtures.

2.5.2 Maximizing resolution by optimizing column efficiency

When adjusting k_2' to the optimal range fails to achieve the desired separation, the best solution is usually an increase in N. This can obviously be achieved by increasing column length, but the resolution is increased only by the square root of the increase in length at the expense of a linear increase in analysis time, although the increased length can be compensated for example, by increasing the column inlet pressure in HPLC. Nevertheless, in general, increasing plate number by varying column length with other parameters held constant can be expensive in terms of analysis time. Alternatively, the increase in plate number can be accomplished by a reduction in plate height as discussed under rate theory. In such circumstances, the selection of another column and/or mobile phase or even another chromatographic technique may be more suitable.

Equation 2.54 can be rearranged and used to calculate the number of plates required for a given separation:

$$N = 4R_s^2 \cdot \left[\frac{\alpha + 1}{\alpha - 1} \right]^2 \cdot \left[\frac{k_2' + 1}{k_2} \right]^2 \qquad 2.59$$

As an illustration of this relationship, Table 2.6 lists the number of theoretical plates required to achieve a specified resolution for various capacity factor values; a resolution of 1.5 represents baseline resolution, while a resolution of 1.0 corresponds to about 98% separation. The calculated values show a rapid decrease in the required N value as the capacity factor is increased from 0.2 to 5 and the effect of a small change in selectivity factor, from 1.05 to 1.10, corresponding to the practical case of moderately difficult separations. The reduction in required plate number also affects other column characteristics. Assuming a constant H value, then, because less plates are needed, the same resolution can be obtained on a shorter column. Conversely, a shorter column means a shorter retention time (assuming the same average linear mobile phase velocity) since $t_r = L/u \, (1 + k')$.

2.5.3 Maximizing resolution by optimizing column selectivity

An increase in column selectivity, α, is a powerful technique for improving resolution because resolution can be increased while analysis time is decreased. However, determining the necessary operating conditions is time consuming, difficult to treat from a theoretical viewpoint and often becomes a matter of trial and error. This reflects the general lack of understanding of intermolecular interactions between complex molecules that pervades the physical sciences in general [78]. In many instances, specialty texts and the literature on technical data from instrument manufacturers or chromatography supply companies may be helpful.

Table 2.6 Plate number required to achieve a given resolution.

Capacity factor k_2'	$R_s = 1.0$		$R_s = 1.5$	
	$\alpha = 1.05$	$\alpha = 1.10$	$\alpha = 1.05$	$\alpha = 1.10$
0.2	242 064	63 504	544 644	142 884
0.5	60 516	15 876	136 161	35 721
1.0	26 896	7056	60 516	15 876
1.5	18 678	4900	42 024	11 025
2.0	15 129	3969	34 040	8930
5.0	9683	2540	21 786	5715
10.0	8136	2134	18 306	4802
50.0	6996	1835	15 740	4129

If the capacity factor and efficiency terms are held constant, equation 2.55 may be written in the form

$$R_s/Q = \frac{\alpha - 1}{\alpha + 1} \qquad\qquad 2.60$$

The variation of resolution with column selectivity is then obtained (Fig. 2.13). When $\alpha = 1$, it follows that the distribution coefficients of the two analytes are identical and separation is not possible. When selectivity is less than about 1.1, a small change in α produces a large change in resolution and this can be a useful strategy for improving resolution of analytes with low α values. In other circumstances changes in resolution by selectivity changes are desirable, but unfortunately may lead merely to a reshuffling of peaks with no real improvement in the overall separation.

For complex mixtures containing several components, the same principles can be applied to optimize the separation of the solute pair which is most difficult to resolve. The often time-consuming trial and error process of selecting and optimizing chromatographic conditions makes the practice of chromatography something of an art. The burden of development and optimization of chromatographic methods is being reduced as more sophisticated optimization schemes are developed. These approaches generally involve established procedures such as SIMPLEX, response-surface methodology or multilevel factor analysis. Modern HPLC instruments, where the problems of optimization are more acute than in GC, can be programmed to handle these mathematical approaches and in the future smart chromatographs will increasingly become available.

The series of hypothetical chromatograms in Fig. 2.14 illustrate the effect of changes in the partition ratio, efficiency and selectivity on the resolution. Chromatogram (a) shows an initial separation in which the two analytes are only partially resolved. In chromatogram (b) plate number has been increased but with no change in resolution as a result of the reduced capacity factor. The increased capacity factor in (c) results in improved resolution. Altering the column selectivity by increasing α

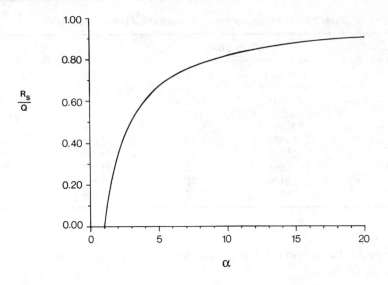

Fig. 2.13.   Effect of selectivity on resolution.

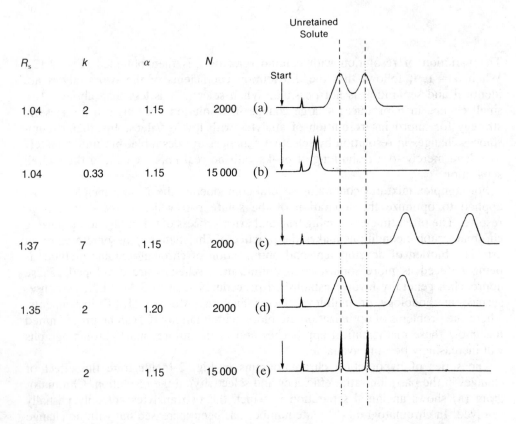

Fig. 2.14.   Hypothetical chromatograms illustrating changes in resolution as a result of changes in the capacity factor, column selectivity and efficiency.

results in a displacement of one zone centre relative to the other and a rapid increase in resolution, as shown in (d). Finally, an increase in plate number (see e) because of greater efficiency results in sharper better-resolved peaks without altering retention times.

2.6 Overall System Performance

In the early stages of development of chromatography, separation efficiency, as measured by parameters such as plate number, plate height and Trennzahl was the most important quality performance parameter. Now, separation efficiency is only one of many performance criteria and the full characterization of a system should include [79, 80] assessment of efficiency, retention (k), relative retention (α), column activity, sample capacity, and bleed rate. In the case of HPLC, column pressure drop or permeability (at a specified flow-rate and temperature) should be included [81]. Although time consuming, regular testing of column performance should be a routine part of quality assurance. In many instances complete evaluation of the system is not essential but should concentrate on those aspects of performance critical to the particular application. The requirements of a system used routinely for qualitative analysis, for example, may differ substantially from those required in a quantitative analysis at the ultratrace level.

Procedures used to assess column activity and bleed rate are somewhat specific to the individual technique and more appropriately treated in later chapters. Column activity refers to any interactive effects observed in a column that are additional to the desired process. The assessment of this activity has assumed increasing importance as the performance of other parts of the chromatographic system (e.g., detector sensitivity) has improved. Column bleed refers to the elution of material from a column which is not attributable to sample. This arises from leaching effects of the mobile phase in partition systems using coated phases or, more commonly, from stationary phase degradation of bonded phases. The problem is obviously magnified as detector sensitivity is improved and in ultratrace analysis. It is more prevalent in GC because of elevated temperature operation.

References

1. Martin A.J.P. and Synge R.L.M. (1941). *Biochem. J.*, **35**: 1358.
2. Craig L.C. (1950). *Anal. Chem.*, **22**: 1346.
3. Glueckauf E. (1955). *Trans. Faraday Soc.*, **51**: 34.
4. van Deemter J.J., Zuiderweg F.J. and Klinkenberg A. (1956). *Chem. Eng. Sci.*, **5**: 271.
5. Giddings J.C. (1959). *J. Chem. Phys.*, **31**: 1462.
6. ASTM (1977). *Standard Recommended Practice for Gas Chromatography Terms and Relationships.* ASTM E 355. American Society for Testing Materials, Philadelphia, 1977.
7. ASTM (1979). *Standard Practice for Liquid Chromatography Terms and Relationships.* ASTM E 682, American Society for Testing and Materials, Philadelphia, 1979.
8. Irving H.M.N.H., Freiser H. and West T.S. (1978). *Compendium of Analytical Nomenclature. Definitive Rules. 1977* (IUPAC Orange Book), Pergamon, Oxford.
9. Ettre L.S. (1981). *J. Chromatogr.*, **220**: 65.
10. Ettre L.S. (1981). *J. Chromatogr.*, **220**: 29.

11. Ettre L.S. (1979). *J. Chromatogr.*, **165**: 235.
12. Wainwright M.S. and Haken J.K. (1980). *J. Chromatogr.*, **184**: 1.
13. Smith R.J., Nieass C.S. and Wainwright M.S. (1986). *J. Liquid Chromatogr.*, **9**: 1387.
14. Berendson G.E., Schoenmakers P.J., De Galen L., Vigh G., Varga-Puchony Z. and Inczedy J. (1980). *J. Liquid Chromatogr.*, **3**: 1669.
15. Geiss F. (1987). *Fundamentals of Thin Layer Chromatography* Huethig, Heidelberg, (p. 91).
16. Conder J.R. (1982). *J. High Resol. Chromatogr., Chromatogr. Comm.*, **5**: 341, 397.
17. Kirkland J.J., Yau W.W., Stoklosa H.J. and Dilks C.H. (1977). *J. Chromatogr. Sci.*, **15**: 303.
18. Bidlingmeyer B.A. and Warren F.V. (1984). *Anal. Chem.*, **56**: 1583A.
19. Belenkij B.G., Nesterov V.V., Gankina E.S. and Smirnov M.M. (1967). *J. Chromatogr.*, **31**: 360.
20. Guiochon G., Siouffi A., Engelhardt H. and Halasz I., *J. Chromatogr. Sci.*, **16**: 152.
21. Guiochon G. and Siouffi A. (1978). *J. Chromatogr. Sci.*, **16**: 470.
22. Guiochon G., Bressolle F. and Siouffi A. (1979). *J. Chromatogr. Sci.*, **17**: 368.
23. Guiochon G. and Siouffi A. (1979). *Analusis*, 7: 316.
24. Siouffi A., Bressolle F. and Guiochon G. (1981). *J. Chromatogr.*, **209**: 129.
25. Guiochon G. and Siouffi A. (1982). *J. Chromatogr.*, **245**: 1.
26. Said A.S. (1983). *Theory and Mathematics of Chromatography*. Huethig, Heidelberg.
27. Giddings J.C. (1965). *Dynamics of Chromatography, Part 1*. Marcel Dekker, New York, pp. 13–25.
28. Sternberg J.C. (1966). In Giddings J.C. and Keller R.A., Editors, *Advances in Chromatography*, Vol. 2. Marcel Dekker, New York, p. 205.
29. Perry J.A. (1981). *Introduction to Analytical Gas Chromatography. History, Principles and Practice*. Marcel Dekker, New York.
30. Huber J.F.K., Lauer H.H. and Poppe H. (1975). *J. Chromatogr.*, **112**: 377.
31. Horvath C. and Lin H.-J. (1976). *J. Chromatogr.*, **126**: 401.
32. Horvath C. and Lin H.-J. (1978). *J. Chromatogr.*, **149**: 43.
33. Knox J.H. (1977). *J. Chromatogr. Sci.*, **18**: 453.
34. Batu V. (1983). *J. Chromatogr.*, **260**: 255.
35. Knox J.H. and Scott H.P. (1983). *J. Chromatogr.*, **282**: 297.
36. Meyer, V.R. (1985). *J. Chromatogr.*, **334**: 197.
37. Barber W.E. and Carr P.W. (1981). *Anal. Chem.*, **53**: 1939.
38. Foley J.P. and Dorsey J.G. (1983). *Anal. Chem.*, **55**: 730.
39. Yau W.W. (1977). *Anal. Chem.*, **49**: 395.
40. Foley J.P. and Dorsey J.G. (1984). *J. Chromatogr. Sci.*, **22**: 40.
41. Bidlingmeyer B.A. and Warren F.V. (1984). *Anal. Chem.*, **56**: 1583A.
42. Dezaro R.A., Floyd T.R., Raglione T.V. and Hartwick R.A. (1986). *Chromatogr. Forum*, **1(1)**: 34.
43. Higgins J., *Technical Notes 912*. Brownlee Labs.
44. Pazdernik O. and Schneider P. (1981). *J. Chromatogr.*, **207**: 181.
45. Said A.S, Al-Ali H. and Hamad E. (1982). *J. High Resol. Chromatogr., Chromatogr. Comm.*, **5**: 306.
46. Conder J.R., Rees G.J. and McHale S. (1983). *J. Chromatogr.*, **258**: 1.
47. Chesler S.N. and Cram S.P. (1971). *Anal. Chem.*, **43**: 1922.
48. Vidal-Madjar C. and Guiochon G. (1977). *J. Chromatogr.*, **142**: 61.
49. Huber J.F.K. and Hulsman J.A. (1967). *Anal. Chem.*, **38**: 305.
50. Ettre L.S. and Purcell J.E. (1974). In Giddings J.C. and Keller R.A., Editors, *Advances in Chromatography* Vol. 10. Marcel Dekker, New York, pp. 1–97.
51. Hawkes S.J. (1983). *J. Chem. Ed.*, **60**: 393.
52. Golay M.J.E. (1958). In Desty D.H., Editor, *Gas Chromatography*. Butterworths, London.
53. Bristow P.A. and Knox J.H. (1977). *Chromatographia*, **10**: 279.
54. Chen J.-C. and Weber S.G. (1982). *J. Chromatogr.*, **248**: 434.
55. Laird G.R., Jurand J. and Knox J.H. (1974). *Proc. Soc. Anal. Chem.*, **11**: 310.
56. Grushka E., Snyder L.R. and Knox J.H. (1975). *J.Chromatogr. Sci.*, **13**: 25.
57. vanWasen U., Swaid I. and Schneider G.M. (1980). *Angew. Chem., Int. Ed.*, **19**: 575.
58. de Ligny C.L. (1970). *J. Chromatogr.*, **49**: 393.
59. Guiochon G. (1978). *Anal. Chem.*, **50**: 1812.
60. Casper G., Vidal-Madjar C. and Guiochon G. (1982). *Chromatographia*, **15**: 125.
61. Schutjes C.P.M., Vermer E.A., Rijks J.A. and Cramers C.A. (1982). *J. Chromatogr.*, **253**: 1.
62. Gonnord M.F., Guiochon G. and Onuska F.I. (1983). *Anal. Chem.*, **55**: 2115.

63. Bottari E. and Goretti G. (1978). *J. Chromatogr.*, **154**: 228.
64. Lauer H.H. and Rozing G.P. (1981). *Chromatographia*, **14**: 641.
65. Gill W.N. and Sankasubramanian R. (1970). *Proc. Roy. Soc. London*, **A316**: 341.
66. Snyder L.R. (1972). *J. Chromatogr. Sci.*, **10**: 200.
67. Giddings J.C. (1967). *Anal. Chem.*, **39**: 1027.
68. Kaiser R. (1966). *Chromatographie in der Gasphase*, 2nd edn, Vol. 2. Bibliographisches Institut, Mannheim, pp. 47–48.
69. Ettre L.S. (1975). *Chromatographia*, **8**: 291 & 355.
70. Grob K. and Grob K. (1981). *J. Chromatogr.*, **207**: 291.
71. Said A.S. (1979). *J. High Resolut. Chromatogr., Chromatogr. Comm.*, **2**: 193.
72. Foley J.P. (1991). *Analyst*, **116**: 1275.
73. Guiochon G. (1979). *J. Chromatogr.*, **185**: 3.
74. Guiochon G. (1980). *Anal. Chem.*, **52**: 2002.
75. Snyder L.R. (1969). *J. Chromatogr. Sci.*, **7**: 352.
76. Snyder L.R. (1968). *Principles of Adsorption Chromatography*. Marcel Dekker, New York.
77. Snyder L.R. (1970). *J. Chromatogr. Sci.*, **8**: 692.
78. Poole C.F. and Poole S.K. (1989). *Anal. Chim. Acta*, **216**: 109.
79. Pauls R.E. and McCoy R.W. (1986). *J. Chromatogr. Sci.*, **24**: 66.
80. Walker J.Q., Spencer S.F. and Sonchik S.M. (1985). *J. Chromatogr. Sci.*, **23**: 555.
81. Bristow P.A. and Knox J.H. (1977). *Chromatographia*, **10**: 279.

Bibliography

Retention time and retardation factor

Dallas M.S.J. (1968). The effect of layer thickness on RF values in TLC. *J. Chromatogr.*, **33**: 193.
Dhont J.H. (1980). The RF value as a constant in TLC. *J. Chromatogr.*, **202**: 15.
Dhont J.H., Vinkenborg C., Compaan H., Ritter F.J., Labadie R.P., Verweij A. and de Zeuw R.A. (1972). Application of RF correction in TLC by means of two reference RF values. II. Results obtained with a polar multi-component solvent system. *J. Chromatogr.*, **71**: 283.
Dhont J.H., Vinkenborg C., Compaan H., Ritter F.J., Verweij A. and de Zeuw R.A. (1977). Application of RF correction in TLC by means of two reference RF values. III. Results obtained in reversed-phase TLC. *J. Chromatogr.*, **130**: 205.
Nurok D. (1981). Controlled migration in TLC. *Anal. Chem.*, **53**: 714.
vanWendel de Joode M.D., Hindriks H. and Lakeman J. (1979). RF correction in TLC. *J. Chromatogr.*, **170**: 412.

Peak shape

Ettre L.S. (1965). Remarks to the question of peak symmetry calculation. *J. Gas Chromatogr.*, **3**: 100.

Separation efficiency

Bidlingmeyer B.A. and Warren F.V. (1984). Column efficiency measurement. *Anal. Chem.*, **56**: 1583A.
Ettre L.S. (1975). Separation values and their utilization in column characterization, I & II. *Chromatographia*, **8**: 291 & 355.
Grushka E., Snyder L.R. and Knox J.H. (1975). Advances in band spreading theories. *J. Chromatogr. Sci.*, **13**: 25.
Guiochon G., Ghodbane S., Golshan-Shirazi S., Huang J.-X., Katti A., Lin B.-C. and Ma Z. (1989). Nonlinear Chromatography: Recent theoretical and experimental results. *Talanta*, **36**: 19.
Hayashi Y. and Matsuda R. (1993). Optimization theory of chromatography. *Chemom. Intell. Lab. Syst.*, **18**: 1.
Hurell R.A. and Perry S.G. (1962). Resolution in gas chromatography. *Nature*, **196**: 571.

Kaliszan R. (1992). Quantitative structure-retention relationships. *Anal. Chem.*, **64**: 619A.

Kirkland J.J., Yau W.W., Stoklosa H.J. and Dilks C.H. (1977). Sampling and extracolumn effects in high performance liquid chromatography; influence of peak skew on plate count calculations *J. Chromatogr. Sci.*, **15**: 303.

Knox J.H. (1977). Practical aspects of liquid chromatography theory. *J. Chromatogr. Sci.*, **15**: 352.

Knox J.H. and Scott H.P. (1983). B and C terms in the van Deemter equation for liquid chromatography. *J. Chromatogr.*, **282**: 297.

De Ligny C.L. (1970). The contribution of eddy diffusion and of the macroscopic mobile phase velocity profile to plate height in chromatography. A literature investigation. *J. Chromatogr.*, *Chromatogr. Rev.*, **49**: 393.

Meyer V.R. (1985). High-performance liquid chromatographic theory for the practitioner. *J. Chromatogr.*, *Chromatogr. Rev.*, **334**: 197.

Novotny M. (1985). Capillary separation methods: a key to high efficiency and improved detection capabilities. *Analyst*, **109**: 199.

Tchapla T. (1992). Optimization software in chromatography. *Analusis*, **20**: M71.

Holdup volume

Berendson G.E., Schoenmakers P.J., de Galen L., Vigh G., Varga-Puchony Z. and Inczedy J. (1980). On the determination of the hold-up time in reversed-phase liquid chromatography. *J. Liquid Chromatogr.*, **3**: 1669.

Smith R.J., Nieass C.S. and Wainwright M.S. (1986). A review of methods for the determination of hold-up volume in modern liquid chromatography. *J. Liquid Chromatogr.*, **9**: 1387.

Gas Chromatography 3

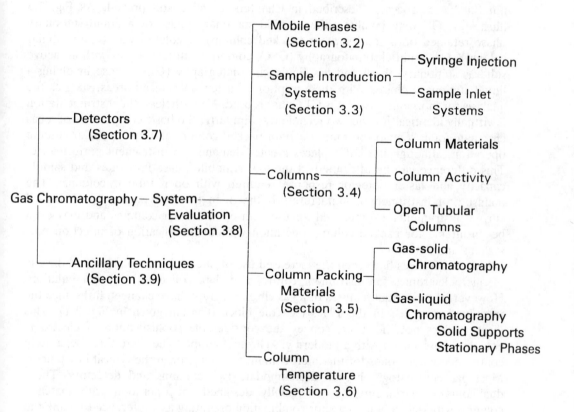

Gas Chromatography — System Evaluation (Section 3.8)

- Mobile Phases (Section 3.2)
- Sample Introduction Systems (Section 3.3)
  - Syringe Injection
  - Sample Inlet Systems
- Columns (Section 3.4)
  - Column Materials
  - Column Activity
  - Open Tubular Columns
- Column Packing Materials (Section 3.5)
  - Gas-solid Chromatography
  - Gas-liquid Chromatography Solid Supports Stationary Phases
- Column Temperature (Section 3.6)
- Detectors (Section 3.7)
- Ancillary Techniques (Section 3.9)

3.1   Introduction

Gas chromatography (GC) describes all chromatographic methods in which the mobile phase is a gas. It is a well-established analytical technique used routinely in most industrial and academic laboratories because of its capability of high resolution, selectivity and sensitivity. From the 1950s and until the advent of high-performance liquid chromatography (HPLC), it dominated separation methods. Instrumentation and techniques have become increasingly sophisticated but the basic processes described in Chapters 1 and 2 still prevail. As Fig. 1.3 illustrates, GC may involve either a solid stationary phase or a liquid stationary phase retained on a solid sorbent (packed column) or column wall (open tubular column). Gas–solid chromatography (GSC) comprises all techniques with an active solid as stationary phase and gas–liquid chromatography (GLC) those involving a liquid stationary phase. With the exception of a few specialized areas, such as the analysis for inorganic gases, it is GLC that is used. Nevertheless, the instrumentation is virtually identical for the two techniques. Similarly, the basic components of a gas chromatograph remain the same for both packed column and open tubular column operation, although the latter places greater demands on instrument performance. This difference can be attributed to the lower mobile phase flow-rates and sample capacity and faster detector response required with open tubular columns. The mainstay of instrument manufacturers is the open tubular system and almost all current instruments are intended for use with open tubular columns and may even be unsuitable for packed column operation without modification of injection port systems and detectors.

Recent advances have seen the increased use of open tubular columns and the use of microelectronics for instrument control and data collection and manipulation. However, the essential features of a gas chromatograph have changed little since the introduction of GC in the 1950s (see the block diagram given in Fig. 3.1). This diagram does not, however, convey the considerable sophistication of electronic equipment associated with a modern gas chromatograph. There are three separately controlled heated zones for the inlet, column and detector in the typical instrument. Most gas chromatographs can accommodate two columns and detectors. These dual-column instruments were originally designed for operation with matched columns with one column/detector combination operating as a reference to blank the analytical column. Because of the greater stability and lower column bleed (*vide infra*) with open tubular columns, dual column systems are now usually operated independently, using two different columns and separate detector types (e.g. a selective detector and a universal detector). Data on commercial instruments have been reviewed by Bayer [1].

In the most common approach (elution development) the sample is introduced into the chromatograph via the sample inlet into a continuous flow of mobile phase which, in GC, is referred to as the carrier gas. The sample is vaporized in the inlet system and transported by the carrier gas to the thermostatted column, where separation occurs. The individual components give rise to an electrical signal in the detector, which may have provision for the inlet of additional make-up gas. This is necessary to permit separate optimization of gas flow through the column and detector. After suitable amplification the detector signal is conducted to a recording

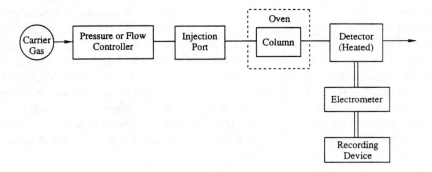

Fig. 3.1. Block diagram of a gas chromatograph.

device. Sample components can be identified from their characteristic retention times. With proper calibration, the amounts of the components of a mixture can be measured accurately also. Thus, both qualitative and quantitative data can be obtained, provided accurate and precise control of both temperature and carrier gas flow-rate is achieved. With isothermal operation, the column temperature remains constant throughout the analysis, whereas temperature programmed gas chromatography involves a controlled rise in temperature.

3.1.1 Organization of this chapter

GC and HPLC are frequently seen as competitors rather than complementary techniques. The approach taken in this chapter on GC differs from that of Chapters 5 and 6 on HPLC in one important aspect. The selection of a suitable system for HPLC is examined in Chapter 6 from the perspective of the analyte, whereas in the corresponding Section 3.5 on GC the emphasis is placed on column packing rather than the analyte. This difference can be attributed to the greater diversity of separation mechanisms in HPLC. In contrast, separations by GC are restricted to essentially only two mechanisms, namely, adsorption and partition/bonded phase which simplifies the process of selecting a suitable system.

Factors which must be considered in choosing a system for GC are then:

- The carrier gas.
- Sample introduction system.
- The type of column and stationary phase.
- The column temperature.
- The detector type.

A more detailed discussion of these aspects is given in the following sections.

3.1.2 Sample type

The basic requirement with respect to sample is that it is thermally stable and has an appreciable vapour pressure at the column temperature. This allows the sample components to vaporize in, and move with, the gaseous mobile phase. This requirement is not as severe a restriction as it appears as column temperatures as high as 450°C (300°C is more common) are used in GC. Thus, GC can be applied to all permanent gases, most non-ionized small or medium-sized organic molecules (typically up to C_{30}) and many organometallic compounds but it cannot be used for macromolecules or salts. In some instances, non-volatile compounds can be converted into more volatile and stable derivatives before chromatography. In a typical sample containing a mixture of volatile and non-volatile components, care must be taken that the non-volatile solutes are not deposited in the system where they can interfere with subsequent analyses. The subject of sample preparation is discussed in Chapter 8.

3.2   Mobile Phases

Ideally, the mobile phase for GC will be nonreactive toward the analyte, nonflammable and cheap since it is vented at the end of the instrument. The use of gases or vapours (either as mobile phase additives or alone) capable of interacting with the analyte and influencing selectivity [2] is limited by practical problems related to detector compatibility, safety and the limited temperature range for the desired effect to apply. Moreover, chemical reaction of analyte and gas is problematical [3, 4] at the elevated temperatures usually encountered in GC. The usual mobile phase in GC is therefore a noninteractive gas that does not influence selectivity. Hence, the choice of a mobile phase or carrier gas is determined by practical constraints of cost, availability, inertness, detector compatibility, etc., rather than its ability to effect a particular separation. However, the carrier gas can influence resolution through its effect on column efficiency because of differences in solute diffusion rates for various gases. Moreover, it can affect analysis time as discussed below, and plays a role in pressure-limiting situations because of differences in gas viscosities (see Table 3.1).

Taking these considerations into account, hydrogen, helium and nitrogen, are the most popular carrier gases in GC. With packed columns, there is little to recommend either one of these gases over another, although in certain situations [5, 6] one may be favoured. For example, in cases where the pressure drop is a limiting

Table 3.1   Physical properties (at 273 K and 101 kPa) and applications of gases used in gas chromatography.

Gas	Thermal conductivity $(10^8\,W\,m^{-1}\,K^{-1})$	Viscosity $(10^{-7}\,Pa.s)$	Density $(kg\,m^{-3})$	Application
Hydrogen	16.75	84	0.0899	Carrier and burner gas
Helium	14.07	186	0.1785	Carrier gas
Nitrogen	2.39	166	1.2505	Carrier gas
Argon	1.67	212	1.7839	Carrier gas
Neon	4.56	298	0.8999	Carrier gas
Oxygen	2.43	192	1.4289	Burner gas
Air	2.39	171	1.2928	Burner gas

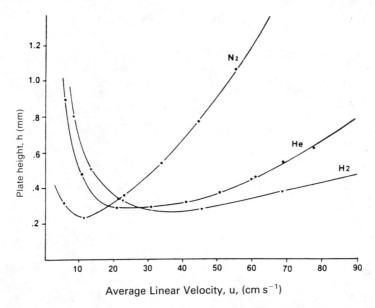

Fig. 3.2. Experimental van Deemter curves for *n*-octadecane showing the effect of mobile phase on efficiency. Data obtained with a 25 m × 0.22 mm (d_f = 0.25 μm) BP1 WCOT column.

feature, hydrogen should be used because of its lower viscosity. Conditions with open tubular columns are somewhat different and careful selection of the most appropriate carrier gas is important in optimizing the separation.

The most efficient separations (smallest value of *h*) are achieved with nitrogen as carrier gas, as shown by the van Deemter curves (Fig. 3.2) for a 25 m × 0.22 mm open tubular column operated with nitrogen, helium or hydrogen. This can be attributed to the higher molecular mass of nitrogen and smaller diffusion coefficients, i.e. lower *B* term in the rate equation. However, in order to achieve this efficiency there is a considerable sacrifice in analysis time, since the optimum mobile phase velocity for nitrogen is 8–10 cm s^{-1}. The use of helium or hydrogen as mobile phase entails a small sacrifice in efficiency, but analysis time is reduced since the optimum mobile phase velocity is 16–20 or 35–40 cm s^{-1} for helium and hydrogen, respectively. Furthermore, helium and hydrogen perform better (see Fig. 3.2) at higher carrier gas velocities, exceeding the optimum values where the *C* term in the rate equation dominates. The practical consequences, in terms of resolution, are illustrated in Fig. 3.3. Here, separation of a two-component mixture is compared using all three gases at their individual optimum average linear velocity, with maximum resolution being achieved for nitrogen. However, when a common velocity exceeding the optimum is used, the maximum resolution is achieved with hydrogen, as demonstrated by the flattest van Deemter curve above the optimum velocity. This enables use of higher velocities to shorten analysis time with minimal sacrifice in efficiency. For this reason, the use of hydrogen has gained in popularity. The effect of shorter analysis time is particularly noticeable in temperature-programmed operation [7]. Furthermore, any changes in carrier gas velocity during temperature-programmed operation do not significantly alter efficiency under these

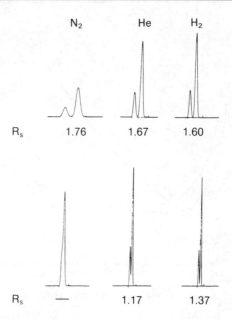

Fig. 3.3. Variation in resolution of *n*-pristane and *n*-heptadecane using different carrier gases at (i) optimum average linear mobile phase velocity (nitrogen, 11.7 cm s^{-1}; helium, 23.2 cm s^{-1}; hydrogen, 25.0 cm s^{-1}) and (ii) constant average linear mobile phase velocity (58 cm s^{-1}). Copyright 1981 Hewlett Packard Company. Reproduced with permission. R.R. Freeman, Editor, High Resolution Gas Chromatography, 2nd edn.

conditions. Nevertheless, hydrogen is highly flammable and its mixtures with air can be explosive. Precautions are therefore necessary when using hydrogen and a leak sensor in the column oven is highly desirable. Moreover, in some situations hydrogen may react with sample components to produce hydrogenated artefacts [8] that may or may not interfere in the analysis. Additional considerations apply to the choice of carrier gas when using some detectors.

3.2.1 Carrier gas purification—oxygen and moisture traps

It is critical that the highest purity gas be used to reduce deterioration of the stationary phase and lessen detector noise. Nevertheless, it is usual to include oxygen and moisture traps in the carrier gas lines, even though high-purity gases are employed. These traps are commercially available and contain activated carbon (to remove organic impurities) or molecular sieves or Drierite (to remove moisture and oxygen). The traps must be monitored and periodically regenerated. Contamination can also arise from compressed air supplies and from oil, soldering flux or solvent vapours on newly installed gas lines, which should be of metal construction to avoid ingress of moisture and oxygen which is common through nylon tubing. The electron capture detector is particularly susceptible to traces of oxygen or moisture (see Section 3.7). Inclusion of traps is especially important with polar stationary phases, such as polyols and polyethers which are easily degraded at typical column temperatures by traces (10 p.p.m.) of oxygen or moisture. Porous polymers (see later) and solid stationary phases may also be deactivated by moisture. Any newly fitted open tubular column should be purged with carrier gas (15 min for a 50-m column) prior to heating to remove air from the column. Similarly, the column should be cooled prior to stopping the flow of carrier gas if it is to be removed from the instrument at the conclusion of an analysis. No open tubular column should be exposed to air at temperatures exceeding 50°C.

3.2.2 Carrier gas regulation

The flow of carrier gas is described by two variables, the flow-rate as measured in ml min^{-1} and the pressure drop between the injection port inlet and detector outlet. A constant carrier gas flow-rate is desirable so that retention times will not vary and flow sensitive detectors will not become nonlinear. Carrier gas is usually supplied from a cylinder with an initial pressure of 17 MPa and reduced using a double-stage pressure regulator to 50–300 kPa. The pressure regulators are generally of the diaphragm type with a polymeric membrane, which can be a source of carrier gas contamination [9]. Similarly, delaying the changing of gas cylinders too long can cause problems. Running an instrument with a cylinder at low outlet pressure increases the probability of introducing contamination (especially water) from the cylinder into the GC flow system. When changing cylinders, it is important to ensure that all fittings are free of dust and dirt particles before connection to gas lines.

At the instrument, the carrier gas is regulated by either a pressure regulator or flow controller. Differential flow regulators maintain a constant volumetric flow of carrier gas during changing pressure drops across the column as caused, for example, by temperature programming where the increasing column temperature increases the resistance to flow in the column, causing a greater pressure drop across the column. Volumetric flow-rates with 4-mm internal diameter (i.d.) packed columns are usually 40–80 ml min^{-1} whereas much lower flow-rates of 1–4 ml min^{-1} are used with open tubular columns (see Tables 2.4 and 3.2). Flow is generally pressure regulated with these columns in order to accommodate split injectors, with the result that volumetric flow decreases during a programmed temperature analysis owing to increased gas viscosity. At the start of a programmed analysis, it is usual to use a

Table 3.2 Typical values of flow rate, pressure and average linear gas velocity for different sized open tubular columns using hydrogen as carrier gas.

Column i.d. (mm)	Film thickness (μm)	Film length (m)	Flow rate (ml min^{-1})	Average linear velocity (cm s^{-1})	Pressure (kPa)
0.10	0.10	12	0.2–0.5	38	80
0.22	0.5	12	0.8–2.0	36	35
0.32	0.5	12	1.7–4.0	34	17
	2.0	12	1.7–4.0	28	17
	0.5	25	1.7–4.0	28	35
0.53	1.0	12	3–50	28	7
	5.0	12	3–50	13	7
	5.0	25	3–50	13	14

higher mobile phase flow-rate than that used for an isothermal run to allow for this decrease. Hydrogen has an advantage over nitrogen and, to a lesser extent, helium since the plate number varies less for hydrogen than for other gases as mobile phase velocity changes.

Conventional flow controllers are inadequate for the low flow rates encountered in open tubular systems and electronic programmable flow controllers, which are not based on diaphragm technology, are increasingly used in modern instruments. With these devices, the flow rate and pressure are specified via the instrument keyboard. The inlet system for open tubular systems contains several controlled static (split vent, septum purge) leaks and dynamic leaks (inlet purge) and for split and splitless injection, pressure regulation is superior to flow control, whereas on-column injection is compatible with flow control.

When an open tubular column is installed in an instrument it should be checked for carrier gas flow before connecting the detector end of the column, to avoid the possibility of heating a column with no flow. This is conveniently done by dipping the end of the column into a volatile solvent such as pentane, from which a strong stream of bubbles should then emerge. When the column connection has been completed, the system should be checked for gas leaks. A convenient method is to apply soap solution to all connections while the instrument is cold. Any leaks will appear as a soap bubble formation. Once any leaks have been eliminated and the column purged with carrier gas, the volumetric flow-rate (ml min^{-1}) can be measured and controlled electronically or measured with a soap bubble flow-meter, a rotameter or a hot wire flow-meter. A soap bubble flow-meter can be used to measure flow at the detector end of the column. The flow is calculated from the time taken for the carrier gas to carry a soap bubble up a calibrated tube of known volume. If necessary the flow can be corrected to normal conditions of temperature and pressure (0°C and 101 kPa). In practice, measurement of the low flow-rates encountered in open tubular columns can be a problem with a soap bubble flow-meter. Measurement of the theoretically more useful average linear gas velocity (cm s^{-1}) is a common alternative. This is obtained from the known column length and column dead time, which is measured by injecting an unretained solute [10, 11]. Methane is usually suitable for this purpose. In either case, the second parameter can be calculated from the measured value using equation 3.1.

$$F = uA \qquad\qquad 3.1$$

where F is the volumetric flow-rate (ml min^{-1}), u is the average linear gas velocity (cm min^{-1}) and A is the column cross-sectional area available to the mobile phase.

3.3   Systems for Sample Introduction

Sample introduction has always been a problem area in GC for a number of reasons. The inlet system or injection port must receive the sample and deliver the correct amount of material to the column so as not to exceed the sample capacity of the column or the linear range of the detector in use. In addition, the sample has to be evaporated rapidly and delivered to the column as a sharp band. In normal practice, the material entering the column must have the same composition as the original sample. There are some specialized techniques in which the sample entering the column is different from the composition of the original sample. These techniques are treated separately but include equilibrium headspace sampling, purge and trap sampling, pyrolysis GC and multidimensional chromatography.

3.3.1 Syringes

The method of sample introduction will depend on the state of aggregation of the sample and the nature of the separation procedure. Samples for GC can be solids, liquids or gases, although solids and liquids are usually introduced as dilute solutions in a volatile solvent. A universal method of sample introduction is by means of a microsyringe through a septum made of elastomer or rubber which seals the inlet system as the syringe needle is withdrawn. Despite the seeming simplicity of this system, many variants exist, as discussed below. This can be attributed to the practical difficulties encountered with syringe injection, which are caused by selective vaporization from the syringe needle [12] during introduction of the needle into the hot vaporization chamber and from the residue remaining in the needle after the bulk of the sample has been expelled. Reproducibility of syringe injection with gases (50–1000 μl) is relatively poor because the volume of a gas is so temperature dependent. In this case, higher precision can be achieved with a sampling valve. Solids can be injected with a specially designed syringe with a cavity in the tip of the needle. However, slow vaporization can cause problems and for this reason solids are invariably injected as dilute solutions.

Syringes are available from a number of manufacturers in various configurations; needle point style, length of needle, fixed or replaceable needle. Most needles are constructed of stainless steel but specialty fused silica needles are available for on-column injection (*vide infra*). An important consideration in choosing a syringe is the correct needle length to ensure delivery of the sample at the correct position in the injection zone. For quantitative work, the syringe should be gas-tight at the column back-pressure to prevent loss of sample during the injection. Syringes with replaceable or cemented needles are available. The latter have a smaller dead volume but generally shorter lifetimes. The method of introducing the syringe into the inlet can be a source of poor reproducibility and requires considerable care and practice. The best procedure for this will depend on the injection mode.

Syringe handling techniques

With packed columns, the sample solution is introduced via a syringe (1, 5 or 10 μl) into the injection port (sealed with a septum), which is heated to a higher temperature (by 20 to 40°C) than the column in order to assist vaporization. Alternatively, the sample solution may be deposited, via a syringe, directly onto the end of the column housed inside the injection port. The latter approach is more efficient and there is less chance of sample decomposition. The first few centimetres of column packing may require periodic replacement because of adsorbed sample components or because stationary phase has been washed from the solid support material. Sample discrimination is not a problem, although septum bleed may cause

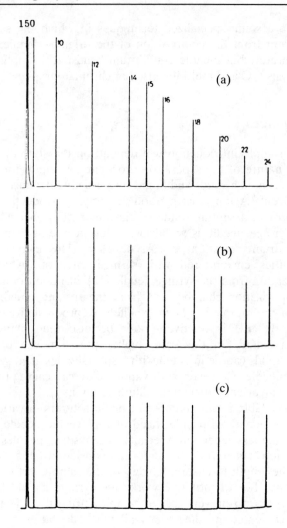

Fig. 3.4. Discrimination of *n*-alkanes illustrated for different injection techniques. (a) Filled needle; (b) hot needle; and (c) cold on-column. Reproduced with permission from Grob and Grob (1979). *J. High. Resol. Chromatogr., Chromatogr. Comm.*, **2**: 109.

irregularities. Packed columns are relatively forgiving of poor technique because of the relatively large sample sizes.

Conversely, open tubular columns are more demanding on technique and considerably more attention has been given to detailed investigation of various syringe handling methods. The syringe [12, 13], the septum [14–17] and the configuration of the inlet system [18] can be sources of operational problems, the extent of which are greatly magnified with open tubular columns because of the much smaller sample capacity and carrier gas flow-rates. These problems manifest as sample discrimination, which can be regarded as a measure of how well the detected peak areas reflect the original sample composition. The syringe can contribute to discrimination in at least two ways. First, sample vaporization often occurs from

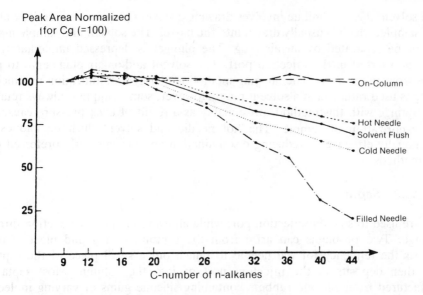

Fig. 3.5. Discrimination of *n*-alkanes obtained by different injection techniques using a split injection (1 μl; split ratio 1:15) and injection port temperature of 350°C. Methods of syringe handling are described in the text. Reproduced with permission from Grob and Neukom (1979). *J. High Resol. Chromatogr., Chromatogr. Comm.*, **2**: 15.

the metal surface of the syringe needle and, second, selective prevolatilization of the solvent and more volatile sample components from the needle is a common occurrence. The effects of discrimination are shown in Fig. 3.4 for different injection techniques. There is obviously considerable loss of the higher homologues with the filled needle technique in comparison with cold on-column injection. The metal surface of the syringe can also cause adsorption or catalytic conversion of labile solutes [19].

A variety of individual techniques have evolved for injecting a sample with a syringe. With a vaporizing injector (i.e. heated injection port) the poorest method is to retain sample in the needle when introduced into the heated injection port [20]. Nevertheless this is probably the most common injection method because of its simplicity. There are at least two variants of filled syringe needle injection which consist simply of filling the flushed syringe needle with sample, which may also be drawn into the barrel. The needle is inserted into the heated injection zone and sample is allowed to evaporate or it is expelled by depressing the plunger of the needle. This technique may show significant discrimination for high boiling-point components as shown in Fig. 3.5.

In cold syringe needle injection, the sample is sucked into the barrel of the syringe leaving no sample in the needle (in theory at least) and after insertion through the septum the plunger is depressed rapidly to deposit the sample in the heated zone without time for the needle to equilibrate to the injection port temperature; any sample remaining in the needle leaves by evaporation. Hot needle injection is a variation in which the needle is allowed to equilibrate to the injection port temperature before depressing the plunger.

The solvent flush technique involves drawing a solvent plug into the syringe ahead of the sample, which is usually drawn into the barrel. The solvent and sample may or may not be separated by an air plug. The plunger is depressed immediately the sample is inserted in the injection port. The solvent and/or air plug serve to push sample from the syringe. The air plug method is merely a modification in which an air plug is used rather than a solvent plug. However, some sample is always retained in the syringe with this method, presumably as a result of back pressure caused by rapid vaporization of sample. The hot needle and solvent flush techniques are about equally effective in reducing discrimination and are currently preferred over other methods.

3.3.2 Septa

Septa are used to seal the injection port while allowing sample introduction through a syringe. Two problems can arise from the septum: coring and bleed. Coring describes the tendency for the needle to punch out a small piece of the septum, which then deposits in the injection port or into the column. Most septa are manufactured from silicone rubbers containing silicone gums of varying molecular weights which are polymerized with a catalyst. However, unpolymerized oils and previously retained solutes can diffuse out of the septum giving rise to the phenomenon of septum bleed, which is exacerbated at higher column temperatures and by temperature programming. The consequences of septum bleed, are reduced column efficiency and in many cases a high, unsteady detector baseline resulting in a reduced signal-to-noise ratio. When temperature programming is used, the oils will elute according to their boiling points resulting in spurious peaks. The extent of such problems depends on a number of factors, including the composition of the septum, the type of detector (selective detectors give a higher response for certain compounds), the sensitivity setting of the detector amplifier and the injection port temperature. This is illustrated in the chromatograms shown in Fig. 3.6. Various solutions are available; low bleed septa or a septum purge device in which a portion of the carrier gas flows across the face of the septum and exits via an adjacent orifice can be used. Furthermore, a good operating principle is to perform a separation with a solvent blank, particularly with a new batch of septa. The face of the septum should not be touched as finger oils can appear in chromatograms. For high-sensitivity work, the septa should be stored in a metal or glass container to avoid contamination by plastic containers. Septa have an upper temperature limit above which excessive degradation and/or bleed occurs. Some septa have a moulded layer of PTFE; this reduces the area that can bleed, but also limits the maximum temperature to 225°C. The ultimate solution may be the elimination of the septum in favour of a valve injector, an approach adopted by Alltech who produce the 'Jade' valve.

Septa have a limited lifetime (20–100 injections) dependent on the skill of the operator, the injector temperature and the septum quality. A skilful operator is able to penetrate the septum in the same place on each injection thereby prolonging septum life. It is also for this reason that lifetimes are usually much improved with automatic injectors in which the sample is sealed in a sample tube closed with a septum and placed in a carousel or rack of samples. The syringe is repeatedly flushed

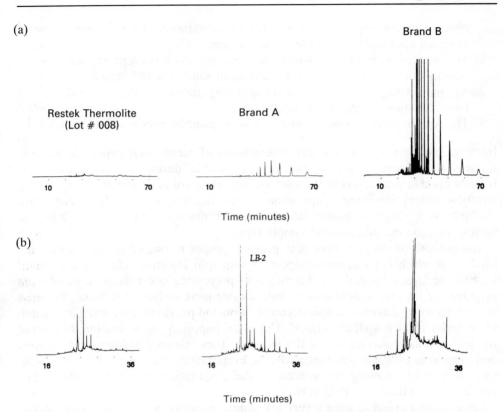

Fig. 3.6. Bleed comparison of three unconditioned low-bleed septa using (a) flame ionization detection and (b) electron capture detection. A piece of septum was installed into a clean splitless sleeve inserted in the instrument injection port which was heated to 250°C for 15 min and then turned off. Analysis was performed on a 15 m × 0.53 mm Rtx-1 column (1.0 μm film) programmed from 40°C (15-min hold) to 280°C at 15°C per min using hydrogen carrier gas (40 cm s^{-1}). Reprinted with permission from *Restek Corporation Catalog*, pp. 148, 149.

with solvent, filled with sample and turned to the inlet system, where the syringe needle pierces the septum and the preset volume of sample is injected. As well as enabling high sample throughput, automatic injection increases precision by eliminating much of the variability caused by operator injection. Recently, a long-life septum has been released (Microseal septum, Activon) for use with an autosampler or manual injection using a syringe having a blunt-tipped needle. The claimed lifetime for this septum is 25 000 injections at an injection port temperature of 300°C.

3.3.3 Sample inlet systems

The inlet system should ideally meet the following requirements [21] but in practice a compromise is typically necessary:

1. The sample must be introduced into the column in as small a volume as possible because the peak width at the column outlet is the initial band width

plus the broadening that occurs during the separation process. Thus, a poorly designed inlet system can defeat a separation.

2. The heat capacity of the system must be such as to prevent any appreciable cooling during fusion and/or vaporization of solid or liquid samples.
3. Sample decomposition must not occur during the vaporization process.
4. Discrimination effects should be absent.
5. The solvent peak should not interfere with quantification of solute peaks.

Initially, sample inlet systems were constructed of metal, thus providing metallic surfaces where sample decomposition was possible during sample evaporation. Interchangeable glass liners in the inlet are now standard in practically every sample injection system involving evaporation of the injected sample. Inlet liners are available in a range of configurations to meet the different needs of injection techniques (split or splitless) and sample types.

The purpose of the inlet liner is to provide: proper mixing of sample vapour and carrier gas which is particularly important with split injection; efficient transfer of heat to the injected sample; and a means of preventing non-volatile material from reaching the column. Liners also provide a more inert surface than metal, reducing the possibility of catalytic sample decomposition and providing easy interchangeability in order to give a clean surface. This is an important consideration as charred residues gradually accumulate in the injector. These charred residues may catalyse decomposition of labile substances during evaporation. The liner also enables an easy method of changing the volume of the evaporation chamber by the proper selection of the liner wall thickness.

Open tubular columns have a very low sample capacity, and to avoid overloading the stationary phase specialized injection systems have evolved. The more important of these are listed in Table 3.3 but there are many others, each having its own advantages and disadvantages [22]. As stated in the preface of Sandra's text [22] 'all these systems have a place in capillary chromatography and a knowledge of their possibilities and limitations . . . is therefore a must. . . . On the other hand, we now have a much better understanding of injection phenomena and can easier select the appropriate injection technique for a given separation problem.'

Split injection

Split injection was the first open tubular column injection technique and was used almost exclusively for a decade up to about 1968. It remains the simplest method of injection and is suitable for many applications. The sample is injected, preferably using the solvent flush or hot needle technique and, after evaporation and homogeneous mixing with the carrier gas, is split into two unequal portions, the smaller one passing to the column, while the rest is vented to waste. Dynamic splitting involves the evaporation, mixing and splitting of the sample in a flowing stream and accounts for the majority of applications. With such systems the split ratio is determined by the relative magnitude of the two flow-rates which, in turn, are determined by the pneumatic resistances of the column and vent.

In a typical split system, two valves are fitted to the injection port, one functions as the septum purge and allows a small flow of carrier gas (2–5 ml min^{-1}) from just

Table 3.3 Open tubular column injection modes.

Injection mode	Concentration (ng per component)	Type of injector	Inlet temperature	Column temperature	Reference
Split	>50	Vaporizing	Hot	Hot or cold	23
Splitless	<50	Vaporizing	Hot	Cold	24, 25
Cold on-column, direct on-column or on-column syringe	<100	Non-vaporizing	Cold	Cold	22
PTV split/splitless option	Versatile	Versatile	Cold with ballistic heating	Hot or cold	26–28

below the septum to eliminate any contaminants being released by septum bleed. The second valve can be adjusted to control the ratio of carrier gas being vented to waste in the atmosphere and the flow onto the column. In a properly designed system, this valve should be at the same temperature as the rest of the system, thus preventing sample condensation which could possibly alter the split ratio. In the extreme, the vent might become completely blocked. With the valve maintained at the inlet temperature, changes in the splitter temperature will alter the split ratio, since flow through the vent will be affected while column flow will remain unaltered. The split ratio is conveniently calculated from the measured volumetric flow-rate at the vent outlet and the column exit using

$$\text{split ratio} = \frac{\text{split vent flow} + \text{column flow}}{\text{column flow}} \qquad 3.2$$

In some texts, the split ratio is incorrectly reported as the ratio of split vent flow to column flow. The difference between the two definitions will only be significant with low split ratios. Depending on column characteristics and the analytical problem, split ratios range from 10:1 to 500:1. With a pressure-regulated flow system, the split ratio decreases as column temperature is increased, owing to the decreased column flow caused by increasing carrier gas viscosity.

The injection port volume must be sufficient to accommodate the vaporized sample without causing severe pressure surges during injection, leading to nonreproducible split ratios [29]. Under such conditions sample vapours may be forced upstream into the carrier gas inlet lines leading to peak tailing as the vapours slowly re-enter the inlet. It is obvious that in a system with a fixed volume injector, the sample size must be appropriate to this volume.

The term linear splitting is often applied to split systems. This simply means that each sample component is split in precisely the same ratio regardless of its chemical nature or concentration, i.e. sample discrimination is absent. Linear split is always sought but reproducible nonlinear behaviour is accepted in practice. Split injection is a flash vaporization technique and sample discrimination is a distinct possibility. Many interrelated variables can affect the fidelity of the split, including the molecular size and polarity of components, injected volume, inner diameter of the

INLET SYSTEMS

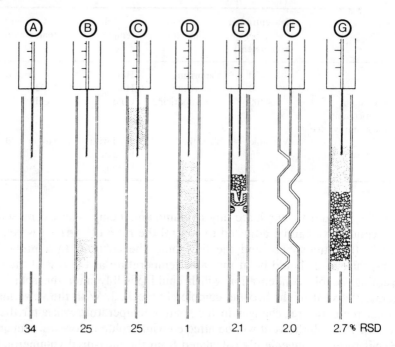

Fig. 3.7. Effect of inlet liner on repeatability of peak area ratios of methanol:2-ethyl-1-hexanol. (A) empty tube, (B) short glass wool plug in the splitting region, (C) short glass wool plug in the injection region, (D) long and tight glass wool plug, (E) Jennings tube, (F) deformation of cross-section, (G) chromatographic support packing. Reprinted with permission from Schomburg *et al.* (1977). *J. Chromatgr.*, **142**: 87.

split liner and viscosity of the column and vent streams. Any change in the viscosity of the vent stream, such as caused by sample vapours during injection, will alter the split ratio. In addition to the use of packed liners to eliminate this problem, the effect of any viscosity changes can be minimized by designing an injector with an appropriate volume. A number of parameters affecting linearity are a function of splitter design and can be minimized by choosing an appropriate inlet liner, see Fig. 3.7 [30–32]. Assuming a well-designed splitter is available, three operational factors appear to exercise major interrelated effects on the linearity of the split. These are the degree of mixing of sample and carrier gas, the inlet temperature and sample type (especially solvent volatility). For example, split injection of a wide boiling temperature range mixture [12, 33] showed discrimination and poor precision with the inlet temperature too high (300°C) or a too low boiling temperature solvent (*n*-octane). Mass discrimination was reduced by changing the solvent to *n*-dodecane or by lowering the inlet temperature to 210°C. Of course lowering the inlet temperature can be carried too far as it must be sufficiently high to volatilize the sample rapidly.

A typical flow-rate through the inlet might be 200 ml min^{-1} and consequently the sample residence time in the inlet is short. Thus, for high boiling temperature solutes, a packed liner is most appropriate as this also increases the liner's heat

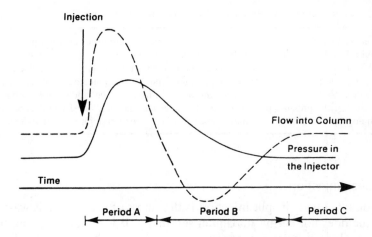

Injection

Flow into Column

Pressure in
the Injector

Time

Period A | Period B | Period C

Fig. 3.8. Effect of split injection on pressure and column flow. Reproduced with permission from Grob and Newton (1979) *J. High Resol. Chromatgr., Chromatogr. Comm.*, **2**: 563.

capacity and assists in sample vaporization. Care must be exercised with labile solutes as a packed liner can contribute to solute decomposition. In the case of very volatile solutes, a packed liner will increase band broadening because of the multipath diffusion through the packing and an unpacked liner is appropriate. With a wide boiling temperature range mixture of hydrocarbons, narrow liners (1.7 mm versus 3.9 mm) contribute greater mass discrimination and give poorer reproducibility [18], in contrast to splitless injection, where narrower liners give better results [21]. This difference in behaviour highlights the difficulty in attempting to generalize such effects because of the large differences in splitter design and injection conditions.

The amount of sample actually entering the column is related to the split ratio times the amount of sample injected, but it is difficult to determine accurately. This can be explained as follows. Sample vaporization causes a pressure surge at the moment of injection which, in turn, increases the column flow (see Fig. 3.8), thereby increasing the amount of sample entering the column more than would be expected from the steady-state split ratio. Changes in the split ratio might not produce equal changes in the amounts of different components entering the column because of discrimination effects.

In practice, the absolute value of the split ratio is usually unimportant. Moreover, the discrimination effects are generally reproducible and can be compensated for by use of standards [34] and appropriate split conditions. Nevertheless, Poole and Schuette [35] state 'The apparent simplicity of the split sampling method conflicts with the many problems that arise in obtaining quantitative data for all but the simplest of mixtures.' This view is not shared by Ettre [36] in whose opinion a well designed and properly used split system can provide quantitatively reliable data [37]. Criticisms of split systems seem to relate more to shortcomings in design, incorrect operation, such as viscosity changes in the vent restrictor, or use of split injection to analyse a sample more appropriately treated by an alternative injection technique. Recommended conditions for using split injection are given in Table 3.4.

Table 3.4 Recommended procedures for split injection.

1. Sample volume should be reproduced precisely to minimize variations of pressure pulse.
2. Use a rapid injection with the hot needle or solvent flush technique.
3. The length of penetration of the syringe needle into the injector should be reproduced precisely.
4. The same solvent should be used for all samples and standards to eliminate any solvent-related effects. The distribution of sample between vapour and droplet phases can be influenced by solvent volatility. The magnitude of the pressure pulse is related to molecular mass and density of the solvent.
5. Choose an appropriate inlet and column temperature, liner configuration and size by experiment.

Splitless injection

The main disadvantage of split injection is that most of the sample is wasted. Thus, the technique does not permit maximum sensitivity. It is therefore inappropriate for ultra-trace analysis requiring maximum sensitivity, or for very complex samples containing components with a wide range of boiling points. In many cases, it is solvent rather than sample components which would overload the column and an injection technique which places the entire sample on the column is desirable. There is, in fact, a bewildering array of such techniques but they fit into two basic categories: (a) evaporation and trapping (referred to as splitless injection with a recent variant, cold splitless injection with programmed-temperature vaporizer; (b) direct introduction of the (cold) sample into the column without evaporation—cold on-column injection and direct injection termed direct on-column and on-column syringe injection, respectively, by Schomburg [22]. In splitless mode (an unfortunate choice of name as it is not the only splitless technique) [38–43], a relatively large volume (1–5 μl) of dilute sample is introduced by syringe injection into the same (but with the split vent fully closed) or a similar device as used for split injection. The carrier gas velocity is much lower in splitless mode than in split mode. Therefore, the residence time of the sample in the injection port liner is much longer in the splitless mode (15 s or more compared with less than 1 s for split injection), permitting lower injection port temperatures for effective sample vaporization. Nevertheless, the inlet temperature must be sufficiently high to produce flash vaporization of sample onto the column.

Unnecessary mixing of sample with carrier gas must be avoided during the injection process and to achieve this, the column inlet should be 8–10 mm from the tip of the syringe needle. Because of the relatively large injection volume and the time required for transfer of sample components from the injection port onto the column, the sample must be reconcentrated at the head of the column. This is achieved via the solvent effect [18, 24, 41–44] or thermal focus (cold trapping). Splitless injection [21] is a relatively slow injection (up to 20 s) and, consequently, an open glass tube is a suitable injection port liner. In order to avoid a long solvent tail as the last portion of solvent enters the column in an exponential decay fashion [37], the injection port is back-flushed 30–60 s after injection, by opening the septum purge. If properly timed [45], this back-flushing vents mainly solvent (perhaps 10% of total solvent with a similar amount of sample) preventing its slow diffusion onto the column. The maximum volume of various solvents that can be accommodated using splitless injection with a 250 μl injection liner is given in Table 3.5. Inlet liners

Table 3.5 Maximum solvent injection volumes using splitless injection with a head pressure of 101 kPa., injection temperature of 275°C and an injection liner volume of 250 μl.

Solvent	Density (g cm^{-3})	Expansion factor*	Volume (μl)†
Hexane	0.66	172.5	1.45
Iso-octane	0.69	136.1	1.84
Dichloromethane	1.33	351.8	0.71
Chloroform	1.48	278.4	0.90
Methanol	0.79	555.0	0.45
Acetone	0.79	306.2	0.82
Water	1.00	1249.0	0.20

*Expansion factor

$$= 22\,400 \times \frac{\text{density}}{\text{molecular mass}} \times \frac{15}{15 + \left(\dfrac{\text{column pressure (kPa)}}{6.89}\right)} \times \frac{\text{injection port temperature (K)}}{273}$$

†Under the specified conditions this volume will expand to 250 μl.

are available with different internal diameters; Hewlett Packard, for example, supply three different sized injection liners with internal diameters of 1.5 mm, 2.0 mm and 4.0 mm and respective volumes of 135 μl, 245 μl and 980 μl.

With the thermal focus technique, sample components are focused at the head of the column by holding the column temperature at least 100°C (usually 150–200°C) below the analyte boiling points. This initial low temperature causes the sample components to have large capacity factors that effectively prevent their migration from the head of the column. During this time, solvents (and other highly volatile components) are eluted. Increasing the temperature of the column decreases the capacity factors and the sample components begin their migration along the column. This method is applicable to temperature-programmed operation only and is unsuitable for analysis of samples containing very volatile components, unless sub-ambient temperature control of the column is used.

Successful operation of the solvent effect requires an initial column temperature 10–30°C below the boiling point of the solvent. This ensures that solvent condenses at the head of the column to create a temporary thick film on the column wall, which provides a very low value phase ratio and thus a highly retentive zone at the head of the column. Sample components exhibit large capacity factors in this region which prevents their migration. The solute bands become narrower during this process since their front moves on an increasingly thick film (i.e. decreasing phase ratio which forces a change in solute partition ratio) while their back moves on a thinner film and therefore at a faster rate. The solvent soon revaporizes, returning the column to its original retentive properties. Detailed investigations have demonstrated that conditions such as solvent boiling point and column temperature must be carefully chosen for successful operation of the solvent effect [37, 40, 46–48]. The polarity of the solvent should be similar to that of the stationary phase. The use of polar solvents with non-polar phases and vice versa generally will not produce good results. The solvent effect is intensified with a retention gap [43, 46, 49, 50] (a short length, typically 60 cm, of uncoated column joined to the inlet end of the column which provides negligible retention compared with the coated section of column) because, on evaporation, the solutes are focused onto the start of the coated column.

Splitless injection is not a universal panacea for all sample types, being limited to samples at trace levels. Moreover, disadvantages appear when it is extended to samples containing components of high molecular mass, low volatility or low thermal stability [18]. Difficulties arise from a number of sources [37, 40], including the flooding effect [39] involving condensation of sample in the column inlet and its spreading along the early sections of the column. Splitless injection places a high solvent load on the column and is therefore recommended for use with bonded phases only. A retention gap is essential when using splitless injection with coated nonbonded phases.

Cold on-column injection

Non-split systems of introduction, such as splitless injection, suffer a number of limitations: a syringe with a metal needle is used to inject the sample into a hot vaporizing chamber. Poor precision and accuracy are often associated with such procedures because of discrimination effects and catalytic decomposition on the metal surfaces. These problems can be eliminated if the sample is deposited directly on the open tubular column [39, 42, 51, 52]. In principle, this technique is not new, having been used widely with packed columns. Desty [53] advocated the use of direct sample introduction into the column in 1965 with the statement 'directly into the column where it can dissolve in the stationary phase in the top few plates and thus be subjected to the lowest possible temperatures—that of the column itself.'

Grob and Grob [54, 55] introduced the technique of cold on-column injection for conventional-bore open tubular columns. The sample is deposited directly onto the column at a temperature equal to the boiling point ($\pm10°C$) of the sample solvent, following which the column is heated to the desired temperature. An advantage of the technique is the absence of a septum, a modified injection port being used in which a valve replaces the septum. Solute decomposition and discrimination effects [56] are reduced significantly, provided the injection is made as rapidly as possible [57]. The original technique was modified [58, 59] and, as now practised, employs a special syringe with a fine needle made of fused silica. In this way, sample contact with metal is completely eliminated.

Both Freeman [60] and Onuska and Karasek [61] caution that cold on-column injection should not be confused with direct injection. As applied by Freeman, the latter describes a system of injection used with megabore open tubular columns in which the sample is first vaporized in a glass inlet and then passes directly to the column. Conversely, Onuska and Karasek describe direct injection as a technique 'based on introduction of the sample into the WCOT column without previous vaporization outside the column.' This highlights the confusion that has arisen in the use of terminology applied to injection systems. Indeed, there is an apparent inconsistency in the terminology used by Schomburg and Ettre in the same text [22].

At least one company (Hewlett Packard) produce electronic pressure control for split/splitless and on-column injection. This modification, which can be operated in several modes (constant pressure, constant flow, vacuum compensation, pressure programming, etc.) enables sample to be swept rapidly from the inlet and into the column. This is achieved by using a programmed high carrier gas flow-rate (e.g. 50 ml min^{-1}) during injection, followed by a rapid reduction in inlet pressure to a normal value (e.g. 2 ml min^{-1}). Preliminary results with this technique show [62] reduced sample decomposition for thermally sensitive compounds (e.g. endrin) and reduced discrimination.

Programmed temperature vaporizer

Direct on-column injection techniques have one major drawback: non-volatile material present in the sample will enter the column and remain there. The programmed temperature vaporizer (PTV) is a recent and versatile development [26, 63] which can be operated in several modes. When maintained at constant temperature, it can be used as a conventional split/splitless system, although there is little to be gained from using it for isothermal flash vaporization. Alternatively, it can be used with wide-bore columns for on-column injection or programmed-temperature vaporization using cold injection. In this case, the sample is injected into a cold split/splitless injection port which is then rapidly heated (programmed from ambient up to 400°C in about 10–20 s) to transfer the sample to a cool column before temperature programming is commenced. The systems described by Schomburg et al. [27] and Vogt et al. [64] are based on the same principle but employ a different practical approach. The feature that distinguishes these systems from split and splitless injection is the fact that the sample is injected into a cold vaporizer. The advantages include reduction in sample decomposition by avoiding a high inlet temperature and that any non-volatile material remains in the inlet and will not contaminate the column. Furthermore, high temperatures are avoided in the injection port until the syringe needle is withdrawn and selective vaporization of that portion of sample remaining in the syringe needle is avoided. Thus, needle discrimination is minimized. One disadvantage is that double peak formation may be observed for early eluting very volatile components owing to vaporization from the cold injector prior to heating.

PTV injection is perhaps the closest approach to a universal injection system and appears the best approach for the chromatographer with a diverse range of separation problems. As with any of these techniques, optimization is dependent on several factors. In the split/splitless mode with solvent elimination, factors which must be considered are the design of the inlet liner, inlet pressure, initial liner temperature, purge flow, speed of sample introduction (related to the saturated vapour volume of the solvent), sample volume and physicochemical properties of the solvent. Thus, the user should consult trade or specialist literature for recommendations on the use of the various inlet systems.

3.4 Columns

The column is the heart of the chromatographic system. It determines the selectivity and efficiency of the separation. Even the most sophisticated electronics cannot compensate for an inadequate column. Columns for GC are reusable, as in HPLC and SFC, and with proper care will last a long time, which is fortunate since columns

Table 3.6 Comparison of packed and open tubular columns.

Parameter	Packed	Microbore open tubular	Open tubular	Megabore open tubular
Length (m)	0.5–3	5–50	5–100	5–100
Internal diameter (mm)	2–4	>0.1	0.18–0.32	0.53–1.00
Permeability (10^{-7} cm^2)	1–50		300–20 000	
Film thickness (μm)	1–10	0.1	0.2–2	1–5
Carrier gas average linear velocity (cm S^{-1})*	4–8	70–90	60–80	20–50
Flow rate (ml min^{-1})	40–80	0.2–0.5	0.6–4	2–50
Phase ratio, β	5–35	300–1500	80–250	25–130
Pressure drop (kPa)	70–275	70–100	14–35	7–14
Effective plates per metre	1000–2000	8000–12 000	3000–5000	1400–1800
Sample capacity (ng)	20 000	<5	20–500	1000–15 000

*These values are optimum for hydrogen. For nitrogen the values would be about 0.3 times those shown and for helium, the values would be about 0.55 times those shown.

may be very expensive. This contrasts with classical column chromatography and TLC where the stationary phase is usually discarded after a single use.

From its inception, up to the 1980s, almost all separations in GC were performed on conventional packed columns in which the liquid stationary phase was coated on an inert support packed in a metal or glass tube. This was despite the demonstration by Golay in 1957 of much greater efficiency obtainable with open tubular columns. Typical sizes of packed columns (Table 3.6) are now 0.5–3 m long with a ⅛ inch (3 mm) or, more commonly, ¼ inch (6 mm) outside diameter and 2–4 mm inside diameter. This mixture of metric and imperial units is a result of the preference for standard pipe fittings. In most laboratories, precoiled glass columns are either purchased or fabricated in a glass workshop and packed in-house. Alternatively, most chromatography suppliers have standard prepacked columns available and also offer a customized service for preparation of specially packed columns. The diameter of the coil is unimportant provided that it is at least 10 times the diameter of the tubing in order to avoid significant differences in the path length of the carrier gas flow on the inner and outer edges of the coil. In general, packed columns are not interchangeable between instruments from different manufacturers and even between models because of different shapes and/or dimensions.

There were several reasons for the continuing use of packed columns; the most important being their lower cost and the lesser technical expertise required to obtain reproducible results. However, the obvious advantages of open tubular columns in terms of higher resolution (Fig. 3.9), greater sensitivity (despite injection of less solute), reduced analysis time (to achieve equivalent resolution) and greater chemical inertness were gradually recognized. More recently, polymer-clad flexible fused silica open tubular columns with chemically bonded and/or cross-linked immobilized stationary phases have become commercially available at reasonable

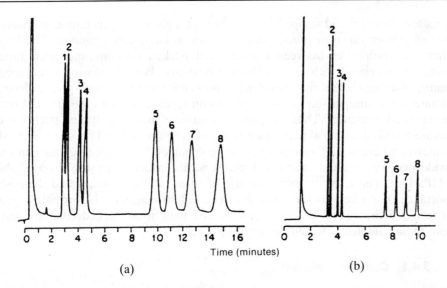

Time (minutes)

(a)   (b)

Fig. 3.9. Efficiency comparison of (a) a packed column (1.8 m × 4 mm i.d.) and (b) an open tubular column (15 m × 0.32 mm) for the separation of chlorinated pesticides. Conditions: (a) 1.3% OV-17 + 2.1% QF-1 on Chromosorb 750, 100–120 mesh, 200°C, nitrogen carrier gas, 60 ml min^{-1}; (b) DB-5, 0.25 μm film thickness programmed from 150°C to 240°C at 5°C min^{-1}, helium carrier gas. Pesticides are: 1, α-BHC; 2, lindane; 3, B-BHC; 4, heptachlor; 5, o,p'-DDD; 6, endrin; 7, p,p'-DDD; 8, endosulfan.

cost, and this has led to the current popularity of open tubular columns. These columns now routinely provide high efficiency, inertness and reproducibility. Some separation efficiency can be sacrificed by using shorter columns to achieve very rapid analyses (see Section 3.9.5).

With few exceptions, open tubular columns are purchased ready for use. They are available from several manufacturers in a wide range of column internal diameters (0.1–1.0 mm), column lengths (5–50 m) and stationary phase film thicknesses (0.1–5.0 μm). Generally, sample capacity increases but the efficiency decreases as the internal diameter or film thickness increases. The larger bore open tubular columns, with internal diameters between 0.53 mm and 1.00 mm, are termed wide-bore or megabore open tubular columns and these have similar capacities, but greater efficiencies, than packed columns (see Table 3.6). These columns represent an excellent compromise for the analyst not possessing the instrumentation necessary for the narrower bore open tubular columns. Columns with thicker films (1–3 μm) have completely replaced the older surface-coated open tubular or SCOT columns. In the earliest open tubular columns the stationary phase was coated on the internal wall of the column giving rise to the name wall-coated open tubular column. Nowadays, most open tubular columns contain a bonded phase in which the stationary phase is chemically bonded to the column wall by cross-linking. Depending on the column manufacturer, more or less binding to the column wall may occur. However, regardless of this consideration the stationary phase is permanently held in the column and such columns can even be washed with solvents to remove contaminants (see Section 3.5.2).

Greater efficiency and sample detectability for a given analysis time can always be achieved [65] on an open tubular column than on any packed column. The largest variation in properties between conventional packed columns and open tubular columns is associated with the column permeability. For this reason, open tubular columns offer much less flow resistance and can be used in much greater lengths. Ultimately, the comparison of different column types is between the efficient use of column head pressure. Thus, a packed column containing $10 \mu m$ particles can generate 50 000 theoretical plates per metre but requires a head pressure of 20 MPa per metre, whereas a 70 m open tubular column of $50 \mu m$ internal diameter can provide over one million theoretical plates with a column pressure drop of about 2.2 MPa [66]. The trend to increased use of open tubular columns and diminished importance of packed columns can be expected to continue. Nevertheless, many practical and theoretical developments in GC have used packed columns and to ignore them completely would be inappropriate.

3.4.1   Column materials

The column tubing must support the stationary phase and direct carrier gas from the point of injection to the point of detection but play no part in the actual separation process. It follows that the tubing must be chemically inert to prevent sample decomposition and also be thermally stable. It should also be robust and flexible so that it can be formed into a coil or, less commonly, a U-tube that can fit inside a thermostatted oven. The earliest columns, both packed and open tubular, were made of metal (stainless steel, copper, aluminium or nickel). However, as GC developed and detectors of greater sensitivity were employed, it soon became apparent that the metal surface was too reactive, particularly for reactive analytes. On-column decomposition of say, $1 \mu g$, was tolerable (if observed at all) when using an injection of milligram quantities, but with the advent of the flame ionization detector, injection quantities of less than $1 \mu g$ became commonplace and the same level of decomposition now caused total loss of sample. Consequently, glass columns became very popular despite their greater frailty. Glass remains the preferred construction material for packed columns and with a little care, glass columns are relatively easy to handle. Glass-lined metal tubing has also been used, as it has the inertness of glass with the mechanical strength of metal, but has not proved popular. Passivated nickel [67] has been regarded as being inferior to glass for chemical inertness although some recent data suggests [68] that this is not necessarily correct. Stainless steel, copper and aluminium columns are generally too reactive for high-performance analyses. Polytetrafluoroethylene (PTFE) tubing is very inert and flexible but cannot withstand high temperatures and is best restricted to the separation of corrosive compounds. Moreover, columns made of PTFE are subject to molecular diffusion through the column walls.

Borosilicate glass open tubular columns achieved some popularity during the 1970s and early 1980s and many individual laboratories possessed capillary column glass drawing machines. However, such columns were extremely brittle and caused problems in routine use. An important development was the preparation of open tubular columns from fused quartz and soon thereafter, fused silica tubing. Fused silica columns are weak and friable and subject to atmospheric corrosion [69]. For

Table 3.7 Bulk chemical composition (%) of glasses used for open tubular column construction.

Component	Soda lime (soft)	Borosilicate (hard, Pyrex, Duran)	Fused quartz	Fused silica
SiO_2	68.0	81.0	99.9	99.9
Na_2O	15.5	4.0		
CaO	6.0	0.5		
Al_2O_3	3.0	2.0	$100 \ \mu g \ g^{-1}$	$<1 \ \mu g \ g^{-1}$
B_2O_3		13.0		
MgO	4.0			
BaO	1.0			
K_2O	0.5			
Fe_2O_3			$100 \ \mu g \ g^{-1}$	$<1 \ \mu g \ g^{-1}$

this reason fused silica open tubular columns are protected by an outer polymeric or aluminium sheath. An advantage of fused silica columns is their ability to be bent, thereby simplifying installation in an instrument. Furthermore, the same columns fit all instruments. Fused silica columns are much more chemically inert than comparable open tubular columns prepared from other glasses [60, 70].

3.4.2 Column activity

The effects of chemical activity of a column are observed as tailing of a peak or, in extreme cases, by total sorption of the analyte. Chemical interactions causing column activity are more pronounced on thin film columns where minimal shielding is provided by the stationary phase. Freeman [60] reported the separation of mercaptans and phenols on columns prepared from different glasses. A number of compounds totally retained on columns prepared from other materials were eluted as sharp symmetrical peaks on the fused silica columns. Such comparisons should not be misinterpreted. Columns prepared from fused silica are not automatically more inert than borosilicate glass columns. A poorly prepared fused silica column may, in fact, be less inert and even the most inert fused silica column may show activity following use owing to retention of sample components, or if improperly handled.

Differences in activity between columns prepared from different glasses (e.g. soda lime glass, borosilicate glass, quartz and fused silica) can be attributed to the chemical composition of the glasses (see Table 3.7). The major component of all three glasses is silica. More important than the bulk composition, however, are the surface properties of the glass, for it is at the surface that any catalytic or sorptive effects will arise [71–74]. Activity effects are attributed to the silica surface structure where various groups have been identified:

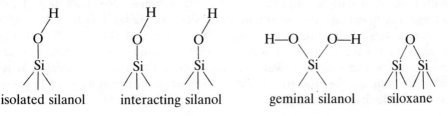

isolated silanol   interacting silanol   geminal silanol   siloxane

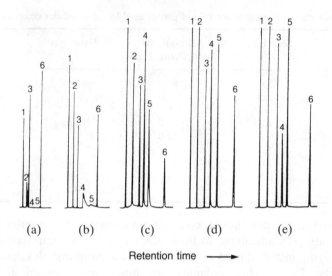

Fig. 3.10. Intermediate activity test of a borosilicate glass open tubular column (a–d) and a soda lime glass (e) open tubular column after various deactivation treatments. (a) bare borosilicate glass; (b) after etching with hydrogen chloride gas; (c) column (b) after deactivation with BTPPC; (d) column (b) after deactivation with Carbowax 20M; (e) soda lime glass column treated as in (d). Components of test mixture are identified: 1, n-decane; 2, dibutylketone; 3, undecane; 4, 2-propylcyclohexanol; 5, 2,6-dimethylaniline; 6, n-dodecane. Reproduced with permission from Sandra and Verzele (1977). *Chromatographia*, **10**: 419.

Activity effects have also been attributed to the presence of trace metal ions at the surface.

The surface hydroxyl groups can act as proton donors in hydrogen bonding interactions and can act as very strong sorptive sites for molecules with localized high electron density. Conversely, surface siloxane bridges give rise to significant van der Waals interactions and can act as proton acceptors functioning as sorptive sites for molecules such as alcohols. In Fig. 3.10(a), dibutylketone exhibits tailing while 2-propylcyclohexanol and 2,6-dimethylaniline are completely adsorbed on the bare Pyrex surface. The activity is slightly reduced by etching the glass with hydrogen chloride (Fig. 3.10b). Column activity is further diminished by deactivation with benzyltriphenylphosphonium chloride (BTPPC) (Fig. 3.10c), although some tailing of the test solutes is still evident. Treatment with Carbowax 20M is more effective (Fig. 3.10d). The greater activity of borosilicate glass columns compared to soda lime glass columns is seen by comparing Fig. 3.10(d) and (e).

Both soda lime and borosilicate glasses contain appreciable quantities of metal ions, whereas fused silica is essentially pure SiO_2 containing less than $1 \mu g \ g^{-1}$ of metal impurities. Metal ions can function as Lewis acid sites [75–77] adsorbing molecules that have regions of localized high electron density such as olefins, aromatic compounds, alcohols, ketones and amines. The lower chemical activity of fused silica columns [78] is attributed to the reduction in the number of Lewis acid sites (see Table 3.7) in fused silica.

Column deactivation

Various procedures have been suggested [79, 80] to reduce the activity of both Lewis acid sites and silanol groups. One common treatment involves silanizing or silylating the inner surface of the column, a process [81] of deactivation of reactive hydroxyl groups by chemical reaction with a silanizing reagent, for example, dimethyldichlorosilane. The reaction involving interacting silanol sites is depicted in equation 3.3.

$$
\begin{array}{c}
\equiv\!Si\text{–}OH \quad Cl \quad CH_3 \\[4pt]
\qquad\qquad \diagdown \;\; \diagup \\
+ \qquad Si \qquad \rightarrow \\
\qquad\qquad \diagup \;\; \diagdown \\
\equiv\!Si\text{–}OH \quad Cl \quad CH_3
\end{array}
\qquad
\begin{array}{c}
\equiv\!Si\text{–}O \quad CH_3 \\[4pt]
\diagdown \;\; \diagup \\
Si \qquad + \; 2HCl \\
\diagup \;\; \diagdown \\
\equiv\!Si\text{–}O \quad CH_3
\end{array}
\qquad 3.3
$$

The effect of silylation on column activity is shown in Fig. 3.11, for four acid-leached soda lime glass open tubular columns which have been silylated under various conditions. The chromatograms clearly demonstrate the importance of using the correct silylation process. Intermediate tests have been described [82–86] for assessing the activity of uncoated columns and determining the effectiveness of various surface preparation procedures and, in particular, column deactivation processes. This is of particular relevance to column manufacturers and individual research laboratories involved in column preparation.

Column deactivation procedures cause modification of the surface, which in some instances, may be left either acidic or basic. Moreover, some stationary phases are intentionally non-neutral. An example is the free fatty acid phase (FFAP) produced by refluxing the polyethylene glycol, Carbowax 20M, with terephthalic acid. This acidic phase is designed for the analysis of acidic solutes. In such cases, the method of surface treatment and/or stationary phase preparation determines the range of solutes that can be analysed on a column, particularly at ultra-trace levels. It also partly accounts for the differences in column performance between nominally equivalent columns from different manufacturers.

3.4.3 Open tubular columns

Fused silica columns represent the state-of-the-art in GC. They are inherently straight as well as being extremely flexible and virtually unbreakable in normal use, providing the outer surface is protected from surface damage. Traditionally, this protection has been provided by a polyimide coating on the outer surface. However, the thermal instability of the coating above 370°C limits the maximum column temperature and aluminium-coated fused silica columns with a temperature limit of 450°C have been prepared. The maximum temperature limit of the stationary phase must still be observed. Moreover, these columns have a limited life because of differences in thermal expansion between the fused silica, and aluminium coating which leads to formation of micro-cracks in the fused silica, especially when temperature programming is used. Chrompack [87] have recently released a metal column which is claimed to be more inert than fused silica due to a completely new deactivation technique. To date there is insufficient data to judge the claims of

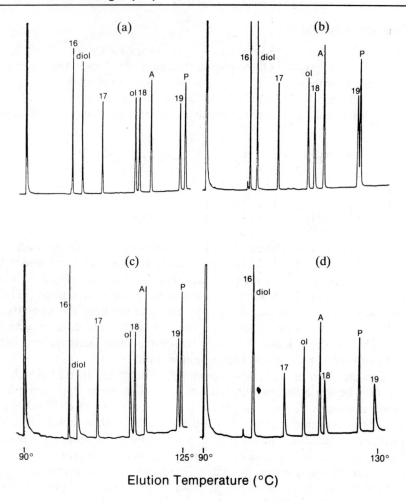

Fig. 3.11. Intermediate activity test of open tubular columns following various silylation procedures: (a) Pre-column; (b) proper silylation; (c) insufficient silylation; (d) excessive silylation. Components of test mixture are identified: 16, 17, 18 and 19 are *n*-alkanes; diol, butane-2,3-diol; A, 2,6-dimethylaniline; P, 2,6-dimethylphenol; ol, decan-1-ol. Reproduced with permission from Grob *et al.* (1979). *J. High Resol. Chromatogr.*, *Chromatogr. Comm.*, **2**: 31.

superiority, but certainly the absence of silanol groups can be expected to produce better peaks for hydroxy-compounds, such as alcohols. Metal columns have advantages in situations where high temperature stability and strength are required.

Column installation

The primary aim when connecting the column to the injector and detector is to obtain leak-free connections with zero dead volume. Connecting fused silica columns is very simple compared with the old type of rigid glass columns. The inherently straight and flexible nature of the material allows it to be ideally positioned in the injector and detector. The column is connected to the instrument with nuts that are

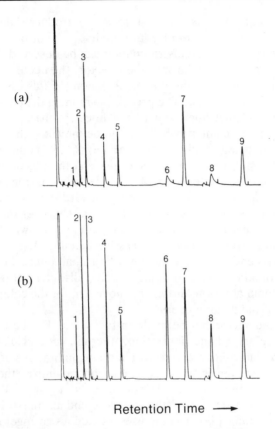

Retention Time ⟶

Fig. 3.12. The effect of a contaminated flame jet on performance. Chromatogram (a) was recorded with the column terminating at the jet causing contamination. For chromatogram (b) the column terminated 3 mm from the flame jet. An SP-2100 fused silica column 20 m × 0.20 mm with flame ionization detection was used in both cases. Components of the test mixture are identified as: 1. octan-1-ol; 2. 2,6-dimethylphenol; 3. 2,6-dimethylaniline; 4. 3,4-methylphenol; 5. n-tridecane; 6. DCHA; 7. acenaphthylene; 8. dodecan-1-ol; 9. n-pentadecane. Reproduced from Freeman R.R., Editor (1981). *High Resolution Gas Chromatography*, 2nd edn, Copyright 1981 Hewlett Packard Company. Reprinted with permission.

sealed by ferrules made of either graphite, vespel (a polyimide) or a composite of the two. In sliding the ferrules over the column, extreme care is necessary to avoid any material from the ferrule entering the column. It is a good practice to break off a small length of column after it has been pushed through the ferrule. Special care must be taken to ensure that the column ends are cut perfectly square and that pieces of polyimide coating, which cause activity, are not allowed inside the column. Jagged column ends can cause excessive carrier gas flow resulting in tailing, split or broadened peaks. Jagged ends may be avoided by first scratching the outside of the column where the break is required using a diamond tipped pencil or a piece of carborundum. The cut can be effected by grasping the scribed column on each side of the scribe and by bending and pulling the column at the point of the scribe. After cutting, it is essential that the column is examined under ×10–20 magnification for jagged ends.

Connection to the injector is usually best done by introducing the column directly into the injector. If the column is too far into the inlet, the distance for good mixing of the sample is reduced and some discrimination can be observed. It is also possible that if the column is inserted too far into the injector, the needle of the syringe will be past the end of the column at the time of injection. In this case, poor sensitivity or no peaks at all will be observed. The precise positioning depends on the injection mode and design of the inlet liner. For a split injector, there will be an optimum position for the end of the column, which should be indicated in the instructions for the particular system being used. In general terms, the column end should be positioned so that the split of the sample takes place at the end of the column, which must be located inside the heated part of the injector. Positioning the end of the column in other injection modes is more straightforward than in split mode, and the instructions accompanying the injector should adequately cover this area.

The column must be inserted in the column oven in such a way that the tubing is not subjected to undue stress caused by excessive bending. Larger diameter tubing cannot tolerate sharp bends as well as tubing with a small internal diameter. Contact of the tubing with column identification tags, oven walls, etc., must be eliminated in order to prevent column abrasion during any movement of the column caused by the forced air currents of the column oven.

Connection to the detector can be made in two ways. The best technique in the case of flame detectors (flame ionization, thermionic and flame photometric detectors) is to insert the column end to within a few millimetres of the detector jet. This ensures that the system is truly 'all glass' and minimizes the possibility of a contaminated area affecting the sample components. Figure 3.12 compares data obtained by different positioning of the detector end of the column. The other possibility is to use a capillary connecting tube and add a scavenger (make-up) gas to the column eluate in order to speed up the passage of the eluted components, so that additional band broadening is minimized. Assuming a typical flow-rate of 2 ml min^{-1} and a peak width of 5 s then the gas volume corresponding to this peak is 0.167 ml compared with the volume of a typical electron capture detector of 0.5 ml. Thus, the use of a scavenger gas cannot be avoided in the case of detectors having a large volume, such as the electron capture detector.

The methane test

Following installation of an open tubular column, it should be tested to ensure that this has been done correctly. This can be achieved by injection of an unretained solute such as methane[1] with a flame ionization detector. With the carrier gas flowing, and the injector and detector at the required operating temperature, the column is set to about 50°C. Methane is injected and the chromatogram recorded at a high chart speed of about 5 cm min^{-1}. The high speed of the recording device facilitates diagnosis of the resulting peak. The methane peak should be very sharp and symmetrical if the column has been installed correctly. Any tailing of the methane peak is an indication of dead or unswept volume in the system. The

---

[1] Strictly speaking methane is not an unretained solute. However, the extent of retention is minimal and can be ignored in all but the most exacting work.

retention time of the methane peak, which corresponds to t_m, is used to calculate the average linear mobile phase velocity using $u = L/t_m$, where L is the length of the column.

Column diameter

The column diameter is one of six interrelated parameters that determine the effectiveness (degree of separation and analysis time) of a separation. Three of these parameters (column temperature, carrier gas and carrier gas velocity) are operational parameters that are easily changed, whereas the remaining three (column length and internal diameter, film thickness) are characteristic of a given column. The internal diameter of the column has a significant effect in that it determines the maximum obtainable efficiency (as predicted from the rate equation); the smaller the column internal diameter the greater the efficiency. Table 3.6 shows that the number of effective plates per metre decreases from 8000–12 000 for a microbore column to 1400–1800 for a megabore column. But how profound are these differences when translated to real samples? Figure 3.13 compares chromatograms of a lemon oil on 0.25 mm and 0.53 mm columns. Resolution is very similar, except for the resolution of p-cymene and limonene (peaks 13 and 14), linalool and nonanal (peaks 18 and 19) and an unknown from β-bisabolene (peak 38). The smaller diameter column certainly produces the better resolution, as predicted from rate theory, but the differences are subtle. For this reason. the number of theoretical plates should not be a major factor in choosing the internal diameter of a column for a particular separation.

Column diameter influences the phase ratio and hence capacity factor and column operating temperature. For example, decreasing the internal diameter of a column causes a decrease in the phase ratio for a constant film thickness. The inner surface area of an open tubular column (which governs V_s at constant film thickness) varies directly with column diameter, while the volume of the column (which governs V_m) varies directly with the square of the column radius, r_o. Hence, both the column diameter and stationary phase film thickness influence the β value of open tubular columns as shown in equation 3.4

$$\beta = r_o/2d_f \qquad\qquad 3.4$$

In Chapter 2, it was noted that K, the product of $k'\beta$, is a thermodynamic constant whose magnitude depends only on the solute, the nature of the stationary phase and the column temperature. Thus, as β decreases as a result of reducing the column diameter, there must be a corresponding increase in the partition ratio, k', which is manifest as a longer retention time. Conversely, a larger diameter column will result in a shorter analysis time (see Fig. 3.14) and can therefore be operated at a lower column temperature, which is an advantage with thermally labile solutes. However, the greater efficiency of the smaller diameter column means that the same efficiency can be achieved with a shorter column at the same carrier gas velocity. Furthermore, advantage can be taken of the flatter van Deemter curve of the smaller-diameter column to operate at over twice the optimum gas velocity. Thus, the analysis time required to produce a given resolution can be reduced on a smaller diameter column owing to the increased column efficiency.

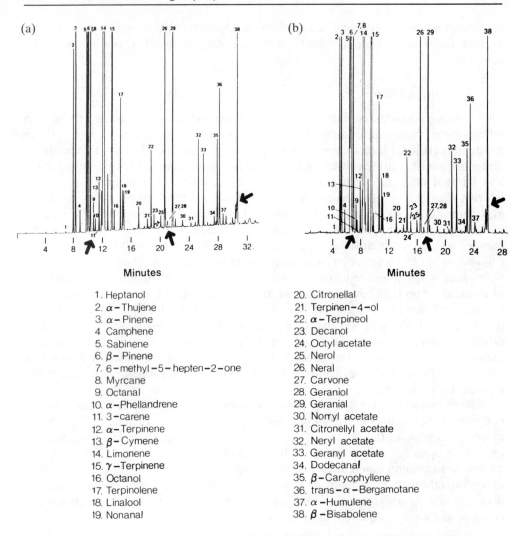

Fig. 3.13. Comparison of the resolving power of (a) a 0.25 mm column and (b) a 0.53 mm column for the separation of a lemon oil. Differences between chromatograms are highlighted by arrows. Conditions: 30 m RTx-5 fused silica column, 1.0 μm film thickness programmed from 75°C (4-min hold) to 200°C at 4°C per minute (*Restek Catalog* (1992) p. 16). Reprinted with permission.

1. Heptanol
2. α – Thujene
3. α – Pinene
4. Camphene
5. Sabinene
6. β – Pinene
7. 6 – methyl – 5 – hepten – 2 – one
8. Myrcane
9. Octanal
10. α – Phellandrene
11. 3 – carene
12. α – Terpinene
13. β – Cymene
14. Limonene
15. γ – Terpinene
16. Octanol
17. Terpinolene
18. Linalool
19. Nonanal
20. Citronellal
21. Terpinen – 4 – ol
22. α – Terpineol
23. Decanol
24. Octyl acetate
25. Nerol
26. Neral
27. Carvone
28. Geraniol
29. Geranial
30. Norryl acetate
31. Citronellyl acetate
32. Neryl acetate
33. Geranyl acetate
34. Dodecanal
35. β – Caryophyllene
36. trans – α – Bergamotane
37. α – Humulene
38. β – Bisabolene

Column internal diameter has a third effect on a separation. The sample capacity of a column is a function of the amount of stationary phase in the column. In practice, the sample capacity is the amount of solute that can be chromatographed without peak distortion caused by overloading. Overloaded (fronting) peaks are characterized by an Anti-Langmuir isotherm (Section 2.3). Most integrators can accurately quantify peak areas that are slightly overloaded. However, severe overloading causes spurious on/off integration cycles and false peak maxima locations that affect quantitative accuracy. In general, sample capacity has been neglected in column comparisons. This can be attributed, in part, to the absence of an accepted definition. It is sometimes specified as the amount that creates a 10%

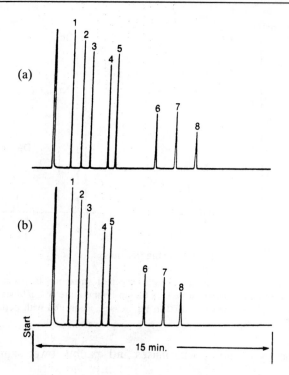

Fig. 3.14. Effect of column diameter on retention. Chromatogram (a) [$k_{Peak7} = 6.3$] was obtained on a 0.25 mm column and chromatogram (b) [$k_{Peak7} = 5.7$] on a 0.32 mm column. Other conditions were constant for both chromatograms: 30 m DB-5 fused silica column at 105°C, 0.25 μm film thickness with helium carrier gas, 40 cm s^{-1}; 1 μl split injection, 1:100; detector, FID with nitrogen make-up gas. Compounds are identified: 1, decane; 2, octan-1-ol; 3, 2,6-dimethylphenol; 4, 2,6-dimethylaniline; 5, naphthalene; 6, decan-1-ol; 7, tridecane; 8, methyl decanoate. Reproduced from *J & W Catalog* (1992) p. 9 with permission from J & W Scientific, Division of Curtin Matheson Scientific, Inc.

reduction in the number of theoretical plates. However, there are a number of problems associated with this definition [88].

A more accurate expression of sample capacity of a column is the amount of solute when the resolution of two specified compounds is reduced to a certain level, e.g. $R_s = 1.5$ representing baseline resolution. Sample capacity depends on column diameter, film thickness, k and solute solubility in the stationary phase. It is increased by increasing internal diameter of the column. Figure 3.15 compares a 0.53 mm and a 0.25 mm column to demonstrate the effect of column diameter on sample capacity. The 0.25 mm column shows some evidence of peak overload with the elution of octane. Nonane is badly overloaded and decane exhibits severe peak fronting. In contrast, the 0.53 mm column shows symmetrical peaks for octane and nonane and the decane peak shows only slight overloading. Megabore columns have a similar capacity to packed columns but produce better resolution in less time and are more inert. The main factor determining the larger sample capacity of megabore columns is that standard film thicknesses are 5–10 times higher than for standard open tubular columns [89]. Thicker films (see below) and better solute/stationary phase solubility also increase sample capacity. For instance, hydrocarbons exhibit

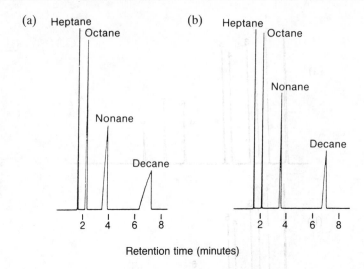

Fig. 3.15. Sample capacity effects. Chromatogram (a) was performed on a 0.25 mm column at 70°C and chromatogram (b) on a 0.53 mm column at 53°C. Column temperature was adjusted to yield similar k values. A Rtx-1 column, 0.25 μm film was used in both cases. Reprinted with permission from *Restek Catalog* (1992) p. 14, Fig. 13.

higher sample capacity on non-polar phases and exhibit lower sample capacity on polar phases.

The sample capacity of any column can be enhanced by forcing the solute to elute faster, but a drastic loss of resolving power will occur at very low values of the capacity factor. This can be achieved for late eluting solutes by temperature programming. Increasing the temperature programming rate will further increase sample capacity. However, programme rates that are too fast will decrease the resolving power of the column.

In summary, the nature of the sample and the available equipment dictate the selection of the column diameter. General recommendations on column diameter are:

1. Columns with internal diameters of 0.10 mm are suitable when very high column efficiencies are needed as, for example, in the case of very complex samples, i.e. greater than 100 components. Alternatively, short column lengths may be used for very high-speed analyses. The use of these columns requires very specialized state-of-the-art equipment if full column performance is to be realized. They are generally considered difficult to use.
2. Columns with internal diameters of 0.18 mm are ideal for use with GC–MS systems with low pumping capacities.
3. Columns with internal diameters of 0.23–0.32 mm offer a good compromise between column efficiency and speed of analysis. These are the open tubular columns used in a majority of laboratories and require some dedicated 'capillary' instrumentation.
4. Columns with internal diameters of 0.53 mm up to 1.00 mm are suitable replacements for packed columns and for separating dirty samples because they are less affected by non-volatile sample residues. They are also useful for

samples containing a wide dynamic range of analyte concentrations and for relatively simple separations, i.e. samples with fewer than 30 components. They require no specialized equipment and are more forgiving of operator deficiencies. Sample valves and purge and trap inlet systems function better with wide-bore columns because of the higher operating flow-rates. This minimizes dead volume effects and induces better sample transfer efficiencies from the adsorbent trap.

Column length

Column length has less impact on the resolution of a solute pair than either column internal diameter or stationary phase film thickness. Increasing column length has three effects; the number of theoretical plates and hence resolution, column backpressure and analysis time are all increased. The choice of column length is then a compromise between efficiency, column operating pressure and analysis time. The shortest column capable of generating the required separating efficiency should be chosen.

Resolution achieved on a column is proportional to the square root of the column length whereas retention time is directly proportional to column length for isothermal analysis. Thus, if resolution must be improved by a factor of two and a 30 m column is in use, then a 120 m column is necessary, resulting in an increase in analysis time by a factor of four. Figure 3.16(a) compares the isothermal separation of fatty acid methyl esters on a 30 m and 60 m column under otherwise identical conditions. The expected increase in resolution is observed, but at the cost of an increased analysis time. Better means of increasing the resolution are achieved by changing column temperature, film thickness or column internal diameter. The position is somewhat different with temperature-programmed operation. In this case, the compounds elute according to column temperature. Using a longer column and the same temperature profile, compounds elute at higher temperatures on the longer column which, in turn, reduces the analysis time. This is illustrated in Fig. 3.16(b). Areas where resolution was markedly improved on the 60 m column are denoted by arrows, and the analysis time only increased from 22 to 28 min. The maximum gain in column efficiency is achieved on the 60 m column by reducing the temperature programme rate by one-half. However, the analysis time would double using these conditions and the gain in resolution is not sufficient to justify this increase.

The small loss in column length from required end-trimming during installation, or by breaking off the front portion of a contaminated column, has little effect on subsequent separations because of the square root relationship between efficiency and column length. Shorter columns (0.2 m–4 m) are useful for screening samples and for samples containing a relatively small number of components. Most analyses are performed on columns of intermediate length (e.g. 15–30 m × 0.25 mm i.d.).

3.5 Column Packing Materials

The stationary phase in gas–solid chromatography (GSC) is a solid adsorbent, whereas in gas–liquid chromatography (GLC) it is a liquid either coated on a solid

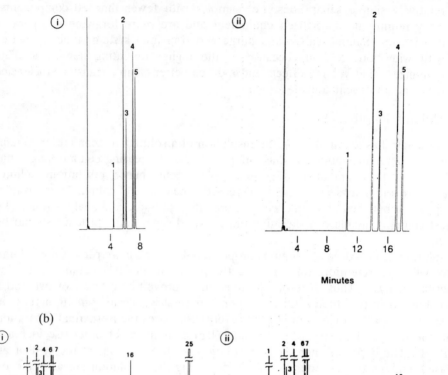

Fig. 3.16. Chromatograms showing the effect of column length on (a) isothermal separation of myristic, palmitic, palmitoleic, stearic and oleic acid methyl esters, and (b) temperature-programmed separation of volatile flavour compounds in spearmint oil. Chromatograms in (a) were obtained on (i) 30 m and (ii) 60 m BP-X70 columns (0.22 mm; 0.25 μm) at 150°C. The chromatograms in (b) were obtained on (i) 30 m and (ii) 60 m Supelcowax 10 columns (0.25 mm i.d.; 0.25 μm), temperature programmed from 75°C (4-min hold) to 200°C at 4°C min⁻¹. Carrier flow was 25 cm s⁻¹ and flame ionization detection was used in all cases. Compound identification (b): 1, α-pinene; 2, β-pinene; 3, sabinene; 4, myrcene; 5, α-terpinene; 6, L-limonene; 7, 1,8-cineole; 8, cis-ocimene; 9, γ-terpinene; 10, para-cymene; 11, terpinolene; 12, 3-octyl acetate; 13, octan-3-ol; 14, trans-sabinenehydrate; 15; L-menthone; 16, β-bourbonene; 17, linalool; 18, terpinene-4-ol; 19, β-caryophilene; 20, dihydrocarvone; 21, trans-didihydrocarvyl; 22, trans-β-farnesene; 23, α-terpineol; 24, germacrene-D; 25, carvone, 26, cis-carvyl acetate. (b) Reprinted with permission from Supelo Inc., Bellefonte, PA 16823, Fig. A, p. 42, *Catalog 30* (1992).

support (packed column) or deposited directly on the column walls. It is therefore convenient to examine the stationary phases for the two techniques separately.

3.5.1 Gas–solid chromatography

GSC preceded GLC but has never achieved the same prominence. There are a number of reasons for this. First, adsorption isotherms are frequently nonlinear, leading to several abnormal phenomena such as asymmetric peaks and retention times that are dependent on sample size (see Section 2.3). Second, retention times are excessively long because of the high surface areas of adsorbents. Thus, it is restricted to relatively low molecular mass solutes. Third, adsorbents are difficult to standardize and prepare reproducibly and many active solids are efficient catalysts. Nonetheless, GSC enjoys some advantages over GLC and it has some important application areas such as the separation of isomers, where it exhibits greater selectivity and also in the separation of inorganic gases and low molecular mass hydrocarbons for which GLC shows little selectivity. Furthermore, adsorbents are stable over a wide temperature range with virtually nonexistent column bleed, thereby allowing the use of high column temperatures. Other general properties required of an adsorbent for GSC are a large, homogeneous surface and no catalytic activity.

Adsorbents

GSC of inorganic gases is the one area where packed columns are still used almost exclusively. It is a fact of contemporary gas analysis, however, that complex column systems (see Section 3.9.4 on multidimensional chromatography) may be required to separate and quantify even relatively simple mixtures [90]. The main adsorbents are based on silica, charcoal, alumina or molecular sieves (Table 3.8). The development of new adsorbents is, however, continuing.

Porous silica is available in a variety of surface areas and pore diameters [91] which have been used for separation of low molecular mass saturated and

Table 3.8 Selected adsorbents and molecular sieves for gas–solid chromatography.

Chemical type	Commercial name	Specific surface area ($m^2 g^{-1}$)	Pore diameter (nm)
Silica	Porasil B	185	15
	Porasil C	100	30
Alumina	Various	–	–
Graphitized carbon black	Carbopack C	12	–
	Carbopack B	100	–
	Carbosieve	1000	1.3
	Spherocarb	1200	1.5
Carbon molecular sieve	Carbosphere	1000	1.3
Sodium aluminium silicate	Molecular sieve 13X	700–800	1.0
Calcium aluminium silicate	Molecular sieve 5A	700–800	0.5

unsaturated hydrocarbons (e.g. methane, ethane, ethene). Alumina shows similar retention properties to silica of comparable specific surface area but has different selectivities. These selectivities are due to Lewis acid sites associated with aluminium ions on the surface of alumina. However, the heterogeneous nature of the surface of both silica and alumina [92] often leads to peak tailing. To overcome such problems and to reduce retention volumes, modify selectivity and improve efficiency, both adsorbents can be coated with a small quantity of non-volatile liquid [93]. As the film thickness of modifying liquid is increased, retention times reach a minimum and then increase slowly, reflecting the change from an adsorption to partition process. It is assumed that adsorption dominates at the low levels of liquid phase commonly used in GSC and that the modifying liquid functions by selectively adsorbing to the most energetic sites, thereby reducing the heterogeneous nature of the adsorbent. As a result, both retention and peak asymmetry are reduced. Alternatively, the surface of silica or alumina can be modified by addition of an inorganic salt [94]. In similar fashion, a dramatic reduction in the retention of non-polar solutes can be obtained by the uptake of water vapour by the adsorbents or by use of humidified carrier gas. The selectivity of graphitized carbon permits some unique separations. Nevertheless, these adsorbents are not very popular and have been replaced by molecular sieves and porous polymers.

Molecular sieves

Molecular sieve is a general term, but when unqualified, usually refers to artificially prepared zeolites which are sodium, potassium or calcium aluminosilicates. Of the range of molecular sieves available, only 5A and 13X are in common use and have pore diameters (Table 3.8) that are of the same order as the dimensions of small molecules. For this reason both molecular sieves 5A and 13X are used primarily for the separation of low molecular mass inorganic gases and hydrocarbons which can penetrate the pores. The reason for choosing one or the other will depend on the precise nature of the separation. The molecular sieves exhibit variable retention behaviour, depending on the extent of activation [95] and the carrier gas [96] and must be treated to remove active sites to achieve the successful separation of reactive gases [97]. There may also be significant differences in behaviour between batches of the same molecular sieve. The separating power of molecular sieves can slowly deteriorate because of adsorption of water from carrier gas or from samples.

Porous polymers

In contrast to molecular sieves, porous polymers are excellent for separating polar gases because of near-linear sorption isotherms [98]. These materials are usually copolymers of divinylbenzene (DVB) with another aromatic olefin or acrylates and are used in preference to adsorbents because of their ease of use, superior performance and unique separating capabilities. The precise nature of the separation mechanism on porous polymers remains unclear but adsorption is undoubtedly involved, particularly at lower temperatures [99, 100]. Porous polymers are marketed under various tradenames but the best known are Porapak, Chromosorb Century Series and Tenax (Table 3.9). The commercial materials typically contain

Table 3.9   Physical properties of selected porous polymers.

Porous polymer		Type*	Surface area† $(m^2\,g^{-1})$	Pore diameter (nm)	Temperature limit (°C)
Porapak	N	VP	250–350	–	200
	P‡	PS-DVB	100–200	–	250
	Q‡	EVB-DVB	500–600	7.5	250
	R	VP	450–600	7.6	250
	S	VP	300–450	7.6	250
	T	EGDMA	225–350	9	200
Chromosorb	101	PS-DVB	50	300–400	275
	102	PS-DVB	300–500	8.5	250
	103	PS	15–25	300–400	275
	104	ACN-DVB	100–200	60–80	250
	105	Acrylic ester	600–700	40–60	250
	106	PS	700–800	500	250
	107	Acrylic ester	400–500	800	250
	108	Acrylic ester	100–200	250	250

*VP, vinylpyrollidone; PS, polystyrene; DVB, divinylbenzene; EVB, ethylvinylbenzene; EGDMA, ethylene glycol dimethacrylate; ACN, acrylonitrile.
†Surface areas quoted in the literature vary widely.
‡Also available in a silanized version, PS or QS.

some monomer and before use overnight conditioning at 250°C (190°C for Porapak N and T) with carrier gas flow is necessary.

Porous polymers find many applications in the analysis of volatile organic and inorganic compounds, such as water, permanent gases, and low molecular mass compounds containing halogens or sulfur. For example, Porapak N provides excellent separations of low molecular mass hydrocarbons, including methane, ethane, ethylene and acetylene, whereas nitrogen, carbon dioxide and nitrous oxide are well resolved on Chromosorb 101. The retention behaviour of gases on the various series is similar, but subtle changes in relative retention are observed and hence difficult separations often can be effected by selection of an appropriate material. Alternatively, custom-made porous polymers have provided unique selectivities [101]. For example, a material has been described which selectively retains oxygen by virtue of an anchored cobalt(II)–Schiff base complex. The best source of information regarding the uses of these polymers is the trade literature produced by the various chromatography supply companies. Some of the porous polymers have exhibited specific reactivities to certain compounds [102]. For example, Porapak Q and Chromosorb 102 are nitrated by nitrogen dioxide.

Porous layer open tubular columns

Developments in open tubular column technology including megabore and PLOT columns coated with Porapak Q and molecular sieve 5A or alumina, may ultimately provide an answer to the difficult separation problems encountered in gas analysis and separation of low molecular mass hydrocarbons. The usual advantages of open tubular columns, namely, higher efficiency and sensitivity combined with shorter analysis times are obtained. PLOT columns based on molecular sieve 5A can separate permanent and noble gases, whereas the alumina-based columns are ideal for hydrocarbon analysis. Alumina PLOT columns are available [103] in two polarities exhibiting different selectivities for unsaturated hydrocarbons, depending on the column deactivation process (sodium sulfate versus potassium chloride). PLOT columns containing alumina or molecular sieve 5A exhibit the usual problems of deactivation caused by adsorption of water and/or carbon dioxide. However, in comparison with packed columns, the newer PLOT columns are easily regenerated by heating at elevated temperatures for as little as 10 min, although it is more usual to reactivate by heating at 250°C overnight or at 300°C for 3–4 h.

3.5.2   Gas–liquid chromatography

The stationary phase for GLC is a liquid which is retained either on a solid support (packed columns) or the column wall (WCOT or BPOT columns). Before proceeding to a consideration of stationary phases it is appropriate to examine the nature of the materials used as solid supports.

Solid supports

The support [81] retains the liquid stationary phase and provides a large interface between the mobile and stationary phases. It should be inert, have a large surface area, be thermally stable and be mechanically strong to avoid fragmentation during the column packing process. The best quality support can be ruined by mishandling, as this can cause fracturing which produces fines (particles of much smaller size than the bulk material) and also exposes untreated surfaces. Optimum column efficiency requires that the material be of uniform pore and particle size to enable it to be packed uniformly. The supports in most common use (see Table 3.10) are based on diatomaceous earths. These are diatom skeleton deposits, consisting mainly of silica with minor metallic impurities, that are specially processed by calcinating, fluxing, crushing and grading to produce materials of high surface area and uniform particle size. Non-diatomite supports which have found occasional use include PTFE beads, glass beads, charcoal (e.g. Carbopak B and C) and porous silica (Porasil). The loading of the stationary phase on the support material is expressed as a percentage by mass (i.e. % w/w) of the solid support.

Although the support should be totally inert and unreactive with sample components, this cannot be achieved in practice, as the surface must have sufficient energy to both hold the liquid phase stationary and to cause it to wet the support surface as a thin film. A possible exception are PTFE beads (e.g. Chromosorb T) which are, in consequence, very difficult to coat with a uniform layer of liquid phase. In the absence of a liquid phase, most supports would act as an adsorbent and GSC could be carried out. Coating the support does not totally eliminate the support activity, although the effectiveness increases with the thickness of the liquid phase. Various techniques are available to reduce the activity of the support. These include acid washing (AW) to eliminate metallic impurities and silanization with dimethyldichlorosilane (DMCS) or hexamethyldisilazane (HMDS) to eliminate surface activity caused by silanol groups. Commercial support materials are

Table 3.10 Physical properties of selected solid supports.

Trade name	Specific surface $(m^2 g^{-1})$	Pore diameter (μm)	Packed density $(g\, ml^{-1})$	Maximum loading (% w/w)
Chromosorb W	1.0	0.9	0.24	15
Chromosorb P	4.0		0.47	30
Chromosorb 750	0.5–1.0		0.36	7
Anakrom	1.0–1.4	1.0		
Gas Chrom Q	Data unavailable			
Supelcoport	Data unavailable			
Glass beads	0.04–0.36			0.5
Chromosorb T	7–8		0.49	20

Table 3.11 Support treatments.

AW or A	Acid washed
NAW or U	Non-acid washed (or untreated)
DMCS or S	Dimethyldichlorosilane treated
AW-DMCS	Acid washed and dimethyldichlorosilane treated
HMDS	Hexamethyldisilazane treated
HP or Q	High-performance (high quality AW DMCS treated)

Table 3.12 Mesh sizes of solid supports in gas–liquid chromatography.

Mesh size (ASTM sieve)	Nominal screen size (mm)
40–60	0.42–0.25
60–80	0.25–0.18
80–100	0.18–0.15
100–120	0.15–0.125

frequently designated according to their treatment (Table 3.11). Some supports such as Gas Chrom Q are also base-washed to remove organic impurities.

Support materials are available in a range of particle sizes (Table 3.12) as well as chemically treated and untreated materials. The particular grade of support chosen depends on the column length and the maximum column pressure drop that can be tolerated. The use of fine grades will give high pressure drops, because such particles are more difficult to pack uniformly, but will also produce higher column efficiencies. For preparative columns, either 40–60 or 60–80 mesh supports are usually used, and for analytical columns 80–100 or 100–120 mesh supports are used.

Ideally, the support material is inert, and for this reason one support can be replaced, at least in theory, with another of similar quality without altering the separation. However, subtle differences in the residual activity of even the best quality supports are sometimes observed which can influence the behaviour of a chemically reactive solute such as a steroid; for example, Gas Chrom QII and Gas Chrom Q are both high-quality diatomite supports, but the latter is more inert to thermally labile and sensitive compounds, whereas Gas Chrom QII generally gives higher efficiency, lower back pressures and better coating characteristics for polar liquid phases. Thus, in changing from one support to another, it is best to confirm its suitability for the particular separation by careful examination of standard samples. In cases where the support can be substituted, the loading of stationary phase may require adjusting because of the different packed densities (supports are also characterized by the free-fall density) of the support materials to produce the same retention time. For example, Chromosorb P is twice as dense as Chromosorb W and, if the same loading is applied to both materials, the column packed with Chromosorb P will contain twice as much stationary phase and so retention times will be correspondingly longer.

Stationary phases

The liquids used as stationary phases in packed and open tubular columns are closely related. Nevertheless, liquid phases in open tubular columns are usually cross-linked and bonded and may exhibit slight differences in selectivity to nominally equivalent packed column materials. The selection and comparison of stationary phases is confusing for the newcomer as some 300 phases are available and approximately 1000 have been described in the literature. In general, selection of a phase [104–106] is based on trial and error. Many of the phases are very similar, differing only in trade name. In practice, a limited number are in common use for packed columns and even fewer in open tubular columns. Moreover, two forces have combined to contain the proliferation of phases. First, the high efficiency of open tubular columns has reduced the necessity for many selective liquid phases and second, theoretical studies have aided [107] in phase selection. It is interesting to note that as open tubular column technology has matured, the search for selective phases has begun once again.

Properties desirable of an ideal liquid phase are a low vapour pressure, thermal and chemical stability, low viscosity, nonreactivity towards sample components and a wide operating-temperature range, extending from −80°C to 450°C. The phase must exhibit reasonable solvent properties (i.e., dissolving power) for the solutes in order to ensure symmetrical peaks. The lower temperature limit is set by the temperature at which the phase begins to behave as a solid, since efficient operation requires a liquid phase (consider the effect of viscosity on the C term of the van Deemter equation). However, operating below this temperature should not cause permanent damage to the column. The upper temperature limit of the operating range is determined by the volatility or, more commonly, the thermal stability of the phase. This temperature is referred to as the maximum allowable operating temperature and exceeding this value will cause accelerated degradation of the phase. The upper limit is not a precise threshold and exceeding this value for short periods will not cause instant phase deterioration. Hence, separate upper limits are set for isothermal and temperature-programmed operation. Major effort has been directed to finding phases of increased thermal stability and low volatility in order to extend the usefulness of GLC to higher and higher boiling temperature solutes. Unfortunately, no phase meets all the requirements and a compromise must be reached.

Precoated support materials for packed columns can be purchased as loose packing material or already packed in columns. Alternatively, uncoated solid supports can be purchased and coated before packing in the column. The latter is probably not cost-effective for routine applications because of column to column variation. With the exception of a few research laboratories, virtually all open tubular columns are purchased ready coated.

Stationary phases can be divided into non-polar, polar and speciality phases. These differ in their ability to interact with solutes of different structure, i.e. their selectivity. The non-polar phases contain no functional groups capable of specific interaction (e.g. hydrogen bonding or dipole interactions) with the sample. Here, interaction between solute and stationary phase is limited to dispersive forces, and components therefore separate according to their volatility with the elution order following their boiling points. Compounds that cannot be differentiated on the basis of their boiling points (i.e. they have similar or equal boiling points) require a different stationary phase for separation. To obtain the differentiation of solutes by forces other than dispersion, a polar phase containing groups capable of specific interactions with sample components is required. The elution order now depends on a combination of volatility and specific polar–polar interactions. The relative magnitude of the various interactions (dispersive, dipole, hydrogen bonding and acid–base) determines the selectivity of the phase toward particular solutes. The selectivity and resolution of a separation can be optimized by choosing a stationary phase that exploits the different interactions.

Polarity and selectivity are often used synonymously but incorrectly. Polarity is best defined as the chemical nature of the stationary phase as determined by the number and type of chemical functionalities present as part of the phase. Selectivity is defined as the degree of interaction between the stationary phase and a solute. A stationary phase exhibiting high selectivity for a particular solute may exhibit lower (different) selectivity toward a solute with different functional groups.

Some selected stationary phases

With the large number of phases available, it is reasonable to wonder whether it is possible to be certain of choosing the correct phase for a particular separation. One of the challenges facing the gas chromatographer is the standardization of existing phases, while developing newer specialty materials (e.g. high temperature, low bleed rate phases, chiral phases). The following list of materials is not meant as an exhaustive list of stationary phases but rather as an indication of the diversity.

A. Non-polar phases

Non-polar phases are excellent solvents for non-polar solutes. Alkanes are selectively retained as compared with polar solutes of similar boiling points. Perfluoroalkanes are the obvious choice as a non-polar phase. However, they have limited applicability because they are poor solvents for most solutes as a result of their extremely low polarity:

Hydrocarbons

A number of hydrocarbons have been used as non-polar stationary phases. The requirement of low volatility restricts suitability to high molecular mass discrete hydrocarbons, such as squalane or Apolane C87, or mixtures of long-chain *n*-alkanes such as Apiezon L. None of this group has found wide application in open tubular columns but they are, nonetheless, important as reference phases. Squalane, for example, has a low upper temperature limit, is a single compound and therefore is easily characterized. It is the least polar of phases in common use and is used as a reference point to define the polarity of other phases. It is also useful for the separation of non-polar solutes such as complex hydrocarbons.

Alkylsilicone phases

Polymers based on a silicon–oxygen–silicon backbone (Fig. 3.17) form the basis of the most widely used group of stationary phases. These linear polysiloxanes differ in their average molecular mass, thermal stability and viscosity. The chemical difference lies in the substituent and degree of substitution on the silicon backbone. Most of these phases can be used at temperatures up to 300–350°C without significant volatilization or decomposition. The dimethylsilicone phases (e.g. SE-30, OV-1) are available as closely related materials from several manufacturers (Table 3.13) and have very similar selectivities. Higher efficiencies are obtained on packed columns with the less viscous liquids, such as OV-101, while viscous gums (e.g. OV-1) produce a more even coating on fused silica or glass and are used for WCOT columns. Coated phases are being replaced increasingly by the cross-linked bonded phase dimethylsilicones. The non-polar dimethylsilicones are very popular as general purpose phases. They usually exhibit longer column life and lower bleed rates than substituted polysiloxanes.

The methylphenylsilicones are a closely related series of phases in which different proportions of the methyl groups are replaced by phenyl groups. The substitution can be random, or the phase can be a copolymer of dimethylsilicone and diphenylsilicone groups. These phases often give better peak shapes than the dimethylsilicones, particularly for polar solutes or those containing aromatic rings. The polarity of the phase increases with increasing degree of phenyl substitution and retention of compounds containing aromatic or polarizable groups is also increased.

Table 3.13   Trademarks and abbreviations.

Trademark	Company
BP	Scientific Glass Engineering (SGE)
DB	J & W
OV	Ohio Valley
CP	Chrompack
SE	General Electric
DC	Dow Corning
SP	Supelco
SPB	RSL Belgium/Alltech
Superox	RSL Belgium/Alltech

Chemical structure	Classification	Uses

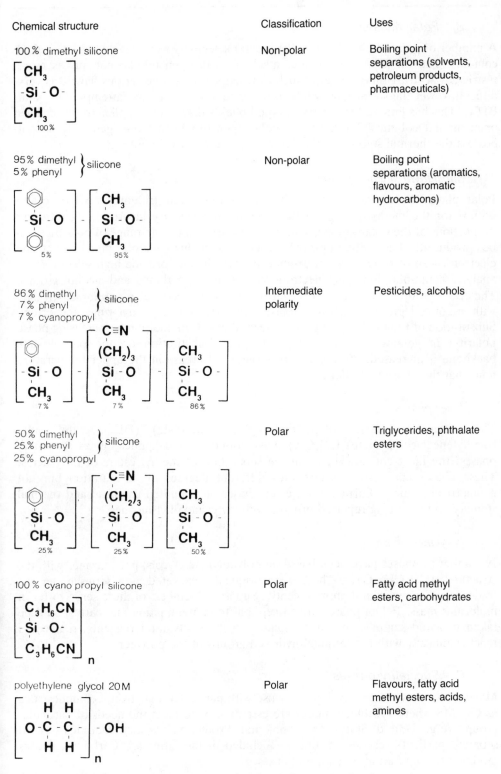

100% dimethyl silicone — Non-polar — Boiling point separations (solvents, petroleum products, pharmaceuticals)

95% dimethyl / 5% phenyl silicone — Non-polar — Boiling point separations (aromatics, flavours, aromatic hydrocarbons)

86% dimethyl / 7% phenyl / 7% cyanopropyl silicone — Intermediate polarity — Pesticides, alcohols

50% dimethyl / 25% phenyl / 25% cyanopropyl silicone — Polar — Triglycerides, phthalate esters

100% cyano propyl silicone — Polar — Fatty acid methyl esters, carbohydrates

polyethylene glycol 20M — Polar — Flavours, fatty acid methyl esters, acids, amines

Fig. 3.17. Structure of polysiloxane and polyethylene glycol stationary phases.

B. Polar phases

A number of polar phases that provide greater selectivity were developed for packed columns. The range of polar phases available with open tubular columns is more restricted. A 'truly inert' material such as fused silica has one serious flaw; it is very difficult to wet the surface, as will be obvious to anyone who has attempted to paint PTFE. This has limited the variety of liquid phases that can be applied to fused silica open tubular columns. Furthermore, polar open tubular columns generally do not possess the thermal stability and efficiency of non-polar columns.

Substituted silicone phases

Polar phases have been prepared by substituting polar trifluoropropyl or cyano groups for the methyl groups of the dimethylsilicones. By incorporating different proportions of the polar groups, stationary phases with a wide range of polarities can be produced. The trifluoropropyl group has a high dipole moment, strong electron-acceptor properties and trifluoropropyl silicones provide high selectivity for analytes containing lone-pair electrons such as nitro, carbonyl and alcohol groups. The cyano group strongly attracts electrons and cyanoalkyl silicones interact strongly with π-bonded groups, such as olefins, carbonyl groups, phenyl rings and esters. Substitution of the methyl groups has a second effect, in addition to increasing phase polarity: in general, as the amount of polar substitution on the polysiloxane backbone is increased, there is a corresponding decrease in the upper temperature limit that the phase will tolerate.

Ester phases

The polymeric esters, poly(diethyleneglycol succinate) (DEGS) and poly-(diethyleneglycol adipate) (DEGA), have found limited application as stationary phases; the most common being in the separation of methyl esters of fatty acids. These phases can resolve esters with different degrees of unsaturation but not geometrical isomers. Furthermore, ester phases have limited chemical and thermal stability and are being replaced progressively by cyanoalkylsilicones.

Polyether phases

A number of phases have been based on polyethylene glycols, polyoxiranes with the structure $-(CH_2CH_2-O)_n-$. These were originally marketed as a series under the tradename Carbowax and, more recently, Superox. Members of these series differ in molecular mass. Polyol phases are susceptible to contamination and care should be taken to avoid samples containing solutes, such as silylation reagents, that could react chemically with the terminal hydroxyl groups of the polymer.

C. Speciality phases

Many phases have been developed for use with particular analytical techniques such as GC–MS where low bleed phases are essential, or to meet the needs of particular groups (e.g. United States Environmental Protection Agency methods) or to separate particular classes of solutes. Included in the latter are Carbowax phases modified for separation of acids and bases.

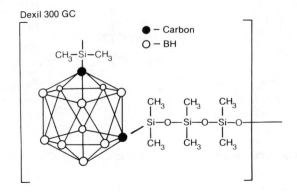

Fig. 3.18. A silicone–carborane copolymer used as a high temperature stationary phase.

High-temperature phases

Extension of the limits of GC has proceeded in four directions; production of more inert columns, increased temperature limits of the chromatographic oven, improvement of column material and stationary phase, enhanced detection limits and column selectivity. The development of phases stable at temperatures up to 500°C is a priority with column manufacturers. The range of analytes that are stable at this temperature is limited but such columns are, nevertheless, highly desirable. Their major advantage is in being able to eliminate stationary phase bleed for use with very sensitive detectors. The most successful high-temperature phases to date have been based on substituted silicone–carborane copolymers (Fig. 3.18). These phases can be used to temperatures as high as 450°C as a result of the stabilizing effect of the carborane.

Chiral phases

In many situations it is important to be able to distinguish between enantiomers of optically active compounds. For many biomolecules, only one optical isomer is biologically active. Drugs are an obvious example. Unfortunately, enantiomers have identical physical properties and most stationary phases are achiral, so that enantiomeric species have the same retention times and are unresolved. One solution to this problem is derivatization of sample components with an optically active chiral reagent to produce a pair of diastereoisomers with different physical properties, which can be resolved on a conventional achiral column. This approach has not been favoured for a number of practical reasons. Alternatively, the stationary phase itself can be chiral and the development of such phases that can separate enantiomers has received particular attention.

The earlier chiral phases such as Chirasil-Val (Fig. 3.19) incorporated amino acid-derived chiral centres into a silicone based polymer. Of more recent origin are the chiral phases based on α-, β- and γ-cyclodextrins, which are formed by the α-1,4 linkage of glucose units with differing numbers of cyclic glucose units. S-hydroxypropyl (hydrophilic properties), dialkyl (hydrophobic properties) and trifluoroacetyl (intermediate properties) derivatives of the three forms have been

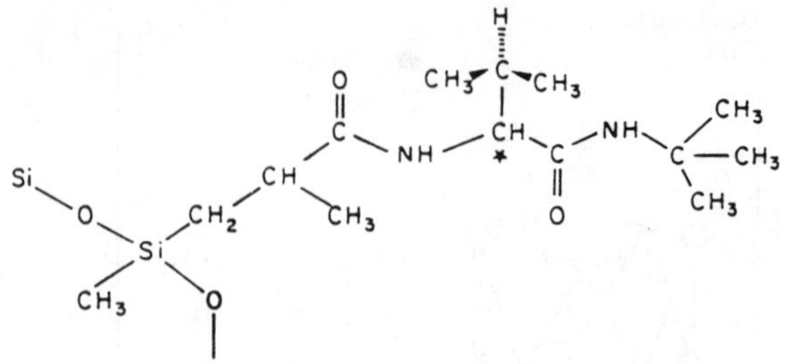

Fig. 3.19.   Structure of the chiral stationary phase, Chirasil-Val.

manufactured which exhibit properties allowing them to be used as stationary phases. These chiral phases separate nonaromatic enantiomers, including saturated alcohols, amines, carboxylic acids, epoxides, diols, polyols, heterocyclic compounds, lactones, amino acids, amino alcohols, pyrans and furans. Each of the phases has a selected area of specificity. Also, reversal of elution order can occur from one series to the next, as well as from one cavity size (β) to another (γ). The acylating reagent (S-hydroxypropyl, dialkyl or trifluoroacetyl) also can contribute to stereo-selectivity. Further more versatile chiral phases are likely to be produced in the near future.

Stationary phase selection

There are several factors to consider in selecting a stationary phase. General considerations include temperature limits of the stationary phase, column efficiency and lifetime and detector compatibility. Since non-polar phases generally provide more efficient columns, which also exhibit superior lifetimes, it is wise to use the least polar phase that provides satisfactory separation. Phases containing the specific element corresponding with element-selective detectors (e.g. cyanopropyl phases with an NPD detector; trifluoropropyl phases with an ECD detector) should be avoided where possible. These selective detectors will be substantially more sensitive to normal column bleed with such phases.

The most difficult factor to assess is the ability of a phase to effect the desired separation. From this perspective, the selection of a stationary phase [104–106, 108, 109] and column is a daunting prospect. In theory, the selection is based upon achieving the maximum selectivity between the phase and solutes of interest. The separation is increased by exploiting solute–stationary phase interactions that retard the progress of some solutes relative to others so as to increase their retentions. The types of interactions to consider are:

- London or dispersion forces which are weak and non-specific.
- Dipole–dipole interactions or dipole induced dipole interactions.
- Acid-base interactions or proton transferring (or sharing) tendencies of either the solute or stationary phase.

In practice, experience, literature data and availability are often the deciding factors. Moreover, practising chromatographers usually develop an intuitive feel for which phase to employ for a particular separation. The reliability of these approaches leaves much to be desired. For instance, if a separation cited in the literature used a lower efficiency system, the stationary phase was selected for its greater selectivity. With a more efficient system, better separation can be achieved, even with a less retentive or selective phase. As a general rule, a stationary phase should be chosen such that it is chemically similar to the substances to be separated: 'Like dissolves like'. For example, a polar phase would be chosen for the separation of alcohols whereas alkanes require a non-polar phase. There are many situations where this generalization holds. However, a stationary phase of opposite polarity might be best to separate solutes that differ little in polarity. The relatively non-polar xylenes, for instance, are best separated on a polar stationary phase (e.g. DB-Wax; polyethylene glycol) that effects the separation by highlighting the slight differences in polarity among the isomers. In similar fashion, non-polar fatty acid methyl esters are well resolved on a highly polar Rtx-2330 phase (90%biscyanopropyl/10% phenylcyanopropylsilicone). In addition, most samples are mixtures of compounds containing a range of functional groups and polarities, in which case the choice of stationary phase is more complicated.

A fairly limited set of packed columns will suffice in most laboratories. Hawkes [107] found that six phases, OV-1, OV-17, DEGS, OV-275, OV-210 and Carbowax 20M were most popular, followed by another 20 that were also common, and 13 that were used for speciality applications. Similar sets have been proposed by other authors [110, 111], while a more limited set of open tubular columns will satisfy the needs of most laboratories. With open tubular columns, a non-polar dimethylsilicone bonded phase column, a polar bonded phase Carbowax column and possibly a cyanomethylsilicone bonded phase column should enable most separations. Open tubular columns supplied by different manufacturers, which are nominally equivalent, are listed in Table 3.14. Further information may be obtained by consulting catalogues of the various chromatography suppliers.

Rohrschneider constants

One of the most logical approaches to characterizing stationary phases was proposed by Rohrschneider [112]. However, a measurement scale is essential to characterization and classification. Retention times are unsuitable for this purpose because of their dependence on operating conditions. The specific retention volume overcomes this objection, but it is not used in practise because it is awkward to calculate. The use of relative retention data is, at present, the accepted method. Under isothermal conditions, a plot of the logarithm of the adjusted retention time versus molecular mass of a homologous series is linear. Therefore, a series of reference compounds are used to establish a linear reference scale into which the retention of other compounds can be interpolated. The most established and useful retention index system (see Section 9.2.1) is the one proposed by Kovats [113–115] in 1958, in which the series of compounds used to establish the linear reference scale were *n*-alkanes. The retention index (I) of a substance is defined as being equal to '100 times the carbon number of a hypothetical *n*-paraffin with the same adjusted retention time as the substance of interest'. According to convention, the *n*-alkanes have an index of 100 times the carbon number in question, e.g. 400 for *n*-butane and 900 for *n*-nonane, at any temperature and on any liquid phase.

Rohrschneider measured the retention indices of five test compounds (see Table 3.15) as a guide to the different interactions of the stationary phase relative to a

Table 3.14 Listing of nominally equivalent open tubular columns and their manufacturers.

Composition	Restek	J&W	SGE	Supelco	HP	Ohio Valley	Alltech	Chrompack	Quadrex
100% Dimethylpolysiloxane	RT$_x$-1	DB-1	BP-1	SP-2100,SPB-1	HP-1,Ultra-1	OV-1	RSL-150,160	CP SIL 5B	007-1
95% Dimethyl:5%-diphenylpolysiloxane	RT$_x$-5	DB-5	BP-5	SPB-5,PTE-5	HP-5,Ultra-2	OV-5	RSL-200	CP SIL 8CB	007-2
80% Dimethyl:20%-diphenylpolysiloxane	RT$_x$-20	–	–	SPB-20	–	OV-7	–	–	007-7
65% Dimethyl:35%-diphenylpolysiloxane	RT$_x$-35	–	–	–	–	OV-11	–	–	007-11
14% Cyanopropylphenyl:86%-dimethylpolysiloxane	RT$_x$-1701	DB-1701	BP-10	–	–	OV-1701	OV-1701	CP SIL 19CB	007-1701
50% Methyl:50%-phenylpolysiloxane	RT$_x$-50	DB-17	–	SP-2250	HP-17	OV-17	RSL-300	–	007-17
Trifluoropropyl-methylpolysiloxane	RT$_x$-200	DB-210	–	–	–	–	–	–	–
50% Cyanopropylphenyl:50% dimethylpolysiloxane	RT$_x$-225	DB-225	BP-225	SP-2300	HP-225	OV-225	RSL-500	CP SIL 43CB	007-225
Carbowax PEG	Stabilwax	DB-Wax	BP-20	Supelcowax-10	HP-20M	Carbowax 20M	Superox-II	CP WAX 52CB	007-CW
Carbowax PEG for amines	Stabilwax-DB	CAM	–	–	–	–	–	–	–
Carbowax PEG for acids	Stabilwax-DA	DB-FFAP	BP-21	SP-1000	HP-FFAP	OV-351	Superox-FA	CP WAX 58CB	FFAP
90% bisCyanopropyl:10% phenylcyanopropylpoly-siloxane	RT$_x$-2330	DB-23	BPX-70	SP-2330	–	–	–	CP SIL 84	CPS-1
100% bisCyanopropylpoly-siloxane	RT$_x$-2340	–	–	2P-2340	–	OV-275	–	CP SIL 88	CPS-2
EPA volatile organic methods	RT$_x$-Volatiles	DB-624	BP-624	VOCOL	–	OV-624	AT-624	CP SIL 13CB	007-624
EPA Volatiles in Methods 502.2, 524.2	RT$_x$-502.2	DB-624	–	VOCOL	–	–	–	–	–
EPA Pesticides Method 608	RT$_x$-35	DB-608	–	SPB-608	–	–	–	–	007-608

Table 3.15   Calculation of Rohrschneider Constants for Carbowax 20M.

Compound	I_{20M}	$I_{squalane}$	$\Delta I/100$	Constant
Benzene	967	649	3.18	x
Ethanol	917	384	5.33	y
Butan-2-one	912	531	3.81	z
Nitromethane	1159	457	7.02	u
Pyridine	1199	695	5.04	s

Table 3.16   Reference substances for the classification of stationary phases.

Reference substances for the Rohrschneider system	Reference substances for the McReynolds system	Compounds with similar behaviour in gas chromatography
Benzene	Benzene	Aromatics, olefins
Ethanol	Butan-1-ol	Alcohols, phenols, weak acids
Butan-2-one	Pentan-2-one	Carbonyls, esters
Nitromethane	1-Nitromethane	Nitriles, nitro compounds
Pyridine	Pyridine	Bases, certain aromatics, N-heterocyclics
	2-Methylpentan-1-ol	Branched chain compounds, certain alcohols
	1-Iodobutane	Halogen-containing compounds
	Oct-2-yne	Acetylene
	1,4-dioxane	Ethers, bases
	cis-Hydrindan	Non-polar steroids, terpenes

standard non-polar phase (squalane). The difference in retention behaviour, which is a measure of the increased polarity of the phase in question over that of squalane, he termed ΔI. The values of ΔI for the five test solutes, benzene, ethanol, butan-2-one, nitromethane and pyridine, divided by 100, yield the Rohrschneider constants, x, y, z, u and s, respectively. Rohrschneider constants for Carbowax 20M, for example, were established as shown in Table 3.15. The Rohrschneider constants provide both qualitative and quantitative data on the selectivity of a stationary phase. Stationary phases with similar Rohrschneider constants are considered to have comparable retention and selectivity properties. Two phases with different constants should demonstrate different selectivity properties [111]. Phases with large x values, for instance, would be expected to exhibit greater retentivity of double-bonded compounds and alcohols would exhibit longer retentions in phases with large y values. However, Rohrschneider constants will not tell which stationary phase is best suited for a given separation.

McReynolds [116] later proposed slight modifications to the Rohrschneider system, including changes to the reference compounds, extension of the number of reference substances to 10 (Table 3.16) and dispensing with division by 100. These changes, with the exception of the extended number of reference compounds, have been adopted and McReynolds constants are now widely used to compare the polarities of different phases. Extensive compilations of McReynolds constants are available for a wide range of stationary phases. McReynolds constants have also been reported for a number of bonded phases on open tubular columns; some small, but significant, differences have been observed between the bonded phases and the nominally equivalent non-bonded phase. Often, the individual McReynolds constants are summed to give an overall polarity value which has been used by some manufacturers to provide polarity classifications of stationary phases. However, phases with identical overall polarities can show marked differences in individual values, reflecting different selectivities. For instance, the bonded phases Rt_x-35 and Rt_x-1701 (Restek Corporation) have identical overall polarities. However, butanol elutes much earlier on Rt_x-35 than on Rt_x-1701 while pyridine elutes much earlier on Rt_x-1701 than on Rt_x-35.

Other methods of stationary phase selection have been described [117–121] but are not used routinely. Nevertheless, more work is required in this area as there are a number of difficulties and criticisms associated with using McReynolds constants [122].

Stationary phase film thickness

The stationary phase in a packed column is coated at a certain loading or concentration, and hence film thickness, on the inert solid support. Typical stationary phase loadings were 20–30% w/w prior to the 1970s, but with the advent of improved quality supports much lower loadings of the order of 1–3% became popular. The stationary phase in open tubular columns is coated or bonded to the capillary wall and the film thickness, d_f, is an important column variable which has a direct effect on the retention, sample capacity and elution temperature for each sample component.

For increased column efficiency (reduced H value), the film thickness is kept as thin as possible in order to reduce the resistance to mass transfer in the stationary phase, i.e. the value of C_s in the van Deemter equation. In general, film thickness has little effect on column efficiency if $d_f < 0.4\,\mu$m. For stationary phases with high solute diffusivities, D_s, which is the case for most non-polar phases, only a slight sacrifice in efficiency occurs when d_f is increased to $1.0\,\mu$m. Greater loss in efficiency occurs with lower diffusivity phases (in general, the more polar phases) when d_f exceeds $0.4\,\mu$m. Hence, columns coated with a thin film generally show higher column efficiencies, as expressed by plate number or H value. However, the resolution of comparable solutes is usually better on thicker film columns [88]. Exceptions are observed when the film thickness is increased to a point where optimum mass transfer is prevented. The maximum thickness for a film is dependent on the column diameter and the stationary phase polarity. The thickest film column supplied by manufacturers for each particular column diameter usually approaches this point.

Thick film columns offer the further advantages of increased sample capacity and greater column inertness [123]. Sample capacity effects are illustrated by the data in Fig. 3.20 for n-octane and hexan-1-ol. Note that the sample capacity is related to the solute structure, but that in both cases it is increased on the column with the thicker $5.0\,\mu$m film. The respective sample capacities (using a 10% increase in peak width as the criterion) for the n-octane and hexan-1-ol are $18\,\mu$g and $5\,\mu$g on the $5\,\mu$m film column, and $5\,\mu$g and $2\,\mu$g on the $3\,\mu$m film column.

Thicker films also increase the column oven temperature required for compounds

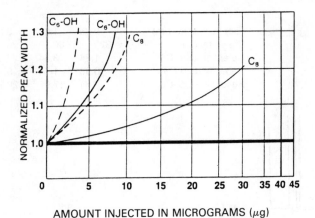

Fig. 3.20. Effect of film thickness on sample capacity for *n*-octane ($k = 5$) and hexan-1-ol ($k = 7$) on a 0.32 mm DB-1 column: —, $5\,\mu$m film thickness; – – –, $3\,\mu$m film thickness. Reproduced from *J & W Catalog* (1992) p. 9.

to elute and retain compounds longer, as shown by the following calculation. Considering the separation of a solute on two open tubular columns differing only in the film thickness of the stationary phase, then the effect of film thickness on capacity factor can be calculated. Assuming a column diameter, d_c, of 0.40 mm and film thicknesses of $0.2\,\mu$m (Column 1) and $1.0\,\mu$m (Column 2), then

for Column 1

$$\beta_1 = d_c/4d_f = 500 \quad \text{and,}$$

for Column 2

$$\beta_2 = d_c/4d_f = 100$$

but, for a given phase and temperature,

$$K = \beta k' = \text{constant}$$

thus,

$$\beta_1 k_1' = \beta_2 k_2'$$
$$k_2'/k_1' = \beta_1/\beta_2$$
$$= 500/100$$
$$= 5$$

Hence, Column 2 is five times more retentive than Column 1, i.e. capacity factors are five times greater on Column 2 and fewer theoretical plates will be required on

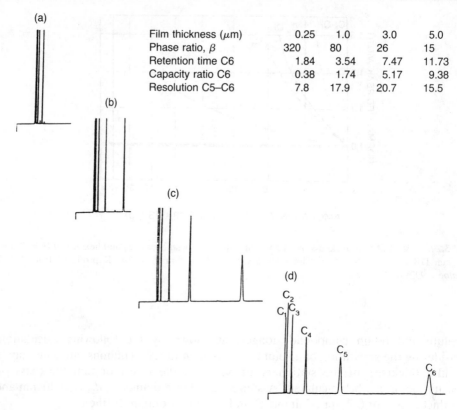

Film thickness (μm)	0.25	1.0	3.0	5.0
Phase ratio, β	320	80	26	15
Retention time C6	1.84	3.54	7.47	11.73
Capacity ratio C6	0.38	1.74	5.17	9.38
Resolution C5–C6	7.8	17.9	20.7	15.5

Fig. 3.21. Chromatograms showing the effect of stationary phase film thickness on the separation of *n*-alkanes. Conditions: 30 m × 0.32 mm DB-1 column at 40°C with helium carrier gas 36 cm s^{-1} with film thicknesses of (a) 0.25 μm, (b) 1.0 μm, (c) 3.0 μm and (d) 5.0 μm. Reproduced with permission from J & W Scientific, Division of Curtin Matheson Scientific Inc., *J & W Catalog* (1992) p. 8.

this column to achieve the same resolution. Alternatively, a higher column temperature can be used with Column 2. In summary, if one assumes a constant column temperature, increasing the film thickness results in a corresponding increase in the capacity factor, as shown by the chromatograms of Fig. 3.21. Carrier gas velocity is kept constant for the series in order to facilitate comparisons. This results in a sacrifice of efficiency with the thicker film columns.

Column bleed and sample induced problems

Column bleed is the normal background signal caused by the elution of stationary phase degradation products. Every column will exhibit some column bleed, although the extent will depend on the type (packed versus coated phase versus bonded phase) and quality of the column and its previous history. The extent of normal open tubular column bleed increases with increasing stationary phase film thickness, column internal diameter, length and temperature and is usually greater for polar than non-polar phases, reflecting the relative thermal stability of polar and non-polar phases. The effect of column bleed is observed as a rising baseline when detector output is monitored during temperature-programmed operation. Column bleed is distinguished from septum bleed by the absence of discrete peaks, which result from the single-point introduction of degradation products from the septum onto the column. The actual magnitude of column bleed at any temperature will depend on the detector sensitivity. Column bleed becomes especially significant for trace analysis, which requires high sensitivity, or for very sensitive detectors. The magnitude will also be enhanced by using selective detectors. For example, baseline elevation for a trifluoropropyl polysiloxane phase will be greater with an ECD detector, which has a larger response to halogens, than with a FID detector due to the presence of fluorine in the phase.

Even new columns exhibit bleed of low molecular mass liquid-phase fractions and residual traces of solvent. For this reason, they require conditioning at a temperature 10–20°C above the maximum value that will be used in subsequent separations, provided this value does not exceed the maximum allowable isothermal temperature limit for the column. Before heating to this temperature the column, once installed, and any carrier gas leaks eliminated, should be subjected to normal carrier gas flow conditions at 70–100°C for 15–30 min. Only after this initial heating should the column be temperature programmed to the conditioning temperature. Conditioning removes from the column or injector any volatile material resulting from careless handling or column preparation procedures or, in the case of a used column, from previous injections. The time required for conditioning varies from several hours (10–30) for a packed column to 30 min for a non-bonded phase open tubular column and a few minutes for a bonded phase open tubular column. During conditioning the detector should be maintained at an elevated temperature to minimize condensation of eluting impurities. With a packed column, the detector connection should not be made until after conditioning is completed. More prolonged conditioning is required for high-sensitivity trace analysis than for work requiring less-sensitive detector settings. If the detector baseline is unstable or excessively high after conditioning, there is probably a contaminated section of the system, which must be identified and decontaminated.

Excessive column bleed results from thermal or chemical damage to the stationary phase and/or system contamination. Thermal damage is caused by exceeding the maximum allowable operating temperature for prolonged periods, whereas chemical damage can result from exposure of the stationary phase to oxygen, particularly at elevated temperatures. Bonded cross-linked phases are chemically very stable to organic solvents and aqueous solutions. However, chemical damage to the phase from the introduction of mineral acids or bases, or some of the multi-halogenated acids in high concentrations, will cause excessive column bleed. The other cause of excessive baseline elevation is contamination of the injector and/or column by non- or semi-volatile residues introduced with samples. Subsequent elution of the residues can simulate the appearance of column bleed. Such residues can also cause adsorption and/or decomposition of sample components in subsequent injections, resulting in ghost peaks, baseline disturbances or peak distortion.

Often the performance of contaminated open tubular columns can be restored. The first few centimetres of a column are prone to damage from non-volatile residues, mechanical damage from the syringe etc. The easiest approach, therefore, to restoring performance is to remove approximately 50 cm from the beginning of the column, following the guidelines for breaking the column and re-installation in

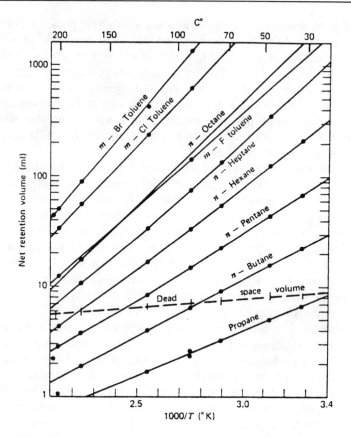

Fig. 3.22. Temperature dependence of retention volume. Reprinted with permission from Harris and Habgood, Copyright (1964). *Talanta*, **11**: 115.

the instrument. If this fails to restore performance, the column should be conditioned (with the detector end of the column disconnected) by programming at $5°C\,min^{-1}$ to its maximum isothermal temperature limit. Depending on the degree of contamination, any time from 1 h to 24 h of conditioning at the maximum temperature may be required. If removing a section of column and conditioning does not restore performance, the next step for bonded and cross-linked phase columns is to rinse them using special kits that are available from most chromatography suppliers. The contaminated column is removed from the instrument and a portion of column (about 0.5 m) is broken from the injector end of the column. Solvent is flushed through the remainder of the column at a flow of $1–2\,ml\,min^{-1}$ using either a low positive pressure or vacuum. Almost any common solvent can be used. However, a good general clean-up can be accomplished with methanol, followed by dichloromethane then hexane. In choosing a solvent for the final rinse, consideration should be given to detector compatibility.

Prevention of column contamination is obviously preferable to curing the problem afterwards. Sample clean-up procedures (see Chapter 8) such as liquid–liquid extraction and solid phase extraction can minimize the extent of residue problems.

However, with some samples (e.g. biological fluids and tissues) complete removal of all non-volatile residues is probably not feasible. In such cases, use of the injection liner as a trapping device may delay the onset of residue problems. The use of a packed liner insert, which can be periodically replaced, is sometimes beneficial. Packing may include a small plug of high quality silanized glass wool or lightly loaded OV-1 or OV-101 packed column material. A further possibility is the use of a guard or pre-column, frequently referred to as a retention gap, to 'trap' non-volatile residues. This consists of a piece of uncoated deactivated fused silica, typically 0.5–1.0 m in length, which is joined to the injector end of the analytical column. Special connectors are commercially available to facilitate joining the two columns. The residues accumulate in the pre-column which can be easily replaced when excessively contaminated.

3.6 Column Temperature

Column temperature is an important variable that must be controlled in GC. Thus, the column is housed in a thermostatted oven. The reproducibility of retention times with modern gas chromatographs is extremely high, with standard deviations of 0.01–0.10%. This can be attributed to the accurate temperature ($\pm0.1°C$) and pneumatic control provided by modern instruments. Temperature stability in early gas chromatographs was ensured by the high thermal mass of the column oven. Uniformity of temperature throughout the oven is now considered more important, particularly for open tubular columns because of their low thermal mass. Thus, most ovens in modern instruments are constructed of low-mass stainless steel to permit rapid heating and cooling when desired. In most designs, circulating air is blown by high capacity fans past the heating coils then through baffles, past the column and back to the blower to be reheated and recirculated. The temperature in the oven is controlled by thermocouples and feedback proportional heaters. The temperature programmer on most instruments can include several timed events, including an initial isothermal period, different rates of temperature rise (from 0.25°C min^{-1} up to 40°C min^{-1}) with pauses at intermediate fixed temperatures and a final isothermal temperature with an automated cooling cycle to return temperature to the initial value ready for a new cycle. Lower programme rates (0.25–5.0°C min^{-1}) are used with open tubular columns and the higher values with packed columns.

The most important result of an increase in column temperature is that it forces a larger fraction of solute molecules into the vapour phase. Other less important results are changes in diffusion coefficients, viscosity, etc. Zone migration depends entirely on the solute vapour (since it is vapour that is carried along by mobile phase), while separation depends on solute in the liquid phase, where selective molecular interactions operate. At very low concentrations of solute in the vapour phase, as occurs at low temperatures, there is so little solute in the form of moving vapour that the solute peak nearly ceases to migrate. In addition to the lengthy waiting times for the emergence of the solutes, the peaks under these circumstances are very broad and difficult to detect. At very high concentrations of solute in the vapour phase, as occurs at high temperatures, essentially all of the solute is removed from the stationary phase. Since interaction with this phase is responsible for separation, very little separation occurs. Thus, temperature must be adjusted to an appropriate value. As it is not possible to calculate the optimum operating temperature from readily accessible data, the temperature behaviour of a sample and column should be determined empirically from a few injections.

Gas chromatographic retention times decrease as column temperature increases because distribution coefficients, K, are temperature dependent in accordance with the Clausius-Clapeyron equation:

$$\log p^0 = -\Delta H/2.3RT + \text{constant} \qquad 3.5$$

where ΔH, the enthalpy of vaporization of the solute, is assumed constant over the range of temperatures investigated. Hence, solute vapour pressure, p^0 increases with temperature, resulting in an increase in equilibrium concentration of solute in the moving gas phase and a decrease in distribution coefficient and retention volume. That the relationship in equation 3.5 is at least approximately true is shown in Fig. 3.22, in which the log of the net retention volume is plotted versus $1/T$. The slope of each line is proportional to the enthalpy of vaporization of the solute; the fact that straight lines are obtained indicates that the assumption of a constant enthalpy is valid. This is correct over narrow temperature ranges only. Close inspection of the lines in Fig. 3.22 reveals that they diverge slightly. Hence, at some temperature the lines can cross, although this may be at a temperature of no practical interest. In some cases, however, crossover occurs at typical column temperatures, as shown here for n-octane and m-fluorotoluene. At a temperature of about 140°C, these solutes cannot be separated on this phase; at a higher temperature, n-octane elutes first, whereas the reverse is true at lower temperatures. Inversions of the elution order with changes in column temperature (or programme temperature rate) are relatively common in open tubular column GC.

3.6.1   Temperature programming

For simple samples containing relatively few peaks, an appropriate column temperature can be determined experimentally to achieve the separation and isothermal analysis is suitable. Isothermal operation (Section 2.5.1) is also preferred when accurate retention data are required for identification purposes, as it is easy to control and reproduce. The cycle time of a programmed run can take much longer than an isothermal separation because of the time required to cool and re-equilibrate the oven at the completion of a cycle. Hence, isothermal operation has an advantage in situations, such as process control, where analysis time and sample throughput are important considerations. Nonetheless, many samples contain components with a wide range of volatility, making temperature programming essential, as shown in Fig. 3.23. The more volatile components are eluted rapidly with no resolution when analysed isothermally at a high temperature. On the other hand, the analysis time is unacceptably long and later eluting peaks are very broad and may be lost as baseline drift when analysed isothermally at a low temperature. Temperature programming is very useful for initial screening of unknown samples. It has the additional advantage that the increasing column temperature also sharpens the peaks of the later eluting

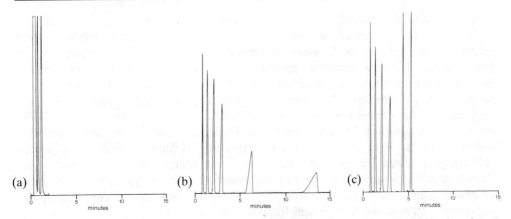

Fig. 3.23. Comparison of isothermal and temperature-programmed separation of an equimolar mixture of ethanol, butan-2-ol, 2-methylpropanol, butan-1-ol, pentan-1-ol and hexan-1-ol using a PTV injection (0.1 μl) in split mode (1:100) and flame ionization detection. Column temperatures: (a) and (b), isothermal at 105°C or 50°C, respectively and (c) 50°C isothermal for 3 min then programmed to 140°C at 10°C min^{-1}. In chromatogram (b) the effects of exceeding column sample capacity are seen as a fronting peak. Note that use of a higher column temperature in chromatogram (a) enhances the sample capacity. Chromatograms courtesy of J. Thompson.

components. The increased peak width with retention time observed for isothermal operation can thus be avoided, giving increased peak heights and sensitivity.

The theory of temperature programming has been treated thoroughly by Harris and Habgood [124] and by Mikkelsen [125]. A simple but adequate treatment has been provided by Giddings [126]. This model leads to the important conclusion that the solute starts moving through the column very slowly and that it speeds up as the temperature is increased, with nearly all the migration occurring at the higher temperatures near the elution temperature. The operation of temperature programming can be envisioned as follows: sample is injected and for a long initial period the components are essentially stationary at the origin; as the temperature increases, the analytes 'boil off' and migrate along the column at increasing rates until they elute.

Consideration must be given to the choice of optimum carrier gas flow for temperature-programmed operation. Open tubular columns are usually operated at constant inlet pressure because the commercially available flow regulators are unsatisfactory at low volumetric flow-rates. Increased temperature causes increased carrier gas viscosity and a corresponding decrease in carrier gas flow and velocity in order to keep the pressure drop constant. This means that a temperature-programmed analysis should be commenced at higher linear velocities than perhaps would be used for an isothermal run so that we operate at or above, rather than at or below optimum velocity during most of the run.

3.7 Detectors

In Chapter 1, chromatography was defined in terms of a separation process. Nevertheless, on-line detection is an integral part of a gas chromatograph. The detector monitors the column effluent and produces an electric signal that is proportional to the amount of analyte being eluted. The output signal is recorded as a continuous trace of signal intensity against time and peak areas (or heights) can be measured either electronically using an integrator, or manually from a chart recorder (see Chapter 9). The widespread application of GC can be attributed to the ability to simultaneously separate and quantify low solute concentrations. In principle, any physical or physicochemical property of the analyte which deviates from the properties of the carrier gas plus analyte can serve as the basis for detection. Thus, over 100 detectors for GC have been described but relatively few are in common use. One of the most sensitive detection systems available exploits the biological selectivity of insect-pheromone response or the human olfactory system. Such detection methods suffer a number of disadvantages such as safety, fatigue and lack of linear response but, nevertheless, the human nose has an excellent record as a detection system in the flavour and fragrance industry.

3.7.1 Detector specifications

The operation and applicability of different detectors can be compared against several performance criteria. These criteria are introduced at this point in order to facilitate further discussion, but are treated in more detail in Chapter 9. These criteria include the sensitivity, noise, minimum detectable quantity (MDQ) or detection limit, detector time constant and response time and the selectivity of the response. For example, a detector is considered to be selective if its response to a certain type of compound differs markedly from that to another type of compound; the usual meaning attached to 'markedly' is that response differs by a factor higher than 10. A detector that does not exhibit selectivity can be termed a universal detector. For purposes of screening a sample of unknown composition, a universal detector has definite advantages whereas a selective detector may aid in the identification of an unknown. Selective detectors are particularly useful for the analysis of complex mixtures, where the selectivity may greatly simplify the chromatogram.

In addition to such performance parameters, there are many ways in which detectors can be classified. Novak [127] has criticized the fundamental principles on which these classifications are based, but nevertheless the distinction between detectors is useful to the practising chromatographer. One method of classification [128, 129] distinguishes between concentration-sensitive and mass-sensitive detectors. The response of concentration type detectors is proportional to the relative concentration of analyte in the carrier gas (i.e. mass of solute per unit volume of carrier gas), while mass-sensitive detectors produce a signal proportional to the absolute mass of solute vapour reaching the detector per unit time. In contrast to concentration detectors, they are independent of detector volume. The response of the two detector types to mobile phase flow changes is very different.

A second classification is that of destructive and nondestructive detectors [130]. With nondestructive detectors, the original chemical form of the analyte persists throughout the detection process. This is an obvious advantage when the analyte is required for further analysis. In destructive detectors, the process of detection involves an irreversible chemical change in the analyte.

A consideration of the characteristics discussed above (and in Chapter 9) and the needs of a particular analytical problem will determine the most appropriate detector for a given problem. A detector with a wide linear dynamic range and low detection limit will be adopted for the determination of trace components in addition to main components in a sample. Conversely, the use of a selective detector is convenient if the trace components belong to a particular class of substance or possess some common functional group.

3.7.2 Make-up gas

Standard detectors on commercially available instruments usually require 30–40 ml min^{-1} total carrier gas flow for optimum sensitivity and peak shape. Carrier gas flow-rates for open tubular columns are well below this range for which detectors exhibit optimal performance. In this situation, make-up gas is added to the column effluent to supplement the carrier gas flow in order to achieve a total carrier gas flow of 30–40 ml min^{-1} at the detector. The make-up gas can be the same as the carrier gas or a different gas may be preferred for some detectors. The instruction manual accompanying the instrument being used should contain recommendations on the need for a make-up gas with details of the type of gas and flow-rates.

3.7.3 Common detectors used in gas chromatography

Of the many available detectors, most work is performed with one or other of the five established detectors (see Table 3.17):

- Thermal conductivity detector (TCD).
- Flame ionization detector (FID).
- Electron capture detector (ECD).
- Alkali flame ionization detector (AFID).
- Flame photometric detector (FPD).

Table 3.17 Classification of the most common gas chromatographic detectors.

Detector	Response	Optimal detection limit	Linear range	Classification
TCD	Organic and inorganic solutes	10^{-9} g ml^{-1}	10^4	Concentration; nondestructive
FID	All organic solutes except formic acid and formaldehyde	10^{-12} g ml^{-1}	10^7	Mass flow-rate; destructive
ECD	Halogenated and nitro compounds	10^{-16} mol ml^{-1}	10^3–10^4 (pulsed)	Concentration; nondestructive
AFID	P- or N-containing solutes	N:10^{-14} g s^{-1} P:10^{-13} g s^{-1}	10^3–10^5	Mass flow-rate; destructive
FPD	P- or S-containing solutes	S:10^{-10} g s^{-1} P:10^{-12} g s^{-1}	S:10^3 P:10^5	Mass flow-rate; destructive

The TCD and FID are usually considered universal detectors as they respond to most analytes. The remaining three detectors are the most useful selective detectors and give differential responses to analytes containing different functional groups. Note that this does not imply that the magnitude of the response of the universal detectors is constant to all analytes.

Thermal conductivity detector

The first detector commercially available for GC, the TCD remains a consideration for situations requiring universal detection. It is also known as the hot-wire detector and katharometer. The principle of the TCD is the reduction in heat loss of a hot wire because of the presence of solute molecules. The TCD consists of four tiny coiled wires or filaments arranged in a Wheatstone bridge configuration. The wires can be supported on holders or be mounted concentrically in a cylindrical cavity. The latter arrangement allows a smaller detector cell volume which minimizes band broadening. The filaments are electrically heated while the column effluent flows past two of the hot filaments and reference gas flows past the other two. The gas flow carries away excess heat at a rate dependent upon the thermal conductivity of the gas in the cavity. A thermal equilibrium is established when the column effluent consists of carrier gas alone. When a peak is eluted and enters the detector cell, the thermal conductivity of the carrier gas stream will change and hence the rate of heat loss from the filament will alter. This will cause the temperature and thus the resistance of the filament to change, altering the balance of the Wheatstone bridge. The signal appears as a difference in current from the reference. In newer instruments, the current flowing through the filaments is adjusted electronically to maintain a constant temperature and the change in applied potential is monitored.

The TCD responds to any compound, irrespective of its structure, whose thermal conductivity differs from that of the carrier gas. Hence, it is the only choice for detection of compounds to which other more-sensitive detectors give a poor or negligible response. In particular, it is the standard detector for determination of inorganic gases such as hydrogen, oxygen, nitrogen, carbon disulphide and water. Sensitivity depends on a number of factors, including detector cell design and the difference in thermal conductivity (Table 3.18) of the carrier gas and carrier gas plus analyte. For this reason, hydrogen or helium are usually used as carrier gas. Nitrogen is not used because its thermal conductivity is similar to that of most organic compounds. However, if hydrogen or helium is an analyte, then nitrogen or argon are used as carrier gas to increase the sensitivity. Sensitivity is also increased by heating the filament to a higher temperature with the power supply.

The TCD requires constant temperature control with good thermal insulation from the column oven, especially when temperature programming is being used. A major problem with the TCD is that the filaments must be protected from oxygen while they are hot. Care must be taken therefore to remove any air from the detector by flushing with carrier gas before the current is applied to the filaments. In many modern instruments, the filaments are protected by a pressure-sensitive switch which shuts filament power off in the event of a drop in carrier gas supply pressure. A TCD can be cleaned by disconnecting from the system and soaking in a series of hot solvents such as methanol, water and acetone. After drying, the detector is re-installed in the instrument and heated for 24 hr with a flow of carrier gas.

Table 3.18 Relative thermal conductivities of selected compounds.

Compound	Relative thermal conductivity
Carrier gases	
Helium	100.0
Nitrogen	18.0
Hydrogen	128.0
Argon	12.5
Carbon dioxide	12.7
Typical analytes	
Ethane	17.5
n-Butane	13.5
iso-Butane	14.0
Benzene	9.9
Ethanol	12.7
Acetone	9.6
Chloroform	6.0
Ethyl acetate	9.9

Flame ionization detector

The FID is the standard workhorse detector in GC. It consists of a stainless steel jet constructed so that carrier gas exiting the column flows through this jet, mixes with hydrogen gas and flows to a microburner tip, which is swept by a high flow of air, for combustion. Ions produced by the combustion are collected at a pair of polarized electrodes, constituting a small background current (10^{-13} to 10^{-14} A) which is the signal. When solutes enter the detector, they are combusted and the signal increases to 10^{-12} to 10^{-6} A. The current produced is then amplified and passed to a recording device. The exact mechanism of flame ionization is uncertain, but the ionization efficiency, while very low, is sufficient to give excellent sensitivity and linearity. Unlike the TCD, the FID gives virtually no response to inorganic compounds. Most organic compounds, however, give similar responses, which is approximately proportional to the total mass of the carbon and hydrogen in the analyte. A reduced response is usually observed with the first members of an homologous series and compounds with a large proportion of oxygen.

An FID utilizes three gases; the carrier gas plus hydrogen and air for combustion. The flow rate of hydrogen and air must be correctly adjusted with respect to one another and to the flow of carrier gas. If the flow ratios differ greatly from the optimum values, the flame will be difficult to ignite and keep alight. In addition, there is an optimum ratio of carrier gas:hydrogen (usually about 1.2:1) for maximum linearity and sensitivity (Fig. 3.24). Large deviations from this ratio can give an unstable flame, which produces a noisy response or is extinguished as the sample solvent peak is eluted. The air flow is less critical. Detector response increases with rising air flow followed by a plateau and all that is necessary is to work in this plateau region. An air flow of 300–600 ml min^{-1} lies in this range for most instruments.

After extended use, an FID can become contaminated by stationary phase and sample residues, producing a noisy baseline or spikes in the chromatogram. In such

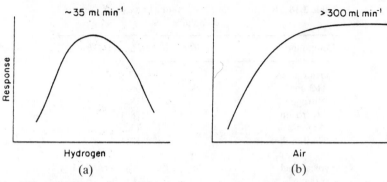

Fig. 3.24. Effect of (a) hydrogen and (b) air flow on the response of the flame ionization detector.

cases the detector should be heated to a higher temperature in order to evaporate any condensed substances. If this treatment does not help, the detector has to be disconnected and cleaned with a small flue brush and solvents. The system should then be reassembled, and with carrier gas flowing, heated to 120°C, and the flame may be ignited.

Electron capture detector

The ECD is the most widely used of the selective detectors (Table 3.17). Its popularity can be attributed to the high sensitivity to organohalogen compounds, which include many compounds of environmental interest, including polychlorinated biphenyls and pesticides. It is the least selective of the 'selective' detectors but has the highest sensitivity of any contemporary detector. As a result of the combination of these properties, operation of an ECD requires considerable care, experience and expertise to achieve consistent results.

The ECD was first reported in 1960 [131] and its design and acceptance has grown steadily. The detector consists of a chamber containing a radioactive source which is usually ^{63}Ni or, less commonly, tritium. Of these, ^{63}Ni has a considerably longer half-life and it can be used at higher temperatures, of up to 380°C, without appreciable loss of radioactive material. ECD chambers are constructed in two forms – the more traditional parallel plate design and the more common concentric tube design. The latter permits a lower dead volume detector cell and, because of its geometric shape, optimizes the electron capture process. The radioactive source emits high energy β particles capable of ionizing the carrier gas to produce secondary thermal electrons. Each β particle may generate 100–1000 thermal electrons by collision with carrier gas before it has reduced its kinetic energy. The thermal electrons are monitored at an anode by applying a potential of about 50 V across the chamber. This background or standing current is amplified and passed to a recorder. If an electrophilic analyte molecule that is capable of capturing electrons is eluted, and collides with a thermal electron, either a dissociative or non-dissociative reaction can occur depending on the analyte.

$$AB + e^- \rightarrow AB^- \text{ nondissociative capture}$$
$$AB + e^- \rightarrow A + B^- \text{ dissociative capture}$$

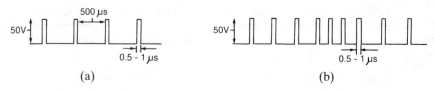

Fig. 3.25. Comparison of the different modes of pulsed operation of an electron-capture detector. (a) constant pulse rate and (b) pulse modulated showing increased pulse rate during analyte elution.

As a result of these reactions, the fast-moving electrons are replaced by slow-moving analyte ions which take longer to move to the anode and, therefore, have a higher probability of recombination with positive carrier gas ions to form neutral molecules. Thus, the system loses electrons and the standing current is reduced. The reduction in standing current is displayed as a positive peak by setting the standing current signal to the baseline and reversing the output from the amplifier.

Ideally, the detector temperature should be optimized for each analyte, as an increase in the detector temperature favours the dissociative reaction, while a lower temperature enhances the nondissociative mechanism. Thus, detector sensitivity is enhanced by a low temperature in the case of nondissociative reactions and by a high temperature for dissociative reactions. The type of reaction involved for a particular analyte can be easily identified by injection of the sample at two different detector temperatures.

The potential can be applied to the detector cell in one of three ways: direct current (d.c.) mode (continuous potential), pulsed mode (constant pulse frequency) or pulse-modulated mode (constant current with variable pulse frequency) (see Fig. 3.25). The last of these is now the accepted method, but understanding of the benefits of this approach can be gained by examining the historical development of the three modes.

A variable d.c. voltage supply is the simplest form of applied potential. This mode is very susceptible to anomalous response, particularly under conditions of detector contamination. This can be attributed to the build-up of a contact potential on the electrode surfaces, which may oppose the applied potential and cause reverse peaking. In pulsed mode, the potential across the cell is applied as short pulses of 0.5–1 μs with a pulse interval of 500 μs. Hence, the detector potential is off for most of the time and so the density of electrons in the cell is greater than under d.c. operation, where the electron concentration is reduced by the constant migration of electrons to the anode. Pulsed operation avoids most problems of charge build-up and increases the sensitivity of the detector by increasing the probability of interaction of free electrons and analyte. The linear concentration range with pulsed operation is still limited. In most modern instruments this is overcome by operating in a constant current mode in which the pulse frequency is varied, so that there is a constant current across the cell. When electron-capturing species enter the detector cell, electrons are removed and the number collected with each voltage pulse is reduced; hence, the voltage must be pulsed more frequently in order to maintain a constant current. In addition to providing an extended linear concentration range, the detector is reported to show less susceptibility to contamination.

The anomalous responses obtained with electron-capture detectors are usually of

Table 3.19   Relative response of the ECD and FID to selected analytes.

Analyte	Relative response of the ECD compared with FID
Halogenated organics	10^4
Organometallic compounds	10^3
Aromatic compounds	10^2
Conjugated unsaturated compounds	10^1
Non-conjugated compounds	10^{-4}

two basic types. Reverse response results from the detector operating in some mode other than the electron capture mode. This is invariably a consequence of detector contamination. Flat-topped peaks result from detector overload. The magnitude of the signal from an electron-capture detector is limited to the standing current. Once the original standing current is fully quenched, any further increase in sample size cannot produce further change and overloading the detector produces a flat-topped peak. This effect will occur for electron-capturing substances at very low concentrations, near the limit of detection of the FID, because of the high sensitivity of the ECD.

In general, nitrogen can be used as carrier gas but sensitivity is increased by using argon containing 5–10% of either methane or carbon dioxide as a quench gas, which reduces the mean speed of the thermal electrons and increases the probability of their interaction with an analyte molecule. Pure argon produces metastable ions capable of interaction with the analyte to give a signal and is therefore unsuitable. Many ECD chambers have a significant internal volume (0.2–0.3 ml) and the quench gas can also function as make-up gas in order to reduce the effective dead volume of the detector by increasing the flow rate through the chamber.

The ECD is easily contaminated; the usual symptoms are loss of sensitivity, baseline drift or a negative peak on the peak tailing edge. Water and oxygen are electron capturing and will quench the standing current. They must therefore be removed from the carrier gas. Low bleed stationary phases are essential and phases containing electron-capturing groups should be avoided. Careful consideration should be given to sample preparation and use of chlorinated solvents must be avoided, as even residual traces left after sample clean-up can saturate the detector for prolonged periods. Contaminated ^{63}Ni detector cells can often be cleaned by heating to 350°C or by removing the detector and rinsing with solvent (usually methanol followed, in order, by acetone, toluene and hexane), while taking necessary precautions and monitoring for any radioactivity. The last solvent should be removed by drying for 1 h in an oven preheated to 120°C. Under no circumstances should the detector cell be dismantled or mechanically cleaned, except by the manufacturer.

Alkali flame ionization detector

The AFID is often also known as the N/P detector (NPD) or thermionic ionization or emission detector (TID). It is a modified FID in which a constant supply of an alkali metal salt, such as rubidium chloride, is introduced into the flame. The AFID has been known for many years, but earlier versions were plagued with instability which made the AFID unsuitable for routine analysis. Improved detector design increasing detector reliability and stability is responsible for the more recent increase in use of the AFID.

Various methods have been used to introduce a supply of alkali metal salt into the flame but most modern commercial detectors are based on a three-electrode design [132, 133]. A ceramic or silica bead coated with a rubidium or caesium salt is placed between the flame jet and the ion collector of an FID. The bead is heated electrically to a dull red colour (600–800°C). The alkali salt is volatilized and passes into the flame where ionization occurs. The ions are then measured in the flame where they create a constant standing current. The selectivity and the response of the detector can be altered by adjusting the potential applied to the flame jet, the bead, and the collector. The passage of the correct compound into the flame causes an increase in the total number of salt ions present in the flame. The increase in numbers of ions is amplified and measured in the usual manner.

In the most useful mode, the flame is not ignited, but rather the flow of hydrogen is reduced to 1–2 ml min^{-1} to give a low temperature plasma. The AFID responds selectively to compounds containing nitrogen or phosphorus atoms, the response being typically 10^4 greater than for hydrocarbons. In phosphorus mode, the hydrogen flow is increased to a level similar to that in an FID. A hot flame is produced and the detector jet is grounded. Under these conditions the detector is selective for phosphorus-containing compounds. However, this mode has limited application. It is also possible to adjust the detector to give selectivity towards other elements, but these modes are rarely used.

Silylated derivatives or silylation reagents can reduce the response of the AFID through formation of silica in the flame, followed by deposition over the alkali salt tip, thereby reducing the amount of volatilization of alkali metal from the tip. The biggest disadvantage of the AFID, however, is that because alkali salt is continually being used up, the response of the detector changes, requiring frequent recalibration. Lifetimes of the alkali salt bead range from 100–1000 h depending on operating conditions.

Flame photometric detector

The operating principle of the FPD is the detection of specific luminescent emission originating from various excited-state species in a flame. The detector consists of a flame ionization chamber in which the column effluent is mixed with hydrogen and burned in a relatively cool hydrogen-rich flame. The gas ratio will affect the relative production of excited species and hence detector performance. Each design of detector will have an optimum gas ratio, and departure from this value will, in most cases, reduce detector performance. The sample luminescence originating in the flame occurs at wavelengths characteristic of different species. This radiation is collected by a mirror and passed to the optical system which is thermally insulated from the detector block. Interference filters permit light of the desired wavelength to be chosen. Although flame photometric detection has been described for a range of species, it is most commonly used for the specific detection of compounds containing either sulphur and/or phosphorus. Under the correct conditions, the combustion of phosphorus and sulphur-containing compounds produces two excited species, HPO*

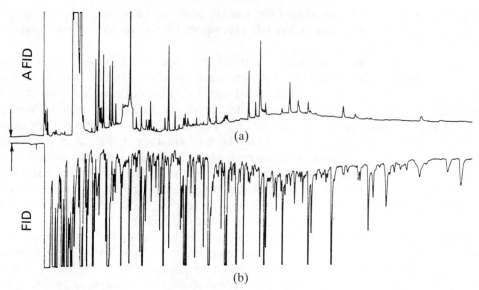

Fig. 3.26 Chromatograms of a serum sample obtained with dual detection. (a), alkali flame ionization detector; (b), flame ionization detector. Reproduced with permission from F. Hsu *et al*. (1980). *J. High Resol. Chromatogr., Chromatogr. Comm.*, **3**: 648.

and S_2^*, which emit at 526 nm and 394 nm, respectively. By using an appropriate filter, only the emission of the particular element of interest will be detected. Nonspecific radiation arising from the flame and stray light gives rise to a background signal which can be minimized by correct detector design, including the quality of the filter. In the sulfur mode the response is nonlinear, and the detector signal varies approximately as the square of the amount of sulfur present in the sample. This nonlinear response can cause problems with calibration.

3.7.4 Dual detection

The simultaneous use of two or more detectors, whose outputs complement each other can aid in compound identification [134] by generating substance-characteristic detector response ratios [135]. In some instances, the detectors are operated sequentially or, alternatively, the column eluate is split and passed separately to the individual detectors. Effluent splitters are usually in the form of a simple T- or Y-splitter for dual detector operation or more complex star-shaped splitters for multiple detection. The splitter must have a low dead volume and should provide a fixed split ratio that is independent of sample volatility, carrier gas flow-rate and column temperature. The combination of a selective with a universal detector [135] can provide information on the whole sample and, at the same time, give greater quantitative sensitivity on specific components. The chromatogram in Fig. 3.26 demonstrates the advantage of using a selective detector in conjunction with a universal detector. The FID chromatogram illustrates the complexity of the sample while the AFID records only those compounds that contain nitrogen.

3.8   System Evaluation

Once the column has been installed, any leaks eliminated, the column conditioned and checked using the methane test (Section 3.4.3) it is ready for use. At this point the performance of the system should be tested and compared with the computer-generated report (Fig. 3.27) supplied with most open tubular columns. Any significant differences suggest a problem in installation or that the column has deteriorated since being tested by the manufacturer. The initial test will provide a reference against which future performance can be measured. It can also serve as a diagnostic tool should problems arise. The testing should be repeated at intervals during use to ensure that performance (efficiency, selectivity, activity) has not deteriorated. Some standard methods may require additional method-specific tests. In Section 2.6, a number of factors were identified which should be measured in order to characterize the performance of a chromatographic system. In this section, further details are presented for the assessment of activity (sorptive and catalytic) and bleed rate. Most tests refer to column activity and column bleed, but it must be borne in mind that the column is part of the overall system and it is really the latter that is being assessed.

Residual chemical activity in modern open tubular columns is inevitably a consequence of reactive sites causing reversible or irreversible sorption of solutes. More recently, catalytic phenomena have also been identified as a cause for concern [136, 137]. All three effects give rise to reduced peak heights for susceptible solutes [138]. Reversible sorption is distinguishable in that it gives rise to tailing or broadened peaks, often with increased retention times, but with correct peak areas [139]. Irreversible sorption and catalysis, which are less easily distinguishable, lead to reduced peak area because of the permanent loss of solute [136, 139]. As the column can only be tested as part of a chromatographic system, extra-column effects such as syringe and injection system or column use can enhance [140, 141] the real or apparent activity of the column. Catalytic activity causes time-dependent, concentration-independent losses of sample components. With increasing column temperature, sorption decreases while catalytic effects and thermal decomposition increase.

3.8.1   Test solutions

The activity of a system can be evaluated by chromatographing a properly designed test mixture. Test mixtures contain compounds [78, 82, 139, 142, 143] that have a range of functional groups which are chosen so as to provide different pieces of information about the system, including column efficiency (number of plates, Trennzahl) and chemical activity. Hydrocarbon peaks are the standard to which all other peaks are compared. Because of the lack of functionality, hydrocarbon peaks should be sharp and symmetrical; malformed peaks indicate gas flow problems (dead volumes, gas leaks, improper column installation, poorly cut column ends, a broken or obstructed injection liner), poor injection technique, an extremely contaminated system or, in very rare cases, an extremely damaged phase owing to oxidation. Alcohols provide a good test of column activity because the hydroxyl group can

OPEATOR   JS
CHANNEL NO.   02
INJECTOR SYSTEM   SGE   Univector split mode
DETECTOR SYSTEM   FID

COLUMN DESCRIPTION

CODE   12QC2/SP10 0.25
PART NO.   052330
SERIAL NO.   35219
LENGTH   12 meter
TYPE   BONDED PHASE
MATERIAL   FUSED SILICA
PHASE   BP10
FILM THICKNESS   0.25 micron
MAX. TEMPERATURE   270 deg. C
I.D.   0.22 mm
O.D.   0.33 mm

COLUMN WAS CONDITIONED AT 240 C. FOR 16 HOURS

TEST CONDITIONS

COLUMN TEMPERATURE   - 115   GAS VELOCITY   35.0 cm/sec
DETECTOR TEMPERATURE - 280   SAMPLE SIZE   0.1 ul
INJECTOR TEMPERATURE - 240   SPLIT RATIO   60:1
CARRIER GAS   - H2   ATTENUATION   32*10E-12
INLET PRESSURE   - 6.0 PSI DETECTOR   FID

RETENTION TIMES

SOLVENT - PENTANE

PEAK	DESCRIPTION	TR (min)	KTH	KOVATS RETENTION INDEX
1.	PENTANE	0.577	0.010	
2.	n-NONANE	0.845	0.480	900.00
3.	2-OCTANONE	1.416	1.480	1064.44
4.	1-OCTANOL	1.992	2.489	1173.17
5.	n-DODECANE	2.167	2.796	1200.00
6.	2,6-(dmp)	3.289	4.762	1333.00
7.	NAPHTHALENE	3.489	5.112	1351.81
8.	2,4-(dma)	3.784	5.629	1377.66

COLUMN PERFORMANCE

EFFECTIVE PLATES (2,4-(dma))	29325
THEORETICAL PLATES	40669
EFFECTIVE PLATES/METER	2444
TZ (Peaks 7-8)	2.50
ACID: BASE (Peak 6: Peak 8)	1.23

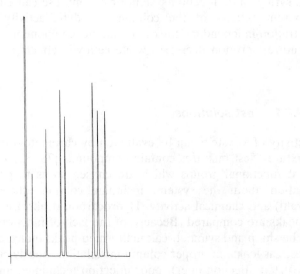

Fig. 3.27. Computer report giving the results of a system evaluation based upon the analysis of a test mixture. Reproduced with permission from Scientific Glass Engineering, Australia.

interact with any material that can hydrogen bond. This is usually due to silanol groups, or alternatively, an oxidized stationary phase, or sample residues from previous injections. The behaviour of the system towards acidic or basic solutes is tested with substituted phenols (acidic) and anilines (basic). Other substances, such as fatty acid methyl esters, are frequently included in test mixtures. Nonetheless, the best indicator of system performance is always the sample being analysed. For example, if only esters are being analysed, the fact that alcohols will tail does not effect the results. In addition, some compounds are notoriously difficult to chromatograph and symmetrical peaks should not be expected. Moreover, the effects of catalytic activity can be quite specific.

The need for more demanding test mixtures has increased with improved column manufacturing techniques. This can be illustrated by the development of tests for acid sites in open tubular columns. Dimethylaniline, once used as the test solute for acid sites is now eluted [144] without any difficulty on most columns. The integrity of test mixtures is often overlooked. Test mixtures contain a range of solutes, some of which are very reactive and subject to oxidation, decomposition or reaction with other components. The use of poor quality test mixtures can obscure the true performance of the system.

3.8.2   Performance tests

Performance tests have been divided [145] into three categories as (i) peak symmetry measurements, (ii) variation in peak height or peak area ratios for a reactive versus inert solute and (iii) variation in retention index resulting from changes in the nature or amount of stationary phase in the column. In each approach, selected test solutes are chromatographed and peak characteristics measured. It must be remembered that the result of any activity test is influenced by the amount of solute injected and the residence time [145]. Type (iii) tests are more appropriate for measuring column bleed, while the tailing factor, as defined in Section 2.3, is a common measure of peak asymmetry used in type (i) tests. Grob [see 145] has indicated, however, that peak shape alone is now no longer sufficient for characterizing column activity. The standardized Grob test [139] is a type (ii) test designed [78] to overcome this criticism and remains the standard against which other activity tests should be measured.

Standardized Grob test

The standardized Grob test is performed under optimized conditions of carrier gas flow-rate and rate of temperature programming as follows:

- The column oven is cooled to 25°C (100°C for thick film columns with d_f exceeding 0.7 μm) and the system dead time measured by injection of methane. In the case of split injection, split flow should be set prior to measurement because changes in the split flow usually change the pressure at the column inlet. The carrier gas flow-rate is adjusted to give the standard dead time and, similarly, the temperature programme rate is set to the appropriate value, as listed in Table 3.20. Some instruments do not allow the temperature pro-

Table 3.20   Standard experimental conditions for performing the Grob Test.

Column length (m)	Hydrogen		Helium	
	CH$_4$ elution (s)	Temperature programme rate (°C min^{-1})	CH$_4$ elution (s)	Temperature programme rate (°C min^{-1})
10	20	5.0	35	2.5
15	30	3.3	53	1.65
20	40	2.5	70	1.25
30	60	1.67	105	0.84
40	80	1.25	140	0.63
50	100	1.0	175	0.50

gramme rate to be changed continuously, in which case the dead time should be selected corresponding to the nearest available programme rate.

- The test mixture, consisting of the compounds listed in Table 3.21, is injected so that approximately 2 ng of each solute enters the column (e.g. 1 μl of the test mixture using split injection with a split ratio of 1:50), the column oven is immediately heated to 40°C and the temperature programme commenced.
- Within the temperature range (usually 110–140°C) in which the third ester is eluted, two temperatures are marked on the chromatogram.
- At the end of the run, the elution temperature of the third ester is interpolated or extrapolated and the '100% line' is drawn over the two alkanes and the three esters expressing the height of other peaks as a percentage of the difference between the baseline and the 100% line.
- The separation number is determined as an average for methyl decanoate/ methyl undecanoate and methyl undecanoate/methyl dodecanoate.

Table 3.21   Composition of the test mixture for performing the standardized Grob test.[†]

Component*	Abbreviation	Concentration (mg per 20 ml)
Methyl decanoate	E10	242
Methyl undecanoate	E11	236
Methyl dodecanoate	E12	230
n-Decane	10	172
n-Undecane	11	174
n-Dodecane	12	176
Octan-1-ol	ol	222
Nonanal	al	250
Butane-2-3,-diol	D	380
2,6-Dimethylaniline	A	205
2,6-Dimethylphenol	P	194
Dicyclohexylamine	am	204
2-Ethylhexanoic acid	S	242

*This composition may be unsuitable for some columns because of co-elution of two or more components. In such cases the test mixture should be modified.

[†]The components of this concentrated test mixture are prepared separately and combined prior to use. Ref. 139 or Grob *et al*. (1981), *J. Chromatogr*., **219**: 13 should be consulted for details.

- The stationary phase film thickness can be determined using a nomogram published by Grob *et al.* [139].

The utility of the Grob Test is shown by the chromatograms in Fig. 3.28. The alkanes and esters are inert and should elute intact, although on very inert, non-polar columns, partial retention of the alkanes may occur. The extent of retention by a hydrogen bonding mechanism is assessed by the alcohols octanol and butane-2,3-diol. Acid–base interactions are measured by the retentive behaviour of 2,6-dimethylaniline, 2,6-dimethylphenol, dicyclohexylamine and 2-ethylhexanoic acid. Solutes with sterically hindered groups are preferred for this purpose in order to avoid retention by hydrogen bonding. Dicyclohexylamine and 2-ethylhexanoic acid provide much more stringent tests of acid–base behaviour than aniline or phenol. Nonanal is used to assess the capacity of the column to retain saturated aldehydes. The upper chromatogram in Fig. 3.28 was obtained with a column following overnight conditioning with the column exit connected to the flame ionization detector. Stationary phase bleed, which accumulated in the column exit, was carbonized when the flame was subsequently lit. The chromatogram shows the typical effect of carbonized contamination: general adsorption is least for the most polar substances (very little for butane-2,3-diol; less for octan-1-ol than nonanal), increasing with decreasing polarity and with decreasing volatility (esters). The lower chromatogram is a repeat test after removing 3 mm of column exit. The column is now weakly acidic (dicyclohexylamine strongly adsorbed, 2,6-dimethylaniline 90%), the inertness good (1-octanol 100%) but not excellent (butane-2,3-diol 84%; 2-ethylhexanoic acid 82%).

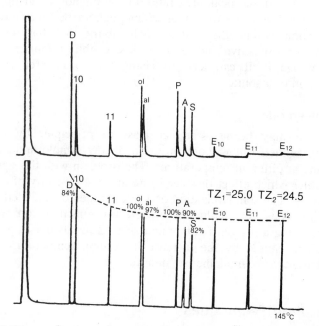

Fig. 3.28. Chromatograms for two columns tested using the procedure described by Grob. For a description of conditions refer to the text. Reproduced with permission from Grob *et al.* (1981). *J. Chromatogr.*, **219**: 13.

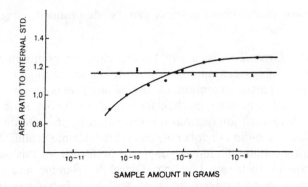

Fig. 3.29. Ratio testing at different dilutions showing irreversible sorption of 3,5-dimethylphenol. ×, dicyclohexylamine to C_{15}; •, 3,5-dimethylphenol to *p*-dimethoxybenzene. Reproduced with permission from Dandeneau and Zerenner (1979). *J. High Resol. Chromatogr., Chromatogr. Comm.*, **2**: 351.

Ratio testing

A second test for column activity is to monitor how the relative response factor for a given test substance varies with concentration [84, 146, 147]. A plot of response factor relative to a hydrocarbon or other solute, which is not subject to sorption, versus concentration (Fig. 3.29) provides an indication of the degree of column retention, which is independent of the peak shape. In Fig. 3.29, as the solute concentrations become smaller, the response ratio for the amine remains constant, whereas there is a shift in the peak area ratio for the phenol, indicating irreversible retention. This provides a means of distinguishing phenomena causing reversible and irreversible retention within the system. Schomburg [84] has shown that *n*-octylamine is a very sensitive indicator of irreversible retention. At extreme dilutions, the syringe itself can lead to changes in peak area ratios through preferential sorption of a solute.

Column bleed rate

System bleed rate may be assessed, at least in principle, by performing a temperature-programmed run without injection of any sample and measuring the change in baseline at different temperatures. The problem with this approach is the number of variables which effect the bleed rate and the difficulty of expressing the result quantitatively. The standardization of a test for measuring bleed rate presents a number of difficulties. The influence of detector sensitivity on results can be eliminated by injecting a standard compound such as an alkane as part of the test [33]. Column bleed caused by the increase in column temperature can then be expressed as micrograms of *n*-alkane per minute.

3.9   Ancillary Techniques

A number of diverse procedures are used to extend the capabilities of gas chromatography. The more important of these are discussed in this section.

3.9.1   Derivatization

It has been estimated that from 10–20% of known organic compounds are directly amenable to GC. The remaining 80–90% are not directly suitable because they are thermally unstable or non-volatile. This is the case with many biological samples, such as proteins, and compounds with polar or ionizable groups (e.g. carbohydrates and amino acids) which have limited volatility. The use of alternative techniques such as LC may seem to be the logical answer but many of these compounds provide different, but still troublesome problems. Furthermore, in some cases, GC remains the preferable technique because of higher sensitivity. A solution to the limited applicability is the preparation of a suitable derivative. While this provided the original motivation, derivatization is now performed for a variety of reasons, which may be identified as:

- Increasing the volatility of sample components that would otherwise require excessive column temperatures for elution.
- Decreasing the volatility of sample components where low molecular mass, very volatile components are present. The reduction of volatility in this case aids in sample recovery, may eliminate the need for cryogenic temperature control at the gas chromatograph, and shifts analyte peaks further away from interference by the sample solvent peak.
- Enhancing thermal stability by removing functional groups which may otherwise encourage isomerization or decomposition.
- Improving detector response by introducing specific functional groups into the sample molecule which permit use of selective detectors. In particular, reagents that introduce an electron-capturing moiety into the sample molecule for detection by the sensitive ECD have been well used. This may also aid in identification of sample components. An example is the formation of deuterated derivatives that can be easily distinguished by their higher molecular mass when analysed by GC–MS.
- Improving separation of compounds that are not easily differentiated in the underivatized sample, such as isomeric compounds.
- Improving peak shape by removing reactive hydrogen atoms which are capable of hydrogen bonding. A simple example is the esterification of carboxylic acids.

Derivatization is not without problems. The major disadvantage is the need for increased sample handling, which can lead to loss of the component of interest. Moreover, excess derivatizing reagent in the sample extract may reduce column life if not removed prior to injection. Hence, certain criteria have been established as guidelines for evaluating a derivatizing reagent. According to such criteria a good reagent will: not cause any rearrangements or structural changes in the sample during formation of the derivative; produce a derivatization reaction that is 95–100% complete; produce a derivative that is stable with respect to time; not contribute to loss of the sample during the reaction; and produce a derivative that will not interact with the gas chromatographic column. To this list could be added the need for the derivatization reaction to proceed smoothly to the formation of a single product at all reasonable concentration ratios of derivatizing reagent to sample. Finally, the derivatizing reagent itself should not interfere in the subsequent detection or,

Table 3.22 Typical derivatizing reagents.

Reaction type	Compound	Derivatizing reagent
Acylation	Alcohols, primary and secondary amines	(Perfluoro)acid anhydrides; (perfluoro)acylimidazoles
Alkylation	Acids	Boron trihalides/alcohol; i.methanolic sodium hydroxide +ii.BF$_3$/methanol;
	Bile acids, acids, primary amines, amino acids	Anhydrous HCl/methanol; N,N-dimethylformamide; dimethyl-acetal
Cyclic alkylboronation	Carbohydrates, Catecholamines, Corticosteroids	n-Butylboronic acid or methylboronic acid
Oxime formation	Aldehydes, ketones	O-alkylhydroxylamine.HCl
Pyrolytic methylation	Fatty acids, barbiturates	Aryltrimethylammonium salts
Silylation	Alcohols, amines, thiols	Various silanes e.g., trimethylchlorosilane, N,O-bis(trimethylsilyl)acetamide, N-trimethylsilylimidazole
Transesterification	Esters such as triglycerides	Sodium methoxide/methanol; aryltrimethylammonium hydroxide in methanol; i.NaOH/methanol + ii.BF$_3$/methanol

alternatively, a means should exist for its easy removal from the sample after reaction is complete.

A derivatization reaction can, in principle, be performed either pre-column, on-column or post-column, however, pre-column has been favoured in GC. This can be attributed, at least in part, to incompatibility of typical derivatizing reagents and the detection system. There are a few exceptions where the reagents are mixed and injected, the reaction occurring on-line in the heated injection port. The following discussion (see Table 3.22) focuses on the basic types of derivatization reactions. Manufacturers trade literature is the best source of detailed information on recommendations for usage of specific products and typical procedures for various reagents. The most favoured reactions are simple one-step reactions, that can be performed at room temperature. Excess reagent is usually used to force the reaction to completion, thus leaving an excess of the reagent in the sample. Reaction vessels are usually very small (0.1–10 ml) glass tapered reaction vials or culture tubes with PTFE-lined plastic screw caps. Slow reactions are generally accelerated by heating the reaction mixture.

Silylation

Silylation is the most versatile and widely used derivatization reaction in GC. It results in replacement of an active hydrogen atom with a trialkylsilyl group. Several factors contribute to the popularity of silylation. Nearly all functional groups which present a problem in GC can be converted to alkylsilyl ethers or esters. With the exception of compounds containing unprotected ketone groups, most silylation

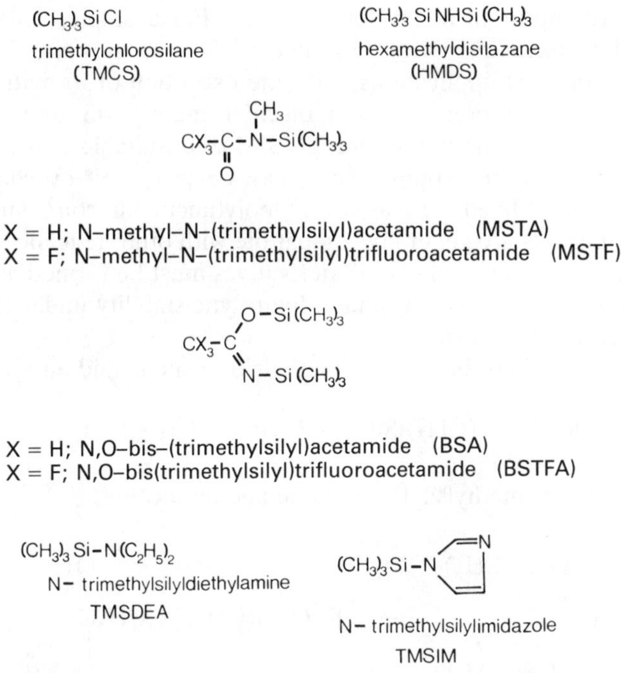

(CH₃)₃Si Cl

trimethylchlorosilane
(TMCS)

(CH₃)₃ Si NHSi (CH₃)₃

hexamethyldisilazane
(HMDS)

$$CX_3-\overset{\overset{\displaystyle CH_3}{|}}{\underset{\underset{\displaystyle O}{\|}}{C}}-N-Si(CH_3)_3$$

X = H; N–methyl–N–(trimethylsilyl)acetamide   (MSTA)
X = F; N–methyl–N–(trimethylsilyl)trifluoroacetamide   (MSTF)

$$CX_3-C\underset{\displaystyle N-Si(CH_3)_3}{\overset{\displaystyle O-Si(CH_3)_3}{<}}$$

X = H; N,O–bis–(trimethylsilyl)acetamide   (BSA)
X = F; N,O–bis(trimethylsilyl)trifluoroacetamide   (BSTFA)

(CH₃)₃ Si–N(C₂H₅)₂

N– trimethylsilyldiethylamine
TMSDEA

(CH₃)₃Si–N

N– trimethylsilylimidazole
TMSIM

Fig. 3.30. Structures of the most commonly used trimethylsilylating reagents.

reactions occur cleanly without artefact or byproduct formation. In general terms, silylation reduces the polarity of the molecule and decreases the possibility of hydrogen bonding which, in turn, enhances the volatility of the compound and reduces its reactivity. The most common group is a trimethylsilyl or TMS group, although derivatives incorporating other alkyl groups are useful in certain conditions to impart greater hydrolytic stability to the derivative, increased sensitivity when used with selective detectors, or improved separation characteristics. TMS ethers are very susceptible to hydrolysis, and therefore water must be rigorously excluded from the reaction mixture. Newer reagents designed specifically for use with GC–MS provide greater diagnostic information. This is the case with *t*-butyldimethylsilyl ether derivatives, where the loss of the *t*-butyl group from the molecular ion gives a very characteristic fragment.

A number of different reagents can be used to synthesize TMS derivatives. The structures of the most widely used reagents are shown in Fig. 3.30. When choosing a reagent, the reactivity of the sample must be considered. In general, the ease of reaction follows the order: alcohols > phenols > carboxylic acids > amines > amides. Within this sequence, primary functional groups react faster than secondary functional groups which, in turn, are more reactive than tertiary functional groups. The relative silylating strength of the reagents is also a factor and this increases in the series:

HMDS < TMCS < TMCS:HMDS (in pyridine) < MSTFA
< BSA < BSTFA < TMSIM < BSTFA:1% TMCS.

Other factors are important in certain situations. For example, TMSIM does not promote enol-ether formation with unprotected ketone groups. TMSIM is the preferred reagent for most applications, with the exception of formation of N-TMS derivatives and the separation of low molecular mass TMS derivatives, where BSTFA is preferred. In situations where TMSIM is unsuitable and the compound contains unprotected ketone groups, these may be protected by conversion to a methoxime derivative. Mixed derivatives of polyfunctional compounds are used frequently. TMS reagents are generally compatible with other reagents for formation of mixed derivatives. However, the TMS derivatives must be formed as the last step in the reaction sequence because of limited hydrolytic stability under the conditions used to prepare most other derivatives.

Typical reactions are those between trimethylchlorosilane and an alcohol:

$$R–OH + (CH_3)_3SiCl \rightarrow R–OSi(CH_3)_3 + HCl \qquad 3.6$$

and between N,O-bis-(trimethylsilyl)acetamide and an alcohol:

$$R–OH + CH_3\underset{\underset{N–Si(CH_3)_3}{\|}}{\overset{\overset{O–Si(CH_3)_3}{|}}{C}} \rightarrow R–O–Si(CH_3)_3 + CH_3\underset{\underset{N–Si(CH_3)_3}{\|}}{\overset{\overset{OH}{|}}{C}} \qquad 3.7$$

A closely related reagent, bis(trimethylsilyl)trifluoroacetamide produces a more volatile reaction byproduct (not a more volatile derivative) which does not interfere with the separation of low molecular mass derivatives.

Silylation reagents are highly reactive and should be stored in a refrigerated desiccator. Most silylation reagents are liquid at room temperature, which allows them to act as solvent for the sample as well as reactant. If a solvent is used, it is frequently pyridine or dimethylformamide. Pyridine does have some undesirable side-effects in certain situations. For example, it promotes the formation of secondary products by reaction with the enol forms of keto compounds (e.g. 17-ketosteroids). Although enol-TMS ethers may be formed, in some instances in quantitative yield, they are undesirable because of low thermal and hydrolytic stability.

$$RH_2C\underset{\underset{O}{\|}}{C}–R \longleftrightarrow R–CH=\underset{\underset{OH}{|}}{C}–R \quad \text{- - -} \rightarrow \quad R–CH=\underset{\underset{O–Si(CH_3)_3}{|}}{C}–R \qquad 3.8$$

keto-enol equilibrium   enol-TMS ether derivative

Frequent use of silyl derivatives and direct injection of the reaction mixture usually leads to a build up of byproducts, such as SiO_2, in the detector of the gas chromatograph, contributing to increasing detector noise levels. Several reagents (e.g. N-methyl-N-trimethylsilylheptafluorobutyrylamide) have been developed to minimize the extent of such deposition.

Acylation

Acylation with acid anhydrides or acyl imidazoles converts alcohols and primary and secondary amines to stable derivatives. The acid anhydrides produce carboxylic acids as by-products, whereas acyl imidazoles produce imidazole as a by-product as illustrated by the following generalized reactions:

$$ROH + (R_1CO)_2O \rightarrow R_1COOR + R_1COOH \qquad 3.9$$

$$ROH + R_1CO(C_3H_3N_2) \rightarrow R_1COOR + C_3H_4N_2 \qquad 3.10$$

Undesirable side reactions, such as dehydration and enolization, are observed with sensitive molecules owing to the strong acid conditions applying in the reaction medium. For these compounds, acylation with an acyl imidazole is preferred because the reaction byproduct is the weakly amphoteric imidazole. Acetylation is the most common acylation reaction. Fluorinated derivatives are used to enhance the sensitivity of detection by ECD.

Alkylation

In alkylation, a reactive hydrogen, such as –OH or –NH is replaced by an alkyl group or occasionally an aryl group. A variety of reagents and methods are used for alkylation, of which the most important are boron trihalide–alcohol, diazoalkanes, N,N-dimethylformamide dialkylacetals and pyrolytic alkylation. Diazomethane is an ideal reagent in many ways:

$$RCOOH + CH_2N_2 \rightarrow RCOOCH_3 + N_2 \qquad 3.11$$

The reaction is conducted in an ice bath and is complete in a few seconds, gaseous nitrogen being the only byproduct of the reaction. However, diazomethane is best avoided because of safety and toxicity hazards and alternative reagents are now available.

Alcohol–boron trihalide reagents convert carboxylic acids to esters. Ethanolic BF_3 is the general purpose reagent, while ethanolic BCl_3 may be used for acids that contain groups which decompose readily. Higher homologues of ethanol are used to reduce the volatility of short-chain carboxylic acids.

Pyrolytic alkylation is useful for the analysis of organic compounds containing acidic OH and NH functional groups. These include carboxylic acids, phenols, barbiturates, sulfonamides, purines, pyrimidines and xanthines. The sample is reacted with an aryltrimethylammonium reagent to form a salt which is injected into the heated injection port where the alkylated derivative and other volatile byproducts are produced by thermal rearrangement.

Oxime formation

The gas chromatographic behaviour of carbonyl compounds is usually entirely adequate. Oxime formation is used under these circumstances to protect the carbonyl group in compounds containing both carbonyl and hydroxyl groups during derivatization of the hydroxyl group. Aldehydes and ketones are conveniently derivatized by forming oximes with O-alkylhydroxylamine hydrochloride.

$$RR_1CO + R_2ONH_2 \rightarrow RR_1C{=}NOR_2 + H_2O \hspace{3cm} 3.12$$

The methyl derivative (R_2 = CH_3) has been used with ketosteroids, saccharides, prostaglandins, aldoacids and ketoacids. Alternatively, O-benzylhydroxylamine hydrochloride forms a less-volatile derivative which is useful in instances where the methyl oximes are not resolved from other components in the sample. O-(pentafluorobenzyl)hydroxylamine hydrochloride produces a derivative with very high sensitivity for detection by ECD.

3.9.2 Pyrolysis gas chromatography

Pyrolysis GC is a specialized sample introduction technique in which a sample is heated in the injection chamber of the gas chromatograph to a temperature at which thermal decomposition of the sample occurs. By chromatographing the pyrolysis products, the structure of the original material can be elucidated, at least in theory. Frequently, no conclusions can be drawn from the pyrolysis products as to the structure of an unknown substance, in which case identification is empirical and based on a fingerprint pattern, which may be characteristic of the particular parent sample.

Many investigations of the optimum conditions for thermal pyrolysis have been reported [148]. Pyrolysis temperatures range from 400–1000°C. The practical problem consists in imparting the desired temperature to the entire sample rapidly and reproducibly. A disadvantage of pyrolysis GC is the relatively poor reproducibility between different types of pyrolysers, as a result of the different temperature behaviour. The duration of the heating also exerts a decisive influence on the pyrogram. There are two types of pyrolysis inlet, pyrolysis chambers or microreactors, and filament systems. Filament systems, in which the sample is deposited as a thin film on the filament and heated using the Curie point principle, may promote catalytic reactions. Another problem is in changes of the filament in the course of time. Using a pyrolysis chamber, the sample is injected directly into the hot furnace. With this type of system, the conditions under which pyrolysis occurs are accurately known. As the temperature is isothermally maintained, the pyrolysis temperature is known and the residence time of the sample in the pyrolysis zone can be controlled.

3.9.3 Headspace gas chromatography

Headspace GC is a sampling technique that involves the indirect determination of volatile constituents in liquid or solid samples by analysing the vapour phase that is in thermodynamic equilibrium with the sample in a closed system [149, 150]. It is used predominantly for the determination of volatile constituents present in trace concentrations. Traditional extraction and enrichment procedures, such as solvent extraction, steam distillation and vacuum distillation, suffer inherent drawbacks relating to the co-extraction of matrix components and introduction of extraneous compounds from the extracting medium. In contrast, headspace analysis necessarily provides an extract limited to volatile components which are ideally suited to GC. The volatile nature of the headspace sample reduces the need for cleaning and maintenance of the injector, column and detector, with savings in instrument down-time. Examples where headspace analysis is especially useful are the determination of contaminants and migratory volatiles in packaging and food, pesticide residue determinations and flavour and drinking water analyses. Headspace analysis has been extremely well documented in several reviews and monographs [151–153].

Headspace techniques are so extensive and diverse that any attempt at classification must be incomplete. However, with this in mind it is hardly possible to avoid a superficial classification. Thus, Nunez *et al*. [153] distinguish according to the method of extraction of volatiles, i.e. between static and dynamic headspace methods. Dynamic methods include a variety of techniques variously referred to as strip-trap and purge, gas phase stripping analysis and purge and trap. With static methods, the sample and vapour are brought to equilibrium in a closed vessel and the headspace is sampled either manually by syringe or automatically. In dynamic methods, an inert gas is stripped through or passed over the sample in order to recover the analyte(s). In either case, the gaseous effluent, following stripping, is generally passed through a suitable trapping medium such as a cold trap, activated charcoal or porous polymer where the volatiles are trapped and subsequently desorbed, either thermally or with a solvent, into the gas chromatograph. This approach provides a concentration effect by which sensitivity is increased.

The sensitivity of static methods can be increased in a number of ways. Headspace sampling is usually performed at elevated temperatures in order to increase the sample vapour pressure and obtain a more favourable distribution constant. This technique is of limited experimental application, owing to the risk of bursting the sample container. Thermal stability of the sample must also be considered. Alternatively, an increase in sensitivity can be effected by increasing the activity coefficient of the sample. This can be achieved by changing the pH of the sample, or by adding inorganic salts (salting out) such as potassium carbonate, sodium carbonate and sodium chloride to aqueous solutions, or water to organic solvents. The magnitude of the salting-out effect often depends on the nature of the salt employed. As an illustration of these effects, sensitivity can be increased in the analysis of low molecular mass carboxylic acids by methylating the acids and lowering the pH of the solution, hence partitioning into the gas phase is enhanced.

Equipment for headspace analysis is very diverse. In its simplest form, the sample is placed in a glass bottle which is sealed with a rubber septum cap and placed in a thermostat. After thermodynamic equilibrium is reached, an aliquot of the vapour is removed using a gas-tight syringe and injected immediately into a gas chromatograph. Even in this simple form there are many variations in procedure which relate to precise details of the sampling (e.g. the use of positive versus negative pressure in the sampling system). At the other extreme, fully automated headspace sampling devices are commercially available. The diversity of headspace techniques causes almost overwhelming evaluation problems. Accuracy of results depends on many factors and, in order to produce reliable data, a thorough understanding of the procedure being used is required. For example, the volume of the headspace vessel should not be too large compared with that of the sample in order to avoid the possibility of total evaporation of the component into the gas phase. The assumption of an equilibrium would no longer be valid under such circumstances. Sampling from containers sealed with rubber septum caps can involve uncertainties affecting both the accuracy and precision of the analysis [154, 155]. Specialist literature on headspace analysis should be consulted for further details.

The chromatographic peak area, A_B of compound B in the gas phase above a sample is proportional to its partial vapour pressure, p_B:

$$A_B = c_B p_B \qquad\qquad 3.13$$

where, for a given sample size, each substance has a specific value for the constant c_B, which also depends on the characteristics of the detector being used. According to Henry's Law, the partial vapour pressure of the analyte above the sample depends on the mole fraction, X_B of the analyte B and the saturated vapour pressure of pure component B at a given temperature, P_B, corrected for any deviation from ideality by the activity coefficient γ_B:

$$p_B = X_B \gamma_B P_B \qquad\qquad 3.14$$

These expressions can be combined to obtain the fundamental equation upon which headspace analysis is based:

$$X_B = \frac{A_B}{c_B P_B \gamma_B} \qquad\qquad 3.15$$

In this equation, the denominator is a constant calibration factor which has to be determined by measurement. The activity coefficient is matrix dependent and unknown for most analyses. It is therefore accounted for by calibrating the headspace sampling system. The calibration must be performed with a mixture, the composition of which corresponds to that of the sample. As a consequence, considerable difficulties often arise with unknown samples. Suitable calibration methods include the use of model systems (e.g. volatile organic compounds in drinking water or vegetable oils), matrix matching, standard addition and multiple headspace extraction. The latter involves repeated analysis from the same container and extrapolation of the data to give the original concentration. It is used for problem samples in which matrix effects are difficult to reproduce. The calibration must be performed under conditions identical to the sample analysis, since at a given temperature, the vapour pressure is characteristic of a compound. The temperature must be precisely controlled because of the temperature dependence of the vapour pressure in accordance with the Clausius–Clapeyron equation. It is also important in static headspace analysis to know the time required to establish the vapour pressure equilibrium, which can vary between compounds, and also be very dependent on the sample viscosity.

3.9.4 *Multidimensional gas chromatography*

Open tubular column GC is generally regarded as a high-resolution technique, but even the best column will separate a rather limited number of peaks. It has been shown by Giddings (see Section 9.3.1) (assuming a random distribution of peaks) that to have a reasonable probability of separating a large proportion of the peaks in a sample, extremely large and impractical numbers of theoretical plates are required, even for a surprisingly low number of components in the sample. The resolving power of open tubular column chromatography can be increased enormously by using multidimensional chromatography, an all-encompassing term which refers to columns connected in series or parallel and configured to perform any of a range of operations. A feature common to all multidimensional techniques in GC is the selective transfer of a gas chromatographic fraction from one column to another. Multidimensional GC requires at least two separation systems with, ideally, two separate ovens capable of independent control and which are equipped with a suitable switching device (e.g. tap, valve). Instrumentation ranges from simple, manually operated arrangements to sophisticated, fully automated microprocessor-controlled systems that include a cold trap between the two columns.

Principal applications include the separation of complex mixtures containing components that are difficult to separate, trace enrichment of dilute samples and process control, employing such techniques as column switching, heartcutting and back-flushing. Heartcutting is the technique where a particular fraction from one column is passed to a second column which may have a different polarity. A second column switching technique, called back-flushing, can be used if only the rapidly eluting volatile components in the sample are of interest. With two columns connected in series, once the components of interest have eluted from the first column, carrier gas flow in this column is reversed. The less-volatile components, which have only travelled a short way down the column, are back-flushed. This process can be used to reduce analysis time, protect the second detector or to protect the second column. Pressure tuning utilizes two or more columns with different polarity and/or selectivity connected in series. By adjusting the relative pressures on the first column, and at the mid-point, large changes in the overall separation are possible. Cryogenic focusing is sometimes necessary in multi-dimensional chromatography to focus all or part of a sample at a point in a column.

The principles of multidimensional chromatography are illustrated by the following example. In Fig. 3.31, limonene and p-cymene are difficult to separate, as shown in chromatogram (a). Complete resolution of this critical pair was achieved by sending the 9 s section of the first column containing the p-cymene/limonene to a more polar second column. The remaining material in the first column was then back-flushed giving a significant reduction in analysis time, as shown in the chromatogram in Fig. 3.31b.

3.9.5 High-speed gas chromatography

High-speed gas chromatography (HSGC) is one of those unfortunate terms which is best avoided. However, the attainment of faster analyses does represent an important direction in which the limits of GC are being extended. For this reason alone, and apart from its descriptive nature, the term high speed is likely to persist and become even more widely used. The immediate question arises of how fast is sufficient to warrant the description of high speed. It is not possible to provide a definitive answer, although analyses requiring more than 2 min for completion

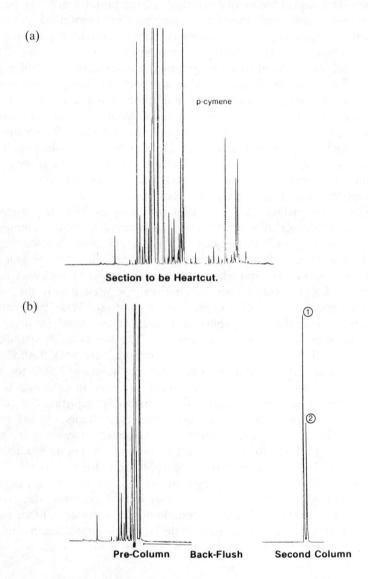

Fig. 3.31. The application of multidimensional chromatography for the analysis of lime oil, showing the advantages of heartcutting and the time-saving of back-flushing. Chromatogram (a) shows the full chromatogram of distilled lime oil with incompletely resolved *p*-cymene and limonene on a 25 m × 0.25 mm BP5 pre-column programmed from 75°C, 1-min hold, to 190°C at 8°C min^{-1}. The lower series of chromatograms (b), showing the complete resolution of *p*-cymene (peak 2) from limonene (peak 1), were obtained by heartcutting the 9-s section of the pre-column chromatogram (a) containing the *p*-cymene to a more polar 25 m CW20M column. The remaining material in the pre-column was then back-flushed. Reproduced with permission from Scientific Glass Engineering, Australia.

probably would not qualify. What is certain is that what currently constitutes high speed will be routine within 10 years.

In some respects HSGC is at the opposite end of the scale to multidimensional chromatography. In HSGC, some resolution is always sacrificed in favour of a reduced analysis time, whereas maximum resolution is the ultimate goal of multidimensional chromatography. To some extent, the loss of resolution in HSGC can be compensated by the choice of an appropriate selective detector. Other factors which can be optimized to minimize the loss in resolution are the column length, carrier gas velocity and a reduced column outlet pressure [156]. With conventional GC, long open tubular columns are used to achieve very high resolution separations, however, this necessarily involves long analysis times which are unacceptable in many situations. The use of relatively short lengths of open tubular columns allows for the possibility of very fast, lower resolution separations. Retention times can be reduced further by minimizing the internal diameter of the column [157].

Peters *et al.* [156] reported the separation of up to 10 compounds in about 3 s using a 2-m long, 0.25-mm i.d. open tubular column operated at an average carrier gas velocity of about $200 \, cm \, s^{-1}$. Early eluting components in HSGC may have bandwidths in the range 30–50 ms, whereas a conventional inlet for GC, equipped with a splitter, generally produces inlet bandwidths of over 100 ms. Hence, conventional inlet systems cause excessive band broadening, even when used with a commercially available auto-injector and are therefore unsatisfactory for most high-speed applications. The wider application of HSGC will follow the commercial availability of suitable inlet systems.

3.9.6 Coupled techniques

It is difficult to provide an adequate definition of a coupled or hyphenated technique which is relevant to all situations. Hirschfeld [158] defined a hyphenated instrument as 'one in which both instruments are automated together as a single integrated unit via a hardware interface . . . whose function is to reconcile the often extremely contradictory output limitations of one instrument and the input limitations of the other'. In other instances, coupled techniques have been defined as those involving the on-line coupling of devices which are analytical methods in their own right. According to either definition, conventional HPLC using an ultraviolet detector would be considered a hyphenated technique, which is contrary to accepted practice. In reality, the inability to provide an adequate definition is not important. Coupled techniques are extremely important in qualitative analysis, a topic examined in Section 9.2.

A gas chromatograph may be coupled with a spectrometric technique such as mass spectrometry (MS), Fourier Transform infrared (FTIR), nuclear magnetic resonance (NMR) or atomic emission spectrometry (AES) to aid in sample identification or, with a second separation technique (e.g. LC or SFC, capillary electrophoresis) to enhance the versatility of the separation process. On-line coupling of two established techniques limits both instruments to operating within conditions that are compatible with one another. The most successful combination to date has been that of GC with MS [159], which represents one of the most powerful analytical techniques available. Other coupled techniques include LC–GC, SFC–GC, GC–FTIR [160, 161] and

GC-AES [162]. The tandem operation of chromatographic and spectrometric techniques provides more information [158] about a sample than the sum of the information gathered by the instruments independently [163].

Coupled techniques have depended on the development of adequate data systems, interfacing and measuring techniques with very high-speed scanning. Although the combinations of GC with the different spectrometric techniques have individual requirements, a few generalizations are possible. The data system must be capable of acquiring and handling the data generated by a fast-scanning spectrometer and also, ideally, of instrument control. The data handling should include background subtraction, normalization, formatting, display and possibly library searching. Interfacing difficulties concern incompatibility of the chromatographic eluant with the inlet of the measuring device, and the states of matter most easily handled by the various techniques [164, 165]. For example, the mass spectrometer operates at approximately 10^{-2} Pa, whereas the outlet from a column in GC is slightly above atmospheric pressure (10^5 Pa). In order to couple these instruments it is necessary to reduce this pressure. A third requirement is for fast scanning of the column eluate by the measuring device so that full spectra of solutes can be obtained as they elute. For peaks obtained with open tubular columns this means that scans must be completed in the order of milliseconds.

Coupled column chromatography

This is an extension of multidimensional GC whereby sample fractions are transferred on-line from one column to another with no restriction on the nature of the system. LC–GC is a potentially important area of chromatography [166]. Significant advantages are gained by LC–GC which allow the isolation of an LC fraction of interest for direct introduction into a GC column. The wider application of LC–GC has been slowed by the difficulty of interfacing two chromatographic systems which operate with different mobile phases. The physical nature of the LC mobile phase and the volume of LC eluant transferred into the gas chromatograph are key factors of successful LC–GC interfacing. For reasons of mobile phase compatibility, coupling of supercritical fluid systems with GC has more promise, at least in the short term.

Gas chromatography–mass spectrometry

The successful operation of a gas chromatograph–mass spectrometer requires a dedicated operator. Operators generally fall into one of two schools: those who consider the mass spectrometer as the ultimate selective detector for their gas chromatograph and those who consider the gas chromatograph as an expensive inlet system for their mass spectrometer. Practitioners equally knowledgeable in both areas and dedicated to the overall process are less common. Settlage and Jaeger [167] point out that regardless of the cause, GC–MS technology falls far short of its potential. Nonetheless, GC–MS has developed into one of the most fruitful techniques in analytical chemistry. GC–MS is capable of providing both quantitative and qualitative data by means of spectral interpretation procedures. Figure 3.32 is a block diagram that shows how the components of a GC–MS–computer system are interrelated.

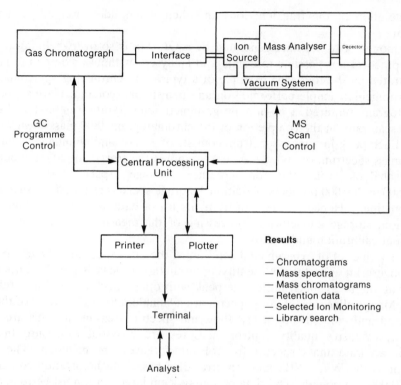

Fig. 3.32. Block diagram of a GC–MS–computer system.

The mass spectrometer receives the separated chromatographic zones which enter the ion source, maintained at a pressure of 10^{-2} Pa to ensure efficient production of ions from the neutral molecules. Ion production can be achieved by electron bombardment (electron ionization), chemical ionization (CI), field ionization or field desorption techniques. The various ions must then be separated in the mass analyser unit and recorded as a spectrum of ions according to their mass-to-charge (m/z) ratios and relative abundances. Various types of mass separation methods, including quadrupole and magnetic/electrostatic sector spectrometers with high scanning rates, have been used but, in general, the quadrupole instrument is preferred for GC–MS because of its relatively high sensitivity, resolution of about 1000 amu, and ability to operate at pressures up to 10^{-2} Pa, with scan rates of 10^{-1} s per mass decade. The only disadvantage in using quadrupole analysers is their relatively low resolving power. With applications requiring high resolution as, for example, in the case of suspected interference, high resolution magnetic sector or double-focusing instruments are needed. More recently, Fourier Transform mass spectrometers and ion trap detectors have been introduced.

The mass spectrum is a line spectrum showing the peak of the molecular ion and the signals corresponding to the individual fragments. Figure 3.33 depicts the mass spectrum of 1,1,1-trichloroethane. The molecular ion indicates the molecular mass of the parent compound. Lines grouped around this peak result from the natural abundances of various isotopes. Further structural information can be obtained from a consideration of the fragment ions and their magnitude observed in the mass spectrum.

The amount of useful data generated by a GC–MS is so overwhelming that it is neither practical nor possible to obtain and analyse it without a computer. This can be illustrated by the following example of a typical gas chromatographic analysis of an environmental sample performed on an open tubular column. If we suppose the chromatogram required a 60-min programmed temperature run and a 2-s mass spectral scan rate, at the completion of the chromatogram 1800 mass spectra will be stored. Each peak in a mass spectrum consists of a mass and abundance value, so a typical mass spectrum of about 100 peaks requires the storage of 200 numbers. The entire analysis will therefore require storage space for more than $200 \times 1800 = 360\,000$ numbers. Additional storage space is required for such data as retention times. Hence, modern instruments use a data system for the automatic acquisition, storage and efficient processing of the measured data, as well as for instrument calibration and control.

For a specific set of experimental conditions, the mass spectrum of a compound is like a fingerprint. The task of identifying an unknown peak is greatly simplified by comparing the mass spectrum for the peak with other spectra stored in a reference library. Many collections of mass spectra are available, although several of these are specialized and contain only a few thousand spectra. Reasons for this are readily apparent; extensive quality control measures are essential to ensure that only verified, accurate mass spectra free of interferences are included. The largest collection is the Wiley–NBS mass spectral data base, which currently contains less than 100 000 compounds. This data base may seem large but it must be remembered that the number of known organic compounds exceeds 5 million. Manual comparison with reference spectra is tedious and comparison is usually performed with a computer. A useful feature of computer matching is the calculation of factors which are used to distinguish between good, average and poor matches. However, no system can give a completely unambiguous identification of an unknown based on search data only. The need for human interpretation of mass spectral data remains. Evaluating the quality of data obtained from the various commercially available systems requires an understanding of different search strategies [168].

Interface

A high vacuum must be maintained in the mass spectrometer so that molecular reactions can be avoided. However, if carrier gas from the gas chromatograph is admitted directly to the ion source, the pressure will rise. Many methods have been described for interfacing the gas chromatograph and mass spectrometer to ensure that transfer of sample between the two is maximized, while removing most of the carrier gas. A poorly designed interface can compromise the performance of the column, the mass spectrometer, or both. An ideal interface should be such that: band broadening is minimized by a minimum dead volume; the pressure is reduced at the column exit to approximately 10^{-2} Pa at the ion source; there are no discrimination effects against thermally or chemically labile compounds owing to active sites or heating of the interface; and there is efficient transfer of the entire sample to the ionization chamber.

Ideally, the interface should isolate the column from the vacuum in order to

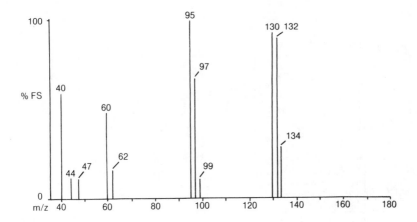

Fig. 3.33.   Electron-impact mass spectrum of trichloroethane at 70 eV ionizing energy.

minimize the effects of the vacuum on the column efficiency [169]. Isolation is essential for packed columns, and in early instruments, an all-glass interface, known as a jet separator, was widely adopted as the most efficient interface. Isolation becomes increasingly less important for narrow bore and long (>15 m) open tubular columns where low carrier gas flow-rates are usual. Hence, because of the popularity of open tubular columns and improvements in vacuum pump design, 'direct' connection of the column to the ion source is now normal practice. The ion source is designed to cope with the low flow encountered with open tubular columns (<5 ml min^{-1}) by using high capacity diffusion pumps. Direct connection can take several forms. The easiest method is direct connection of the column to the ion source, with no attempt to isolate the column from the vacuum. This approach works well for longer narrow-bore columns, but even here there is the disadvantage that the column cannot be removed without venting the mass spectrometer. Alternatively, some isolation can be effected by placing a small-bore capillary tube, either platinum–iridium or fused silica, in the transfer line. Several reports indicate that platinum–iridium alloy is responsible for catalytic decomposition of certain substances. A third and most successful method of direct configuration is the open–split [170].

Scanning mode

The mass spectrometer can be operated in a number of scanning modes with different selectivities and sensitivities. With most systems, data is acquired by continuous repetitive scanning at a predetermined rate (usually 0.1–1.5 s per scan) of the chromatographic eluate. Each scan results in a mass spectrum stored in computer memory. The data system is capable of displaying the stored data in a number of ways to enable real time analysis of the course of the chromatogram and post-run manipulation of the acquired data.

A reconstructed gas chromatogram (RGC) can be generated by summing the ion currents of all ions in each mass spectrum and plotting these values as a function of mass spectrum number or elution time. The RGC is also designated as a total ion

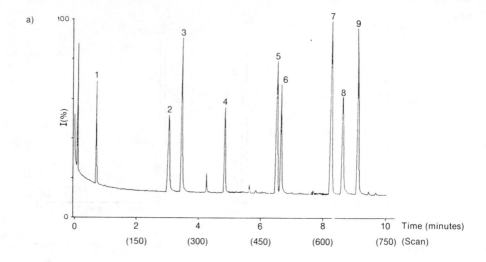

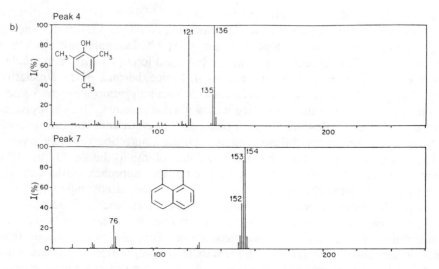

Fig. 3.34. (a) Reconstructed gas chromatogram of a test mixture performed on a 25 m BP-1 open tubular column temperature programmed from 60°C to 130°C at 4°C min^{-1} with direct coupling to the mass spectrometer. Compounds are identified: 1, 1,2-dimethylbenzene; 2, isooctanol; 3, 2,6-dimethylphenol; 4, 2,4,6-trimethylphenol; 5, 2,6-dimethylundecane; 6, nicotine; 7, 1,2-dihydro-acenaphthylene; 8, dodecan-1-ol; 9, 2-methyltetradecane. (b) The mass spectra obtained from peaks 4 and 7.

current chromatogram (see Fig. 3.34a). It is analogous to the chromatogram produced by a flame ionization detector, although the relative sensitivities of the GC peaks will differ between the two traces, because the FID and mass spectrometer respond differently to different compounds. The RGC is also used to identify the mass spectrum associated with a specific chromatographic peak. Individual mass spectra, corresponding to the various peaks (see Fig. 3.34b), can be examined in detail and used to identify the different peaks. In order to obtain adequate

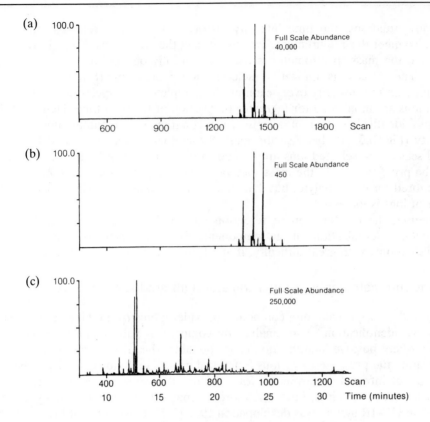

Fig. 3.35. Comparison of (a) selected ion monitor, (b) mass chromatogram, and (c) reconstructed gas chromatogram plots for a real environmental sample containing aromatic compounds.

qualitative data about peak purity, several mass spectra are recorded for each chromatographic peak. The purity of a single peak can be confirmed by comparing the mass spectrum recorded at different points of a peak. The validity of such comparisons depends on using a scan speed that is sufficiently fast that the concentration of the chromatographic peak component does not change enough to distort the normalized mass spectrum. For example, if a peak is 6 s wide and two 3-s mass scans were taken during its elution, one mass spectrum would have higher relative abundances for high-mass ions.

An RGC shows all eluted peaks but provides no mass spectral data. Alternatively, mass chromatograms can be plotted in which the ion abundances of a few selected m/z values are extracted from each stored mass spectrum and plotted versus the spectrum number or retention time. Selectivity can be achieved using this technique by judicious choice of the m/z value which is plotted. An RGC obtained from the analysis of an environmental sample is shown in Fig. 3.35(c). This sample is known to contain a series of aromatic compounds with a characteristic m/z value of 91. Also plotted in Fig. 3.35(b) is a mass chromatogram of m/z 91, clearly showing the presence of the aromatic compounds which were previously barely discernible as minor components in the RGC trace.

In many situations, the time intensity distribution of a few selected ions is of interest. To meet this requirement, cycling through the entire mass range is avoided, but instead the mass spectrometer is focused cyclically on one or a few masses of interest. The process is known as selected ion monitoring (SIM). There is a significant gain in sensitivity over scanning the complete mass spectrum because the selected ions are monitored for a greater proportion of the scan time. Hence a SIM plot looks identical to that of a mass chromatogram (MC) but at much higher sensitivity (Fig. 3.35(a)) because the mass chromatogram was generated from full spectral scans. The selected ions are chosen to be characteristic of the analytes and should be principal ions in the mass spectra of the analytes. More than one ion can be monitored for each analyte, but detection limits become increasingly poor as the number of ions is increased.

Frequently, the combination of retention data and mass spectrum permits the unambiguous identification of a component. However, isomers can often be distinguished only with great difficulty, if at all, by means of a mass spectrum.

Gas chromatography–Fourier transform infrared spectrometry

The infrared (IR) spectrum of a compound provides a fingerprint which can be used for positive identification of an analyte by comparison with standard spectra. In instances where positive identification is not possible, the IR spectrum is still useful for indicating the presence of various functional groups in a molecule. The inherent advantages of infrared spectrometry for structure determination also apply to the coupled technique which is, however, more amenable to quantitative analysis. The first on-line GC–IR system was developed in 1967 [171]. However, until recently the technique has not been used widely as a result of inadequate spectral libraries, low sensitivity and the time required to measure an infrared spectrum with a conventional dispersive (grating or prism) spectrometer. Typically, this time exceeded 2 min, by which time the analyte would have passed through the cell before the spectrum had been recorded. Dispersive instruments could be used as selective single-wavelength detectors for GC, in which case no spectral information beyond the presence of a functional group absorbing at the selected wavelength would be obtained. Dramatic improvements have occurred because of advances in instrument design and the use of Fourier Transform (FT) algorithms to convert a rapid-scan interferogram into an IR absorption or transmission spectrum. Using FTIR, spectral scan times of 0.2 s are achieved, which are compatible with peak widths observed with open tubular columns. A computer system is mandatory for instrument control and data acquisition and manipulation. Improvements in computer software and hardware allow simultaneous display of the spectral and chromatographic data. Sensitivity is greatly enhanced, compared with dispersive instruments, by summing individual spectral scans during the elution of a chromatographic peak.

Early instruments employed a gold-coated glass detector flow cell, called a lightpipe. FTIR spectra of the vapour phase chromatographic fractions were recorded as they passed through this lightpipe. The development and performance of lightpipe systems have been reviewed [172] and indicate that further improvements in detection limits are not practical using the lightpipe design. The relatively low sensitivity of lightpipe-based interfaces has been addressed through the use of

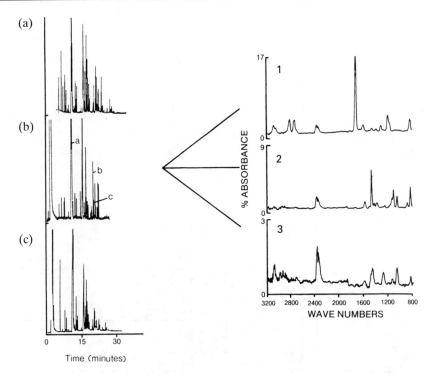

Fig. 3.36. Chromatograms of a hazardous waste sample. (a) reconstructed gas chromatogram using GC–MS; (b) Gram–Schmidt chromatogram with infrared spectra of peaks (1), (2) and (3); and (c) trace obtained with a flame ionization detector. Reprinted with permission from Shafer *et al.* (1984) *Anal. Chem.*, **56**: 237. Copyright (1984) American Chemical Society.

matrix isolation or direct deposition techniques [173]. With matrix isolation interfacing, each chromatographic peak is trapped in a matrix of argon that is condensed on a moving metallic substrate, and the reflection–absorption spectrum is measured. Because the sample is surrounded by the matrix and present in low concentration, intermolecular interaction and molecular rotation are prohibited. This gives rise to very sharp spectra. In the direct deposition interface, chromatographic eluates are condensed on a cooled moving window without dilution in a matrix of any type.

Data may be presented as a Gram–Schmidt chromatogram, which is a plot of the infrared activity versus elapsed time, calculated from the interferograms collected. It is analogous to an RGC or the chromatogram produced by an FID, although the relative sensitivities of the GC peaks will differ between the three traces (see Fig. 3.36). The quality of infrared data that can be obtained is illustrated by spectra obtained for three of the components. The alternative presentation is a functional group chromatogram produced by plotting the detector output at a specific wavelength against time. Functional group chromatograms often allow compounds containing a common functional group to be extracted from a complex run containing many overlapping peaks. This process is analogous in many ways to selective ion monitoring in GC–MS.

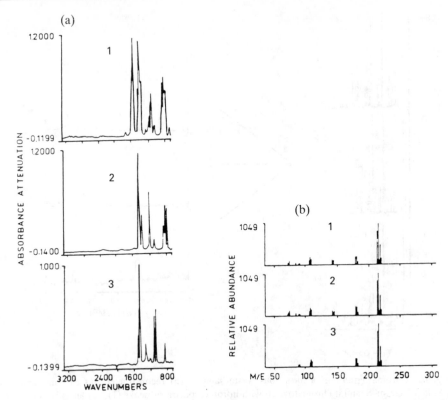

Fig. 3.37. (a) Electron-impact mass spectra and (b) vapour phase infrared spectra of (1) 1,2,3,5-tetrachlorobenzene; (2) 1,2,3,4-tetrachlorobenzene; (3) 1,2,4,5-tetrachlorobenzene. Reprinted with permission from Gurka and Betowski (1982). *Anal Chem.*, **54**: 1819. Copyright (1982) American Chemical Society.

Computerized identification of compounds from their infrared spectra has been performed in several ways. One common method involves comparison of the absorbance spectrum of the unknown with a library of reference spectra after scaling the absorbance of the most intense peak in the sample and reference spectra to the same value. A number of reference spectral libraries are now available and spectral searching has become very fast and can, in some instances, give less ambiguous identifications than GC–MS data. However, unequivocal identification depends on the availability of good reference spectra and there are many pitfalls [174]. Reference libraries of spectra of compounds in the condensed phase generally exhibit less uniformity than their vapour phase analogues. For some collections of condensed phase spectra, samples have been prepared as mineral oil mulls and for others as KBr discs. Some libraries can be purchased in digitized form, whereas others contain only hard-copy spectra. Some data bases were obtained from grating spectra but others contain high-quality spectra measured on FTIR spectrometers.

GC–MS versus GC–FTIR

Several comparisons of GC–FTIR and GC–MS have been reported [175–178]. To date most comparisons have favoured GC–MS for greater sensitivity but have shown

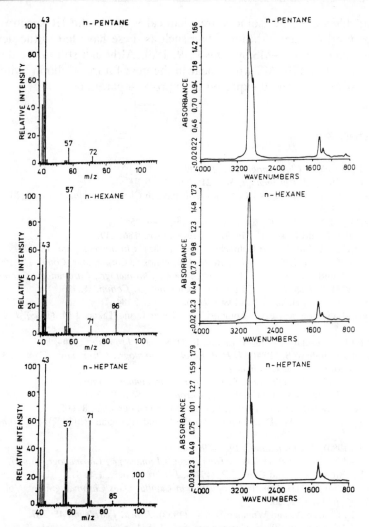

Fig. 3.38. Low resolution GC–MS and gas phase IR spectra of *n*-alkanes. Reprinted with permission from Wilkins *et al.* (1981). *Anal. Chem.*, **53**: 113. Copyright (1981) American Chemical Society.

many cases where identification has been more definitive with GC–FTIR, particularly in the case of isomeric materials. A further advantage is the ability to recognize functional groups. In other instances, such as the identification of members of an homologous series, GC–MS is the preferred technique. Thus, the identification technique of choice, either GC–MS or GC–FTIR, depends to a great extent on the nature of the sample. In many cases, particularly when the compound is a complete unknown, the combined data from both techniques are preferred. The complementary nature of GC–MS and GC–FTIR is well documented [164, 175, 177], as illustrated in Figs 3.37 and 3.38, demonstrating the IR and mass spectra of tetrachlorobenzene isomers and *n*-alkanes, respectively. The IR spectra of the alkanes are identical, whereas the mass spectra differ. In contrast, the mass spectra of the tetrachlorobenzene isomers are identical whereas the IR spectra yield data that can be used for identification. Data such as these have led to the development of directly linked GC–IR–MS systems [179, 180]. Although still in the development stage, routine GC–FTIR–MS will appear in the near future. Column switching when applied to GC–FTIR can be expected to improve separations.

References

1. Bayer F.L. (1986). *J. Chromatogr. Sci.*, **24**: 549.
2. Parcher J.F. (1983). *J. Chromatogr. Sci.*, **21**: 346.
3. Nonaka A. (1975). In Giddings J.C. Editor, *Advances in Chromatography*, Vol. 12. Marcel Dekker, New York, p. 223.
4. Siu K.W.M and Aue W.A. (1980). *J. Chromatogr.*, **189**: 255.
5. Rohrschneider L. and Pelster E. (1979). *J. Chromatogr.*, **186**: 249.
6. Kimperhaus L., Richter F. and Rohrschneider L. (1982). *Chromatographia*, **15**: 577.
7. Grob K. and Grob G. (1979). *J. High Resol. Chromatogr., Chromatogr. Comm.*, **2**: 109.
8. Liddle P.A.P. and Bossard A. (1985). *J. High Resol. Chromatogr., Chromatogr. Comm.*, **7**: 646.
9. Grob K. (1978). *J. High Resol. Chromatogr., Chromatogr. Comm.*, **1**: 173.
10. Wainwright M.S. and Haken J.K. (1980). *J. Chromatogr.*, **184**: 1.
11. Budahegyi M.V., Lombosi E.R., Lombosi T.S., Timar I. and Takacs J.M. (1983). *J. Chromatogr.*, **271**: 213.
12. Schomburg G., Husmann H. and Rittmann R. (1981). *J. Chromatogr.*, **204**: 85.
13. Grob K. and Rennhard S. (1980). *J. High Resol. Chromatogr., Chromatogr. Comm.*, **3**: 627.
14. Oslavicky V.M. (1978). *J. Chromatogr. Sci.*, **16**: 197.
15. Pretorius V. (1978). *J. High Resol. Chromatogr., Chromatogr. Comm.*, **1**: 227.
16. Ottenstein D.M. and Silvis P.H. (1979). *J. Chromatogr. Sci.*, **17**: 389.
17. Purcell J.E., Downs H.D. and Ettre L.S. (1975). *Chromatographia*, **8**: 605.
18. Schomburg G., Behlau H., Dielmann R., Weeke F. and Husmann H. (1977). *J. Chromatogr.*, **142**: 87.
19. Brotell H. (1980). *J. Chromatogr.*, **196**: 489.
20. Grob K. and Neukom H.P. (1979). *J. High Resol. Chromatogr., Chromatogr. Comm.*, **2**: 15.
21. Yang F.Y., Brown A.C. and Cram S.P. (1978). *J. Chromatogr.*, **158**: 91.
22. Sandra P. Editor, (1985). *Sample Introduction in Capillary Gas Chromatography*, Vol. 1. Huethig, Heidelberg, pp. 14, 15 & 56.
23. Desty D.H., Goldup A. and Whyman B.A.F. (1959). *J. Inst. Petroleum*, **45**: 287.
24. Grob K. and Grob G. (1969). *J. Chromatogr. Sci.*, **7**: 584.
25. Grob K. and Grob K. (1974). *J. Chromatogr.*, **94**: 53.
26. Poy F. (1982). *Chromatographia*, **16**: 345.
27. Schomburg G., Husmann H., Behlau H. and Schulz F. (1983). *J. Chromatogr.*, **279**: 251.
28. Vogt W., Jacob K., Ohnesorge A.B. and Obwexer H.W. (1980). *J. Chromatogr.*, **199**: 191.
29. Grob K. and Neukom H.P. (1979). *J. High Resol. Chromatogr., Chromatogr. Comm.*, **2**: 563.
30. Grob K., Neukom H.P. and Hilling P. (1981). *J. High Resol. Chromatogr., Chromatogr. Comm.*, **4**: 203.
31. German A.L. and Horning E.C. (1972). *Anal. Lett.*, **5**: 619.
32. Jennings W.G. (1975). *J. Chromatogr. Sci.*, **13**: 185.
33. Schomburg G., Dielmann R., Borwitzky H. and Husmann H. (1978). *J. Chromatogr.*, **167**: 337.
34. Springer D.L., Phelps D.W. and Schirmer R.E. (1981). *J. High Resol. Chromatogr., Chromatogr. Comm.*, **4**: 638.
35. Poole C.F. and Schuette S.A. (1984). *Contemporary Practice of Chromatography*. Elsevier, Amsterdam, p. 151.
36. Ettre L.S. (1985). In Sandra P., Editor, *Sample Introduction in Capillary Gas Chromatography*, Vol. 1. Huethig, Heidelberg, p. 8.
37. Purcell J.E. (1982). *Chromatographia*, **15**: 546.

38. Pretorius V., Lawson K. and Bertsch W. (1983). *J. High Resol. Chromatogr., Chromatogr. Comm.*, **6**: 185.
39. Grob K. (1981). *J. Chromatogr.*, **213**: 3.
40. Grob K. (1982). *J. Chromatogr.*, **237**: 15.
41. Grob K. and Schilling B. (1983). *J. Chromatogr.*, **260**: 265.
42. Jenkins R. and Jennings W.G. (1983). *J. High Resol. Chromatogr., Chromatogr. Comm.*, **6**: 228.
43. Pretorius V., Phillips C.S.G. and Bertsch W. (1983). *J. High Resol. Chromatogr., Chromatogr. Comm.*, **6**: 232.
44. Deans D.R. (1971). *Anal. Chem.*, **43**: 2026.
45. Alexander G. and Rutten G.A. (1974). *J. Chromatogr.*, **99**: 81.
46. Miller R.J. and Jennings W. (1979). *J. High Resol. Chromatogr., Chromatogr. Comm.*, **2**: 72.
47. Grob K. (1981). *J. Chromatogr.*, **213**: 15.
48. Grob K. and Romann A. (1981). *J. Chromatogr.*, **214**: 118.
49. Grob K. and Grob K. (1978). *J. High Resol. Chromatogr., Chromatogr. Comm.*, **1**: 57.
50. Grob K. (1983). *Chromatographia*, **17**: 357.
51. Yang F.-S., Shanfield H. and Zlatkis A. (1982). *Anal. Chem.*, **54**: 1886.
52. Zlatkis A., Yang F.-S. and Shanfield H. (1982). *Anal. Chem.*, **54**: 2406.
53. Desty D.H. (1965). In Giddings J.C. and Keller R.A., Editors, *Advances in Chromatography*. Vol. 1. Marcel Dekker, New York, p. 218.
54. Grob K. and Grob K. (1978). *J. Chromatogr.*, **151**: 311.
55. Grob K. (1978). *J. High Resol. Chromatogr., Chromatogr. Comm.*, **1**: 263.
56. Grob K. and Neukom H.P. (1980). *J. Chromatogr.*, **198**: 64.
57. Grob K. and Neukom H.P. (1980). *J. Chromatogr.*, **189**: 109.
58. Galli M., Trestianu S. and Grob K. (1979). *J. High Resol. Chromatogr., Chromatogr. Comm.*, **2**: 366.
59. Galli M. and Trestianu S. (1981). *J. Chromatogr.*, **203**: 193.
60. Freeman R.R., Editor (1981). *High Resolution Gas Chromatography*, 2nd edn. Hewlett Packard, pp. 32, 33, 64.
61. Onuska F.I. and Karasek F.W. (1984). *Open Tubular Column Gas Chromatography in Environmental Sciences*. Plenum Press, New York, p. 81.
62. Wylie P.L., Phillips R.J., Klein K.J. and Thompson M.Q. (1991). *J. High Resol. Chromatogr., Chromatogr. Comm.*, **14**: 649.
63. Poy F., Visani S. and Terrosi F. (1981). *J. Chromatogr.*, **217**: 81.
64. Vogt W., Jacob K., Ohnesorge P.B. and Obwexer M.W. (1979). *J. Chromatogr.*, **186**: 197.
65. Ettre L.S. and March E.W. (1974). *J. Chromatogr.*, **91**: 5.
66. Poole C.F. and Poole S.K. (1989). *Anal. Chim. Acta*, **216**: 109.
67. Fenimore D.C., Whitford J.H., Davis C.M. and Zlatkis A. (1977). *J. Chromatogr.*, **140**: 9.
68. *Alltech Catalog* (1993). **300**: 180. Alltech Associates, Deerfield.
69. Michalske T.A. and Freeman S.W. (1982). *Nature*, **295**: 511.
70. Keith L.H., Editor (1981). Advances in the identification and analysis of organic pollutants in water, *Ann Arbor Science*, p. 155.
71. Sanders D.M. and Hench L.L. (1973). *J. Am. Ceram. Soc.*, **52**: 666.
72. Hercules D.M. (1978). *Anal. Chem.*, **50**: 734A.
73. Lipsky S.R. and McMurray W.J. (1981). *J. Chromatogr.*, **217**: 3.
74. Saito H. (1982). *J. Chromatogr.*, **243**: 189.
75. Cant N.W. and Little L.H. (1964). *Can. J. Chem.*, **42**: 802.
76. Filbert A.M. and Hair M.L. (1968). *J. Gas Chromatogr.*, **6**: 218.
77. Chapman I.D. and Hair M.L. (1965). *Trans. Faraday Soc.*, **61**: 1507.
78. Lipsky S.R. (1983). *J. High Resol. Chromatogr., Chromatogr. Comm.*, **6**: 359.
79. Sandra P. and Verzele M. (1977). *Chromatographia*, **10**: 419.
80. Sandra P., Verstappe M. and Verzele M. (1978). *J. High Resol. Chromatogr., Chromatogr. Comm.*, **1**: 28.
81. Ottenstein D.M. (1973). *J. Chromatogr. Sci.*, **11**: 136.
82. Schomburg G., Husmann H. and Weeke F. (1977). *Chromatographia*, **10**: 580.
83. Schomburg G., Husmann H. and Behlau H. (1981). *J. Chromatogr.*, **203**: 179.
84. Schomburg G. (1979). *J. High Resol. Chromatogr., Chromatogr. Comm.*, **2**: 461.

85. Grob K. and Grob G. (1979). *J. High Resol. Chromatogr. Chromatogr. Comm.*, **2**: 302.
86. Bertsch W., Pretorius V. and Lawson K. (1982). *J. High Resol. Chromatogr., Chromatogr. Comm.*, **5**: 568.
87. Chrompack (1992). *Chrompack News* **19** (2): 7
88. Ettre L.S. (1984). *Chromatographia*, **18**: 477.
89. Grob K. and Frech P. (1988). *Int. Lab.*, **18** October 18.
90. Robards K., Kelly V.R. and Patsalides E. (1992). In Giddings J.C., Grushka E. and Brown P.R., Editors, *Advances in Chromatography*, Vol. 32. Marcel Dekker, New York, Chapter 3, pp. 53–86.
91. Smolkova E., Feltl L. and Zima J. (1979). *Chromatographia*, **12**: 463.
92. Rowan R. and Sorrell J.B. (1970). *Anal. Chem.*, **42**: 1716.
93. Naito K., Kurita R., Moriguchi S. and Takei S. (1982). *J. Chromatogr.*, **246**: 199.
94. Naito K., Endo M., Moriguchi S. and Takei S. (1982). *J. Chromatogr.*, **253**: 205.
95. Aubeau R.N., LeRoy J. and Champeix L. (1965). *J. Chromatogr.*, **19**: 249.
96. Karlsson B.M. (1966). *Anal. Chem.*, **38**: 668.
97. Clay D.T. and Lynn S. (1975). *Anal. Chem.*, **47**: 1205.
98. Stetter J.R., Sedlak J.M. and Blurton K.F. (1977). *J. Chromatogr. Sci.*, **15**: 125.
99. Hertl W. and Neumann M.G. (1971). *J. Chromatogr.*, **60**: 319.
100. Bayer E. and Nikolson G. (1970). *J. Chromatogr. Sci.*, **8**: 467.
101. Gillis J.N., Sievers R.E. and Pollock G.E. (1985). *Anal. Chem.*, **57**: 1572.
102. Castello G. and D'Amato G. (1983). *J. Chromatogr.*, **254**: 69.
103. Chrompack (1992). *Chrompack News* **19**(2): 4.
104. Haken J.K. (1977). *J. Chromatogr.*, **141**: 247.
105. Ettre L.S. and DiCesare J.L. (1984). *Int. Lab.*, **14**: 44.
106. Jennings W.G. (1982). In Giddings J.C., Editor, *Advances in Chromatography*, Vol 20. Marcel Dekker, New York, p. 197.
107. Hawkes S., Grossman D., Hartkopf A., Isenhour T., Leary J., Parcher J., Wold S. and Lancey J. (1975). *J. Chromatogr. Sci.*, **13**: 115.
108. Klee M.S., Kaiser M.A. and Laughlin K.B. (1983). *J. Chromatogr.*, **279**: 681.
109. Boksanyi L. and Kovats E. (1976). *J. Chromatogr.*, **126**: 87.
110. Yancey J.A. (1986). *J. Chromatogr. Sci.*, **24**: 117.
111. Supina W.R. and Rose L.P. (1970). *J. Chromatogr. Sci.*, **8**: 214.
112. Rohrschneider L. (1966). *J. Chromatogr.*, **22**: 6.
113. Kovats E. (1958). *Helv. Chim. Acta*, **41**: 1915.
114. Ettre L.S. (1964). *Anal. Chem.*, **36**: 31A.
115. Ettre L.S. (1974). *Chromatographia*, **7**: 39.
116. McReynolds W.O. (1970). *J. Chromatogr. Sci.*, **8**: 685.
117. Weiner P.H. and Parcher J.F. (1972). *J. Chromatogr. Sci.*, **10**: 612.
118. Laub R.J., Purnell J.H. and Williams P.S. (1977). *J. Chromatogr.*, **134**: 249.
119. West S.D. and Hall R.C. (1976). *J. Chromatogr. Sci.*, **14**: 339.
120. Hartkopf A., Grunfeld S. and Delumyea R. (1974). *J. Chromatogr. Sci.*, **12**: 119.
121. Chong E., deBriceno B., Miller G. and Hawkes S. (1985). *Chromatographia*, **20**: 293.
122. Kersten B.R., Poole C.F. and Furton K.G. (1987). *J. Chromatogr.*, **411**: 43.
123. Grob K. (1977). *Chromatographia*, **10**: 250.
124. Harris W.E. and Habgood H.W. (1966). *Programmed Temperature Gas Chromatography*. Wiley, New York.
125. Mikkelsen L. (1966). In Giddings J.C., Editor, *Advances in Chromatography*, Vol. 2. Marcel Dekker, New York, p. 337
126. Giddings J.C. (1962). *J. Chem. Educ.*, **39**: 569.
127. Novak J. (1988). *Quantitative Analysis by Gas Chromatography*, Chromatographic Science Series, Vol. 41, 2nd edn. Marcel Dekker, New York.
128. Halasz I. (1964). *Anal. Chem.*, **36**: 1428.
129. Novak J. (1974). In Giddings J.C., Editor, *Advances in Chromatography*. Vol. 11. Marcel Dekker, New York, p. 1.
130. Dal Nogare S. and Juvet R.S. (1962). *Gas Liquid Chromatography*. Wiley Interscience, New York, p. 188.
131. Lovelock J.E. and Lipsky S.R. (1960). *J. Am. Chem. Soc.*, **82**: 431.

132. Kolb B. and Bischoff J. (1974). *J. Chromatogr. Sci.*, **12**: 625.
133. Patterson P.L., Gatton R.A. and Ontiveros C. (1982). *J. Chromatogr. Sci.*, **20**: 97.
134. Ettre L.S. (1978). *J. Chromatogr Sci.*, **16**: 396.
135. Nutmagul W., Cronn D.R. and Hill H.H. (1983). *Anal. Chem.*, **55**: 2160.
136. Grob K. (1980). *J. High Resol. Chromatogr., Chromatogr. Comm.*, **3**: 585.
137. Grob K. (1981). *J. Chromatogr.*, **208**: 217.
138. Ahnoff M. and Johansson L. (1983). *J. Chromatogr.*, **279**: 75.
139. Grob K., Grob G. and Grob K. (1978). *J. Chromatogr.*, **156**: 1.
140. Moncur J.G. (1982). *J. High Resol. Chromatogr., Chromatogr. Comm.*, **5**: 53.
141. Patsalides E. and Robards K. (1985). *J. Chromatogr.*, **350**: 353.
142. Cram S.P., Yang F.J. and Brown A.C. (1977). *Chromatographia*, **10**: 397.
143. Verzele M. and Sandra P. (1978). *J. Chromatogr.*, **158**: 111.
144. Blomberg L.G. (1984). *J. High Resol. Chromatogr., Chromatogr. Comm.*, **7**: 232.
145. Lee M.L. and Wright B.W. (1980). *J. Chromatogr.*, **184**: 235.
146. Dandeneau R. and Zerenner E. (1979). *J. High Resol. Chromatogr., Chromatogr. Comm.*, **2**: 351.
147. Grob K. and Grob G. (1979). *J. High Resol. Chromatogr., Chromatogr. Comm.*, **2**: 527.
148. Berezkin V.G. (1981). *CRC Crit. Revs. Anal. Chem.*, **11**: 1.
149. Kolb B., Editor (1980). *Applied Headspace Gas Chromatography*. Heyden, London.
150. Hachenberg H. and Schmidt A.P. (1977). *Gas Chromatographic Headspace Analysis*. Heyden, London.
151. Grob K. and Habich A. (1985). *J. Chromatogr.*, **321**: 45.
152. Ioffe B.V. and Vitenberg A.G. (1984). *Headspace Analysis and Related Methods in Gas Chromatography*. Wiley, New York.
153. Nunez A.J., Gonzalez L.F. and Janak J. (1984). *J. Chromatogr.*, **300**: 127.
154. Davis P.L. (1970). *J. Chromatogr. Sci.*, **8**: 423.
155. Maier H. (1970). *J. Chromatogr.*, **50**: 329.
156. Peters A., Klemp M., Puig L., Rankin C. and Sacks R. (1991). *Analyst*, **116**: 1313.
157. Shutjes C., Vermeer E., Rijks J. and Cramers C. (1982). *J. Chromatogr.*, **253**: 1.
158. Hirschfeld T. (1980). *Anal. Chem.*, **52**: 297A.
159. Karasek F.W. and Viau A.C. (1986). *J. Chem. Educ.*, **61**: A233.
160. McDonald R.S. (1986). *Anal. Chem.*, **58**: 1906.
161. Smith S.L. (1984). *J. Chromatogr. Sci.*, **22**: 143.
162. Seeley J.A. and Uden P.C. (1991). *Analyst*, **116**: 1321.
163. Sevcik J. (1979). *J. Chromatogr.*, **186**: 129.
164. Griffiths P.R., deHaseth J.A. and Azarraga L.V. (1983). *Anal. Chem.*, **55**: 1361A.
165. Schreider J.F., Demirian J.C. and Stickler J.C. (1986). *J. Chromatogr. Sci.*, **24**: 330.
166. Davies I.L., Raynor M.W., Kithinji J.P., Bartle K.D., Williams P.E. and Andrews G.E. (1988). *Anal. Chem.*, **60**: 683A.
167. Settlage J.A. and Jaeger H. (1986). In Nikelly J.G., Editor, *Advances in Capillary Chromatography*. Huethig, Heidelberg, pp. 117–124.
168. Karasek F.W. and Clement R.E. (1988). *Basic Gas Chromatography–Mass Spectrometry: Principles and Techniques*. Elsevier, Amsterdam, pp. 106–117.
169. Vangaever F., Sandra P. and Verzele M. (1979). *Chromatographia*, **12**: 153.
170. Hurley R.B. (1980). *J. High Resol. Chromatogr., Chromatogr. Comm.*, **3**: 147.
171. Low M.J.D. and Freeman S.K. (1967). *Anal. Chem.*, **39**: 194.
172. Cooper J.R. and Taylor L.T. (1984). *Anal. Chem.*, **56**: 1989.
173. Bourne S., Haefner A.M., Norton K.L. and Griffiths P.R. (1990). *Anal. Chem.*, **62**: 2448.
174. Shafer K.H. and Griffiths P.R. (1986). *Anal. Chem.*, **58**: 3249.
175. Shafer K.H., Hayes T.L., Brasch J.W. and Jakobsen R.J. (1984). *Anal. Chem.*, **56**: 237.
176. Gurka D.F. and Betowski D. (1982). *Anal. Chem.*, **54**: 1819.
177. Chiu K.S., Blemann K., Krishnan K. and Hill S.L. (1984). *Anal. Chem.*, **56**: 1610.
178. Wilkins C.L. (1987). *Anal. Chem.*, **59**: 571A.
179. Williams S.S., Lam R.B., Sparks D.T., Isenhour T.L. and Hass J.R. (1982). *Anal. Chim. Acta*, **138**: 1.
180. Wilkins C.L., Giss G.N., White R.L., Brissey G.M. and Onyiriuka E.C. (1982). *Anal. Chem.*, **54**: 2260.

Bibliography

General texts

Berezkin V.G. (1991). *Gas Liquid Solid Chromatography, Chromatographic Science Series, No. 56*. Marcel Dekker, New York.

Braithwaite A. and Smith F.J. (1985). *Chromatographic Methods*. Chapman and Hall, London.

Grob R.L. Editor (1985). *Modern Practice of Gas Chromatography*, 2nd edn. John Wiley, New York.

Guiochon G. and Guillemin C.L. (1988). *Quantitative Gas Chromatography for Laboratory Analyses and On-line Process Control, Journal of Chromatography Library Series, No. 42*. Elsevier, Amsterdam.

Heftmann E. Editor (1992). *Chromatography: Fundamentals and Applications of Chromatography and Related Differential Migration Methods*, 5th edn., *Journal of Chromatography Library Series, No. 51 A & B, Part A, Fundamentals and Techniques, Part B, Applications*. Elsevier, Amsterdam.

Jennings W. (1987). *Analytical Gas Chromatography*. Academic Press, New York.

Miller J.M. (1988). *Chromatography: Concepts and Contrasts*. John Wiley and Sons, New York.

Perry J.A. (1981). *Introduction to Analytical Gas Chromatography: History, Principles and Practice, Chromatographic Science Series, No. 14*. Marcel Dekker, New York.

Poole C.F. and Poole S.K. (1991). *Chromatography Today*. Elsevier, Amsterdam.

Roedel W. and Woelm G. (1987). *A Guide to Gas Chromatography*. Huethig, Heidelberg.

Smith R.G. (1988). *Gas and Liquid Chromatography in Analytical Chemistry*. John Wiley and Sons, Chichester.

Instrumentation

Bayer F.L. (1986). Gas chromatographic equipment. *J. Chromatogr. Sci.*, **24**: 549.

Hurrell R.A. (1992). European gas chromatography instrumentation. *J. Chromatogr. Sci.*, **30**: 86.

Injection

Grob K. (1986). *Classical Split and Splitless Injection in Capillary Gas Chromatography*. Huethig, Heidelberg.

Grob K. (1987). *On-column Injection in Capillary Gas Chromatography*. Huethig, Heidelberg.

Sandra P. Editor (1985). *Sample Introduction in Capillary Gas Chromatophly*, Vols. 1 and 2. Huethig, Heidelberg.

Mobile phases

Parcher J.F. (1983). A review of vapor phase chromatography: gas chromatography with vapor carrier gases. *J. Chromatogr. Sci.*, **21**: 346.

Columns and packings

Bertsch W., Jennings W.G. and Kaiser R.E. Editors (1981). *Recent Advances in Capillary Gas Chromatography*, Vols. 1–3, Huethig, Heidelberg.

Ettre L.S. (1985). Open tubular columns: evolution, present status, and future. *Anal. Chem.*, **57**: 1419A.

Ettre L.S. (1992). Open tubular columns: past, present and future. *Chromatographia* **34**: 513.

Jennings W. (1978). *Gas Chromatography with Glass Capillary Columns*, Academic Press, New York.

Jennings W. (1981). *Comparisons of Fused Silica and Other Glass Columns in Gas Chromatography*. Huethig, Heidelberg.

Jennings W. and Nikelly J. (1991). *Capillary Chromatography: The Applications*. Huethig, Heidelberg.

Kersten B.R., Poole C.F. and Furton K.G. (1987). Ambiguities in the determination of McReynolds stationary phase constants. *J. Chromatogr.*, **411**: 43.

Klee M.S., Kaiser M.A. and Laughlin K.B. (1983). Systematic approach to stationary phase selection in gas chromatography, *J. Chromatogr.*, **279**: 681.

Nikelly J.G. Editor (1986). *Advances in Capillary Chromatography*. Huethig, Heidelberg.

Rood D. (1991). *A Practical Guide to the Care, Maintenance, and Troubleshooting of Capillary Gas Chromatographic Systems*. Huethig, Heidelberg.

Rotzsche H. (1991). *Stationary Phases in Gas Chromatography*, Journal of Chromatography Library Series, No. 48. Elsevier, Amsterdam.

Schoenmakers P.J. (1986). *Optimization of Chromatographic Selectivity: A Guide to Method Development*, Journal of Chromatography Library Series, No. 35. Elsevier, Amsterdam.

Unger K.K. Editor (1989). *Packings and Stationary Phases in Chromatographic Techniques*, Journal of Chromatography Library Series, No. 47. Marcel Dekker, New York.

Detection and quantification

Volume 121 (1992) of *Chem. Anal.* contains a series of articles on various detectors.

Berezkin V.G. (1983). *Chemical Methods in Gas Chromatography*, Journal of Chromatography Library Series, No. 24. Elsevier, Amsterdam.

Cardwell T.J. and Marriott P.J. (1982). Some characteristics of a flame photometric detector in sulphur and phosphorus modes. *J. Chromatogr. Sci.*, **20**: 83.

Dirkx W.M.R., Van Cleuvenbergen R.J.A. and Adams F.C. (1992). Speciation of alkyllead compounds by GC–AAS: A state of affairs, *Mikrochim. Acta*, **109**: 133.

Dressler M. (1986). *Selective Gas Chromatographic Detectors*, Journal of Chromatography Library Series, No. 36. Elsevier, Amsterdam.

Driscoll J.N. (1985). Review of photoionization detection in gas chromatography: the first decade. *J. Chromatogr. Sci.*, **23**: 488.

Drozd J. (1981). *Chemical Derivatization in Gas Chromatography*, Journal of Chromatography Library Series, No. 19. Elsevier, Amsterdam.

Dyson N. (199?). *Chromatographic Integration Methods, RSC Chromatography Monographs Series*. Royal Society of Chemistry, London.

Ebdon L., Hill S. and Ward R.W. (1986). Directly coupled chromatography – atomic spectroscopy. Part 1. Directly coupled gas chromatography – atomic spectroscopy. A review, *Analyst*, **111**: 1113.

Edman D.C. and Brooks J.B. (1983). Review. Gas liquid chromatography – frequency pulse – modulated electron – capture detection in the diagnosis of infectious diseases. *J. Chromatogr.*, **274**: 1.

Farwell S.O. and Barinaga C.J. (1986). Sulfur-selective detection with the FPD: current enigmas, practical usage, and future directions. *J. Chromatogr. Sci.*, **24**: 483.

Jaeger, H. Editor (1987). *Capillary Gas Chromatography – Mass Spectrometry in Medicine and Pharmacology*. Huethig, Heidelberg.

Karasek F.W. and Clement R.E. (1988). *Basic Gas Chromatography–Mass Spectrometry: Principles and Techniques*. Elsevier, Amsterdam.

Krull I.S., Swartz M.E. and Driscoll J.N. (1984). *Multiple Detection in Gas Chromatography, Advances in Chromatography*. Marcel Dekker, New York, Vol. 24, p. 247.

Novak J. (1987). *Quantitative Analysis by Gas Chromatography*, 2nd edn, Journal of Chromatography Library Series, No. 41. Marcel Dekker, New York.

Uden P.C. (1984). Inorganic gas chromatography, *J. Chromatogr.*, **313**: 3.

Uden P.C. *Element-Specific Chromatographic Detection by Atomic Emission Spectroscopy*, ACS Symposium Series No.479, American Chemical Society, Washington.

Verner P. (1984). Photoionization detection and its application in gas chromatography. *J. Chromatogr., Chromatogr. Rev.*, **300**: 249.

Zlatkis A. and Poole C.F. (1981). *Electron Capture: Theory and Practice in Chromatography*, Journal of Chromatography Library Series, No. 20. Elsevier, Amsterdam.

System evaluation

Grob, K. and Wagner, C. (1993). Procedure for testing inertness of inserts and insert packing materials for GC injectors. *J. High Resol. Chromatogr.*, **16**: 464.

Walker J.Q., Spencer S.F. and Sonchik S.M. (1985). Forum for chromatographers: testing gas chromatographic instruments. *J. Chromatogr. Sci.*, **23**: 555.

Derivatization

Ahuja S. (1986). Derivatization for gas and liquid chromatography. In Ahuja S., Editor, *Ultratrace Analysis of Pharmaceuticals and Other Compounds of Interest*. John Wiley, New York, pp. 19–90.

Berezkin V.G. (1981). *Chemical Methods in Gas Chromatography. Journal of Chromatography Library Series, No. 24*. Elsevier, Amsterdam.

Blau K. and King G.S. Editors (1977). *Handbook of Derivatives for Chromatography*. Heyden, London.

Drozd J. (1981). *Chemical Derivatization in Gas Chromatography, Journal of Chromatography Library Series, No. 19*. Elsevier, Amsterdam.

Poole C.F. and Zlatkis A. (1980). Derivatization techniques for the electron capture detector. *Anal. Chem.*, **52**: 1002A.

Pyrolysis gas chromatography

Irwin W.J. (1982). *Analytical Pyrolysis: Comprehensive Guide, Chromatographic Science Series*, Vol. 22. Marcel Dekker, New York.

Liebman S.A. and Levy E.J. (1983). Advances in pyrolysis GC systems: application to modern trace organic analysis. *J. Chromatogr. Sci.*, **21**: 1.

Liebman S.A. and Levy E.J. Editors (1985). *Pyrolysis and GC in Polymer Analysis, Chromatographic Science Series*, Vol. 29. Marcel Dekker, New York.

Headspace gas chromatography

Drozd J. and Novak J. (1979). Headspace gas analysis by gas chromatography. *J. Chromatogr.*, **165**: 141.

Hachenberg H. and Horst A.P. (1977). *Gas Chromatographic Headspace Analysis*. Heyden, London.

Ioffe B.V. and Vitenberg A.G. (1984). *Head-space Analysis and Related Methods in Gas Chromatography*. John Wiley, New York.

Kolb B. Editor (1980). *Applied Headspace Gas Chromatography*. Heyden, London.

Vitenberg A.G. (1984). Theory of gas chromatographic headspace analysis with pneumatic sampling. *J. Chromatogr. Sci.*, **22**: 122.

Wampler T.P., Bowe W.A. and Levy E.J. (1985). Dynamic headspace analyses of residual volatiles in pharmaceuticals. *J. Chromatogr. Sci.*, **23**: 64.

High-speed gas chromatography

Peters A., Klemp M., Puig L., Rankin C. and Sacks R. (1991). Instrumentation and strategies for high-speed gas chromatography. *Analyst*, **116**: 1313.

Multidimensional chromatography

Davies I.L., Markides K.E., Lee M.L., Raynor M.W. and Bartle K.D. (1989). Applications of coupled liquid chromatography – gas chromatography: a review. *J. High Resol. Chromatogr., Chromatogr. Comm.*, **12**: 193.

Grob K. (1991). *On-line Coupled LC–GC*. Huethig, Heidelberg..

Gas chromatography–mass spectrometry

Budde W.L. and Eichelberger J.W. (1980). *Organic Analysis using Gas Chromatography/Mass Spectrometry: A Techniques and Procedures Manual*. Ann Arbor Science, Ann Arbor.

Jaeger, H. Editor (1987). *Capillary Gas Chromatography–Mass Spectrometry in Medicine and Pharmacology*. Huethig, Heidelberg.
Wishnok J.S. (1992). Environmental carcinogens: monitoring *in vivo* using GC/MS, *Anal. Chem.*, **64**: 1126A.

Gas chromatography – infrared spectrometry

Griffiths P.R., deHaseth J.A. and Azarraga L.V. (1983). Capillary GC/FTIR. *Anal. Chem.*, **55**: 1361A.
Griffiths P.R., Pentoney S.L., Giorgetti A. and Shafer K.H. (1986). The hyphenation of chromatography and FT-IR Spectrometry. *Anal. Chem.*, **58**: 1349A.
Hanna A., Marshall J.C. and Isenhour T.L. (1979). A GC/FT-IR compound identification system. *J. Chromatogr. Sci.*, **17**: 434.
Herres W. (1987). *HRGC-FTIR: Theory and Applications*. Huethig, Heidelberg.
Seelemann R. (1982). GC-FTIR coupling – a modern tool for analytical chemistry. *Trends Anal. Chem.*, **1**: 333.

Applications

Berezkin V.G. and Drugov Y.S. (1991). *Gas Chromatography in Air Pollution Analysis, Journal of Chromatography Library Series, No. 49*. Elsevier, Amsterdam.
Bobbitt D.R. and Ng K.W. (1992). Chromatographic analysis of antibiotic materials in food. *J. Chromatogr.*, **624**: 153.
Chen E.C.H. and Morrison N.M. (1983). Application of gas chromatography to brewing research and quality control: a review. *J. Am. Soc. Brew. Chem.*, **41**: 14.
David F. and Sandra P. (1992). Capillary gas chromatography–spectroscopic techniques in natural product analysis. *Phytochem. Anal.*, **3**: 145.
Fodor-Csorba K. (1992). Chromatographic methods for the determination of pesticides in foods. *J. Chromatogr.*, **624**: 353.
Grob R.L. and Kaiser M. (1982). *Environmental Problem Solving Using Gas and Liquid Chromatography, Journal of Chromatography Library Series, No. 21*. Elsevier, Amsterdam.
Guiochon G. and Guillemin C.L. (1988). *Quantitative Gas Chromatography for Laboratory Analyses and On-line Process Control, Journal of Chromatography Library Series, No. 42*. Elsevier, Amsterdam.
Haken J.K. (1974). *Gas Chromatography of Coating Materials*. Marcel Dekker, New York.
Holcomb M., Wilson D.M., Trucksess M.W. and Thompson H.C. (1992). Determination of aflatoxins in food products by chromatography. *J. Chromatogr.*, **624**: 341.
Jack D.B. (1984). *Drug Analysis by Gas Chromatography*. Academic Press, Orlando.
Krull I.S. Editor (1992). *Trace Metal Analysis and Speciation, Journal of Chromatography Library Series, No. 47*. Elsevier, Amsterdam.
Middleditch B.S. (1992). *Analytical Artifacts GC, MS, HPLC, TLC and PC, Journal of Chromatography Library Series, No. 44*. Elsevier, Amsterdam.
Mosandl A. (1992). Capillary gas chromatography in quality assessment of flavors and fragrances. *J. Chromatogr.*, **624**: 267.
Onuska F.I. and Karasek F.W. (1984). *Open Tubular Column Gas Chromatography in Environmental Sciences*. Plenum Press, New York.
Rizzolo A. and Polesello S. (1992). Chromatographic determination of vitamins in foods, *J. Chromatogr.*, **624**: 103.
Robards K. and Whitelaw M. (1986). Chromatography of monosaccharides and disaccharides. *J. Chromatogr.*, **373**: 81.
Robards K. and Towers P. (1990). Chromatography as a reference technique for the determination of clinically important steroids. *Biomed. Chromatogr.*, **4**: 1.
Robards K., Patsalides E. and Dilli S. (1987). Gas chromatography of metal β-diketonates and their analogues. *J.Chromatogr.*, **411**: 1.
Shantha N.C. and Napolitano G.E. (1992). Gas chromatography of fatty acids. *J. Chromatogr.*, **624**: 37.

Planar Chromatography 4

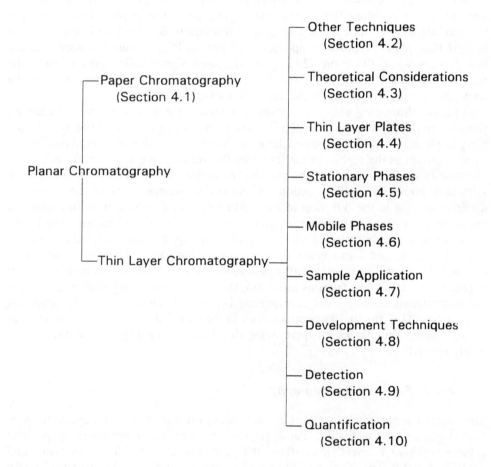

Planar Chromatography

- Paper Chromatography (Section 4.1)
- Thin Layer Chromatography
  - Other Techniques (Section 4.2)
  - Theoretical Considerations (Section 4.3)
  - Thin Layer Plates (Section 4.4)
  - Stationary Phases (Section 4.5)
  - Mobile Phases (Section 4.6)
  - Sample Application (Section 4.7)
  - Development Techniques (Section 4.8)
  - Detection (Section 4.9)
  - Quantification (Section 4.10)

4.1   Introduction

Planar chromatography is a form of liquid chromatography in which the stationary phase is supported on a planar surface rather than in a column. There are two planar techniques: paper chromatography (PC) and TLC. PC preceded TLC by some 10–15 years but has now been largely superseded by the latter owing to greater speed, versatility and reproducibility. The availability of thin layer plates coated with microcrystalline cellulose further contributed to the decline in the use of PC. Separations on cellulose thin layers have separation characteristics similar to paper but provide sharper spot shape and greater resolution. Sherma and Fried [1] noted in 1984 that 'papers reporting important advances in PC continued to decline in the last 2 years while those on TLC again increased significantly.' Nevertheless the tremendous achievements of PC should not be ignored and a brief treatment of the technique is provided at the beginning of this chapter.

In planar chromatography, the sample is applied as a spot or streak of minimum size to a marked position on the planar surface. After evaporation of the solvent, the plate is placed in a suitable sealed tank or chamber with one end immersed in the solvent chosen as the mobile phase (but with the applied spot not covered, otherwise the sample would dissolve off the plate). The mobile phase percolates through the stationary phase by capillary action and moves the components of the sample to different extents in the direction of flow. After the mobile phase front has migrated the necessary distance, the chromatogram is removed from the tank and dried. The point which the mobile phase reaches when it flows up the plate is called the solvent front. The separated components are located and characterized by the R_f value (see Chapter 2). Planar chromatography is commonly regarded as a qualitative technique. However, almost from its inception, the possibility of using planar techniques for quantitative analysis was recognized. Indeed, Kirchner *et al.* [2] described quantitative TLC for the determination of biphenyl in citrus fruits and products as early as 1954. Techniques for quantifying thin layer chromatographic data are now highly refined.

4.1.1   Paper chromatography

The original separations by PC were performed on ordinary filter papers but now special papers are available. These papers are prepared from specially purified cellulose (98–99% α-cellulose, 0.3–1.0% β-cellulose, 0.4–0.8% pentosans and <0.01% mineral ash). They provide a highly purified and reproducible surface with respect to porosity, thickness and arrangement of cellulose fibres. Chromatography papers are characterized by their thickness, weight per unit area and flow-rate index. On this basis, papers may be classified as slow, standard or fast grades corresponding to the degree of coarseness of the cellulose fibres and their packing density. Standard papers represent the best compromise between resolution and the time required to develop the chromatogram. Whatman No. 1 paper or its equivalent is suitable for most separations. Paper has a slight grain, known as the machine direction, and separations may differ according to whether they are made parallel to, or perpendicular to, this grain. Chromatography paper should be stored away from any source of fumes; for example, ammonia has a high affinity for cellulose and modifies the separation process.

The surface of paper can be modified by chemical treatment (e.g., acylation), or by application of other liquids. Impregnating agents include silicone oils for reversed-phase separations, adsorbents (e.g., silica gel and alumina) and liquid ion-exchangers. Chemically modified papers are prepared mainly from ion-exchange celluloses such as carboxymethyl- and diethylaminoethyl-cellulose.

From the practical viewpoint, the main influence that the paper has on the separation is its effect on the rate of flow of the mobile phase. The flow-rate increases with decreasing viscosity of the mobile phase and, therefore, with increase in temperature. However, at fixed temperature, the flow rate of a given mobile phase is controlled by the density and thickness of the paper. Lowering the density or increasing the thickness, gives a higher flow-rate. Thicker papers, up to 3 mm, are useful for separations on a larger scale, as they can accommodate more of the sample without increasing the area of the original and final spots.

Chromatography paper contains sufficient sorbed water to classify PC as a partition process with a 'water–cellulose complex' formed between chemically and physically bound water and the cellulose matrix of the paper. This was originally considered to be the sole operative mechanism. However, the role of the paper is more complex than merely acting as a support for the stationary phase. Adsorption effects due to the polar hydroxyl groups are probably the most important interactions. Furthermore, bleaching processes used during manufacture of the paper increase the number of carboxyl groups in the cellulose. These mildly acidic groups confer weak ion-exchange properties on the paper and ion-exchange processes also contribute to the separation. The properties of the paper may be further modified by sorption of constituents of the mobile phase.

Mobile phases for paper chromatography

The choice of mobile phase is essentially empirical, although so many separations by PC have been reported that it is usually not difficult to find a suitable mobile phase. The usual mobile phase is a mixture, consisting of an organic component, water and various additives, such as acids, bases or complexing agents to modify the solubility of sample components. The selection of a mobile phase can be illustrated by the following example. Polar organic substances, which are more soluble in water than in organic liquids, show very little migration if an anhydrous mobile phase is used, but by adding water to the mobile phase, the substances will be made to move over the paper. For example, n-butanol is not a good solvent for polar amino acids unless it is saturated with water (with which it is partially miscible). If acetic acid is also added to the solvent mixture, then a higher proportion of water may be incorporated, and the resulting ternary solvent mixture consisting of water–butanol–acetic acid greatly enhances the solubility and mobility of the amino acids. Such binary and ternary mixtures are common in PC.

A saturated atmosphere should be maintained in the chromatographic chamber, which should be gas tight and protected from temperature variations. There has been considerable debate on the need to establish an equilibrium in the developing chamber between the mobile phase molecules on the paper and those in the surrounding atmosphere. The extent of equilibration needed depends on the size of the developing chamber, the mobile phase system and the nature of the separation. There is a marked tendency for mobile phase to evaporate from the paper into the surrounding atmosphere. In such cases, the supply of mobile phase to the solvent front is inadequate to ensure uniform flow of mobile phase through the paper by capillary action. The problem can be avoided by preventing surface evaporation of the mobile phase from the paper. This is most easily achieved by saturating the atmosphere in the developing chamber with the mobile phase vapour. In practice, this means adding a large excess of the mobile phase to the closed developing chamber for a period prior to placing the chromatography paper in the tank.

Further information on PC may be obtained in a number of sources [3, 4].

4.1.2   Thin layer chromatography

Thin layer chromatography was discovered by Izmailov and Shraiber in 1938. However, as practised today, it was first described in 1951 [5] for the isolation and identification of the flavouring components in orange and grapefruit juice. The historical development of TLC has been closely related to the analysis of foodstuffs [6]. TLC is also the workhorse of the pharmaceutical industry for determining drug purity, it is widely used in clinical and forensic laboratories and in the chemical industry.

TLC can be performed very simply. The steps involved are:

- Selection of a suitable layer.
- Sample application.
- Selection of the mobile phase.
- Development (the name used for the actual running of the chromatogram).
- Visualization and detection.
- Quantification.

These steps are discussed in the remainder of this chapter and, as each step is discussed, the apparatus required will also be introduced.

4.2   Why Thin Layer Chromatography?

Since the areas of application for the various chromatographic techniques overlap to some extent, there is always interest in comparing their performance. Moreover, the question of which technique is better is frequently asked. However, this question is ill-considered as none of the chromatographic techniques can fully replace any of the others. They are, in fact, mutually complementary. There are situations in which one technique has obvious advantages, whereas in other situations more than one technique may be entirely suitable, and it therefore becomes purely a matter of personal choice. In order to differentiate, the analytical chemist must possess knowledge of each of the techniques and their respective strengths and weaknesses.

At this point, it is worth re-emphasizing that TLC is a variant of liquid chromatography, although the column variant in its abbreviation LC or HPLC, seems to have usurped the name for its exclusive use. Historically, TLC preceded

Table 4.1   Comparison of conventional thin layer and high-performance thin layer chromatography.

Parameter	Conventional TLC	HPTLC
Plate size (cm)	20 × 20; 10 × 20	10 × 10; 20 × 10
Layer thickness (μm)	100–250	100–250
Particle size (μm)		
average	20	between 5 and 15
distribution	10–60	narrow
Number of usable theoretical plates	<600	<5000
Plate height (μm)	30–60	5–20
Separation number	7–10	10–20
Sample volume (μl)	1–5	0.01–0.2
Diameter of starting spots (mm)	3–6	1.0–1.5
Diameter of separated spots (mm)	6–15	2–6
Solvent migration distance (cm)	10–15	3–6
Development time (min)	30–200	3–20
Sample lanes per plate	10	18–36

HPLC but rapid advances in GC technology relegated TLC to the background. At about the time that the initial research interest in GC waned as it became a routine technique, modern column LC (HPLC) emerged and TLC was once again relegated to a secondary role. The greatest overlap in areas of application is, not unexpectedly, between TLC and column LC and the strongest competition for analytical application has come from the column variants of LC, with TLC slowly losing ground. Nonetheless, there is a continued and strong interest in TLC for both routine and research purposes. The latter is evidenced by the approximately 2000 papers published annually on the theory, techniques and applications of TLC.

4.2.1   Comparison of conventional TLC and HPTLC

Modern TLC originated in the mid 1970s with the commercial introduction of fine-particle layers optimized for fast and efficient separations. This version of TLC is often described as high-performance thin layer chromatography (HPTLC). The term HPTLC was originally coined for the use of 5 μm particles in centrifugal TLC, but is now used in a broader sense for any TLC procedure that combines state-of-the-art techniques. HPTLC resulted from the sum of improvements in many of the operations comprising TLC rather than any specific breakthrough in materials or instrumentation. Both conventional TLC and HPTLC continue to exist side-by-side and are compared in Table 4.1. Despite the trend toward increasing use of modern TLC with expensive instrumentation, most notably in western Europe, it is still undoubtedly true that most work with TLC is being performed with relatively inexpensive basic equipment and manual techniques [7].

An important feature of any comparison between TLC and HPTLC is the small particle size of the stationary phase used in the latter technique but, equally as important for the performance of high-performance plates, is the tightness of the particle size distribution. In many cases, HPTLC was not well received when first introduced. The author well remembers a colleague dismissing HPTLC with the advice that the technique had nothing to offer in order to justify its price tag. These criticisms were unfounded in many instances and resulted from the use of methods which were more appropriate to conventional TLC.

The small particle size and more uniform nature of the particles in HPTLC results in much greater chromatographic efficiency than found in conventional TLC and enables the same separations to be achieved much more rapidly and also separations that are not possible by conventional TLC. Although the mobile phase velocity is lower for HPTLC than for conventional TLC, shorter migration distances are required to obtain an equivalent separation. Typical migration distances in HPTLC are 3–6 cm, while development times are an order of magnitude lower than in conventional TLC.

Detection limits in HPTLC are 10 times better than in conventional TLC. Two factors contribute to this improved sensitivity: the layer itself in HPTLC is usually somewhat thinner and, more importantly, the surface is more uniform than that of earlier TLC plates. More uniform surfaces reduce background noise with instrumental detectors. The second factor is the more compact spots on high-performance layers as a result of the reduced band spreading per plate length.

The size of the applied sample spot must be as small as possible (<1.5 mm) in order to attain optimum resolution and sensitivity in HPTLC. The compact starting spots also allow an increase in the number of samples which may be applied along the edge of the plate; as many as 36 samples on a 10×20 cm plate can be simultaneously separated depending on the method of development. Automated micropipettes are necessary to attain the small sample volumes (<200 nl). Following development, separated spots with 2–6 mm diameters are obtained. Smaller developed spots also place greater demands on the optical system of scanning densitometers used for automated detection. Beam broadening by light scattering at the layer surface is probably the limiting factor for resolution. Scanning densitometers remain the weakest link in high-performance systems.

The claims made above and in the literature for the advantages of HPTLC over TLC, including increased resolution, greater sensitivity, better reproducibility, greater number of samples per plate, combined with performance approaching that of HPLC have not been obtained without cost. In particular, the simplicity and robustness of ordinary TLC are not so evident in HPTLC. Instrumentalization of the process is essential to obtain the full benefit and is inferred in HPTLC. Conventional TLC is a labour-intensive discontinuous batch process whereas many of the steps, including spot application, chromatographic development, detection and chromatogram recording have become completely instrumentalized, and often automated, procedures in HPTLC. The slow growth of HPTLC can probably be attributed to the reluctance of analytical chemists to accept the price-tag of instrumentation despite their desire for high performance.

4.2.2 Comparison of TLC with other techniques

In comparisons of TLC with other techniques such as HPLC, the former is often characterized as being inexpensive. This is certainly true of conventional TLC which, at its most basic, required only the plate itself, a mobile phase and a suitable developing chamber. Evaluation of the plate is achieved with the aid of the human eye, aided if necessary by any one of the numerous spray reagents (of varying specificity) which are available. In this form TLC is simple, robust and has a large sample throughput. It is probably this simplicity that has caused TLC to be labelled, by some, with the stigma of a low-resolution technique of poor sensitivity which is, at best, semiquantitative. In contrast, a complete system for automated HPTLC requires a densitometer and automated spotter and can cost as much as an HPLC system. Nevertheless, excellent quantitative results can be achieved with much simpler systems at lower cost, but even with the highest-priced equipment HPTLC is often advantageous because of its simplicity, high sample throughput and versatility [1].

The separation mechanisms for TLC and HPLC are fundamentally the same. Physically, however, the separation processes differ in several ways. In HPLC, the stationary and mobile phases are usually in equilibrium prior to introduction of the sample, while in TLC the mobile phase usually encounters a dry layer at the start of the separation. In column techniques, such as HPLC, the flow of mobile phase results from the pressure difference between column entrance and column exit. Furthermore, the mobile phase velocity is electronically controllable up to a limit established by the maximum pressure gradient that can be maintained across the system. For conventional TLC and HPTLC, the flow of mobile phase is a result of capillary forces and cannot be controlled as it can in column methods.

One of the most important distinctions between TLC and column techniques is in the nature of the separation. Column techniques are time-limited whereas there is a spatial limitation in TLC. Thus, in HPLC all components migrate the same distance and become separated in time, while in TLC all components have the same separation time but are separated in space. In theory, HPLC separations are not limited by time but practical considerations limit such separations to less than 30 min. Assuming a peak width of 10 s at baseline, the HPLC chromatogram can accommodate approximately 200 peaks. On the other hand, the spot capacity or separation number in TLC is defined as the number of spots that could be separated between $R_f = 0$ and $R_f = 1$ with a resolution of unity. Spot capacity represents the maximum number of sample components that can be separated in a system. This is more difficult to calculate than the equivalent value for a column system because the value in TLC is a complex function of the characteristics of the chromatographic system [8, 9]. Solving this problem for different sets of conditions of initial spot size, particle diameter and migration distance indicates that it is very easy to achieve a spot capacity between 10 and 20, but extremely difficult to reach 25 by uni-dimensional development under capillary-controlled flow conditions. Theoretical calculations indicate that under forced-flow development conditions, it should be relatively easy to generate spot capacities well in excess of 500. Layers with very fine particles [ref. 8, p. 81] can ideally separate 13 spots on 1 cm. However, considerable practical difficulties remain before these separation capacities are realized. The primary difficulty is in the detection process until optical scanners can resolve such hyperfine chromatograms.

It is patently absurd to claim that TLC is superior to HPLC or GC (or any other technique) in all applications, but there are many circumstances where it offers a viable alternative. Both TLC and HPTLC have an important role in analytical chemistry which is complementary to the other chromatographic techniques. Among the unique advantages of TLC are the ability to chromatograph several samples simultaneously, thereby reducing the time per analysis, and the fact that the chromatogram provides a permanent record. TLC also gains certain advantages because it is an open bed system which offers a visual display of the entire sample, while column techniques are closed bed systems. In the latter case, the chromatographer may be unaware (at least in the short term) of sample components that are strongly retained. The processes of separation and detection are usually physically separated in TLC, unlike HPLC. Consequently the selection of the mobile phase does not limit the choice of detector. For example, u.v.-absorbing solvents cannot be used with u.v. detectors in HPLC, whereas this is not a problem in TLC because the plate is dried prior to detection. Hence, the separation and detection processes can be optimized independently in TLC.

4.3   Theoretical Considerations

The movement of the solvent front with time in TLC is governed by a quadratic law:

$$z_f^2 = \chi t \qquad 4.1$$

where z_f is the distance from the immersion line to the solvent front, t is the time elapsed since the start of the development and χ is the flow constant or velocity coefficient. For example, on a silica gel 60 TLC precoated plate with toluene as developing solvent, the flow constant has a value of $6.5 \, \text{mm}^2 \, \text{s}^{-1}$ while on a silica gel 60 HPTLC precoated plate with the same solvent the value is $5.0 \, \text{mm}^2 \, \text{s}^{-1}$.

After differentiation of equation 4.1, an explicit relationship is obtained for the migration velocity of the solvent front, u_f at any position of z_f:

$$u_f = \chi/2z_f \qquad 4.2$$

This expression describes the fact that the front velocity decreases hyperbolically with increasing migration distance of the front. The data in Table 4.2 are calculated using equations 4.1 and 4.2 for a hypothetical HPTLC silica plate, with linear development and no preadsorption on the layer.

The measurement of χ is not easy and requires some experimental precautions, such as choosing the correct developing chamber (S type in this case, see below), sealing it and maintaining a constant solvent level. Nevertheless, Geiss [8] has described a convenient method for its measurement and has collected relevant values. The factors influencing the velocity coefficient, χ, have been studied in detail. It is dependent on the surface tension, γ and the viscosity, η of the mobile phase as well as on the nature of the stationary bed:

$$\chi = 2k_o d_p(\gamma/\eta) \cos\theta \qquad 4.3$$

This expression describes the migration when an equilibrium exists between the adsorbent layer, the mobile phase and the vapour phase. If true equilibrium does not exist during the development process, then complex correction factors must be

Table 4.2 Data for the front velocity on silica layers using different solvents.

Solvent	χ (mm^2 s^{-1})	t(min)	z_f(cm)	Momentary velocity, u_f (cm s^{-1})
Ethanol	3.1	0.5	0.96	0.016
		1	1.36	0.011
		5	3.05	0.0051
		10	4.31	0.0036
Dichloromethane	8.7	0.5	1.62	0.027
		1	2.28	0.019
		5	5.1	0.0085
		10	7.2	0.0060
Acetonitrile	11.4	0.5	1.84	0.031
		1	2.62	0.022
		5	5.8	0.0097
		10	8.3	0.0069

Calculated values are illustrative only. Development distance for HPTLC plates rarely exceeds 10 cm with 3–6 cm being normal.

Table 4.3 Changes in the velocity coefficient as a result of variation in wetting angle of RP-18 reversed-phase plates with water–ethanol mixtures [11].

Concentration of water (%v/v)	$\cos\theta$	$\chi \times 10^3$ (cm^2 s^{-1})
0	0.87	16.6
4	0.78	13.4
10	0.76	12.3
20	0.61	7.8
30	0.48	5.5
40	0.34	3.5
50	0.14	1.5

included in the equation [10]. The value of χ also depends on whether or not the layer has been exposed to solvent vapours prior to development. As soon as pre-adsorption occurs, χ increases because the solvent front moves faster. In equation 4.3, k_o is a permeability function reflecting the external porous structure of the layer material [11], d_p is the particle diameter and θ is the contact or wetting angle. For all organic solvents examined on non-bonded phases, θ is less than 10° and hence cos θ is larger than 0.985; that is, for all practical purposes, equal to unity. For aqueous organic solvents on reversed-phase plates, the contact angle may be much larger and therefore the rate of solvent migration can be quite slow. Table 4.3 illustrates the dramatic reduction in the migration rate with increasing water content of an ethanol–water solvent mixture, owing to the corresponding increase in the wetting angle. It should be noted that high water contents in the solvent may result in zero flow rates because of the problems of wettability and, in some cases, to swelling and/or flaking of the precoated reversed-phase layer. This drawback could be partly overcome by adding inorganic salts, principally sodium chloride, to the solvent. Manufacturers are constantly improving their products and the production of a reversed-phase plate that can tolerate high water contents has been a high priority. Wetting and front velocity are improved when a surfactant is added to the aqueous solvent [8].

As implied by the above description, the solvent velocity is not constant and decreases the further the solvent has migrated. Eventually, the velocity drops to zero (for ascending development) which effectively limits the bed length available for the separation and puts a limit on the maximum size for TLC plates.

In addition to the fact that solvent velocity is not constant, it cannot be easily controlled and thus may not be optimum for good chromatography. This is not the only variable in TLC which is difficult to define accurately or to control experimentally. Indeed, under the usual experimental conditions in (HP)TLC, both the stationary and mobile phases remain ill-defined and exist in a state of flux with the vapour phase. The stationary phase may be modified by contact with vapours both prior to, and during, the chromatographic separation. Once in the developing chamber, the sorbent layer progressively sorbs the vapour phase prior to being wetted by the mobile phase. In addition, the solvent does not penetrate the sorbent layer evenly. The solvent flowing in fills the narrower interparticle channels first because of stronger capillary forces. The larger pores below the solvent front remain empty for some time and fill progressively with the liquid flowing from the smaller pores behind. This flow differentiation leads to a frontal volume gradient; that is, the concentration of solvent per unit layer volume continuously decreases towards the solvent front, finally reaching zero. Giddings *et al.* [12, 13] termed the phenomenon 'concentration gradient' but the term 'volume gradient' is preferable to distinguish it from the concentration gradients of solvent components which can form along the plate when developing with solvent mixtures. The two phenomena are of entirely different origin and the volume gradient is present even with single-component solvents. If the vapour and mobile phase are not in equilibrium, evaporation will cause a loss of mobile phase from the surface of the layer. With mixed solvents, preferential evaporation and sorption effects can result in concentration gradients as distinct from volume gradients.

It is difficult to operate TLC by the best theoretical principles because of the number of factors that are difficult to control. For this reason, the operation of TLC has been more empirical than column techniques. Nevertheless, a comprehensive theory of TLC has been developed [8], although most theoretical treatments have been borrowed from GC. Performance in TLC can be evaluated in terms of such parameters as the number of theoretical plates, height equivalent of a theoretical plate and separation number (see Chapter 2). These parameters have been more widely adopted for column techniques and it is therefore necessary to highlight some of the special features of the TLC process. In TLC, the chromatogram is developed for a fixed time and all substances have the same diffusion time but different migration distances. Thus, measures of chromatographic performance must be correlated to the migration distance of the substance. Their numerical values are evaluated for a specific value of R_f and are thus dependent on their position on the layer. The plate height can, in principle, be calculated from a complex formula obtained after integration of the variable 'local plate heights' along the separation distance [8]. However, the practising chromatographer is rarely concerned with the calculation of performance parameters for TLC and thus equations for their calculation are not presented. Nevertheless, the following conclusions should be noted [10, 14–17]:

- The measured plate height in TLC is an average value.
- Average plate height increases with particle diameter, with increasing diffusion constants of the solute and, in particular, with increasing development time, especially for layers with fine particles ($d_p < 5 \, \mu m$). This difference is a result of the fact that the local front velocity never reaches the optimal velocity on fine particle layers. For example, with a $5 \, \mu m$ particle layer the optimal velocity is 0.035 cm s^{-1}. However, normal flow never reaches this value; at $z_f = 0.5$ cm its front has a velocity of 0.025 and at $z_f = 5$ cm it has fallen to 0.005 cm s^{-1}. With a coarser $15 \, \mu m$ layer, the solvent velocity at the start exceeds the optimum value, but then, although progressively falling below this value, it stays fairly close to it up to $z_f = 5$ cm [8]. Hence, the smaller particle layer would require a separation path shorter than $z_f = 1$ cm for the optimal velocity. Under normal operating conditions, the layer cannot effectively use its potential efficiency.
- Layers with wider particle diameter distribution are undesirable, since they have the high plate height values of the coarse fractions and the low mobile phase velocity of its fine fractions.
- Molecular diffusion (van Deemter B term) controls the average plate height for layers with particles of small diameter and long front migration distances while for coarse particles it is the packing (A term) below about $z_f = 10$ cm. The mass transfer term is negligible for small particles.
- The average plate height increases almost proportionally with migration distance in the B term of the van Deemter equation, while the other two terms decrease steadily with increasing development length because of the decreasing solvent velocity. Eventually, the B term always becomes dominant and when this occurs the total plate number no longer increases with development length and no further gain in resolution is achieved. This occurs typically when $5 \, \mu m$ plates are developed beyond 5 cm, in agreement with accepted practice.
- The classical resolution equation (Chapter 2) can be modified for TLC [14] and differentiated to show that optimum resolution is obtained when R_f is about 0.3 and that a solvent should be chosen to achieve this performance. A consequence of this optimum is that solute pairs with R_f values significantly different from 0.3 need increased selectivity to be resolved to the same degree as at the optimum. However, the resolution does not vary significantly with R_f values ranging between 0.2 and 0.5; within this range, the resolution is greater than 92% of the maximum value (Fig. 4.1). Outside this range, the resolution declines rapidly, illustrating the strong correlation between separating power and the position of the spots in the chromatogram.

4.3.1 Forced flow conditions

The discussion thus far has assumed movement of the mobile phase is under the influence of capillary forces. The main drawbacks are varying flow-rates which are not at optimum most of the time and relatively long running times for comparatively short separation lengths. A newer method of development, called overpressured

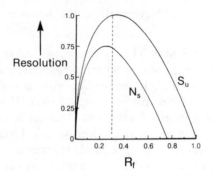

Fig. 4.1. Change in resolution of two closely migrating spots as a function of the R_f value of the faster-moving spot for a saturated N-chamber (N_s) and unsaturated S-chamber (S_u). The optimum is at $R_f \approx 0.3$ for the S_u chamber.

TLC (OTLC), forces the mobile phase through the stationary phase bed. A close examination of OTLC performance reveals the following. Under forced flow conditions, the distance travelled by the solvent front is proportional to time and does not fall hyperbolically as in capillary-driven layers. Moreover, the mobile phase velocity is constant and independent of the migration distance. Equally important is the fact that the velocity can be controlled and hence optimized (about 1.5 cm min^{-1}) for the minimum average plate height values of 15 μm for HPTLC and 35 μm for older TLC plates. There is no optimum front length as for traditional capillary-driven TLC. Moreover, in contrast to ascending TLC, there is a fairly regular linear increase in the number of theoretical plates, the separation number and resolution with solvent migration distance [8]. This is illustrated in Fig. 4.2. The difference in behaviour between normal and overpressured TLC is attributed to the fact that spot broadening by diffusional processes plays a minor role at the relatively high linear mobile phase velocities commonly employed in OTLC.

4.4   Thin Layer Plates

Following the work of Stahl [18] in the 1950s, TLC became a popular technique. In the initial years, manufacturers sold adsorbents with widely varying characteristics and quality. Most users accepted this situation which did not change greatly following the introduction of precoated plates in 1965. Although some workers may still prepare homemade thin layer plates [19, 20], the vast majority of separations today are performed on commercial, precoated plates. This is because the preparation of plates is time consuming. Furthermore, manufacturers now supply a wide variety of precoated plates including reversed-phase layers and the product is usually more consistent than that made in the laboratory. The exacting conditions required for the preparation of high-performance plates are difficult to achieve and their preparation is best left to plate manufacturers. Most TLC plate manufacturers now produce plates for HPTLC but the quality of some high-performance plates leaves much to be desired and care is required in the selection of these plates.

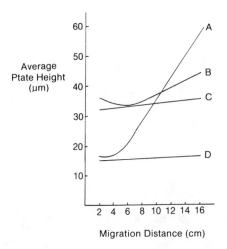

Migration Distance (cm)

Fig. 4.2. Comparison of normal and overpressured TLC showing the variation in average plate height with migration distance. A and D, HPTLC plate; B and C, conventional TLC plate: A and B, normal development; C and D, overpressured development. Reprinted with permission from Minscovics *et al.* (1980) *J. Chromatogr.*, **191**: 293.

Selectivity and efficiency can vary significantly between nominally equivalent materials from different manufacturers.

Commercial TLC plates currently available have layers ranging in thickness from 100 to 2000 μm coated onto glass, aluminium or plastic (usually polyethylene terephthalate) backing sheets. Glass plates are most popular because glass provides an inert, relatively easy to handle support. Plastic plates are unbreakable, lightweight and usually the least expensive because they can be manufactured in rolls. Moreover, plastic sheets are typically 0.2 mm thick, require little storage space and can be cut to any size desired with scissors or a guillotine. However, some solvents interact with the plastic sheet and oligomers which interfere with fluorescent indicators used for detection can be present as a result of the manufacturing process.

The majority of plates are sold with a binder such as gypsum (designated G, e.g., silica gel G), starch or polyesters to impart the desired mechanical strength, durability and abrasion resistance to the sorbent layer. Plates with polyvinyl alcohol binder are very stable and will withstand rather rough handling. It must be remembered, however, that binders will modify the sorption behaviour of the stationary phase and can produce somewhat different separations. Furthermore, a binder used for coating glass plates may have different separation characteristics when used as a binder for coating plastic or aluminium plates. Other options available with precoated plates usually include 'with fluorescent indicator' (F or F254) or without. The 254 used in this designation signifies that maximum fluorescence will be observed at an excitation wavelength of 254 nm. When such plates are viewed under u.v. light, many analytes appear as dark spots against a fluorescent background. The indicators used as phosphors (usually inorganic sulfides) become a part of the stationary phase, but at the levels used do not normally alter the separation.

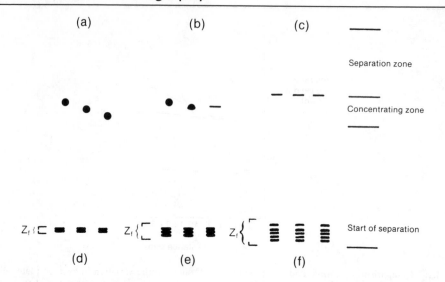

Fig. 4.3. Hypothetical chromatogram showing different phases of separation and concentration on an HPTLC precoated layer with concentration zone. Phases of development are: (a–c) concentration showing the initial application of three samples followed by sequential concentration; (d–f) separation at different migration distances.

A number of specialized designs of plates have been introduced, including plates divided into a series of parallel lanes or channels (generally approximately 1 cm in width) that restrict lateral zone diffusion. Most manufacturers also supply precoated plates with 'concentration zones' designed to increase separation efficiency when large sample volumes are applied. The sorbent materials used for the concentration zone are inert kieselguhr preadsorbent, a large-pore silica gel or a reversed-phase material. The concentration zone extends about 25 mm in the direction of development and merges into the main layer at a sharply defined border, which offers no resistance to the flow of solvent. The concentration zone simplifies the process of sample application since relatively large sample volumes can be applied to the zone (Fig. 4.3). Alternatively, the entire zone can be immersed in a dilute solution of the sample. During development, the sample migrates out of the concentration zone and becomes focused at the interface of the two layers. These plates are also very useful for crude sample extracts, as any insoluble residues are left in the concentration zone and do not interfere with the separation.

4.5　Stationary Phases

TLC has traditionally been performed on a limited range of stationary phases; namely silica gel, alumina, kieselguhr, polyamide and cellulose. Silica gel has been by far the most popular phase and is likely to remain so in the immediate future. However, in the last decade the range of available phases has been extended greatly to include normal and reversed-phase bonded layers comparable to the materials in use for HPLC. Stationary phase materials can be classified in several ways; a

Table 4.4   Classification of stationary phases for thin layer chromatography.

General class	Typical examples
Polar inorganic (hydrophilic)	Silica, alumina, magnesia
Non-polar inorganic	Graphite, charcoal
Non-polar bonded phases	C2, C8, C18 bonded phases
Polar bonded phases	Aminopropyl, cyanopropyl, diol phase
Polar organic	Cellulose, polyamide, chitine

reasonable system seems to be that of Table 4.4. The term sorbent is used as a generic descriptor for the layer material.

4.5.1   *Polar inorganic materials*

Straight phase separations on polar inorganic materials can be adequately described by adsorption processes. With polar adsorbents, non-polar solutes exhibit little affinity for the adsorbent surface and are not retained. In contrast, polar or polarizable (e.g. aromatic compounds) solutes are strongly retained because of relatively strong dipole-dipole and dipole-induced dipole forces. Thus, the order of solute elution on a polar adsorbent is, predictably, in the order of the polarity/polarizability of the solute functional groups, that is, alkanes < alkenes < aromatics, halides < ethers,   polynuclear   aromatics,   nitriles,   nitro   compounds < esters < ketones < aldehydes < amines < alcohols, phenols < amides < carboxylic acids. As an illustration of solute polarity effects, in the separation of lipids on thin layers of silica gel (see Table 4.5), the retention increases in the series cholesteryl oleate, methyl oleate, triolein, oleic acid, cholesterol, lecithin reflecting the increasing polarity from cholesteryl oleate to lecithin.

Silica gel

In addition to being the most popular adsorbent for TLC, microparticulate silica gel is the usual base material for the synthesis of most chemically bonded phases. Silica gel is prepared by dehydration of silicic acid, generated by the reaction of sodium silicate with a mineral acid. The material produced by this process is an amorphous

Table 4.5   Retention data of lipids on silica gel.

Lipid	R_f
Cholesteryl oleate	0.89
Methyl oleate	0.64
Triolein	0.47
Oleic acid	0.31
Cholesterol	0.06
Lecithin	0.0

Mobile phase: hexane + diethyl ether (90 + 10).

Table 4.6   Physical properties of silica gels used to prepare high-performance plates.

Property	Whatman HP-K	Merck Si 60
Mean particle size (μm)	5	5
Mean pore diameter (nm)	8	6
Specific pore volume (ml g^{-1})	0.70	0.82
Specific surface area (m^2 g^{-1})	300	550
Stable pH range	1–8	1–8

porous solid with a surface area that can vary from 200 m^2 g^{-1} to more than 1000 m^2 g^{-1}. The surface contains a number of different species which function as active sites. These species consist of (i) siloxane groups (–Si–O–Si–), (ii) silanol groups, –Si(OH)–, (iii) water, hydrogen bonded to the silanol groups and (iv) nonsorbed capillary or bulk water. The silanol groups on most TLC silica gels are free, with a relatively small population of self-hydrogen bonded adjacent silanol groups. Activation of silica gel plates by heating at 150–200°C removes the physically and reversibly bound water, whereas heating to higher temperatures (400°C) converts adjacent silanol groups to siloxane groups via the abstraction of chemically bound water. Such irreversible changes can alter the structure and lead to a sorbent with different properties.

The physical properties of two silica gels used to prepare high-performance plates are summarized in Table 4.6. The pore system (and hence the specific surface area), the skeletal texture (the ratio of silanol groups to siloxane groups) and the deactivation of the silica surface by adsorption of polar substances (e.g., the uptake of moisture) govern the particular separation characteristics of silica gel. Silica types are frequently denoted by their mean pore size (expressed in angstrom units). The mean pore sizes of silicas used in adsorption chromatography lie mostly within the range 4–12 nm (40–120 Å). The parameters used to characterize the pore structure, that is, mean pore diameter, specific pore volume and specific surface area, are mutually dependent. These parameters determine the particular selectivity of a sorbent whereas separation efficiency is determined more by two secondary parameters: the mean particle size and particle size distribution.

Alumina

Alumina or aluminium oxide is the second most popular adsorbent for TLC. It is prepared from hydrated aluminium hydroxide by thermal removal of water. A variety of crystalline forms exist, depending on the starting material and the dehydration process. These forms differ in their chromatographic properties as a result of differences in surface area, surface energy and pore size. The material for TLC is obtained by low-temperature (200–600°C) dehydration and has a specific surface area of 50–250 m^2 g^{-1} with specific pore volumes between 0.1 and 0.4 ml g^{-1}. Table 4.7 shows the ranges of physical parameters relevant to the chromatographic behaviour of aluminas.

Aluminas are available for TLC both as bulk sorbents, and as precoated layers with variously adjusted pH values. Basic alumina signifies a pH of about 9.0–10.0; neutral signifies a pH of about 7.0–8.0, while acidic aluminas are adjusted to a pH

Table 4.7 Physical parameters of aluminas for thin layer chromatography.

Mean pore diameter (nm)	Specific pore volume (ml g^{-1})	Specific surface area (m^2 g^{-1})
2–35	0.1–0.4	50–250

Table 4.8 Physical parameters of kieselguhrs for thin layer chromatography.

Mean pore diameter (nm)	Specific pore volume (ml g^{-1})	Specific surface area (m^2 g^{-1})
1000–10 000	1–3	1–5

within the range of approximately 4.0–4.5. The chromatographic properties of the sorbent will vary for specific substances between the different aluminas. For example, acids with a pK_a lower than about 13 are strongly adsorbed on basic alumina, whereas the selective adsorption sites of such acids are eliminated on acidic aluminas.

Others

The diatomaceous earth material called kieselguhr is composed of the skeletal material of siliceous algae. The material for chromatography is extensively purified. Nevertheless, kieselguhrs can differ significantly in their chemical composition, as some impurities stay within the bulk phase of the sorbent, possibly influencing the chromatographic behaviour of the material. The range of physical parameters relevant to the chromatographic behaviour of kieselguhrs are shown in Table 4.8. Numerous other inorganic adsorbents have been used to achieve specialized separations. Of these materials, magnesium and calcium silicates and phosphates are worth noting.

Modified adsorbents

TLC on silica gel or alumina is still the method of choice for many separations. The flexibility in choosing a developing solvent provides enormous variation in sample retention and selectivity. Nevertheless, in some instances, polar adsorbents may lack the necessary selectivity for a particular separation (e.g., a series of homologues), or very polar sample components may bind too strongly to the surface and remain too close to the origin. Poor migration characteristics, such as streaking, may also be a problem. The simplest solution is to modify the chromatographic properties of the adsorbent by applying a liquid stationary phase to the sorbent surface.

There are basically two methods of applying a liquid stationary phase to a sorbent in TLC: (i) Impregnation of the sorbent layer before chromatographic development. Substances used to prepare hydrophilic layers by impregnation include dimethyl-formamide, dimethylsulphoxide, and ethylene glycol while lipophilic layers are prepared with, for example, silicone oils, ethyl oleate and undecane. Preparation of these layers is achieved by immersion of the sorbent layer in a solution of the impregnating reagent. The plate is removed, excess solution is allowed to drain and

Table 4.9   Physical parameters of Silica 50 000 for thin layer chromatography.

Mean pore diameter (nm)	Specific pore volume (ml g^{-1})	Specific surface area (m^2 g^{-1})
5000	0.6	0.5

the volatile solvent is evaporated. (ii) A liquid stationary phase can be generated, *in situ*, by sorption on the sorbent layer during chromatographic development.

Separations on liquid stationary phases, whether generated by impregnation or *in situ*, are best described as partition. Unlike adsorption chromatography, effects caused by the gas phase play only a minor role. This is because the active sites responsible for gas phase effects are covered with the liquid stationary phase producing a partition system. Separations on silica using polar developing solvents incorporating an alcohol and/or water are also best explained by partition processes. The materials used for partition chromatography are virtually the same products used for straight phase separations. There is a tendency in partition chromatography to favour the slightly larger pore size silicas (pore size above 6 nm) with a correspondingly lower specific surface area. An example of a separation effected by partition chromatography is the analysis of underivatized amino acids on silica gel 100 using *n*-propanol–water–ammonia solution as developing solvent. This is an instance of the *in situ* application of a liquid stationary phase to the sorbent layer. Kieselguhrs and the synthetic Silica 50 000 (Table 4.9) are widely used as supporting sorbents for partition chromatography. Being synthetic, Silica 50 000 has a high chemical purity. It has similar chromatographic properties to kieselguhr which it is designed to replace.

An alternative method of modifying the chromatographic properties of a sorbent is the incorporation of a chemically selective reagent in the sorbent layer. There are many examples described in the literature, but the best known are silver nitrate impregnation for the separation of saturated and unsaturated compounds by selective bonding of the silver ions with the π electrons of double bonds, and boric acid impregnation for the separation of vicinal bifunctional isomers by chelation. Chelation effects are also exploited in the separation of metal ions on layers impregnated with complexing agents such as salicylic acid.

4.5.2   Non-polar inorganic materials

Among the non-polar inorganic sorbents, charcoal and graphite are worthy of note, but even these two materials have found only limited use, partly because of the difficulty of locating the separated spots on the adsorbent layer.

4.5.3   Non-polar bonded phases

The popularity of bonded phase materials in HPLC has led to their application in TLC. Bonded phases are particularly useful for separating compounds that differ in molecular mass (e.g., homologues). Non-polar bonded phases are mostly used in the reversed-phase mode, which normally involves a relatively non-polar stationary phase used in conjunction with very polar solvents. The operation of TLC in a reversed-phase mode is not new, but the bonded phases make it more practical. Nevertheless, it appears unlikely that this mode will become as popular in TLC as it is in HPLC. The reasons for this are not hard to find. Problems of poor reproducibility and rapid loss of performance in separations on silica gel in HPLC result from the activation or deactivation of the silica gel in the column by the continuous flow of the mobile phase. Over a period of time, this can result in changes in peak shape and retention. The solution to this problem in HPLC has been the use of reversed phases. In contrast, changes in the chromatographic properties of the sorbent are of little consequence in TLC as the plate is only used once.

Nevertheless, there are incentives for using bonded phase layers in TLC. A notable advantage of reversed-phase systems, for example, is the wide range over which a linear relationship exists between R_f and the eluotropic strength of the solvent. This simplifies method development greatly and is in marked contrast to the situation with silica gel where the linear range is narrower and the slope steeper. Under these circumstances, minor changes in solvent composition can have dramatic effects on R_f and spot shape. Moreover, humidity is not a parameter with bonded phase layers whereas for reproducible results on silica and alumina, humidity must be controlled. There are a number of instances where reversed-phase separation is superior [21] to other mechanisms, particularly with compounds of limited stability on silica gel. Moreover, there is much interest in using TLC as a screening procedure and transferring the optimized conditions to a column separation, where bonded phases are the norm.

Almost all bonded phase materials have used silica as the base material. The various methods for attaching a bonded phase to a siliceous support rely on the silanization reaction of surface silanol groups. Non-polar phases involve silanization of alkyl or phenyl groups to the silica surface via –Si–O–Si–R bonds. As the chain length of the bonded alkyl group increases, the sample retention likewise increases (R_f values decrease). The extent of reaction or coverage of surface silanol groups by the hydrocarbon can vary and nominally equivalent materials can display different chromatographic properties. Macherey–Nagel, for example, supplies octadecyl-coated plates prepared from silica gel with 50% and 100% silylation. Catalogues do not specify what the percentage silylation refers to, but it is unlikely that it represents the percentage of silanol groups reacted.

The application of reversed-phase TLC has been limited by the hydrophobic nature of commercially available phases. This makes the use of mobile phases containing large amounts of water impractical. Recently, a fully wettable non-polar phase has been introduced which can be used without restrictions on the solvent composition.

4.5.4   Polar bonded phases

Bonded phases with polar ligates (e.g., amino, cyano and diol functional groups attached by alkyl chains to the silica skeleton) can be used in the normal-phase mode in much the same manner as polar adsorbents. These materials are not available as bulk sorbents for manual spreading but only as high-performance precoated plates.

4.5.5   Other phases

Microcrystalline (e.g., Avicel) and native, fibrous (e.g., Machery–Nagel) precoated cellulose plates provide separations comparable to those obtained by PC. Microcrystalline celluloses offer a number of advantages with respect to development times and separation performance as a result of their rod-like molecular structure. Unlike silica gel and alumina, which are most effective for the separation of lipophilic substances, cellulose and other organic sorbents are used almost exclusively for the separation of very polar hydrophilic compounds, where the process is best described as partition. This is consistent with the low specific surface area of up to approximately $2 \, m^2 \, g^{-1}$.

In addition to the celluloses, polyamide sorbents are also used in partition TLC. Bulk polyamides and precoated polyamide layers are supplied by a number of manufacturers.

Empore TLC sheets have been available for several years but these layers have not yet found wide favour. They are prepared from silica or silica bonded phases entrapped in an inert matrix of PTFE microfibrils. The average particle size of Empore TLC sheets is $8 \, \mu m$. A larger sample capacity has been achieved by using silica with a mean pore size of 6 nm and a surface area of approximately $500 \, m^2 \, g^{-1}$ without any binder. Empore sheets are easier to spot than conventional layers and provide easier sample recovery from the developed layer. However, the performance of Empore sheets in terms of separation number is inferior to that of conventional layers using capillary flow. This can be explained by the very slow solvent migration velocity, which suggests that forced flow development is the application area for these sheets.

Reversed-phase silica gel plates impregnated with a chiral reagent and copper(II) ions have been used to separate optically active isomers based on ligand exchange. Applications described for these precoated plates include the separation of lactones, amino acids and their derivatives. The effectiveness of the plates for pharmaceutical applications is demonstrated by their use in analysis of optical purity of L-DOPA and D-penicillamine.

Various ion-exchange materials are available as precoated layers. In addition to a limited number of polymer-based materials, a number of chemically modified or impregnated celluloses have been designed as ion-exchange sorbents for precoated layers. Impregnated ion-exchange celluloses are: for anion-exchange, PEI (polyethyleneimine); and for cation-exchange, Poly-P (polyphosphate). Surface modified materials are: for anion-exchange, AE (aminoethyl), DEAE (diethylaminoethyl), and ECTEOLA (reaction product of epichlorhydrin, triethanolamine and alkaline cellulose); and for cation-exchange, CM (carboxymethyl) and P (phosphorylated). In addition, amino-modified silica phases can act as weak ion-exchangers with aqueous developing solvents.

4.5.6   Activity

The activity of the sorbent is an important parameter in straight phase TLC. Activity is understood as a surface property of the sorbent which, together with the parameters of adsorbent, solvent and temperature, give rise to a particular R_f value for a given substance. It has important practical consequences since a solute is adsorbed more strongly by an active sorbent, that is, with all other parameters held constant, an increase in layer activity will lead to a lower R_f, and a decrease to higher R_f.

Two components contribute to the activity of a sorbent. First, there is an energy contribution and second, a surface area contribution. The greater the surface energy per unit surface, the greater the adsorptive strength of the sites because of the stronger interactive forces between the sorbent and solute. The surface area determines the relative number of active sites per unit surface and hence the number of sorptive processes that can occur simultaneously.

Preloading

The chromatographic development process is usually regarded as a two-dimensional process, but this is an oversimplification in TLC which is an open system, unlike column techniques. The effect of the gas phase on TLC results must be considered. The term preloading is used to describe sorption of substances from the gas phase by the dry sorbent prior to chromatographic development. There is an additional gas phase effect during development in that the liquid rising in the layer capillaries can interact with the gas phase.

The activity of a sorbent can be varied by controlling the size of the surface on which the sorption phenomena occur. In practice, this is achieved by allowing the sorbent to sorb controlled quantities of a deactivator during preloading which reduces the 'free' surface of the sorbent. Deactivators for hydrophilic adsorbents such as silica and alumina are strongly polar substances (e.g., water, glycol) whereas lipophilic molecules such as toluene and stearic acid are used for non-polar adsorbents (charcoal, graphite). Complete coverage of the surface with a deactivator will completely deactivate an adsorbent. Sample molecules no longer have access to the adsorbent surface and adsorption chromatography is no longer possible.

Water is the most important activity regulator in adsorption TLC on silica gel and alumina. Activation/deactivation is performed almost exclusively via the gas phase where relative humidity is the controlling factor. Relative humidity is defined by equation 4.4,

$$\% \text{ Relative humidity} = \frac{\text{partial pressure of water (at } T°C)}{\text{equilibrium vapour pressure of water (at } T°C)} \qquad 4.4$$

This represents the degree of water vapour saturation of the atmosphere at a given temperature. Consider a practical example. The atmosphere may contain 10 g m^{-3} water with a vapour pressure of 1.33 kPa. At 30°C, where the equilibrium vapour pressure of water is 4.24 kPa, this corresponds to a relative humidity of 31%. By cooling the air to 15°C (equilibrium vapour pressure of water is 1.71 kPa), the relative humidity rises to 78%. Note that the absolute amount of water vapour in the atmosphere remains constant. However, exchange processes between the atmospheric water vapour and TLC adsorbents are controlled by the relative humidity and not by the partial pressure of the water.

Water and other adsorbed substances may be removed from the surface of an adsorbent by controlled heating; the process is called activation of the adsorbent. Thus, the activity of an adsorbent can be adjusted by taking the most active, totally dehydrated form, made by heating the layer to a suitable temperature (different for different adsorbents), and allowing the cooled layer to equilibrate in a specially designed tank with a prescribed relative humidity. How long does the layer retain this imposed activity when removed from the tank and exposed to another relative humidity? In answering this question, the kinetics of the activation–deactivation process must be considered. Unfortunately, considerable re-equilibration to the new activity occurs with many layers within 15 s and is 70% complete for almost all layers within 5 min. The practical consequences are:

- A plate stored in a desiccator at controlled relative humidity (e.g., over a salt solution) will assume the activity corresponding to the prevailing relative humidity of the laboratory atmosphere unless interaction of the layer with ambient air is prevented both before and during development.
- The widespread practice of activating plates at 110°C has no real merit.
- If special developing chambers are unavailable, more reproducible results are generally obtained on air-dry plates than on those stored in conditions of controlled humidity.

As a result of such considerations, many chromatographers use commercial TLC plates directly from the box, although the availability of developing tanks with humidity control may bring about a change in this practice.

In adsorption TLC, retention always decreases and R_f values rise with relative humidity because the adsorbent activity diminishes (Fig. 4.4). This dependence decreases with increasing polarity of the solvent and has become insignificant with methanol as a solvent. Of more importance, the magnitude of any R_f changes for different substances is not constant and, as a consequence, changing activity can affect both the selectivity of a separation and, in the extreme case, even the elution order.

Although water is the most common deactivator on straight phase polar sorbents (e.g., silica), other adsorbed molecules are capable of changing the chromatographic properties of the sorbent. This can be readily achieved by sorption of developing solvent via the gas phase in developing chambers specially designed to permit total saturation of the gas phase with solvent vapour. The properties of the sorbent can also be modified by preconditioning with substances not present in the developing solvent. An example of this type is given in Fig. 4.5, which shows reduced retention on treated layers relative to the equivalent nontreated layer.

Standardization of activity

Brockmann has devised an activity scale for adsorbents in column LC, based on the migration of test dyes in a short column after the movement of a defined amount of mobile phase. The procedure for determining Brockmann activity is readily available in a number of older texts. For the same type of adsorbent, Brockmann activity data from column LC are not related to a distinct R_f value. Activity scales have been defined for TLC adsorbents but are not widely used. Perhaps this failure to measure, control and standardize activity accounts for the occasional complaints about the alleged poor reproducibility of TLC data.

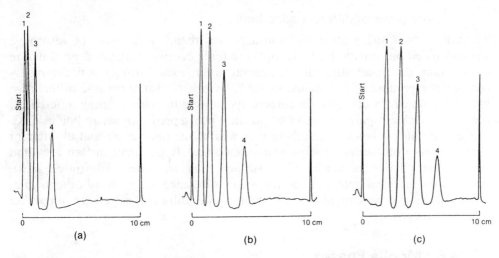

Fig. 4.4. Influence of relative humidity on the retention and resolution of some *m*-oligophenylenes: 1, *m*-quinquephenyl; 2, *m*-quaterphenyl; 3, *m*-terphenyl; and 4, biphenyl. Precoated silica gel 60F 254 plate developed with cyclohexane in a Camag Vario KS chamber. Camag TLC/HPTLC scanner, u.v. 254 nm used for detection. Plates were preconditioned at (a) 20% relative humidity, (b) 50% relative humidity or (c) 80% relative humidity. Reprinted from Hauck and Jost. In Unger K. K., Editor, *Packings and Stationary Phases in Chromatographic Techniques, Chromatographic Science Series, 1989 Vol. 47*, Chapter 5, Fig. 2. By courtesy of Marcel Dekker, Inc.

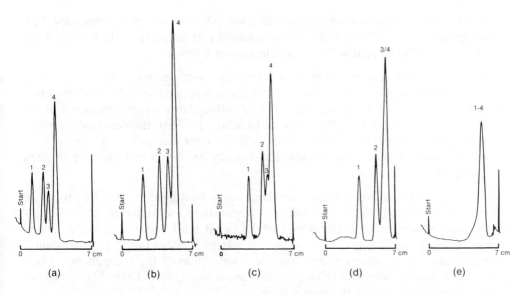

Fig. 4.5. Influence of preconditioning with different substances on the retention and resolution of some insecticides: 1, *p,p'*-DDD; 2, *p,p'*-DDT; 3, *o,p'*-DDT; 4, *p,p'*-DDE. Developing solvent: *n*-heptane. Preconditioning agents (a) without; (b) 25% ammonia; (c) 37% HCl; (d) methanol; (e) dichloromethane. Other conditions as for Fig. 4.4. Reprinted from Hauck and Jost. In Unger K. K., Editor, *Packings and Stationary Phases in Chromatographic Techniques, Chromatographic Science Series, 1989 Vol. 47*, Chapter 5, Fig. 4. By courtesy of Marcel Dekker, Inc.

Activity order of different adsorbents

There have been many attempts to arrange adsorbents in an order of adsorptive strength based on activity. Such attempts are futile because of the ill-defined nature of the term 'adsorbent strength'. A general order, established even under rigidly controlled experimental conditions, would have little relevance because of the large variations in activity of a given adsorbent type owing to manufacturing differences. Moreover, such comparisons are of no practical use in predicting separation in TLC. The activity of two types of adsorbent may well be the same as a result of different contributions from surface energy and surface area (e.g., small surface area and strong adsorptive sites versus large surface area and weak adsorptive sites), manifesting in equal retention of the same substance on both adsorbents but, nevertheless, the adsorbents may exhibit quite different selectivity for other substance groups.

4.6 Mobile Phases

Both the mobile and stationary phase evolve in TLC only during the course of the chromatographic process, through equilibria involving the layer, solvent and gas phase. Hence, solvent is not equivalent to mobile phase since the latter usually has a different composition, which frequently varies along the length of the plate. The liquid in the developing chamber is more aptly described as the solvent or developing solvent and the terms solvent and mobile phase should not be used interchangeably.

There is an enormous choice of possible solvents and solvent mixtures for TLC although a suitable solvent will possess a number of properties which restrict the choice. A suitable solvent will be characterized as follows:

- It should be available in a state of high purity at reasonable cost.
- It must be unreactive towards both sample components and the layer material.
- It should possess a suitable viscosity and surface tension (see equation 4.3)
- It should have a low boiling point to facilitate drying of the developed plate.
- Miscibility is obviously an important consideration with solvent mixtures.
- Properties such as flammability and toxicity are important because of their impact on safety and disposal.

The list of possible solvents is still very large, even bearing these general properties in mind. However, there are other considerations. Chromatographically, the developing solvent must perform two vital jobs: it must transport sample components through the layer; and it must by selective interaction with the sample components and stationary phase produce a separation of the components. These aspects of solvent selection relate to the solvent strength and selectivity, respectively. They are treated in the next section.

4.6.1 Mobile phase optimization

The sample, stationary phase and mobile phase represent the key parameters of LC. Of these, the mobile phase produces the most dramatic changes in a chromatogram and it is the most easily varied, within a wide array of strengths and 'selectivities'. The secret to success in TLC often depends on choosing the developing solvent correctly. Thus, optimizing a separation usually centres on the selection of the best developing solvent by adjusting solvent strength and selectivity to maximize separation. Adjusting the solvent strength so that the R_f values lie in the optimum 0.2–0.5 range is relatively easy, while adjusting selectivity is more complex and remains almost entirely empirical. It is not surprising that the developing solvent has often been selected by trial and error because results are obtained so quickly and easily in TLC. Nevertheless, there are alternative strategies for optimization of retention and selectivity and these can be classified into three groups:

1. Empirical approaches based on experience and intuition.
2. Statistical analysis of published data to permit informed successive trial-and-error changes of conditions.
3. Development of a comprehensive model or theory of retention that is valid for the greatest possible number of separation systems. Ultimately, this approach is the most promising, although a single comprehensive model has not yet been developed. The trend has been toward an ever increasing number of model approaches of increasing refinement and complexity. To date, retention processes for the majority of TLC separations can be adequately described by distinct models that consider the nature of the stationary phase and the mobile phase.

Before discussing the optimization of resolution, we must define what is meant by optimization. The simplest case involves a pair of substances whose separation is to be improved. This is relatively easy to achieve by trial-and-error guided, perhaps, by solvent strength data. The usual case is, however, more complicated in that there are several sample components whose resolution must be optimized simultaneously. In this situation, some compromise in the resolution of individual pairs of compounds may be necessary.

Normal-phase systems

Normal-phase systems involve the use of polar stationary phases in combination with nonaqueous mobile phases. Typical stationary phases include silica gel, alumina and various polar bonded phase materials (e.g., cyano-silica, amino-silica and diol-silica) where adsorption processes play a significant role in the separation. The developing solvent is usually a single-component non-polar solvent or a mixture of a non-polar solvent, such as hexane, with a polar modifier (for example, dichloromethane, various ethers, ethyl acetate, methanol or acetonitrile) for control of solvent strength and selectivity. With silica gel and alumina, small amounts of a third component, such as acetic acid, are often added to partially deactivate the surface and decrease tailing of polar sample components.

The competitive displacement model and solvent interaction model (Chapter 1) of solute retention provide a quantitative description of solvent strength. However, the calculations are somewhat involved and a more empirical approach will be adopted. Hildebrand's solubility parameter is a measure of the forces between molecules and, as such, has been used in an attempt to predict chromatographic retention. In normal phase systems, the strength of the solvent increases with increasing polarity

Table 4.10   Properties of common solvents used in thin layer chromatography.

Solvent	Solubility parameter (δ)	Polarity index, P'	Solvent strength (ε°)*
n-Hexane	7.3	0.1	0.00
i-Octane	7.0	0.1	0.01
Toluene	8.9	2.4	0.22
Ethoxyethane	7.4	2.8	0.29
Chloroform	9.3	4.1	0.31
Dichloromethane	9.7	3.1	0.32
Tetrahydrofuran	9.1	4.0	0.39
2-Butanone (MEK)	9.3	4.7	0.39
Dioxane	10.0	4.8	0.43
Acetone	9.9	5.1	0.43
Ethyl acetate	9.6	4.4	0.45
Acetonitrile	11.7	5.8	0.50
n-Propanol	11.5	4.0	0.63
Ethanol	12.7	4.3	0.68
Methanol	14.4	5.1	0.73
Acetic acid		6.0	Large
Water	21	10.2	Very large

*Compare Table 6.2.

whereas there is an inverse relationship for reversed phase systems. Thus, in normal phase systems, R_f can be increased (less resolution) by increasing the polarity of the developing solvent. The polarity index, P', introduced by Snyder measures the intermolecular attraction between a solute and a solvent whereas the solubility parameter is defined for pure solvent. The polarity index allows for different molecular interactions in the form of solvent proton acceptor, proton donor and strong dipole interactions. For example, ethoxyethane and n-hexane have similar Hildebrand values (Table 4.10). However, ethoxyethane is a proton acceptor through hydrogen bonding to its nonbonding electron pairs and, consequently, its polarity index is 2.9 compared with 0.1 for hexane.

The solvent strength, ε°, which is a dimensionless number ranging from -0.25 to about $+1.2$ is an even better index of solvent strength for normal-phase systems. It is determined by the adsorption energy of a solvent molecule for a given solvent–adsorbent combination. This implies that there are variations in the relative solvent strength among the various adsorbents. For example, more basic solvents are generally stronger when used with acidic silica or alumina while acidic solvents are stronger with amino-silica as the adsorbent. The term eluotropic series was introduced in 1940 for a series of solvents arranged in the order of increasing elution strength. A shortened series for silica and alumina is listed in Table 4.10. In such a series, the greater the solvent strength the larger is R_f, i.e. the weaker the retention, of a given solute.

The general strategy for optimization is to adjust the solvent strength to optimize R_f values in the desired 0.2–0.5 range by replacing a pure solvent by another one or, more commonly, by varying the proportions of a weak solvent and a strong solvent in a binary mixture. While maintaining the optimized solvent strength constant, the solvent composition is then modified to obtain the selectivity necessary to achieve the required separation. The change in solvent strength as a function of the volume

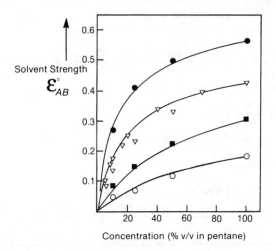

Fig. 4.6. Solvent strengths of binary solvent mixtures in pentane on alumina. Second components are: ○, tetrachloromethane; ■, 1-chloropropane; ▽, dichloromethane; ●, acetone.

per cent of the more polar component is not a linear function, as shown by the representative examples in Fig. 4.6. Particularly notable are the large increases in solvent strength produced by small increases in concentration of the polar solvent when the latter is present at small concentrations. Solvent selection is facilitated by the use of a nomograph such as in Fig. 4.7. For example, if 50% ether in hexane provides the right solvent strength, $\varepsilon^\circ = 0.30$, then any of the solvents falling on the vertical dashed line for $\varepsilon^\circ = 0.30$ are likely alternatives.

Additional information on the selection of an alternative developing solvent from the range of solvents with equivalent strength is necessary. The principal forces responsible for the chromatographic properties of a solvent are dispersion, dipole and hydrogen bonding. Dispersion forces are weak and nonselective and, therefore, not of any real chromatographic significance, except for hydrocarbons where they are the only forces operating. By selecting appropriate substances as probes, the strength of the other forces can be measured relative to a suitable hydrocarbon. The polarity index is then defined as the sum of the three experimentally determined forces or solvent selectivity parameters (χ_c, χ_d, χ_n) which reflect the strength of the interaction due to solvent proton acceptor, proton donor and dipole interactions, respectively. Solvents of similar selectivity can be grouped together by comparing their selectivity parameters. This is usually done diagrammatically in a selectivity triangle (see Fig. 6.15) which shows the various solvents grouped into eight classes (see Table 6.6). In choosing a replacement solvent to enhance selectivity, it is unlikely that a second solvent in the same selectivity group will enhance the separation. However, choosing another solvent with similar strength from a different selectivity group can alter the separation.

The process of selectivity enhancement can be illustrated by a hypothetical example. A separation was performed on silica using the solvent mixture ethoxyethane:hexane (48:52) in the R_f range about 0.4. The resolution was inadequate and

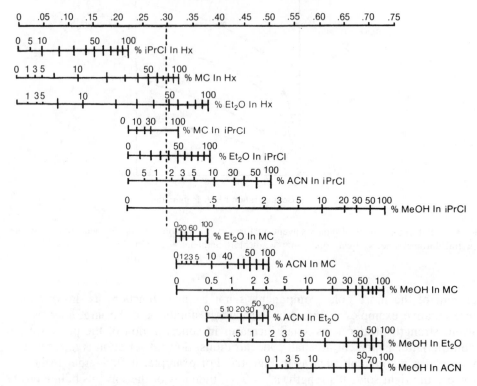

Fig. 4.7. Solvent strengths of some binary solvent mixtures on silica. Hx, hexane; iPrCl, isopropyl chloride; MC, dichloromethane; Et$_2$O, ethoxyethane; ACN, acetonitrile; MeOH, methanol. Reprinted with permission from Saunders (1977). *J. Chromatogr. Sci.*, **15**: 372.

a change in solvent selectivity was therefore necessary. This was most readily achieved by a greater change in the selectivity group of the polar component ethoxyethane belonging to group I (Table 6.6). From Fig. 6.15, groups V and VIII are most distant from group I, and therefore most likely to enhance selectivity. In this case, a reasonable choice might be dichloromethane and from Fig. 4.7, the required concentration is 77% v/v of dichloromethane in hexane to maintain solvent strength.

Binary solvent mixtures provide a simple method of controlling solvent strength but only limited scope for controlling solvent selectivity. Maximum selectivity is achieved with binary solvent mixtures of roughly equal solvent strength when one component is very strong and present only in very small concentrations (e.g. 0.05%). Alternatively, if both components are relatively weak, the stronger component must be present in high concentrations for maximum selectivity (e.g. hexane:toluene, 20:80). However, with complex multicomponent samples, it may not be possible to achieve the required separation with a binary solvent mixture. In such cases, ternary and quaternary mixtures provide more opportunities to fine tune the selectivity while maintaining constant solvent strength.

Various methods [22] have been designed to improve the optimization process in column LC but have not gained wide acceptance for TLC. One approach [23, 24] combines the above process with a mixture–design statistical technique to define the optimum developing solvent for a particular separation. This method of data analysis was called overlapping resolution mapping. A feature of this mixture–design technique is that it leads to the selection of a quaternary developing solvent system for most separations. The recommended solvent mixture for normal-phase separations on silica gel comprises methyl t-butylether, dichloromethane and acetonitrile with hexane as a nonselective solvent to control solvent strength. Further discussion of this approach is beyond the scope of this chapter and is probably not justified for routine purposes.

Polar bonded phases with cyano, amino or diol functional groups, when used with a mobile phase of low polarity, behave in a manner similar to the solid adsorbents. That is, the retention of the sample increases with solute polarity, and increasing the polarity of the mobile phase reduces the retention of all solutes. Polar bonded phase materials offer some notable advantages relative to adsorbents. For example, they are generally less retentive and are relatively free of catalytic activity and the problems of chemisorption.

Reversed-phase systems

Reversed-phase chromatography is characterized by the use of a polar mobile phase in conjunction with a non-polar stationary phase. The common developing solvent is a mixture of water and a water-miscible organic modifier, usually methanol, acetonitrile or tetrahydrofuran. The solvent strength increases with the concentration of organic modifier, and also generally increases with decreasing polarity of the modifier. Different eluotropic scales have been obtained, depending on the type of solute and stationary phase, but the following scale provides a general guide: methanol, acetonitrile < ethanol = acetone = dioxane < tetrahydrofuran, isopropanol.

Retention on reversed-phase materials is of greater complexity than TLC on polar adsorbents partly because of the broader range of operating conditions, but also as a result of the fact that the underlying retention mechanism has not been fully elucidated. With silica-based materials, because of the heterogeneous nature of the surface, a different mechanism can be operative at different regions of the surface. Furthermore, the mechanism is not necessarily the same over the whole range of mobile phase composition since the solvation of the various binding sites on the stationary phase surface is strongly affected by the mobile phase composition. For example, the residual silanols at the stationary phase surface play an increasingly significant role in the retention behaviour of polar solutes as the water content of the mobile phase increases.

With aqueous mobile phases, which are low in organic modifier content, the solvophobic theory [25] offers a valid interpretation of retention based on the assumption that the stationary phase is a uniform layer of a non-polar ligate. The model assumes that the solute binds to the stationary phase surface, which reduces the surface area of solute exposed to the mobile phase, and the solute is sorbed as a result of this solvent effect; that is, the solute is sorbed because it is solvophobic.

Retention in such systems is a function of nonspecific hydrophobic interactions of the solute with the stationary phase, whereas the selectivity results almost entirely from specific interactions of the solute with the mobile phase. This is not so with water-lean, organic-rich solvents where silanophilic interactions predominate. When solvophobic and silanophilic interactions act jointly, retention is determined by the spatial arrangement of the polar and hydrophobic sites on the stationary phase surface and in the solute molecule.

Systematic optimization in reversed-phase chromatography is not developed to the same extent as in normal-phase chromatography because of the complexity of reversed-phase systems. Nevertheless, the approach to solvent optimization discussed for normal-phase systems can be adapted to reversed-phases. The recommended quaternary solvent mixture comprises methanol, acetonitrile and tetrahydrofuran, with water as a nonselective solvent to control solvent strength. However, reducing the water content of the developing solvent not only reduces its strength, but is likely to alter selectivity as well. At this stage of development, trial-and-error is perhaps the best approach to solvent optimization for reversed-phase TLC systems.

Effect of solvent quality

Impurities in solvents can have a significant impact on a separation. Classical cases are the addition of stabilizers by manufacturers to chloroform, ethoxyethane (ether) and tetrahydrofuran and denaturants to denatured ethanol. Solvent mixtures should not be used more than 3–4 times because of the danger of changes in composition as a result of preferential evaporation of the most volatile component or preferential sorption of one or other component on the layer.

4.7 Sample Application

For TLC, a solution of the sample is applied to the layer as a spot or streak sufficiently small so as not to degrade the efficiency of the layer. The sample aliquot must not be so large as to cause overload of the layer. It is important that the application solvent wets the stationary phase layer, otherwise the sample will not penetrate the layer.

Sample application is still performed manually in most instances using micro-pipettes (e.g. Calibra Digital Micropipette) or glass capillaries made specifically for this purpose. Templates are available to facilitate the process of application. The disposable microcaps marketed by Drummond are available in sizes ranging from 0.5 to 200 μl. They are filled by capillary action when dipped in a solution of the sample and dispense when touched gently to the layer. Hard organic bound layers are less likely to be disturbed by the spotting process. Micropipettes which contain plungers for drawing in and expelling variable volumes of sample solution have the advantage of positive displacement and, therefore, do not deform the plate surface even with softer layers. With these devices, the needle is brought sufficiently close to the layer that the convex sample drop of ejected liquid touches the plate surface.

When it is necessary to apply larger sample volumes it is common practice to use repeated application of smaller aliquots to the same origin at intervals of several seconds. This approach ensures a compact initial spot. The sample solvent is dried with a stream of warm air from a hair drier, for example, between applications. Alternatively, plates with a concentration zone are easy to use and facilitate the process of sample application. In some situations, the application of the sample as streaks instead of spots can be advantageous. This is used primarily in preparative TLC because larger amounts of sample can be applied. Improved separation has also been achieved with sample application in bands. Manual application of bands is possible, although not easily performed. Commercial automatic applicators facilitate this operation and are essential for high performance.

Sample volumes for HPTLC must be in the nanolitre range to ensure spot sizes are sufficiently small (<1.5 mm) so as not to degrade layer performance. Various mechanical devices employing dosimeters, microsyringes, spray-on techniques and solid-phase transfer devices are used to deliver these small volumes. A dosimeter marketed by Camag, the Nanomat, automatically delivers 100–200 nl volumes through a platinum–iridium capillary sealed into a glass support capillary of larger bore. The capillary tip is polished to provide a smooth planar surface of small area which, when used with a mechanical applicator, does not damage the layer surface; manual use invariably damages the layer surface and causes blockage of the dosimeter. Dosimeters are suitable for most solutions of average viscosity, although some solvents do not wet the capillary adequately.

Micropipettes for HPTLC application use a micrometer screw gauge to deliver a variable volume (50–250 μl) or a fixed lever mechanism to dispense repetitively a selected constant sample volume. An advantage of the microsyringe over the dosimeter is that it delivers the sample volume by positive displacement rather than capillary action and thus does not deform the layer surface. The Transpot applicator is unique in that it provides for simultaneous solventless application of evaporated sample residues at precise locations on the plate. The sample (up to 100 μl) is placed in depressions in the applicator and the solvent is evaporated by gentle heat under a flow of nitrogen. The TLC plate is then positioned over the applicator and the spots are all transferred simultaneously to the plate. Problems caused by the nonquantitative transfer of crystalline samples are overcome by the addition of a small amount of a non-volatile solvent such as methyl myristate to the solution. This system is particularly adaptable for problem samples such as spotting dilute, viscous or nonhomogeneous samples.

There is scope for much improvement in the area of sample application. The sample application step is certainly the most tedious and probably the least reliable in the TLC procedure. Several companies have addressed the needs of small sample application and automatic sample spotters are commercially available that can apply exact volumes to multiple spots simultaneously [26]. Moreover, microprocessor-controlled automated sample applicators, which combine variable volume application as a spot or band with automatic applicator rinsing between samples, are available to minimize sample application time. Spot application with constant initial zone areas is a requirement for successful quantitative densitometry (Section 4.9). However, to date, no completely automated system is available that can perform the entire process, including sample application in a continuous fashion.

4.8 Development Techniques

Development in TLC refers to the process by which the developing solvent moves through the sorbent layer. Development in closed bed or column techniques is limited to the linear mode. This restriction does not apply to an open bed technique, such as TLC, and both radial and linear development are possible. Within this general framework there are a large variety of development techniques. These methods include continuous development, multiple development and its instrumentalized variants of programmed multiple development (PMD) and automated multiple development (AMD), continuous multiple development (which combines elements of continuous and multiple development), circular and anticircular development, overpressured TLC and two-dimensional TLC. The essential features of the more important of these techniques are presented in this section.

4.8.1 Linear development

Linear development represents the simplest situation and is the most popular mode. The samples are spotted about 1–2 cm from one edge of the plate and the developing solvent is allowed to migrate to the opposite edge. The chromatogram consists of compact symmetrical spots of increasing diameter as R_f increases (Fig. 4.8). Some distortion of spots near the solvent front is relatively common. Linear development has traditionally involved ascending movement of the developing solvent. Because there is no theoretical advantage to ascending development, horizontal developing chambers are becoming more common. These chambers have a number of advantages as described below.

4.8.2 Circular and anticircular development

Centrifugal and centripetal development have been used as alternative names for circular and anticircular development, respectively. Centrifugal is an unfortunate term here because of its connotations with circular motion. Furthermore, there is a variant of centrifugal TLC where the plate rotates. The term anticircular has been criticized as making little sense, but it is descriptive. The following hierarchical tree diagram summarizes the terminology used in this text:

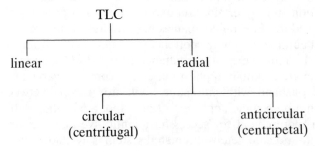

For circular mode, the sample is applied at, or near, the centre of the plate and the point of entry of the developing solvent is here also. This mode provides

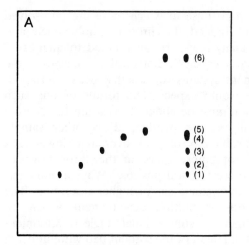

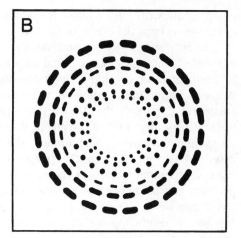

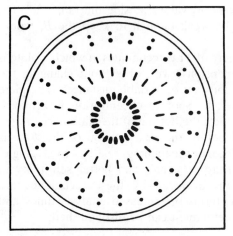

Fig. 4.8. Chromatograms showing appearance of spots following (A) linear, (B) circular and (C) anticircular development. Reprinted with permission from Fenimore and Davis (1981). *Anal. Chem.*, **53**: 252A. Copyright (1981) American Chemical Society.

considerable enhancement of resolution, for comparable development time, in the low R_f range (<0.5) compared with the linear technique. Thus, it is a powerful technique for separating samples where the components exhibit low R_f values. In anticircular development, the sample is applied along the circumference of an outer circle and developed towards the centre of the plate. The unique features of anticircular development are high-speed development (4 min for a migration distance of 45 mm) and the largest sample capacity per plate of any thin layer technique. Unfortunately, the anticircular mode also has the poorest separation performance per unit bed length of all TLC techniques. It is the method of choice for quantitative analysis of samples where the individual components are easily resolved.

4.8.3 Continuous and multiple development

There are many situations where the separating ability of linear and radial development is inadequate. Two situations commonly arise; samples containing components with very similar R_f values, and those in which there are unresolved components near the origin or solvent front. Indeed, the chromatographic uncertainty principle states 'No chromatographic analysis can be guaranteed to quantify a particular chemical substance and only that substance'. Statistical approaches to the problem of peak overlap in multicomponent systems suggest that peak overlap is, discouragingly, more common than one would expect. The results of one such treatment [27] imply that the chromatogram must be about 95% vacant to ensure a 90% probability that a given peak will not be overlapped by other sample components. With simple samples, the probability of peak overlap is lower as a result of the increase of the proportion of empty space in the chromatogram. However, contamination of the spot of interest is still possible. Ways around this problem can be the use of specific detection systems and/or higher resolution. The latter can be achieved by using continuous or multiple development, which are capable of producing higher resolution than the conventional single development modes. Both techniques make use of the full length of the sorbent bed without being subject to the deleterious influence of extreme R_f values outside the range 0.2–0.5. Using these techniques, the solvent must be weak enough to keep the R_f values relatively low (about 0.05 to 0.2).

Multiple development encompasses a variety of TLC techniques in which a plate is repeatedly redeveloped in one or two dimensions, with or without changing the composition of the developing solvent, for constant or variable distances. The plate is dried between each development sequence. Spot reconcentration is achieved during each development as the solvent contacts the bottom of the stationary sample zone, which commences to move forward before the next portion of the zone is contacted. This produces a reduction in zone broadening relative to a single, unidimensional development, thus giving sharper signals and better sensitivity in instrumentalized detection. The separations achieved by these techniques are certainly impressive but this is paid for in terms of increased analysis times and equipment cost. Three types of multiple development can be distinguished:

1. Repetitive development with the same solvent in the same direction, a technique designed to improve the separation of two neighbouring spots.
2. Repetitive development with different solvents in the same direction, a technique designed to cope with samples containing components with a wide polarity range. One common approach involves an initial development with a strong solvent to half the migration distance; more strongly retained substances are relatively well separated while weakly retained substances migrate at the solvent front. After drying, development is repeated with a weaker solvent, only this time using the entire migration distance. The substances separated in the first run remain relatively immobile while the unresolved components from the first run are separated on the upper part of the sorbent layer.
3. Two-dimensional development involving development with two different solvents in perpendicular directions. The sample is developed along one edge of the plate, the solvent is evaporated prior to developing a second time in a direction perpendicular to the first. The use of a different solvent for the second development permits the separation of one group of sample components in the first separation and another group of components in the second.

Two-dimensional development (3 above) is used when sample throughput is not the main concern. Unidimensional development (1 and 2 above) has the advantage of preserving the ability of the plate to handle multiple samples.

PMD is an instrumentalized high-performance technique in which a plate is repeatedly developed with the same solvent in the same direction for gradually increasing migration distances. Solvent is removed from the sorbent layer between developments by controlled evaporation while the plate remains in contact with the solvent reservoir. A typical separation sequence generally involves 10–25 individual developments and each consecutive run goes over a further 3 mm. This corresponds to a total developing time of 0.5–3 h and a total bed length of 3–10 cm. The full benefit of this technique is only achieved when successive runs are made with decreasing solvent strength, that is, by using a solvent gradient. For silica gel, development starts with the most polar solvent and concludes with the least polar solvent. Thus, solvent gradient PMD is suitable for the identification of isocratic solvent systems for poorly resolved substances, and for optimization of separation conditions for samples with a wide variety of component polarities.

In continuous development, the solvent is allowed to traverse the sorbent to a predetermined distance, at which point the developing solvent is allowed or induced to continuously evaporate. Thus, as a result of the complete evaporation of solvent, development does not stop at the conventional solvent front and components with low R_f values can be made to traverse the entire length of the sorbent bed. There are a number of advantages of this approach. Development proceeds undisturbed to completion, in contrast to multiple development where intermediate drying is necessary. Moreover, layer conditioning, if necessary, needs to be done only once thereby saving time. However, improved resolution is achieved at the cost of analysis time. Hence, it is essential that the solvent is optimized in order to minimize the plate length required for the separation.

4.8.4 Development chambers

To obtain reproducible results, the development process in TLC must take place in a developing tank or chamber designed to enable the plate, developing solvent and gas phase to reach equilibrium. Saturation was originally used in reference to saturation of the developing tank atmosphere, but the important role of preloading of the sorbent layer is now recognized. Ignoring the role of the solute, the various equilibrium processes occurring in the chamber may be summarized as:

$$
\begin{aligned}
\text{sorbent} &\rightleftharpoons \text{developing solvent} \\
\text{sorbent} &\rightleftharpoons \text{water} \\
\text{sorbent} \rightleftharpoons \text{developing solvent} &\rightleftharpoons \text{water} \\
\text{water} &\rightleftharpoons \text{developing solvent} \\
\text{developing solvent}_{(g)} &\rightleftharpoons \text{developing solvent}_{(l)}
\end{aligned}
$$

Terminology in relation to saturated systems has not always been consistent; in particular, a distinction should be made between chamber saturation and the different forms of layer preloading and saturation. Before proceeding it is therefore necessary to define [8] various terms relating to saturation and chamber type.

- Chamber saturation. Chamber saturation is achieved when all components of the developing solvent are in equilibrium with all zones of the gas space, before and after development.
- Preloading. Preloading (see Section 4.5.6) refers to any sorptive uptake of gaseous molecules by the unwetted layer from the chamber atmosphere, irrespective of the degree of saturation of chamber and layer. Preloading affects the rate of migration of both the solvent front and solute molecules; the greater the preloading the greater the front migration velocity and the smaller the R_f values.
- Sorptive saturation. Sorptive saturation is the state where the unwetted layer is in equilibrium with all components of the saturated gas space and, as such, represents a special case of preloading, namely its upper limit.
- Capillary saturation. Capillary saturation refers to the process of capillary filling of the free layer volume and the state of the layer after the completion of the development.

Geiss [8] divided developing chambers into two classes if only volume and shape are considered, and into four types (Table 4.11) if saturation is also considered. Normal or N-chambers are characterized by a depth of gas space in front of the layer of greater than 3 mm, whereas chambers with a depth less than this value are named sandwich or S-chambers. The N-chamber is still the most frequently used tank type

Table 4.11 Chamber classification according to Geiss [8].

Depth of gas space/ chamber type	Chamber saturation before starting development	Layer preloading before and during development	Examples
>3 mm Large trough/ unsaturated N-chamber	Practically none	Low	All box and cylinder chambers without paper lining and/or open top. Also SD/CD chamber and Chromatron.
Saturated N-chamber	Medium to high depending on solvent volatility and front velocity	Variable	All box and cylinder chambers with paper-lined walls. Also Camag Twin Trough.
<3 mm Sandwich configuration/ unsaturated S-chamber	None	None	Traditional sandwich chambers without counter layer. Also Camag U-chamber and Chrompres OPTLC.
Saturated S-chamber	Complete	Variable	S-chambers with solvent-wetted counter layer.

in TLC. It is usually a glass tank with a gas space depth of 5–10 cm equipped with a lid. The original tanks were bare-wall, unsaturated N-chambers. Chamber saturation in such tanks probably requires in excess of 2 h. One disadvantage of unsaturated chambers is the uneven movement of the solvent producing a curved solvent front with the spots near the edges travelling faster than those near the centre. Stahl introduced wall lining with filter paper for chamber saturation. In filter paper lined N-chambers, chamber saturation is essentially complete within 10 min of introducing developing solvent.

The twin-trough developing chamber consists of a standard developing chamber with a raised wedge-shaped bottom that divides the tank into two compartments. With this arrangement it is possible to develop two plates simultaneously or to use one compartment for preloading of the sorbent by placing the plate in one compartment and the preloading or developing solvent in the other. After preloading, the chamber can be tilted to allow the transfer of developing solvent into the compartment for development.

Statements that the usual S-chamber is saturated are mostly incorrect, in view of data presented by Geiss [8]. Most S-chambers are unsaturated at the commencement of development, but can easily be converted into a saturated S-chamber just by lining the cover plate with a solvent-wetted filter paper. Nevertheless, environmental characteristics are more easily controlled in an S-chamber because of the smaller chamber dimensions. Most horizontal TLC chambers, including the U-chamber (for circular chromatography) and overpressured tank, are (unsaturated) S-chambers. The linear development chamber is an S-chamber for the horizontal development of high-performance plates. This system allows preloading of the layer and minimal solvent consumption. The samples can be developed from edge to edge or from both ends simultaneously toward the plate centre, depending on whether plate number (efficiency) or sample capacity is the more important consideration. The Vario-KS chamber is a versatile system for horizontal development. The chamber consists of a variety of conditioning trays with up to five developing solvent reservoirs.

Many other chambers are available including the short bed continuous development (SB/CD) chamber, which is designed specifically for use with continuous and multiple development techniques.

4.8.5 Forced flow

The disadvantages of conventional TLC under control of capillary forces, namely variable flow-rates, which are 'off-optimum' most of the time, and relatively long running times, are partly overcome by techniques such as continuous and multiple development. Such techniques generally focus on a narrow band of substances with similar R_f values and sacrifice separation of the rest of the sample mixture. The application of centrifugal force or pressure is a more appropriate remedy for slumping capillary solvent flow. Under forced flow conditions, the flow velocity is constant and can be optimized. The design of instrumentation for this mode is more complicated, but the theoretical advantages of forced flow development outweigh the practical problems of its implementation. A larger number of theoretical plates is available in forced flow TLC, the total number being limited by the available pressure drop and the bed length. Separation times can be dramatically reduced (e.g. separation lengths of 30 cm in less than 30 min are easily achieved) and solvents with poor sorbent wetting characteristics can be used.

Two approaches to forced flow development have been used. In the rotational mode, centrifugal force generated by spinning the sorbent layer about a central axis is used to drive the solvent through the layer. The rate of solvent migration is a function of the rotation speed and the rate at which the solvent is supplied to the layer. Because the layer is not enclosed, the amount of solvent that can be kept within the layer without causing flooding determines the ultimate velocity of the solvent front. The Rotachrom is a commercially available (rotating) centrifugal system which generates a constant flow velocity over a separation path of 10 cm. It can be used in the linear, circular and (by scraping appropriate channels in the layer) anticircular modes.

Overpressured thin layer chromatography (OTLC) was introduced in the late 1970s. The mobile phase velocity is controlled and optimized by adjusting the output of a mechanical pump if the sorbent layer is enclosed using a flexible membrane or an optically flat rigid surface under hydraulic pressure (e.g. Chrompres). The maximum inlet pressure is reached when it equals that of the sealing cushion. The normal mode of OTLC is linear development, with one solvent system in one direction, but circular and anticircular development modes are also possible. In the linear mode, simultaneous development in two directions is also used by placing the solvent inlet in the centre of the plate.

4.9 Detection

After the development is completed and the plate dried, the separated bands must be located *in situ*. Following location and identification of the spots (Section 4.9) they must be quantified (Section 4.10) if that is the purpose of the analysis.

4.9.1 Direct examination

Direct visual examination is usually inappropriate because relatively few compounds are coloured. Notable exceptions are plant pigments, food colourants and dyestuffs. If the analytes are fluorescent, such as polycyclic aromatic hydrocarbons, aflatoxins or riboflavin, they can be located under u.v. illumination. Alternatively, if a plate with a phosphorescent indicator has been used, the plate can still be examined under short-wave (254 nm) u.v. or long-wave (366 nm) u.v. light. With the inorganic phosphors used in most plates the background should fluoresce green, with dark spots where analytes are present which either absorb u.v. light or quench fluorescence. These compounds include most aromatic and conjugated compounds, in addition to some unsaturated compounds.

4.9.2 Spray reagents

If the analytes cannot be located directly, the plate can be treated chemically to produce coloured or fluorescent spots of the analytes. The chemical reagent can be applied by spraying or dipping the plate in either a general purpose or specific reagent. Spraying is generally the more suitable method. The whole area of the plate should be sprayed evenly in a fume hood with a good draught. Occasionally, the reagent is impregnated in the layer prior to development or coloured or fluorescent derivatives are prepared before spotting on the layer. Following application of the detection reagent, the layer is commonly heated in an oven or plate heater (Camag) to complete the reaction and ensure optimum colour development. Prolonged heating or excessive temperature can cause decomposition of analytes or undue darkening of the background. Touchstone and Dobbins [20] describe the use of a total of 207 reagents and Bobbitt [28] lists six universal reagents and 89 other, more specific reagents for TLC. Universal reagents are valuable for characterizing an unknown sample while selective reagents aid in identification. Several compatible reagents can be applied to the layer sequentially by overspraying to maximize the information content of the chromatogram. Selective reagents also simplify the separation process since the need for resolution of the analyte from substances unresponsive to the reagent is less demanding.

The interaction of analyte with many spray reagents involves a chemical reaction to produce a derivative. Such destructive methods are generally unsuitable where recovery of the separated analyte is necessary. A common nondestructive method is visualization with iodine. If a few crystals of iodine are placed in a tank, the vapour pressure of iodine at most laboratory temperatures is sufficient to generate iodine vapour in the tank. When the developed plate is placed in the tank, the iodine dissolves in the solutes which are revealed as brown spots of varying intensity on a pale yellow/brown background. With the exception of a few saturated alkanes, almost all organic compounds are stained by iodine. This method has the advantage that the staining is reversible. After marking the spots the iodine is simply allowed to evaporate.

Another universal method of visualization is the use of highly corrosive charring reagents. This method is restricted to inorganic adsorbents and inert binders. There are many formulae for charring reagents, none of which is especially sensitive. Most are based on sulfuric acid, although a mixture of 3% copper(II) acetate in 8% phosphoric acid is useful. The reagent is sprayed onto the plate, which is heated at 120–130°C for 20–30 min, producing carbonized organic spots on an off-white background. With certain charring reagents some compounds produce characteristic fluorescent spots on light heating, while stronger heating produces brown/black carbon spots.

Selective reagents exhibit varying degrees of selectivity. Carbonyl compounds, for example, are located as orange spots after spraying with 2,4-dinitrophenylhydrazine hydrochloride. Acidic and/or basic analytes can be detected by spraying with an acid–base indicator solution such as bromocresol green. Ninhydrin is generally regarded as a selective reagent for amines and is widely used to detect amino acids, but reducing substances such as ascorbic acid also produce a coloured derivative. The specificity of detection reagents should not be overestimated.

4.9.3 *Miscellaneous*

A number of detection methods have been described which have limited application. Biological activity can be exploited for visualization of biologically active analytes.

Certain pesticides, for example, can be detected by treating the layer with a mixture of indoxyl acetate and cholinesterase, which react to form indoxyl. Subsequent reaction of the indoxyl with molecular oxygen forms indigo blue. The initial reaction is inhibited by enzyme-inhibiting pesticides and so colourless spots are seen against a blue background.

4.9.4   Qualitative analysis

For qualitative analysis, identification is based on R_f values relative to standards run under identical conditions and, preferably, on the same plate. Alternatively, data can be quoted as hR_f values by multiplication of the R_f by 100. Whole numbers are then obtained. Published values of R_f and hR_f are to be regarded as guide values only. Factors like layer thickness, chamber saturation and layer preloading, air humidity, and separation effects of solvent mixtures, which are difficult to reproduce, can exert a marked influence. In some instances, the migration distance of the analyte(s) is compared with that of a simultaneously chromatographed reference substance, which should belong to the same or a similar compound class as far as possible. However, these relative R_f values probably suggest greater accuracy than is possible and can therefore be misleading. For this reason alone, their use is to be discouraged except in situations where the solvent front cannot be measured, such as in continuous development.

One general weakness, shared by all chromatographic techniques, is that they provide insufficient qualitative information to identify positively a substance having a specific R_f value. The colours produced by the visualizing reagent(s) can improve identification, but interfacing the chromatographic system to a detector capable of yielding structural information is the most suitable approach. Instrumental methods of identification such as mass spectrometry are detailed later in this chapter.

4.10   Quantification

Commercial instruments for quantitative evaluation of thin layer chromatograms first appeared in 1967. The evolution of modern TLC has depended on the availability of such instruments; without adequate means of quantifying the chromatogram the resolution achieved in HPTLC would be to no avail. Visual methods of detection show poor precision of no better than 10–30%, with detection limits typically in the microgram range. Similarly, methods based on elution of separated analytes from the sorbent followed by independent measurement are time consuming and relatively insensitive. *In situ* measurement of separated substances on the TLC plate is the only viable method for high-performance systems. Nevertheless, visual scanning of the chromatogram and measurement of spot size and elution require little in terms of sophisticated instrumentation and are useful for this reason.

4.10.1   Elution techniques

Quantitative analysis can be performed by scraping of the spots from the developed chromatogram and elution of the substance from the sorbent. It will be obvious what part of the sorbent should be scraped if the spots can be visualized by direct examination, otherwise spots can be visualized in the usual manner with detection reagents and recovered by scraping and elution. It is more usual, however, in such cases to apply the sample to the plate in duplicate and, following development, the detection reagent is applied to one sample after masking the other. An area at least 10% larger than the visual size of the spot should be scraped to compensate for three-dimensional spreading of the spot below the surface. The scrapings are transferred to either a Soxhlet extractor or a centrifuge tube and homogenized with a suitable eluting solvent. The clear extract or supernatant can be analysed by any convenient procedure, e.g. spectrophotometry or electrochemistry. Standards must be processed in parallel with the sample.

4.10.2   In situ *visual evaluation*

Visual evaluation is performed by applying the sample(s) and a series of reference standards to the plate. The concentration of standards is chosen so as to bracket the expected concentration in the unknown sample. After development and detection, the sizes of the spots are compared to choose the best match between the sample spot and the appropriate standard. The accuracy of this procedure can be improved by repeating the analysis with a narrowed concentration range of standards. Under the best conditions, errors are unlikely to be less than 10% and more probably are closer to 30%.

4.10.3   In situ *optical scanners*

The availability of scanning densitometers and imaging scanners, which can detect visible or u.v. absorbance or fluorescence, has made HPTLC a highly efficient quantitative tool. Unlike column techniques, where the analyte and stationary phase are separated following completion of development, the sample and sorbent remain unseparated at the detection step in TLC. Optical methods of quantification involve measurement of the difference in optical response between blank portions of the sorbent and regions where a separated substance is located. The sorbent is optically opaque and incident monochromatic light is: (i) specularly reflected from the surface; (ii) absorbed by the medium and dissipated, for example, by conversion to heat; and (iii) diffusely reflected or transmitted by the medium.

Specularly reflected light is only important with smooth surfaces but conveys little useful information about any sample distributed within the sorbent layer. The diffusely reflected/transmitted portion is measured for chromatographic quantification, while the specularly reflected component is assumed to be very small. Since the diffusely and specularly reflected components cannot be distinguished, the specularly reflected component does however, contribute to the background noise signal. Light scattering occurs from all particles of the layer, which complicates photometric measurements of separated substances relative to equivalent measurements in solution. Under such circumstances, the Beer–Lambert Law is invalid and an alternative relationship must be sought between the amount of light absorbed (or emitted in fluorescence) and the amount of substance. The most generally accepted theory is derived from Kubelka and Munk [see ref 29].

Spectroscopic detection can be used either directly or after the analytes have been revealed by a spray reagent. Optical scanners are available which are capable of operating in any one of several modes. Scanners typically can detect ultraviolet, visible or near infrared (185–2500 nm) absorbance or fluorescence. Absorbance by a substance in the visible range can be measured in the transmittance or reflectance mode (i.e. diffuse scattering by the sorbent particles diminished by absorbance). Conversely, measurement is restricted to the reflectance mode for substances absorbing only in the u.v. region because silica gel sorbents exhibit a strong u.v. absorbance. The benefits and disadvantages of the different measuring techniques may be summarized as follows:

- Transmittance mode. Layer irregularities caused by variations of layer thickness and particle size inhomogeneities interfere with the signal, resulting in the most unfavourable signal-to-noise ratio of all modes. Moreover, measurement in the transmittance mode is not possible below about 320 nm.
- Reflectance mode. Light scattering in the stationary phase hampers transmission methods, whereas reflectance measurements are easier because scattering of the light beam is largely independent of the plate thickness. Nevertheless, the accuracy and precision of reflectance measurements does require careful control of experimental technique and materials, such as spot shape and size and uniformity of the sample development. The most accurate and precise data are obtained if spot diffusion is consistent. For this reason, the sample and calibration standards should be prepared in different concentrations and applied to the plate using a fixed constant volume of solutions of various concentrations rather than different volumes of solutions. Standards and samples should be run on the same plate, since the deviation between spots measured on the same plate is always less than the deviation between spots on different plates.
- Combined transmittance–reflectance mode. Combined transmission–reflectance mode [8] provides a better signal-to-noise ratio for spots with an absorption maximum greater than 380 nm.
- Decrease of phosphorescence. The substance-induced diminished phosphorescence of an indicator mixed with the sorbent is a special case of absorbance mode measurement. Ultraviolet-absorbing spots absorb a portion of the phosphorescence excitation radiation and thus diminish the phosphorescence emission intensity. This technique suffers from severe background fluctuations, resulting from inhomogeneous distribution of the phosphor in the sorbent layer. The sensitivity of this mode is therefore generally lower than that of the normal reflectance measurement.
- Fluorescence mode. Some compounds exhibit fluorescence after excitation by u.v. light. The absorbed radiation is emitted at a longer wavelength than the absorbed light. The measurement of natural fluorescence is the detection mode of choice, whenever applicable, because of its selectivity and sensitivity. Fluorescence mode has other important advantages: (i) the calibration curve is nearly always linear and (ii) the integrated signal is almost independent of the shape of the spot. Fluorescence quenching and catalytic degradation of the sample [29] may occasionally cause unexpected results; for example, relative to

measurements made in solution, the response may occur at different excitation and emission wavelengths, or the magnitude of the signal may be diminished. Moreover, matrix interferences may cause either enhancement or quenching of the fluorescence. Co-elution of free fatty acids with aflatoxins, for example, increase the fluorescence response [29] of the aflatoxins by 14–36%.

Calibration techniques

Calibration curves for absorption measurements (transmittance and reflectance modes) are generally inherently nonlinear. They usually comprise a pseudolinear region at low sample concentrations curving towards the concentration axis at higher concentrations. The extent of the linear portion is frequently very different for different substances. With fluorescence measurements, the calibration curves are often linear over two to three orders of magnitude, while in some absorption measurements no reasonable linear range may exist. Therefore, different regression techniques are used in calibration. The simplest transformations to linearize the calibration curves involve the conversion of the sample concentration and/or signal into reciprocals, logarithms or squared terms. Reciprocal transformation is probably the best of these linear approximations although, even here, large errors can be introduced if the dependent variable (reflectance data) is transformed. Therefore, nonlinear regression models are to be preferred although, in all likelihood, no single model will be best for all calibration curves. Hence, several models should be tested for each compound–plate type combination and the best model selected using statistical tests.

Optical scanning instrumentation

The essential features of a scanning densitometer are:

- Light source. Different sources are necessary to cover the entire u.v.-visible range; halogen or tungsten lamps for visible wavelengths, deuterium lamps for u.v. wavelengths and high-intensity mercury or xenon sources for fluorescence excitation.
- Wavelength selection device. Filters and monochromators are used to select the measuring wavelength. Grating monochromators with broad spectral sources offer versatility for optimizing sample absorption wavelengths whereas filter densitometers are limited in selectivity to the characteristic line spectrum of the source, which is usually a mercury source. The advantage of filter instruments is their low cost. The excitation wavelength for fluorescence measurement is selected with either a monochromator or filter. A cut-off filter, which attenuates the excitation wavelength but transmits the emission envelope, is placed between the plate and detector. Greater selectivity, but with some loss in sensitivity, can be achieved with interference filters in place of cut-off filters.
- Detection device. Photodiodes or photomultipliers are usually used for signal measurements. The latter are preferred as they provide a linear output signal as a function of excitation energy.

Geometry of the scanning light beam

The position of the beam is fixed in commercial scanners and the plate is scanned by mounting it on a movable stage. With slit scanners, the motor-driven scanning stage transports the plate through an illuminated slit of 4–10 mm in length in the direction perpendicular to the slit length. Each scan represents a lane whose width is specified by the slit setting and whose length is defined by the sample migration distance. A contrasting kind of scanner is the spot scanner, with a small illuminated light spot (small compared with the spot size) which moves in two directions using one of three kinds of motion: meandering scanning, zig-zag scanning or flying spot scanning. There are advantages and disadvantages associated with each type of scanning system and more work is needed in this area. Some instruments have a turntable-type stage as an option for radial scanning (in the direction of development) or peripheral scanning (at right-angles to the direction of development) which are used for circular and anticircular chromatograms. Radial scanning is generally preferred over peripheral scanning.

Optical arrangements

There are three optical arrangements predominantly used in scanning instruments. The simplest and most commonly used technique is the single-beam method. Background noise resulting from fluctuations in the source output, irregularities in the sorbent surface and the distribution of extraneous adsorbed impurities are a problem and other types of scanner have been constructed. Double-beam instruments can partly compensate for background disturbances by exposing the plate to two beams and recording the difference of the two signals. The two beams can be either separated in time or in space. The double beam in space configuration divides a single beam into two beams that scan different positions on the plate. One beam traverses the sample lane while the other scans the region between lanes. A difference signal is thus recorded which compensates for fluctuations in source output, but irregularities in the surface may still pose problems as the two beams scan different areas on the plate.

The double beam in time system uses a rotating chopper to send two wavelengths alternately to the same place on the plate. The two wavelengths are chosen experimentally; one wavelength where the spot absorbs and the second where the spot exhibits no absorption, but experiences the same scatter. Unfortunately, the scatter coefficient is wavelength dependent to some extent, so that the background correction is better when the two wavelengths are as close as possible. This may be difficult to achieve in some situations as absorption spectra are usually broad.

Automation of scanning

Most modern densitometers are designed for automatic scanning of a complete plate. Computer control is used for lane changes, spot detection, optimization of measuring conditions and data collection. Electronic scanning combines fast acquisition of data with simple instrument design and no moving parts, and compatibility with data analysis of two-dimensional chromatograms that are difficult to scan with conventional slit-scanning devices. However, the sensitivity, dynamic signal range and wavelength measuring range are more restricted in electronic scanners than in slit scanning densitometers. Despite the greater cost, electronic scanners are likely to develop rapidly in the future. Scanning takes place electronically while the complete plate is stationary and evenly illuminated with monochromatic light, and the reflected or transmitted light is focused directly onto a vidicon tube which functions as a two-dimensional array of unit detectors. These unit detectors are periodically discharged and the signal digitized for analysis by computer.

Concluding remarks

Variables in the scanning procedure such as mode (absorption or fluorescence, transmittance or reflectance), the size and position of the slit, the direction and speed of the scan and the electronic time constant of the densitometer should be evaluated for each system in order to obtain the maximum precision, linearity of the calibration curve and sensitivity [29]. The principal errors in scanning densitometry have been identified as the reproducibility of sample application, chromatographic separation, positioning the spot in the centre of the measuring beam, and measurement. Good laboratory practices can be used to control errors in sample application and positioning the plate in the beam. Variability in the development process owing to such factors as edge effects, deviations in layer thickness and nonlinear solvent fronts is generally the most significant error. With complete computer control of data acquisition and manipulation the relative standard deviation from all errors in scanning densitometry with HPTLC using absorbance or fluorescence measurements can be maintained below 1% [29].

Because TLC plates are relatively inexpensive and disposable, TLC can be used to separate cruder samples than HPLC and, in particular, samples containing particulates and components that are difficult to elute. Background interference from the sample matrix may complicate quantification under these circumstances [9].

4.10.4 *Other measuring systems*

A number of techniques are being investigated and developed for *in situ* measurement. Such procedures include photothermal deflection densitometers and radioisotope imaging systems [29]. Another type of detection system is provided by the Iatroscan device, which involves chromatographic separation on silica gel coated quartz rods with detection by a flame ionization detector. In another field, that of the TLC of radioactive substances, techniques such as autoradiography and liquid scintillation counting are being replaced by direct scanning with the 'linear analyser'. Autoradiography involves exposure of the chromatogram to an X-ray film, which can then be converted into intensity measurements using a photodensitometer. It is generally slow and has poor accuracy and precision. Scintillation counting is also slow as it requires scraping of the spots from the plate, followed by mixing with a scintillation fluid and counting in the usual manner. Direct isotope scanning is the method of choice. Early instruments with radiation-sensitive detectors (e.g. Geiger counter) and operated in a manner similar to optical scanning densitometers have been replaced by second-generation instruments employing windowless gas-flow proportional counters as imaging detectors.

The emphasis in the treatment of detection systems has been on the quantitative aspects of TLC. However, most scanning densitometers have some provision for *in situ* recording of the spectra of any number of spots and, in these circumstances, can provide qualitative information. Absorption spectra are rarely sufficiently characteristic for substance identification but can be used as an aid to confirm identification. Fluorescence spectra are more characteristic and may provide additional information. Scanning the full spectrum is time consuming and sequential scanning at several characteristic wavelengths (e.g. peak maxima and minima) is usually used as an alternative. The ratios of the signal at these characteristic wavelengths can be used to confirm the similarity between an analyte and a standard or the presence of contamination of a sample spot. For maximum precision, the ratios should be determined for samples and standards run on the same plate. It should be noted that *in situ* spectra and solution spectra may show little correspondence.

TLC has been coupled with other detection devices where the emphasis is on the qualitative information that can be obtained. The techniques of interest are infrared and Raman spectroscopy and mass spectrometry. Various off-line methods following spot elution are used for obtaining infrared spectra of separated components. However, *in situ* methods are essential for high-performance operation. Coupling of these techniques with TLC in an on-line manner has met with various degrees of success. Infrared spectra may be recorded *in situ* by diffuse reflectance Fourier transform infrared spectrometry. The technique has limited sensitivity, and requires scrupulous removal of developing solvent from the plate prior to measurement and background correction for absorption by the plate. However, all common sorbents have regions of strong infrared absorption (e.g., 3700–3100 and 1650–800 cm^{-1} for silica gel) which prevents reliable spectral data being obtained in those regions, even with background correction. Spectral interpretation for an unknown requires reference spectra to be obtained on the same sorbent since compounds bind differently to different sorbents and this changes spectral band positions. In contrast to *in situ* infrared measurements, the capabilities of surface-enhanced Raman spectrometry appear compatible with modern HPTLC. A number of procedural difficulties await solution but the method holds a great deal of promise.

Both off- and on-line MS has been used as an aid in compound identification. Most of these applications of TLC–MS are based on the use of fast atom bombardment from silica gel, although less sophisticated methods of ionization, such as electron impact, can be used. The combination of TLC and mass spectrometry has many of the advantages, as a tool for compound identification, possessed by GC–MS (Chapter 3) without the need for a complicated interface between the chromatograph and spectrometer.

References

1. Sherma J. and Fried B. (1984). *Anal. Chem.*, **56**: 48R.
2. Kirchner J.G., Miller J.M. and Rice R.G. (1954). *J. Agr. Food Chem.*, **2**: 1031.
3. Stewart G.H. (1965). In Giddings J.C., Editor, *Advances in Chromatography*, Vol. 1. Marcel Dekker, New York, p. 93.
4. Sherma J. and Zweig G. (1971). *Paper Chromatography and Electrophoresis*, Vol. 2, *Paper Chromatography*. Academic Press, New York.

 5. Kirchner J.G., Miller J.M. and Keller G.J. (1951). *Anal. Chem.*, **23**: 420.
 6. Laitinen H.H. and Ewing G.W. Editors (1977). *A History of Analytical Chemistry*. American Chemical Society, Washington, pp. 316–321.
 7. Sherma J. (1992). *Anal. Chem.*, **64**: 134R.
 8. Geiss F. (1987). *Fundamentals of Thin Layer Chromatography (Planar Chromatography)*. Huethig, Heidelberg.
 9. Poole C.F. and Poole S.K. (1989). *Anal. Chim. Acta*, **216**: 109.
10. Guiochon G. and Siouffi A.M. (1978). *J. Chromatogr. Sci.*, **16**: 598.
11. Guiochon G., Korosi G. and Siouffi A.M. (1980). *J. Chromatogr. Sci.*, **18**: 324.
12. Giddings J.C., Stewart G.H. and Ruoff A.L. (1960). *J. Chromatogr.*, **3**: 239.
13. Ruoff A.L. and Giddings J.C. (1960). *J. Chromatogr.*, **3**: 438.
14. Thoma J.A. and Perisho C.R. (1967). *Anal. Chem.*, **39**: 745.
15. Guiochon G., Siouffi A., Engelhardt H. and Halasz J. (1978). *J. Chromatogr. Sci.*, **16**: 152.
16. Guiochon G. and Siouffi A. (1978). *J. Chromatogr. Sci.*, **16**: 470.
17. Guiochon G., Bressolle F. and Siouffi A. (1979). *J. Chromatogr. Sci.*, **17**: 368.
18. Stahl E. (1979). *J. Chromatogr.*, **165**: 59.
19. Stahl E. Editor (1969). *Thin Layer Chromatography*. Springer-Verlag, New York.
20. Touchstone J.C. and Dobbins M.F. (1983). *Practice of Thin-Layer Chromatography*, 2nd edn. Wiley, New York.
21. Wilson I.D. (1983). *J. Pharm. Biomed. Anal.*, **1**: 219.
22. Schoenmakers P.J., Droun A.C.J.H., Billiet H.A.H. and de Galan L. (1982). *Chromatographia*, **15**: 688.
23. Glajch J.L., Kirkland J.J. and Snyder L.R. (1982). *J. Chromatogr.*, **238**: 269.
24. Glajch J.L., Kirkland J.J., Squire K.M. and Minor J.M. (1980). *J. Chromatogr.*, **199**: 57.
25. Horvath C. and Melander W. (1977). *J. Chromatogr. Sci.*, **15**: 393.
26. Beesley T.E. (1985). *J. Chromatogr. Sci.*, **23**: 525.
27. Davis J.M. and Giddings J.C. (1983). *Anal. Chem.*, **55**: 418.
28. Bobbitt J.M. (1963). *Thin-Layer Chromatography*. Reinhold, New York.
29. Poole C.F. and Poole S.K. (1989). *J. Chromatogr.*, **492**: 539.

Bibliography

Dallas F.A., Read H., Ruane R.J. and Wilson I.D., Editors (1988). *Recent Advances in Thin-Layer Chromatography*, Plenum Press, New York.
Poole C.F. and Poole S.K. (1989). Modern thin-layer chromatography. *Anal. Chem.*, **61**: 1257A.

Comparison of techniques

Sherma J. (1991). Comparison of thin layer chromatography and liquid chromatography. *J. Assoc. Off. Anal. Chem.*, **74**: 435.

Stationary phases

Grassini-Strazza G., Carunchio V. and Girelli A.M. (1989). Flat-bed chromatography on impregnated layers. *J. Chromatogr.*, **466**: 1.
Hahn-Deinstrop E. (1992). Stationary phases, sorbents. *J. Planar Chromatogr. – Mod. TLC*, **5**: 57.

Development

Tyihak E., Mincsovics E. and Szekely T.J. (1989). Overpressured multi-layer chromatography. *J. Chromatogr.*, **471**: 375.
Vajda J., Leisztner L., Pick J. and Anh-Tuan N. (1986). Methodological developments in overpressured-layer chromatography. *Chromatographia*, **21**: 152.

Instrumentation

Touchstone J.C. (1988). Instrumentation for thin-layer chromatography: A review. *J. Chromatogr. Sci.*, **26**: 645.

Qualitative analysis

Busch K.L. (1992) Mass spectrometric detection for thin-layer chromatography. *Trends Anal. Chem.*, **11**: 314.

Quantification

Pollak V.A. (1989). Sources of errors in the densitometric evaluation of thin layer separations with special regard to nonlinear problems. In Giddings J.C., Grushka E. and Brown P.R., Editors, *Advances in Chromatography*, Volume 30. Marcel Dekker, New York.

Pollak V.A. (1989). Electronic scanning for the densitometric analysis of flat-bed separations. In Giddings J.C., Grushka E. and Brown P.R., Editors, *Advances in Chromatography*, Volume 30. Marcel Dekker, New York.

Poole C.F. and Poole S.K. (1989). Progress in densitometry for quantitation in planar chromatography. *J. Chromatogr.*, **492**: 539.

Shantha N.C. (1992). Thin-layer chromatography – flame ionization detection Iatroscan system. *J. Chromatogr.*, **624**: 21.

High-performance Liquid Chromatography— Instrumentation and Techniques

5

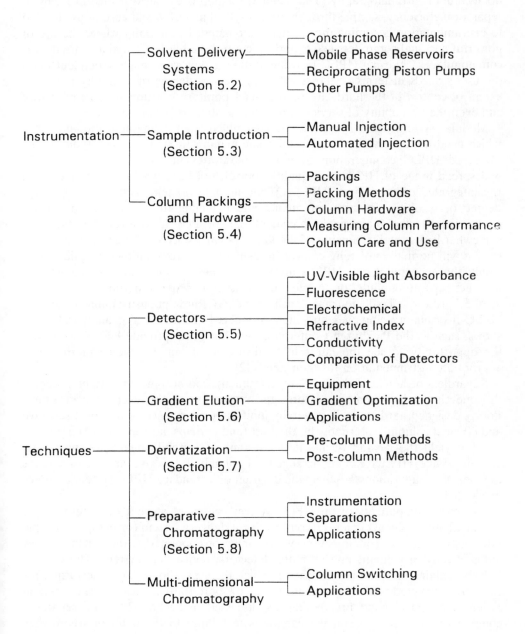

Instrumentation

- Solvent Delivery Systems (Section 5.2)
  - Construction Materials
  - Mobile Phase Reservoirs
  - Reciprocating Piston Pumps
  - Other Pumps
- Sample Introduction (Section 5.3)
  - Manual Injection
  - Automated Injection
- Column Packings and Hardware (Section 5.4)
  - Packings
  - Packing Methods
  - Column Hardware
  - Measuring Column Performance
  - Column Care and Use

Techniques

- Detectors (Section 5.5)
  - UV-Visible light Absorbance
  - Fluorescence
  - Electrochemical
  - Refractive Index
  - Conductivity
  - Comparison of Detectors
- Gradient Elution (Section 5.6)
  - Equipment
  - Gradient Optimization
  - Applications
- Derivatization (Section 5.7)
  - Pre-column Methods
  - Post-column Methods
- Preparative Chromatography (Section 5.8)
  - Instrumentation
  - Separations
  - Applications
- Multi-dimensional Chromatography
  - Column Switching
  - Applications

5.1   Introduction

The instrumentation for LC has evolved constantly over the last two or more decades to the point where today's high-performance (or high pressure) liquid chromatograph (HPLC) bears little resemblance to the original format used for open-column chromatography. The demand for improved performance in terms of separation efficiency, sample throughput, analytical precision and ease of use has led to instrumentation becoming increasingly automated, to the point where the use of powerful personal computers (PCs) both to control the hardware and acquire data is commonplace. Choosing HPLC instrumentation is now very much dependent upon the user's needs in terms of the specific application, degree of flexibility required, extent of control of the hardware (and other laboratory instrumentation) necessary and even the availability of service and technical support personnel.

Modular systems, that is instruments assembled from component modular units which need not necessarily be from the same manufacturer, have been typically the preferred HPLC configuration in most laboratories in the past. However, the widespread usage of HPLC as a quality control tool has seen the re-emergence of the integrated or 'black box' HPLC. While modular systems give the user a high degree of flexibility in terms of mixing and matching the hardware for specific configurations, the increasingly prevalent use of PCs to control the hardware somewhat limits the intermixing of modular systems, since one manufacturer's data system will normally not fully control instrumentation from another supplier. As a result, the majority of new HPLCs are now purchased from a single manufacturer, as either separate modules or an integrated system, in order to permit full control of all the hardware. Trade journals, such as *LC.GC*, now publish annual listings of HPLC instrument companies and their products [1]. Also, major analytical trade shows, such as the Pittsburgh Conference and Analytica, provide HPLC users with the opportunity to see the latest in hardware to aid in the choice of the most appropriate instrumentation for their needs [2].

Regardless of how complex modern instrumentation becomes, the right system is the one that will ultimately perform the separation to the user's specifications. Poorly designed hardware will generally limit the performance of the most selective and efficient column. Alternatively, the best (and perhaps most expensive) hardware will not yield the desired results if the column and mobile phase are incorrectly chosen. While HPLCs can be arranged, or 'plumbed', to give an almost endless number of configurations, a schematic diagram of a standard HPLC system is shown in Fig. 5.1.

The basic components of an HPLC system consist of solvent (also termed the mobile phase or eluent) in a reservoir, a solvent delivery system (or pump), a sample introduction device (or injector) which may be operated either manually or automatically, a column, one or more detectors, and a data system. The column (which contains the stationary phase on which the separation is to be performed) is typically connected to the injector using narrow bore (0.01 inches or less) tubing in order to minimize band broadening, as discussed in Section 2.5. The separated components then pass (again via narrow bore tubing) to the detector where they generate a signal, which is recorded on a data system. More than one detector may be employed in a system and the use of multiple detectors in series, in order to

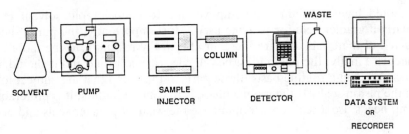

WASTE

COLUMN

SOLVENT   PUMP   SAMPLE   DETECTOR   DATA SYSTEM
                           INJECTOR   OR
                                                                      RECORDER

Fig. 5.1. Components of a basic HPLC system. Diagram courtesy of Waters Chromatography Division of Millipore.

maximize sample information from a single injection, is quite common in HPLC. The detector signal is generated on the basis of some intrinsic property of the solute or mobile phase, such as u.v./visible light absorbance, fluorescence, conductance, etc., while the data system can be as simple as a strip chart recorder or as complex as a VAX-based computer.

The data system may be used not only to acquire the data but it also may be used to control physically the actual hardware parameters, such as pump flow-rate, injection volume or detector wavelength and sensitivity, enabling the HPLC to be operated totally through the computer keyboard. The instrument modules typically communicate via RS-232 serial or IEEE-488 GPIB interfaces [3]; however, some manufacturers are now using fibre-optic cables to communicate between the various modules [4]. In this chapter, the various components of an HPLC system will be discussed in some detail, in addition to a number of specialized HPLC techniques.

5.2   Solvent Delivery Systems

5.2.1   General requirements

The primary function of the solvent delivery system, or pump, is to deliver the mobile phase through the system as reproducibly as possible. The majority of HPLC pumps deliver the mobile phase at a constant flow-rate, although some pumps are also designed to deliver the mobile phase at a constant pressure. However, the constant pressure mode is not generally used for analytical chromatography [5]. For typical analytical HPLC work, a flow-rate range of $0–10\,ml\,min^{-1}$ is satisfactory; for preparative work, flow-rates as high as $45–180\,ml\,min^{-1}$ may be required. Conversely, microbore chromatography utilizes flow-rates in the order of $50–150\,\mu l$ min; hence different pump head sizes are generally required to deliver at these extremes of flow rate.

The small particle size ($3–10\,\mu m$) of modern HPLC column packings demands that the pumps be able to deliver the mobile phase at backpressures up to 6000 p.s.i. at precise flow-rates ($<1\%$ variation). The mobile phase delivery should also be relatively free from pressure pulsations as the baseline from conductivity and refractive index detectors, among others, is very sensitive to pump pulsations. The pump should have a minimal hold-up volume to enable rapid solvent changes, particularly if gradient elution is being used. Most importantly, the pump should be constructed of materials that are inert, or chemically resistant, to the mobile phase components. Finally, the pump must be reliable, easy to prime and it should also be easy to change, on a routine basis, those parts in the pump that are subject to wear, such as the piston seals. If the mobile phase is to contain buffers, it is preferable that the pump heads use a seal wash to minimize the wear of plunger seals and pistons.

5.2.2   Materials of construction

As the mobile phases used in HPLC can be chemically aggressive, i.e. acidic, basic or corrosive, it is essential that the pump, and other components in the system where necessary, be constructed of materials resistant to chemical attack. The majority of the wetted surface in most HPLC pumps is constructed from 316 stainless steel. Once the stainless steel has been passivated, generally with 6 M nitric acid, the surface is resistant to leaching and chemical attack from most solvents, buffers and acids, the exception being HCl, which is corrosive to this type of surface [6]. The major advantages of stainless steel are its relatively low cost, ease of machining and the fact that it can be used at operating pressures up to 6000 p.s.i. [3]. Instrument companies, e.g. Spectra-Physics and Gilson, have also manufactured HPLC pump heads from inert metals such as titanium, however this increases the cost and such materials have not been widely used.

HPLC pump heads may also be constructed from polymers such as PTFE [5] and PEEK (polyethlyethylketone) [7] for separations when corrosive solutions, such as HCl, are to be used as the mobile phase. These materials are also used when there is concern that: (a) the sample and eluent may be contaminated by metal ions produced from corrosion of the stainless steel [8]; (b) proteins may adsorb onto the stainless steel surface [9]. In addition to their higher manufacturing cost, another major disadvantage of polymeric materials is that they can typically be operated only at pressures of 2000–4000 p.s.i. [7]. Considerable discussion over the relative merits of stainless steel and polymeric systems has ensued over the past 10–15 years. Manufacturers market the necessity of inert polymeric systems for many applications when stainless steel would also be equally appropriate [6, 10]. In addition, it has been demonstrated that when recycling eluents typically used for trace metal analysis in ion chromatography, over 99% of metals accumulated in the mobile phase come from the reagents used to prepare the mobile phase rather than from extraction, or complexation, of metal ions from the stainless steel [6, 11].

The piston plungers used in HPLC pumps are normally constructed from sapphire, while the ball used in a ball-and-seat check valve is generally made of ruby and the seat from sapphire [3]. Care must be taken when assembling such valves not to actually touch the ball or seat as this will prevent the correct operation of the check valve. Plunger seals, which typically wear with use, are made from a number of polymeric materials including high molecular weight polyethylene or polypropylene, PTFE or Roulon. Not every type of pump seal is compatible with all solvents and the manufacturer's instructions should be consulted when replacing these seals. The lifetime of plunger seals very much depends upon the type of mobile phase being delivered. Purely aqueous mobile phases, particularly those containing high ionic strength buffers, shorten the life of the pump seals. Alternatively, a mobile phase of methanol:water will typically enhance seal life. Other inert materials which are used for parts, such as washers and spacers, include Kalrez, KelF and various ceramics [4].

5.2.3   Mobile phase reservoirs

The mobile phase reservoir is the container used to store the solvent prior to it being fed into the pump. The container (and the solvent feed line) must be inert, that is, the solvent should not extract organic (or inorganic) compounds from the vessel. Any such extracted compounds would subsequently form part of the mobile phase and, depending upon the type of chromatography, may lead to problems such as baseline noise, loss of resolution, etc. Typically 1–2 l glass bottles, e.g. Schott vessels, are used in most laboratories. The solvent feed line is typically made of nonpermeable teflon and should be fitted with a reservoir filter, which may be either stainless steel or teflon containing 2–10 μm pores to remove particulates from the mobile phase. All mobile phases used in HPLC should be degassed, generally by vacuum filtration through a 0.22–0.45 μm filter, immediately prior to use. Degassing minimizes dissolved gases in the mobile phase and reduces the possibility of bubbles forming in the pump check valves and detectors [5]. Care must be taken when degassing in order to avoid changing the concentration of volatile solvents in the mobile phase. It is also preferable that the reservoir be maintained at a higher elevation than the pump heads in order to maintain a slight positive head pressure and thereby avoid the pump starving at higher flow rates.

If the mobile phase is to be sparged, then the solvent reservoir requires further

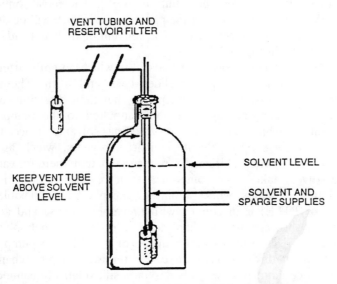

Fig. 5.2.   Typical solvent reservoir for use with a sparged mobile phase. Diagram courtesy of Waters Chromatography Division of Millipore.

consideration. Sparging involves bubbling a gas through the solvent in order to reduce the partial pressure of unwanted, dissolved gases at the surface of the solvent. Typically, an inert gas such as nitrogen or helium is used, although helium is preferred as it is less soluble than nitrogen, and hence has greater displacing power. Sparging will further reduce the background absorbance on a u.v. detector and the quenching phenomenon caused by oxygen on a fluorescence detector [3]. Sparging is essential when carrying out reversed-phase gradient elution, particularly in low-pressure gradients, since it reduces the evolution of air bubbles when mixing water and organic solvents. Sparging will also improve the performance of most piston pumps, even when used in the isocratic mode. The solvent reservoir must have a vent in order to prevent any measurable pressure build-up in the reservoir, as most bottles are not designed to withstand the internal pressures that multisolvent delivery systems equipped with sparging facilities are capable of delivering [3]. Figure 5.2 shows a typical solvent reservoir for use with a sparged mobile phase. Most HPLC manufacturers make accessories to enable mobile-phase sparging [e.g. 12] if the pump does not have the facility built in, as is the case with the majority of isocratic solvent delivery systems.

5.2.4 Reciprocating HPLC pumps

As a result of their overall performance and reliability, the vast majority of pumps used in HPLC applications today are reciprocating pumps, operated either mechanically, using thumbwheels, or more commonly through microprocessor control. These pumps use either a piston (or a diaphragm) to positively displace solvent from small volume (50–250 μl) chambers, via check valves, out of the pump. They may have one, two or even three pump heads and typically incorporate specialized drive mechanisms to minimize solvent flow pulsations [5]. The most common of these pump types is the dual head reciprocating piston pump, which will be described here in some detail. The drive mechanism of a conventional dual head reciprocating piston pump is shown in Fig. 5.3.

Each pump head contains a cavity with a seal in which a tightly fitting, accurately machined plunger (typically sapphire) slides back and forth. The two plungers (usually referred to as pistons) are driven such that their directions are opposing, resulting in one plunger drawing in solvent (supplied to the pump via an inlet manifold assembly) while the other is expelling solvent, hence the pumping chambers alternatively supply solvent. Noncircular gears, which are continually lubricated in an oil chamber, drive the plungers so that their forward motion in expelling the solvent takes more time than their reverse motion in drawing the solvent. Diaphragm pumps work similarly, except that a flexible diaphragm (stainless steel or PTFE) is in contact with the mobile phase and a piston in an oil-filled cavity pushes against the diaphragm in order to expel solvent from the pumping chamber [5]. The major disadvantage of a diaphragm pump is that when the diaphragm wears through, gear oil can enter the pumping chamber and be pumped onto the column; however this design is still relatively commonplace [e.g. 13]. With such pumps, the solvent flow-rate is changed either by varying the piston stroke volume or, most commonly, by varying the piston frequency. The pump

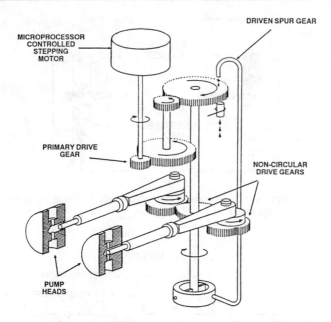

Fig. 5.3.   Drive mechanism of a conventional dual head reciprocating piston pump. Diagram courtesy of Waters Chromatography Division of Millipore.

heads contain two check valve assemblies, namely an inlet and an outlet check valve, which operate according to a push–pull arrangement, as shown in Fig. 5.4.

Since the plungers in the pump chambers are driven alternately, the check valves must also open alternately. Therefore as one pump head expels solvent, its outlet check valve opens and its inlet check valve closes. Simultaneously, the other pump head draws solvent in and its inlet check valve opens and the outlet check valve closes. The valves are arranged to allow solvent flow in one direction only and typically contain a cup filter which prevents any piston or seal wear material from reaching other components in the HPLC system. Inlet check valves are mostly of the ball and seat variety, while there are two major types of outlet check valves—the ball and seat and the spring-loaded soft seat. Of these, the ball and seat is generally preferred as it makes the pump easier to prime, i.e. to expel trapped air bubbles from the pump heads. Figure 5.5 illustrates graphically how the solvent flow from each of the pump heads combines to obtain a steady composite flow [3]. In the piston profile curve in Fig. 5.5, the slope of the curve corresponds to the solvent delivery rate.

In interval A (Fig. 5.5), the right piston is accelerating, increasing the solvent flow from this head, while the left piston decelerates. During interval B, the slope and flow rates are constant with all the flow provided by the right piston. The left pump chamber fills during this period. In interval C, the forward motion of the right piston decelerates, while the left begins to accelerate. During this crossover, the decreasing flow of the right head adds to the increasing flow of the left head to produce a combined steady flow, with only slight variation in flow at the crossover points. If

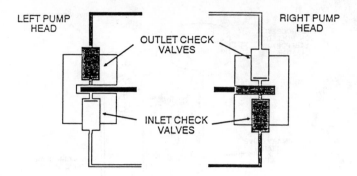

Fig. 5.4. Check valve operation of a conventional dual-head reciprocating piston pump. Diagram courtesy of Waters Chromatography Division of Millipore.

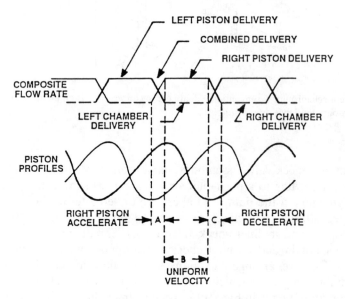

Fig. 5.5. Combined flow produced from dual head reciprocating piston pump. Diagram courtesy of Waters Chromatography Division of Millipore.

circular gears were used, then the flow would momentarily drop to zero during the crossover interval as the slopes of each piston would be zero. This is illustrated in Fig. 5.6, which compares the solvent flow produced from a single-piston pump, a dual-piston pump with circular gears and a dual-piston pump with noncircular (sinusoidal cam) gears.

While pump pulsations have no adverse effect on the actual chromatographic resolution, detectability at trace levels is often limited by baseline noise arising from pump pulsations [5]. Hence, single-piston pumps are rarely used to deliver the analytical mobile phase owing to the fact that they produce pulsed flow, as highlighted by Fig. 5.6. However, such pumps may be used in HPLC to deliver solutions, such as post-column derivatizing reagents [12], where pump pulsations

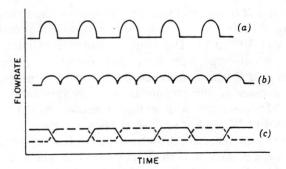

Fig. 5.6. Solvent flow produced from (a) single piston pump, (b) dual piston pump with circular gears and (c) dual piston pump with noncircular gears. Reprinted with permission from Snyder and Kirkland (1979) *Introduction to Modern Liquid Chromatography*, John Wiley & Sons, New York.

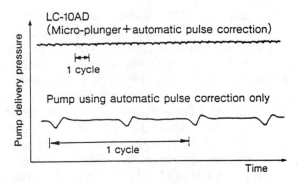

Fig. 5.7. Solvent flow produced from a conventional piston pump and one using micro-plungers. Diagram courtesy of Shimadzu.

may be less critical, as they are typically cheaper than dual-piston pumps. The Dupont company has also manufactured an HPLC pump with three heads which have three pistons 120° out of phase in order to minimize pump pulsations.

HPLC pumps often incorporate additional features in order to minimize pulsations. Many pumps incorporate pulse dampeners, alternatively termed noise filters, which are flow-through devices essentially consisting of lengths of stainless steel tubing compacted by bending them over themselves. By accumulating and discharging solvent, minor flow fluctuations are smoothed out [3]. Another option is to use very small volume pump heads and microplungers coupled with a high-speed drive, as illustrated in Fig. 5.7. This results in reduced pump pulsations; however, the drawback of this approach is that the plungers work faster, potentially reducing seal life. The latest advance in pump technology is to use control algorithms which electronically monitor the pressure output and adjust the pump motor speed so that the instantaneous pump flow is kept constant [3]. Such control algorithms may also compensate for solvent compressibility, which occurs as a result of raising the solvent pressure from ambient to the pump (or column) operating pressure. Alternatively, the pump should have a manual adjustment to allow for accurate delivery of solvents with different degrees of compressibility.

The majority of stainless steel reciprocating pumps operate to approximately 6000 p.s.i., while pumps manufactured of either PTFE or PEEK generally only operate to between 2000 and 4000 p.s.i. The pump pressure is monitored using either a Bourdon tube or strain-gauge transducer and almost all pumps incorporate a high-pressure limit switch which stops the pump flow at a user-specified value in order to prevent damage to the column through overpressuring. Modern HPLC pumps typically deliver solvent with a flow precision of 0.1–0.5% and a flow accuracy in the order of 1–2%. For most work, flow accuracy is less important than the precision as it is generally desirable that a peak elute reproducibly at a known retention time. Flow accuracy is perhaps more critical when performing gradient analysis, which will be discussed further in Section 5.4.

5.2.5   Other types of HPLC pump

Other types of pumps used for HPLC include positive displacement or syringe pumps, pneumatic constant pressure pumps, and gear pumps. The syringe pump is essentially a large (e.g. 10–50 ml) barrel syringe with the plunger connected to a digital stepping motor. As the plunger moves forward, it drives solvent through to the column. The solvent flow-rate is altered by varying the voltage on the motor and most pumps have special fill systems to enable refilling of the empty pump chamber on the return plunger stroke [5]. These pumps have the advantage of being able to deliver completely pulseless flow during the piston stroke, but the maximum run time is limited by the volume of the syringe. Considering the lack of flexibility and the fact that syringe pumps are not compatible with low-pressure gradient elution, this type of pump is not commonly used in HPLC today. Nevertheless, a number of manufacturers have used this approach in particular instruments, e.g. the Waters QA-1 and the ABI 130A Separation System.

Pneumatic pumps utilize gas pressure applied to the mobile phase reservoir to force the solvent through to the column. This simple approach provides pulseless flow, but the operating pressures are governed by the regulators and gas cylinders used to supply the pressure. For this reason, operating pressures are typically only 100–200 p.s.i., hence this type of pump is rarely used to deliver the mobile phase in analytical chromatography. However, such pumps are often used for the post-column addition of derivatizing reagents, where their pulseless flow minimizes baseline noise. A schematic diagram of a pneumatic vessel for delivery of post-column reagent is illustrated in Fig. 5.8. Further discussion on post-column reaction derivatization in HPLC is contained in Section 5.7. Gear pumps have also been used to deliver solvents in LC, but their use is restricted to low pressure (0–100 p.s.i.) applications, such as membrane protein separations [14].

5.3 Sample Introduction in HPLC

The function of the HPLC sample introduction device (or injector) is to introduce the sample into the flowing solvent stream prior to the column so that it may be carried to the column and subsequently separated. The injector may be operated

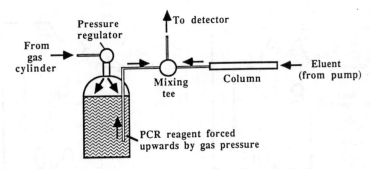

Fig. 5.8. Schematic diagram of a pneumatic vessel designed for the delivery of post-column derivatizing reagent. Reprinted with permission from Haddad and Jackson, *Journal of Chromatography Library Series No. 46*, Elsevier, Amsterdam.

either manually via a hand-held microsyringe or automatically, to enable unattended injection. The injector should be designed to minimize dispersion, or band broadening. It should also allow the sample to be introduced without overly disturbing the solvent flow, as this may result in significant baseline disturbances in flow-sensitive detectors, such as refractive index or conductivity. In some instances, the sample injector may also need to withstand high temperatures (up to 150°C) as certain polymer analyses require that the injector be heated in order to keep the sample soluble [12].

5.3.1 Manual injection

The simplest form of injector is the septum device, commonly used in GC, as described in Section 3.3. This device involves injecting the sample into the pressurized solvent stream through a self-sealing elastic septum using a microsyringe. The greatest drawback of such devices is that they are typically limited to a maximum operating pressure of 1500 p.s.i. [5], hence their use is not widespread in HPLC. Septum injectors have found use in low-pressure LC applications, such as protein separations [14].

Valve-type injectors are by far the most widely used HPLC sample introduction devices. Valve injectors allow reproducible introduction of the sample into the pressurized mobile phase without significant interruption of the flow, even at elevated temperatures. One of the simplest and most common valve injectors is the six-port Valco (or Rheodyne) injector, as illustrated in Fig. 5.9.

In order to inject a sample, the injector valve is turned to the load position as shown in Fig. 5.9 (a). The mobile phase bypasses the sample loop and flows directly to the column, allowing the sample loop to be filled with a microsyringe through the needle port. The valve is then turned to the inject position, as shown in Fig. 5.9 (b) and the mobile phase is used to back-flush the sample from the loop through to the column. The sample loop size is varied depending upon the desired injection volume. It is recommended that the loop be flushed with a volume of sample of at least twice the loop volume when using the injector in the complete-fill mode, as up to 20% of the sample is lost out the vent tube [15]. This also prevents sample

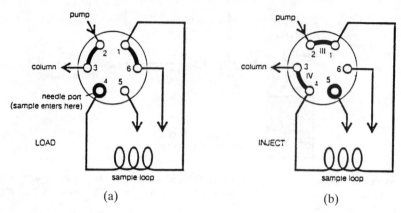

Fig. 5.9. Operation of a six-port high-pressure valve injector: (a) valve in 'load' position; (b) valve in inject position. Diagram courtesy of Rheodyne Incorporated.

carry-over between injections. The injector can also be used in the partial-fill mode, but the volume injected should be <50% of the loop volume. One disadvantage of this injector type is that the accuracy of the injection varies according to the size of the loop, ranging from about 5% for a 2 ml loop to about 30% for a 5 μl loop when used in the complete-fill mode. Hence, while the complete-fill mode offers the best precision, the partial-fill mode must be used in order to know accurately the exact volume injected. In the complete-fill mode, the precision of injection is 0.05% to 1%, depending on the volume loaded; five full loop volumes of sample produces a precision of about 0.1%. The precision of the partial-fill mode varies from 0.2% to 2% relative standard deviation (RSD), depending upon the skill of the operator and the type of syringe used [15].

Except for the minor inconvenience of having to change loops in order to vary the sample injection volume, the six-port valve injector is very reliable and easy to use. Care must be taken to ensure that the valve is switched rapidly from the load to the inject position since the flow to the column is stopped if the rotor is left half-way between these two positions. An example of a more flexible alternative to the six-port valve injector is typified by the Waters U6K universal injector, as illustrated in Fig. 5.10.

This injector has two levers which manually operate the flapper valves A, B and C depicted in Fig. 5.10. With the top lever in the 'load' position, all three valves are closed as in Fig. 5.10(a). The sample loop is isolated and the mobile phase flow passes through the restrictor loop to the column. The lower lever is then turned, opening the vent valve C, as shown in Fig. 5.10(b). This enables the loading plug to be removed and the sample to be loaded at the front of the loop, displacing an equivalent volume of solvent out through the vent valve. The loop is typically 2 ml in volume and the sample must be loaded accurately using a glass microsyringe. Finally, as shown in Fig. 5.10(c), the lower lever is rotated, closing valve C and the top lever returned to the inject position opening valves A and B. The mobile phase then flows through both the sample and restrictor loops with the majority going through the sample loop as a result of its wider bore (therefore lower pressure), back-flushing the sample to the column [16]. This step is also used to trigger the

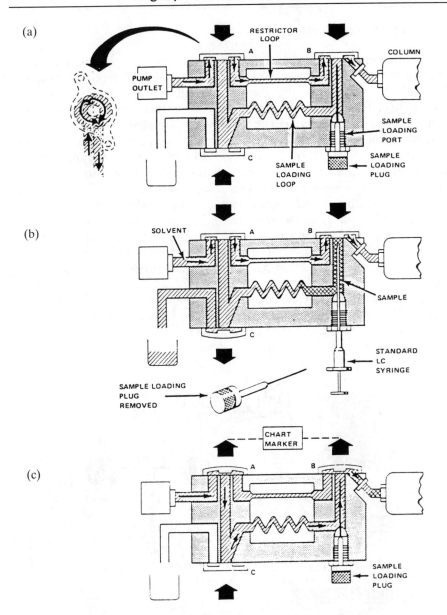

Fig. 5.10. Operation of the Waters U6K universal injector. Diagram courtesy of Waters Chromatography Division of Millipore.

chart marker or data station. This injector type is the most convenient sampling device, as both very large or very small volumes can be easily introduced, although such designs are more expensive and also require more maintenance than the six-port valve-type injectors. The injection precision again depends upon the skill of the operator and the type of syringe used and is similar to that obtained for the six-port valve-type injector when used in partial fill mode, i.e. 0.2% to 2% RSD.

5.3.2   Automated injection

Automated sample introduction devices, commonly referred to as autosamplers or autoinjectors, function in essentially the same fashion as the valve-operated manual injectors described above, except that the sample is introduced from a vial held in a sample carousel (or tray) using a syringe assembly controlled by a stepping motor and the valves are automatically actuated. Early autosamplers used pneumatically actuated valves; however, the majority of new autosamplers available now use electrically actuated valves [17]. In addition to the convenience of allowing continuous unattended sample introduction, autosamplers may also incorporate other features including sample heater–cooler compartments, the ability to perform internal standard additions and even carry out the addition and mixing of pre-column derivatization reagents prior to injection. While autosamplers are typically 5–10 times more expensive than manual injectors, the majority of new HPLC systems purchased contain an autosampler in preference to a manual injector. Figure 5.11 illustrates the normal flow-path of a typical autoinjector, the Waters 717 autosampler.

The flow-path and (electric) valve operation of the Waters 717 autosampler is virtually identical to that previously described for the U6K universal injector (Fig.

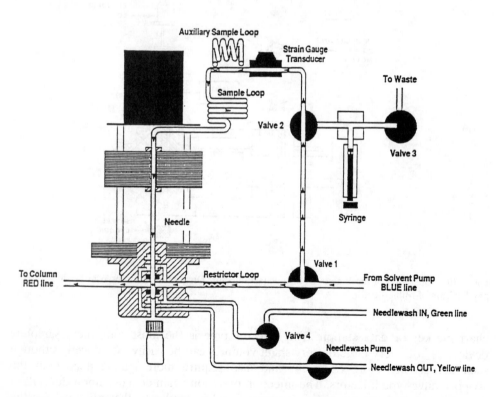

Fig. 5.11. Flow-path of a Waters 717 autosampler. Diagram courtesy of Waters Chromatography Division of Millipore.

5.10), except that an additional valve (V4) is incorporated which allows automatic syringe needle washing to prevent any sample carry-over. The performance of most modern autosamplers, in terms of precision and accuracy, is similar to (or better than) that of manual injectors. Autoinjectors can typically inject volumes from as low as 0.1 μl up to 2.0 ml with a linearity of >99% over the 1–100 μl range [17]. The precision is generally about 0.5% with less than 0.1% carry-over [17]. The sample capacity of most autoinjectors is in the order of 48–100 vials, typically held in a circular rotating carousel. There are exceptions which use an X–Y configuration rather than a rotating carousel and autoinjectors have been produced with a sample capacity as large as 500 vials, e.g. the Gilson 232/401. Most modern autoinjectors allow multiple injections per vial, random access programming, and priority sample interruption routines and usually can be controlled through the data station as well as through the front panel keypad.

Autoinjector vials come in a great variety of shapes, sizes and material of construction, depending upon the type of autosampler. The volume of the standard autosampler vial is in the order of 1–4 ml, although most suppliers offer limited-volume inserts which have low residual volumes (e.g. 5–10 μl) when the sample is limited [12]. The most common material of construction is clear borosilicate glass, but vials are also made from amber glass for light-sensitive samples and from polymers such as polypropylene, for cases where the sample may adhere to glass, e.g. proteins, or when ionic contaminants leached from the glass could interfere with the chromatographic separation, as occurs with ion chromatography [10]. The vial caps may be screw, snap or crimped types and septa, typically PTFE or silicone rubber, are available for single use only or can also be self-sealing to allow sample reinjection [12].

5.4   Column Packings and Hardware

5.4.1   Introduction

The three major contributions to band broadening in LC are eddy diffusion, molecular diffusion and resistance to mass transfer, as described in the van Deemter equation (equation 2.38) in Section 2.4.2. The effect of each of these processes on column performance can be related to a large number of variables including mobile phase flow-rate (or linear velocity), column length, average particle size, particle size distribution, column configuration, porosity of packing, solvent viscosity, shape of packing, integrity of column bed, mass and volume of injection and nature of the chemical interaction. In this section, general aspects of column packings, configuration, hardware and their effect on chromatographic performance are discussed, while the chemical nature of the packing material and its effect on selectivity, retention and resolution is covered in Chapter 6.

5.4.2   Column packings

Modern HPLC packings can be broadly classified according to the type, size, shape, nature and material of the particle. A wide variety of different particle types is used

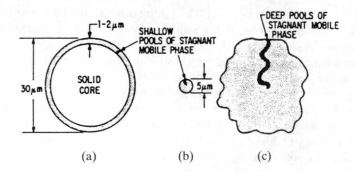

(a)　　　　　　　(b)　　　　(c)

Fig. 5.12. Particle types used in HPLC columns: (a) superficially porous particle; (b) very small totally porous particle; (c) totally porous particle. Reprinted with permission from Snyder and Kirkland (1979), *Introduction to Modern Liquid Chromatography*. John Wiley, New York.

in HPLC columns. Pellicular particles have a solid, inner core and a thin outer surface layer of stationary phase. Alternatively, the outer layer of coated stationary phase can be porous, creating a superficially porous particle. These pellicular materials consist of solid, spherical glass beads of relatively large diameter (e.g. 30 μm), with a thin (approximately 1 μm) layer of porous silica on the surface, as typified by the Zipax range of columns from Du Pont [10]. Porous particles are materials of relatively large diameter (e.g. 30 μm), which are fully porous and can be either irregular or spherical in shape. Microparticulate materials are of smaller diameter (e.g. 3–10 μm), fully porous, and again can be either irregular or spherical in shape. Figure 5.12 illustrates some of the particle types that are used in modern HPLC columns.

Pellicular materials give significantly higher efficiencies (or lower HETP values) than porous particles of the same size as a result of reducing the C term (resistance to mass transfer) in the van Deemter equation. Microparticulate (i.e. smaller diameter) materials give higher efficiencies than larger porous materials as the resistance to mobile phase mass transfer decreases with particle diameter. Smaller particles also reduce the contribution to the A term (eddy diffusion), again resulting in increased column efficiency. Pellicular and microparticulate materials can provide similar chromatographic efficiencies; however, pellicular materials are restricted to small sample loadings because of their low active surface area [10], but are more easily packed into columns than are microparticulate materials. The chromatographic properties of spherical and irregular microparticulate materials are essentially similar in terms of efficiency, but the structure of a bed of irregular particles is less stable than a dense, well-packed bed of spherical particles [5]. This makes preparation of columns packed with irregular particles more difficult and also limits their stability when operated continuously at high pressures. In general, highest column efficiencies are obtained if the stationary phase thickness is minimized, columns are packed with small particles and the packing procedure results in a dense, tightly compacted and uniform column bed. The majority of HPLC columns today are packed with spherical, microparticulate (3–10 μm diameter) materials, which results in stable, high-efficiency columns which can be used for relatively large sample loadings.

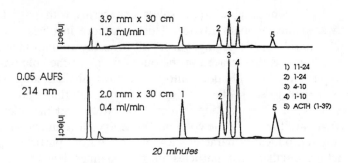

Fig. 5.13. Comparison of sensitivity obtained using 3.9 mm and 2.0 mm i.d. columns. A Waters μBondapak C$_{18}$ column was used with a linear gradient of 10–60% acetonitrile/water over 20 min Two-hundred and fifty picomoles of ACTH peptides were injected. Reprinted with permission from *Developing HPLC Separations, Book One* (1991), Millipore Corporation, Milford.

In addition to the physical nature of the particle (i.e. microporous or pellicular) and its size and shape, the particle material also significantly affects the performance of the HPLC packing. Column packing alternatives include the use of rigid solids (most commonly silica), resins (usually polystyrene divinylbenzene) and soft gels. Rigid solids based on silica are the most common HPLC packings used today, particularly for adsorption and reversed-phase modes of chromatography. Silica packings can withstand the high pressures generated when 10–30 cm columns packed with 3–10 μm particles are used. Silica is abundant and available in a variety of shapes, sizes and degrees of porosity. Most importantly, silica can be readily functionalized and the chemistry of its bonding reactions is relatively well understood. Silica also has a number of disadvantages, largely as a result of its pH instability and its high surface activity. Other rigid particles which have been used as HPLC column packings include alumina, zircon and carbon, although these materials have yet to find widespread acceptance.

Resin-based packings are increasingly widely used in HPLC columns. These packings are predominantly used in gel permeation chromatography (GPC) and ion-exchange chromatography, but resin-based, reversed-phase columns are also available commercially [18]. Resin-based columns have the advantage that they can be used over a wide pH range, although most resin types are limited to moderate operating pressures (1000–2000 p.s.i.) and mobile phases are often restricted to those with low concentrations of organic modifiers [10]. Soft gels, such as agarose and Sephadex, are used almost exclusively for the separation of aqueous proteins; however, they cannot tolerate very high backpressure [5]. More detail on specific column types, applications and their advantages and limitations will be given in Chapter 6.

5.4.3 Column packing methods

The two most common methods for packing HPLC columns are the dry-fill procedure and the wet-fill (or slurry-packing) procedure. The dry-fill procedure is recommended for the packing of rigid solids and resins with particle diameters $>20\,\mu m$, such as pellicular materials [5]. This procedure involves first degreasing then drying the interior of the tubing which will form the column blank. A porous screen (typically $2\,\mu m$) is then placed in the outlet fitting of the column and a small amount of the packing material added into the vertically held column via a funnel. The column is tapped to settle the packing and more material is added and so on until the column is full. The packing is then levelled off and the inlet fitting with screen is screwed onto the top of the column. While dry-fill columns can be prepared to give relatively reproducible characteristics for large porous particles and pellicular materials, packing smaller size particles in this manner leads to poor column efficiency. Small particles tend to agglomerate because of their high surface energy to mass ratio, resulting in a poorly compacted column bed. This results in widely varying flow velocities along the column, creating significant band broadening and hence poor efficiency [5]. The dry-fill method is used routinely for preparing columns for preparative chromatography and also for solid phase extraction cartridges, which are used for sample clean-up, as discussed in Chapter 8.

The wet-fill or slurry packing method of column preparation uses a suitable liquid to suspend the particles, which are then pumped under high pressure into the column blank. The suspending solvent must be chosen to maintain a uniform particle distribution without agglomeration and must also wet the packing thoroughly. High surface energy materials, such as unfunctionalized silica, require polar solvents, while lower surface energy packings, such as C_{18} functionalized silica, may be packed in less polar solvents. The slurrying solvent density should also be considered, particularly when packing particles $>10\,\mu m$ in diameter. For these larger particles, the solvent density should be similar to that of the particles themselves in order to reduce settling, particularly if the particle size distribution is large. In order to pack a column, the packing is slurried at modest concentrations with the solvent and placed in a specially designed reservoir, often called a column bomb, which fits on the inlet end of the column blank. A porous screen ($1–2\,\mu m$) is placed at the outlet end of the column and the solvent is pumped into the bomb at high velocity with a constant-pressure HPLC pump. This forces the slurried packing into the column and produces a compact bed. The packing is complete when a constant flow-rate from the column is finally obtained, after which the bomb is removed, the packing levelled off and the inlet fitting with screen screwed onto the top of the column. The column may be packed in either the upward or downward direction, with the majority of columns being packed in the downward direction.

Both rigid solid and resin-based columns can be prepared in this fashion, although resins must first be allowed to swell in the solvent before packing. Owing to their decreased rigidity, lower packing pressures are normally used for resin-based columns. A standard $25\,cm \times 4.6\,mm$ i.d. HPLC column contains approximately 2–3 g of material if microparticulate silica particles are used. For a more detailed discussion of column packing techniques, see Snyder and Kirkland [5]. While column blanks and packing apparatus are available from column suppliers, such as Alltech, the vast majority of HPLC columns are purchased prepacked by the supplier. The two main reasons for this are that it takes both time and considerable skill to produce reproducible, efficient columns, and most column manufacturers do not sell their analytical scale packings in bulk, preferring to sell prepacked columns.

5.4.4 Column configuration and hardware

The design of the HPLC column has been the subject of much research over many years and reviews on the subject are still published frequently [e.g. 19, 20]. Scott has recently described the development of column theory over the past 30 years of HPLC and its impact on column design [19]. The column must be constructed of materials that withstand the pressures used in HPLC and are also chemically resistant to the mobile phase. As is the case with HPLC pumps, the majority of columns are constructed from 316 stainless steel. Stainless steel columns may be constructed using either a rigid single wall or a double wall. Columns are also available in inert materials, such as glass, teflon and PEEK [18]. Such columns are used in cases where chemically aggressive mobile phases, such as HCl, are used or when the sample may adsorb onto the stainless steel surface, as is the case with some proteins [9]. Stainless steel materials offer great rigidity and can be operated at high pressures, hence are most widely used for silica and bonded phase packings. Polymeric column materials are commonly used for ion-exchange packings, while glass columns are used mostly for protein separations. Both polymeric and glass columns are subject to stringent pressure limitations and care should be taken not to exceed the manufacturer's recommended operating pressures when using such columns.

HPLC columns are available in a wide variety of configurations, shapes and sizes. Most columns are packed in straight sections of tubing fitted with zero dead volume end couplings. Care must always be taken to ensure that compression screws and ferrules being used are compatible with the column end fittings. Generally, compression screws from one manufacturer cannot be used with another's column without permanently damaging the column. Cartridge columns are available as an alternative to the standard steel column, as typified by the Brownlee range of columns. With such columns, the body of the column may be replaced and the old end fittings reused. While the initial purchase price of a cartridge column is usually higher than an equivalent steel column, the replacement cartridges are less expensive, hence they become cost-effective if the user frequently needs to replace columns.

Yet another alternative to the conventional steel column is the radial compression cartridge. In this format, the stationary phase is packed into a teflon cartridge, which is compressed by applying an external pressure radially to the cartridge before use. The pressure is applied by compressing either a liquid or a gas surrounding the cartridge. Reforming the packing bed and removing any voids and channels each time the column is used leads to greater stability, improved durability and efficiency [18]. This type of cartridge reduces the 'wall effects' which result from the fact that there is less resistance to solvent flow near the rigid walls of a conventional column because of the lower density of the packed bed. This wall effect is a significant source of band broadening and the efficiency of radial compression columns has been shown to be up to 25% higher than equivalent steel columns [18].

In addition to choosing the packing material and column hardware, it is also necessary to select the appropriate column dimensions. Column manufacturers provide their packing materials in columns of various lengths and diameters. Generally, the longer the column, the higher its efficiency and resolution; the shorter the column, the faster the separation. Also, the larger the column diameter, the greater is its loading capacity, while the narrower the column, the greater is its mass sensitivity. Most analytical columns range from 50 to 300 mm in length and 2.0 to 5.0 mm in diameter. The use of shorter columns provides rapid separations but limits resolving power; hence, this approach should only be used for simple samples [19]. Some resin-based materials, such as polymethacrylate, are often packed only into short columns in order to keep the pressure drop across the column as low as possible. Longer columns provide greater resolving power, although at the expense of increased chromatographic run times. Narrow bore (or microbore) columns with diameters of 1–2 mm or less are used for applications where high sensitivity is required, where the amount of sample is limited or where solvent (and disposal costs) are significant. Flow rates of 0.05 to 1.0 ml min^{-1} are used with narrow-bore columns. However, to obtain the best results from such columns, it is critical that sources of bandspreading within the system be minimized. Figure 5.13 compares the sensitivity obtained for a mixture of synthetic fragments of ACTH using 3.9 and 2.0 mm i.d. C_{18} reversed-phase columns [21].

Conversely, if larger sample loadings are required than conventional columns allow, larger internal diameter columns should be used. Sample load is proportional to the column length and the square of the column diameter. Large-bore columns are most frequently used in preparative chromatography, as discussed in Section 5.8. In summary, while HPLC columns can be prepared with efficiencies over 1 million theoretical plates, the columns dimensions (7000 × 0.02 cm i.d.) and excessive run times (>7 days) make them unsuitable for routine use [19]. In practical terms, the most commonly used columns are 250–300 mm in length with a diameter of 3.9–4.6 mm, although 150 mm columns are becoming increasingly popular. Columns in this size range offer the best overall compromise in terms of efficiency, sample loading and chromatographic run times, without the need for excessive attention to instrument performance in regard to bandspreading.

5.4.5 Measuring column performance

Column performance can be evaluated using a number of parameters, although traditionally the number of theoretical plates (N) is used as a measure of column efficiency. Other parameters of importance include peak asymmetry, capacity factor, selectivity, resolution of a critical peak pair, pressure drop across the column, etc. Methods for calculating these parameters have been given in Chapter 2. Ideally, all of the above parameters should be monitored continuously and the data recorded. The concept of system suitability is based on careful monitoring and documentation of the column and overall system performance in order to set up control specifications and limits for given assays. This quality approach is becoming increasingly important, particularly in pharmaceutical laboratories. The latest versions of chromatography data software now calculate and automatically document a wide range of system parameters, such as those listed above. In addition to indicating if the system meets the assay specifications, continual monitoring of the column and system performance provides a benchmark to consult when trouble-shooting a chromatographic problem. It also allows the user to gain longer term

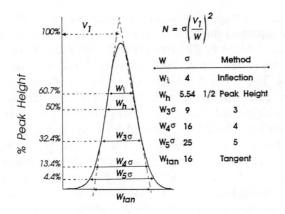

Fig. 5.14. Alternative methods for calculating column efficiency. Reprinted with permission from *Developing HPLC Separations, Book One* (1991), Millipore Corporation, Milford.

data, such as column lifetime, which aids in maintaining stock levels and even deciding if a column has lasted for a reasonable number of injections.

As mentioned previously, determination of N is the most widely used means of 'testing' a column. Almost all column manufacturers provide a set of plate count conditions and a minimum specification of N with their columns. All new columns should be tested according to the manufacturer's specifications when first received to ensure that these specifications are met. If a column fails to meet specifications when new, it can usually be replaced under warranty. The value of N is a measurement of the extent of band broadening in the column and the chromatographic system, as discussed in Section 2.4. The combined effects of each of the processes responsible for this broadening can be described mathematically in terms of the variance (square of the standard deviation) of the Gaussian peak. The value of N can be measured using a number of either manual or statistical moment calculations. Figure 5.14 shows some of the graphical methods for calculating N. The five sigma method is the recommended method [22] since it provides improved accuracy over the more convenient tangent method and is more sensitive to peak distortion so that it gives a more realistic assessment of column performance. Table 5.1 gives the conditions for measuring N for a Waters μBondapak C_{18} reversed-phase column (300×3.9 mm i.d.) and provides a typical example of a manufacturer's recommended conditions.

Table 5.1 Recommended platecount conditions for a Waters μBondapak C_{18} column.

Mobile phase	Flow rate	Test sample	Detector settings
60 : 40 acetonitrile and water	2.5 ml min^{-1}	20 mg uracil and 200 mg acenaphthene in 100 ml methanol	254 nm, 1.0 AUFs*

*Absorbance units full scale.

The chart speed, injection volume and detector sensitivity should be adjusted to give a suitable chromatogram with a chart recorder deflection about 50% of full scale. An injection volume of 10 μl is suitable and a chart speed of 10 cm min^{-1} enables accurate measurement of the peak width and retention volume (V_1), as shown in Fig. 5.14. The uracil is not retained under the above conditions and is used as a void marker peak to enable calculation of k' for the acenaphthene peak. Acenaphthene is widely used as a test (probe) solute for platecounting C$_{18}$ columns. Mobile phase conditions should be adjusted for other columns such that the probe solute has a moderate k' (approximately 5). The effective plate number, as discussed in Section 2.4.1, is a more appropriate means to calculate column efficiency for early eluting peaks. It is also noteworthy that for some column types, such as ion-exchangers, manufacturers will use the half peak height method as these columns generally produce somewhat distorted (fronted) peaks.

5.4.6   Column care and use

HPLC columns are a consumable item and no column will maintain satisfactory performance indefinitely. However, column life (and performance) can be enhanced by taking appropriate care of the column. First, the manufacturer's recommendations regarding mobile phase pH, flow-rates, organic modifier content, temperature, maximum operating pressure, etc., should be followed carefully. For instance, bonded phase C$_{18}$ columns should be operated in the pH range 2–8, as mobile phases <pH 2 will result in hydrolysis of the functional group and loss of retention, while pH values >8 will cause the support to dissolve and create voids in the stationary phase. Usually, operating a column for prolonged periods close to the recommended limits will also reduce its life.

Second, use HPLC grade water (such as Millipore Milli-Q 18 megohm water), solvents and analytical reagent grade chemicals to prepare solvents and standards. Mobile phases should always be filtered through an appropriate solvent compatible 0.45 (or 0.22) μm filter and vacuum degassed before use. Mobile phases for reversed-phase gradient separations should also be sparged with helium. Samples should be filtered through an appropriate 0.45 μm filter to remove particulate matter before injection. In many cases, the use of sample pretreatment, such as solid phase extraction, will also prolong column life.

Third, mobile phase flow-rates should be altered in small increments (i.e. 0.2 ml min^{-1}) to avoid sudden column backpressure changes, particularly for softer stationary phase materials, such as those found in some resin-based columns. Columns should not be back-flushed, that is they should only be used in the flow direction indicated on the column. When not in constant use, the column should be stored according to the manufacturer's recommendations. Columns should normally not be stored in buffers, particularly those which promote microbial growth, such as phosphate buffers. Bonded phase C$_{18}$ columns can simply be stored in pure methanol.

Finally, guard columns should be used wherever possible. Guard columns are very short columns (0.5–3.0 cm) which are packed with the same stationary phase (or equivalent) as that used in the analytical column. These guard columns retain any material which would otherwise permanently bind to the analytical column.

Guard columns are intended to be used only for short periods and should be replaced after a limited number of injections, usually in the order of 50–200. While they may marginally decrease the separation efficiency [5], their use will significantly enhance column life.

Column heaters will also enhance column performance for some applications. Most HPLC instrument manufacturers sell heating devices which will conveniently regulate a column temperature to within 0.1 to 1.0°C, depending upon the particular device [12]. Many separations, such as sugars and amino acids, require elevated temperatures in order to achieve appropriate resolution. Moreover, high-sensitivity analyses, particularly those with conductivity or refractive index detectors, will often be improved if the column is thermally stabilized.

In conclusion, a well-treated column can last for many thousands of injections, but the use of one inappropriate mobile phase or untreated sample may drastically reduce a column's performance. Careful attention to the points listed above will enhance the column life. In addition, a great deal of help is readily available regarding column information and troubleshooting. Many manufacturers of HPLC instrumentation now provide free phone support lines to answer questions regarding column and system performance criteria. Chromatography journals, such as *LC.GC*, include discussions on column troubleshooting and performance [23] and a number of books on HPLC troubleshooting are available [e.g. 24]. Virtually all columns are supplied with a manual which should be followed carefully. Instrument manuals are also a useful source of troubleshooting information.

5.5   Detectors

5.5.1   Introduction

To date, there is no single detector which can be employed for all HPLC separations and a number of different detector types are used commonly. It is convenient to broadly classify these detectors according to the property which forms the basis of the detection method. Thus, solute property detectors respond to a physical property of the solute that is generally not exhibited (to any significant degree) by the mobile phase. These detectors are usually specific in that they respond only to solutes having the particular physical property which is being monitored, and also are often very sensitive. Bulk-property detectors are those which compare an overall change in some physical property of the mobile phase with and without an eluted solute. These detectors give response to a wide range of solutes but are often insensitive. Figure 5.15 shows a schematic chart listing some of the more important examples of each type of detector. This chart forms the basis of the discussion of HPLC detectors included in this chapter.

The normal mode of operation of a detector in HPLC is the monitoring of a signal (due to the analyte) which appears as an increase above the background signal due to the mobile phase alone. This is called direct detection and applies to all separation modes in which the background composition of the mobile phase is not altered by the presence of the analyte. Direct detection will be a useful mode whenever the background detector signal due to the mobile phase alone is small enough to be offset by the zeroing control on the detector. In some chromatographic modes, an alternative detection method, namely indirect detection, is also possible. Indirect detection involves the measurement of a decrease in detector signal when the analyte is eluted and is generally used with mobile phases which give a high background signal. To function correctly, indirect detection requires that the background composition of the mobile phase must alter in the presence of the eluted analyte and it is this change, rather than a specific characteristic of the analyte itself, which is monitored. Ion-exchange chromatography is the most important example of a separation mode which is suited to indirect detection. The reasons for this are discussed more fully in Chapter 6.

Important performance criteria for HPLC detectors

Some of the more important properties of chromatographic detectors were discussed earlier in Section 3.7. These properties include sensitivity (and how it can be measured), linear dynamic range, minimum detectable quantity (and its measurement), baseline noise, dispersion, time constant and response time. Whilst these terms were discussed in the context of detectors for gas chromatography, they are equally valid when applied to HPLC detectors. An ideal HPLC detector might be considered to be one which exhibits the following characteristics [5]:

- High sensitivity.
- Negligible baseline noise.
- Large linear dynamic range.
- Response independent of variations in operating parameters, such as pressure, flow-rate, temperature, etc.
- Response independent of mobile phase composition.
- Low dead volume.
- Non-destructive of the sample.
- Stable over long periods of operation.
- Convenient and reliable to operate.
- Inexpensive to purchase and operate.
- Capable of providing information on the identity of the solute.

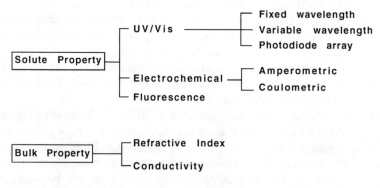

Fig. 5.15. Schematic classification of different types of HPLC detectors.

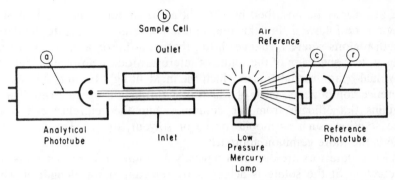

Fig. 5.16. Schematic view of a monochromatic u.v. photometric detector. Reprinted with permission of Du Pont Instrument Products Division.

The detectors listed in Fig 5.15 do not meet all of these criteria but, nevertheless, account for the vast majority of HPLC applications performed routinely. They can therefore be considered to be the most important detectors and will be discussed in descending order of frequency of use. The same general framework will be used for each detector and this will include the principles of operation, operating characteristics and typical applications.

5.5.2   u.v.–Visible light absorbance detectors

Principles of operation of u.v.–visible absorbance detectors

u.v.–Visible light absorbance detectors are by far the most commonly employed detectors in HPLC. This type of detector is normally the first choice when purchasing an HPLC system. The detector operates on exactly the same principles as a spectrophotometer and a general schematic diagram is shown in Fig 5.16. Light from a suitable radiation source is passed through a monochromating device (such as a filter or a grating) and thence to a cell through which the mobile phase flows. The amount of transmitted light is measured using a suitable photodetector or, in some cases, an array of photodetectors.

If light of intensity (or power) P_0 is focused onto the cell and light of power P is transmitted, then from the Beer–Lambert Law we can write:

$$A = \log\left(\frac{P_0}{P}\right) = \varepsilon bC \qquad 5.1$$

where A is the absorbance of the solution in the cell, b is the optical path-length through the cell (in cm), ε is the molar absorptivity of the solute at the particular wavelength used (in $l\,mol^{-1}\,cm^{-1}$), and C is the molar concentration of the solute. At a fixed wavelength a linear relationship exists between absorbance and concentration. This relationship is strictly valid only when truly monochromatic light is used, which does not occur in the case of HPLC detectors which operate typically with a narrow band of wavelengths. If the absorbance was measured in a region where ε changes rapidly with wavelength, then the different wavelengths comprising the light beam may be absorbed by very different amounts and the Beer–Lambert Law may not be followed. For this reason it is important to operate the detector in wavelength regions where ε changes little; that is, at maxima, minima or shoulders in the absorption spectrum of the solute of interest. Modern absorption detectors are generally dual-beam instruments in which the incident light beam is split into sample and reference components, each of which pass through a separate cell. The sample cell contains the effluent from the column, while the reference cell contains a reference solution, such as mobile phase or solvent, or air. The use of air as a reference is the more common approach.

u.v.–Visible detectors are operated typically as solute property detectors, wherein direct detection of the solute is achieved by selecting a wavelength at which the solute exhibits a high value of molar absorptivity. This mode of detection encompasses the use of absorption detectors in almost every form of HPLC. However, indirect detection is employed commonly in ion-exchange chromatography when a strongly absorbing mobile phase is used.

Classification of absorption detectors

Absorption detectors are classified primarily according to the frequency of radiation they measure. Ultraviolet (u.v.) detectors operate in the approximate wavelength range 190–400 nm, whilst u.v.–visible light detectors extend this range up to 700 nm or more. A secondary classification of absorption detectors centres on the range of wavelengths accessible to the detector, so that detectors fall into the following four categories: (i) fixed wavelength detectors; (ii) variable wavelength detectors; (iii) scanning detectors; and (iv) photodiode-array detectors.

Table 5.2 Typical wavelengths for light sources used in absorption detectors.

Source	Wavelength (nm) of emission line(s)
Mercury	254, 313, 365, 405, 436, 546, 578
Cadmium	229, 326
Zinc	214, 308
Magnesium	206
Deuterium	190–350 (continuum)

Fixed-wavelength detectors are the most simple type, with the wavelength used being determined principally by the nature of the light source used. Table 5.2 shows some of the emission lines for common light sources and indicates that mercury, cadmium, zinc and magnesium lamps provide u.v. radiation as sharp lines which can be filtered easily from radiation of neighbouring wavelengths. Deuterium lamps provide a continuum of wavelengths covering most of the u.v. spectrum. Mercury lamps are the most popular in fixed-wavelength detectors and are normally used at a wavelength of 254 nm. It has been estimated that almost two thirds of the organic solutes analysed by HPLC show some absorption of light at 254 nm, especially aromatic compounds which exhibit molar absorptivities of at least 10^4. The molar absorptivity of a compound increases with increasing unsaturation, with conjugated compounds being especially absorptive. While a mercury lamp operated at 254 nm is

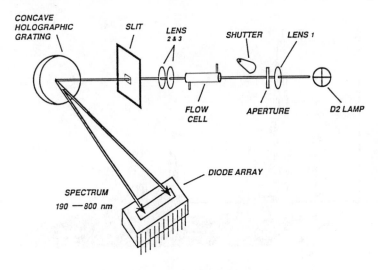

Fig. 5.17. Diode-array detector optics. Reprinted with permission of Hewlett Packard.

suitable for many applications, it is sometimes necessary to use a lower wavelength to detect compounds that have different chromophores. This is particularly true of compounds having the carboxylate functional group, for which a zinc lamp operated at 214 nm is required. It should be noted that most fixed-wavelength detectors are designed to enable the lamp (and filter) to be changed very easily.

Variable-wavelength detectors are those which have the facility to vary the detection wavelength in order to accommodate the absorption characteristics of a particular solute or group of solutes. This facility enables detection sensitivity to be maximized or the choice of a detection wavelength at which interfering solutes do not absorb appreciably. In order to achieve wavelength variability, a continuum source (usually deuterium) is used, together with a suitable variable monochromator (generally a grating type) and the instrument is operated in the dual-beam mode. Scanning detectors form a subset of the variable wavelength type and permit an absorption spectrum of an eluting solute to be obtained as a means of identification of the solute or to check whether two solutes have been co-eluted as a single peak. The absorption spectrum can be obtained by stopping the flow of mobile phase while the solute is in the detector cell and then scanning through the wavelength range accessible to the detector, or by the use of a fast-scanning method in which the spectrum is scanned in a very short time, usually while the mobile phase is flowing.

Much more rapid scanning can be achieved with the aid of a photodiode-array (PDA) detector, a schematic diagram of which is shown in Fig 5.17. It can be seen that the dispersion device is situated after the flow cell, which is the opposite case to that for the other u.v.–visible light detectors discussed above. For this reason, the optical arrangement used in the PDA is sometimes referred to as reversed optics. Light from the continuum source (usually a deuterium lamp) passes through an achromatic lens system which focuses polychromatic light onto the flow cell. The transmitted light then falls onto a holographic diffraction grating where it is dispersed onto a photodiode array consisting of several-hundred photosensitive

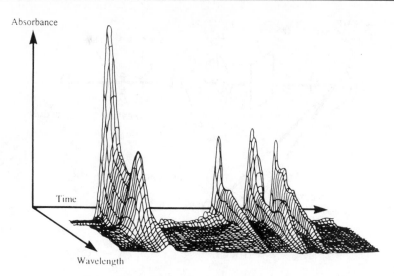

Absorbance

Time

Wavelength

Fig. 5.18. Typical output from a diode-array detector—absorbance data as a function of wavelength and time.

diodes generally configured in a linear pattern which mimics the focal plane of a spectrometer. Each photodiode is connected to a storage capacitor which is charged initially to a known level and is connected to a switch register. Light falling onto the photodiode discharges the storage capacitor to a level dependent on the intensity of the incident light. In each measurement cycle, which takes about 10 ms, the current required to recharge each storage capacitor is measured to provide quantitative data on the intensity of the light transmitted through the cell as a function of wavelength. That is, an absorption spectrum is produced in each measurement cycle. This process can be repeated many times each second so that a large amount of experimental data can be generated during the time that the sample spends in the cell. It is therefore mandatory that the PDA be supported with appropriate computer hardware and software. One form of the output generated by a PDA is illustrated in Fig. 5.18.

The range of wavelengths falling onto a particular photodiode depends both on the number of photodiodes employed to cover the spectral range of the detector and the slit width used. A typical PDA uses about 512 photodiodes to accommodate a spectral range of 190–800 nm, so that each photodiode has a bandwidth of about 2 nm. In practice, PDA detectors are somewhat less sensitive than other absorption detectors and the presence of multiple wavelengths of light in the sample cell increases the possibility of errors arising from secondary fluorescence. However, the advantages offered by the comprehensive spectral data recorded are great. Peak purity can be ascertained by overlaying spectra taken from different regions of the same peak and noting any changes which could be attributed to a co-eluted impurity. Absorbance ratioing techniques can also be used to achieve the same result. Chromatograms at any desired wavelength within the range accessible to the instrument can be displayed without reinjection of the sample and different solutes can therefore be quantified under conditions where they are measured most

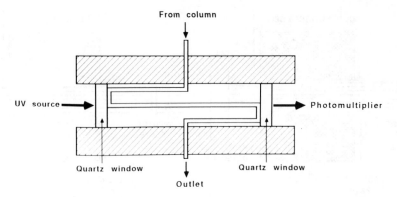

From column

UV source →

← Photomultiplier

Quartz window

Quartz window

Outlet

Fig. 5.19. Z-cell design for the flow-cell of an absorbance detector for HPLC.

sensitively or selectively. Finally, the identity of a solute can often be determined by comparison of its absorption spectrum with standard spectra. This process adds a further dimension to the process of identification using chromatographic retention-time data.

Cell design for u.v.–visible light detectors

The flow cell for an absorption detector should maximize the optical path length in order to increase the absorbance reading for a specified concentration of solute, while also confining the cell volume to the minimum possible size so that dispersion of the sample (and the resultant broadening of the chromatographic peak) is minimized. For this reason, flow-cells are usually constructed with narrow flow paths (approximately 1 mm diameter) and a length of about 1 cm; that is, the same optical path length as that used in a typical spectrophotometer. The geometry of the flow-cell can vary, but a commonly used arrangement is the Z-cell design shown in Fig 5.19.

A problem common to all absorption detectors is the fact that the refractive index of the solution passing through the cell may alter as a result of changes in temperature, flow-rate and mobile phase composition or because of differences in refractive index between the sample and the mobile phase. Changes in refractive index can produce a 'dynamic liquid lens' which will distort the light beam passing through the cell and may reduce the amount of light reaching the photodetector. This, in turn, will appear as an absorbance which may be misinterpreted as an eluted solute. Refractive index effects may be minimized, or even eliminated, by collimating the light beam within stringent limits, but this causes an appreciable loss in the intensity of the light beam reaching the cell. An alternative approach is to use a conical tapered cell to continuously eliminate the effect of the liquid lens as it moves through the cell. The diverging cell walls prevent the distorted light from impinging on the surfaces of the flow-cell. This process is illustrated in Fig. 5.20.

Operating characteristics of u.v.–visible light detectors

While u.v.–visible light detectors are the most rugged and reliable detectors available for HPLC, some factors should be taken into consideration during their

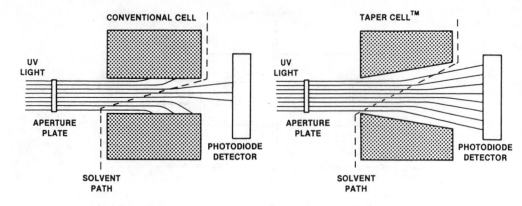

Fig. 5.20. Elimination of the liquid lens effects with a tapered flow-cell. Reprinted with permission of Waters Chromatography Division of Millipore.

use. First, it is imperative that the background absorbance of the mobile phase be kept as low as possible if direct detection is to be used. This requires some knowledge of the u.v. cut-off wavelengths (that is, the wavelength at which an absorbance of unity is recorded for a neat sample using a 1-cm path-length) and Table 5.4 and shows these cut-offs for some commonly employed solvents and buffers used in mobile phase preparation.

It should also be borne in mind when choosing the detection wavelength for a particular solute that the absorption spectrum of that solute will be dependent, to some extent, on the composition of the mobile phase. This is especially true when the mobile phase contains a buffer since many solutes, such as weak acids, show pronounced changes in absorption spectra with pH. For example, the wavelength of maximum absorption for ascorbic acid changes from 265 nm at pH 4.0 to 247 nm at pH 2.0.

While the u.v. region of the spectrum remains the most suitable for a wavelength for sensitive detection of most solutes determined by HPLC, the visible spectrum also offers some opportunities. Relatively few compounds show absorbance in the

Table 5.3   u.v. Cut-offs for some common solvents and buffers.

Solvent	u.v. Cut-off (nm)	Buffer	u.v. Cut-off (nm)
n-Pentane	190	Acetic acid, 1%	230
Carbon tetrachloride	265	Triethylamine, 1%	235
Methanol	205	Sodium citrate, 10 mM	225
Tetrahydrofuran	230	Sodium acetate, 10 mM	205
Chloroform	245	Tris HCl, 20 mM	204
Acetonitrile	190	Potassium phosphate, 10 mM	190
Dioxane	215	Ammonium bicarbonate, 10 mM	190
Ethanol	210	Sodium chloride, 1 M	208
Ethyl acetate	256	EDTA, disodium, 1 mM	190
Petroleum ether	210	Sodium dodecyl sulfate, 0.1%	190

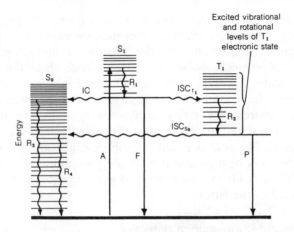

Fig. 5.21. Physical processes which can follow the absorption of a photon by a molecule. S denotes a singlet state and T denotes a triplet state. Solid arrows represent processes involving photons, while wavy arrows denote radiationless transitions. A, absorption; F, fluorescence; P, phosphorescence; IC, internal conversion; ISC, intersystem crossing; R, vibrational relaxation.

visible region, but derivatization reactions (both pre- and post-column) producing coloured products are frequently employed as a means to increase both the sensitivity and selectivity of detection. This type of detection is perhaps the most common application of detection using wavelengths from the visible region. Derivatization reactions are discussed later in this chapter in Section 5.7.

5.5.3 Fluorescence detectors

Principles of operation of fluorescence detectors

Fluorescence involves the instantaneous emission of radiation from a molecule which has attained an excited electronic state after the absorption of radiation. The fluorescence process can therefore be seen to involve two steps: excitation and emission. In the excitation step, absorption of a photon causes the solute molecule to move to an excited electronic state; for example, this absorption may result in the transition of the solute molecule from its ground singlet state to its first excited singlet state. In most cases the absorbed energy will become dissipated through a series of vibrational transitions which will result in the production of minute quantities of heat. This process will occur when the energy levels for the excited and ground states overlap, which is a common occurrence. However, when there is no overlap between energy levels the most convenient method by which the excited molecule can lose energy is through the emission of a photon. This process is termed fluorescence. The energies of both the absorbed and emitted photons may vary according to the actual vibrational energy levels between which the transitions occur, as illustrated in Fig 5.21. That is, both excitation and emission occur within a reasonably broad band of wavelengths, with the emission band being centred at longer wavelengths (lower energy) than the excitation band.

Figure 5.21 also shows that an alternative pathway for energy loss involves intersystem crossing in which transition between the excited singlet state and an excited triplet state occurs. This excited triplet state then returns to the ground state with the emission of a photon and the process is termed phosphorescence. Phosphorescence takes much longer than fluorescence since the transition from triplet to singlet states involves a change of electronic spin. For this reason, phosphorescence may have a life of several seconds after the excitation radiation is removed, whereas fluorescence emission has a very short life and for practical purposes can be considered to cease virtually immediately after the excitation radiation is removed. Fluorescence and phosphorescence are specific examples of the general phenomenon of luminescence. While phosphorescence can be used for detection in HPLC, it is far more common to use fluorescence and the discussion here will be confined to the latter.

Characteristics of fluorescence detection

The magnitude of the fluorescence signal (F) is given by:

$$F = f(\theta)\, g(\lambda)\, I_0\, \theta_f\, \varepsilon\, lc \qquad\qquad 5.2$$

where $f(\theta)$ is a geometry factor related to the positioning of the detector, $g(\lambda)$ is the wavelength response characteristics of the detector, I_0 is the intensity of the incident radiation, ϕ_f is the quantum yield of the analyte molecule, ε is the molar absorptivity of the analyte, l is the optical path length and c is the molar concentration of the analyte.

From this equation it can be seen that the fluorescence signal is proportional to analyte concentration, with the constant of proportionality being dependent, among other things, on the intensity of the excitation source and both the quantum yield and the molar absorptivity of the analyte. The quantum yield (which is a measure of the efficiency of the fluorescence process) is defined as the number of photons emitted per photon absorbed and has a maximum value of 1.0.

The fluorescence radiation is emitted in all directions and in order to avoid problems of differentiating between the excitation and emission radiation, it is usual for fluorescence detectors to operate in a right-angle configuration, such as that shown in Fig 5.22. A lamp of relatively high intensity (often a xenon or deuterium source) passes radiation through a monochromator (the excitation monochromator) into the flow-cell through which the column effluent passes. Fluorescence radiation is collected at right-angles to the excitation beam, where it is passed through a second monochromator (the emission monochromator), and thence to a photomultiplier. In many fluorescence detectors, the monochromators comprise simple filters with moderate spectral bandpasses (e.g. 50 nm). More sophisticated detectors have monochromators with gratings that offer very much smaller spectral bandpasses and hence greatly superior spectral resolution. In addition, some detectors permit rapid scanning of both the excitation and emission monochromators (often after stopping the eluent flow with the analyte trapped in the flow-cell) so that emission and excitation spectra of the analyte can be obtained. Flow-cells are generally constructed from quartz since common excitation wavelengths fall into the low u.v. region.

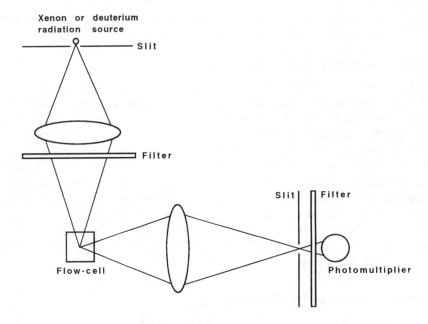

Fig. 5.22.   Schematic layout of a fluorescence detector.

As with all luminescence techniques, fluorescence detection suffers from a number of drawbacks, including a marked dependence of the fluorescence signal on a range of experimental parameters, such as the mobile phase pH, the nature of the components of the mobile phase, the temperature, the concentration of the analyte, and quenching effects. Effects of pH are most pronounced when the solute contains acidic or basic functional groups and with such solutes, it is common for fluorescence detection to be applicable over only a narrow pH range, necessitating the addition of buffers to the mobile phase. The composition of the mobile phase, especially the nature of any organic modifier used, may also exert a major effect on fluorescence output with the dielectric constant of the modifier often governing the wavelength of maximum emission. Temperature is also an important variable since a reduction in temperature can minimize vibrational relaxation processes which are competitive to fluorescence, thereby causing an increase in the fluorescence output. The possibility exists for emitted fluorescence radiation to be reabsorbed by adjacent, unexcited analyte molecules. This process, known as self-absorption, becomes more probable as the concentration of the analyte increases and will result in the appearance of a nonlinear calibration plot of fluorescence intensity versus analyte concentration. In such circumstances care must be taken to ensure that measurements are taken on the linear portion of the calibration plot. Finally, it should be recognized that the fluorescence signal can be reduced by the presence of quenching agents which, through interactions with the excited analyte molecule, enable it to relax to the ground state without emission of radiation. Molecular oxygen is a common quenching agent and in some cases must be removed from the mobile phase if the fluorescence signal is to be maximized. Despite these shortcomings, fluorescence detection affords detection limits which are often 100 times lower than those achieved with u.v.–visible light absorbance detection.

Applications of fluorescence detection

Molecules which are strongly conjugated and have a rigid structure are likely to exhibit fluorescence, with aromatic and polyaromatic compounds being good candidates for fluorescence detection. Nevertheless, the number of solutes which show native fluorescence is relatively small, so that fluorescence detection is often very selective which allows it to be applied to the detection of trace quantities of analyte in complex sample matrices. Examples of analytes detected in this manner include adrenaline, aflatoxins, polynuclear aromatic hydrocarbons, and oestrogen. The applicability of fluorescence detection can be enhanced considerably through the use of derivatization reactions. For example, the introduction of electron-donating groups (such as OH) to aromatic systems increases fluorescence efficiency, as illustrated by the quantum yields of coumarin and 7-hydroxycoumarin, which are approximately 0.0001 and 0.50, respectively.

5.5.4   Electrochemical detection

Definitions

The term electrochemical detection is applied loosely to describe a range of detection techniques involving the application of an electric potential (via suitable electrodes) to a sample solution, followed by measurement of the resultant current. Conductivity detection, which is discussed later, quite properly falls into this category, but is usually treated as a distinct detection method. For the purposes of the present discussion, we will interpret electrochemical detection to embrace the techniques of voltammetry, amperometry and coulometry. The common characteristic of these techniques is that a chemical reaction (e.g. a Faradaic oxidation or reduction) occurs during the measurement.

Voltammetry is a well-established technique which involves the application of a changing potential (measured with respect to a reference electrode) to a working electrode, followed by measurement of the current resulting from the reaction of analysed species at the working electrode. Within the general field of voltammetry, we can identify polarographic techniques as those which measure current–voltage relationships using mercury as the working electrode. The key factor in voltammetry (and polarography) is that the applied potential is varied over the course of the measurement.

The term amperometry describes the technique in which a fixed potential (again measured with respect to a reference electrode) is applied to a working electrode and the current resulting from oxidation or reduction reactions occurring at the working electrode is measured. In the case of chromatographic detection, the working electrode is located in a suitable flow-cell, through which the eluent stream passes. The analyte to be detected undergoes a Faradaic reaction if the applied potential has appropriate polarity and magnitude. However, the surface area of the working electrode in amperometry is generally quite small ($0.5\,cm^2$ or less), so the Faradaic reaction of the analyte is incomplete, causing only a fraction of the total

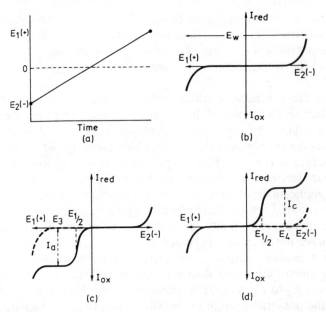

Fig. 5.23. Voltammograms obtained by applying the potential profile illustrated in (a) to a solution of supporting electrolyte alone (b) and to the same electrolyte containing an oxidizable species (c) or a reducible species (d).

analyte to react; less than 10% of the analyte is reacted in a typical amperometric flow-cell at flow-rates around 1 ml min^{-1}. The use of working electrodes of larger surface area can lead to quantitative reaction of the analyte at the electrode, and when this occurs, the technique used is described as high-efficiency amperometry, or coulometry. Thus, amperometry and coulometry can be distinguished by the extent to which the analyte undergoes a Faradaic reaction at the working electrode.

Interrelationships between voltammetry, amperometry, and coulometry

Consider the situation where a working electrode is inserted into a solution containing an electrolyte (e.g. KCl), but with no electrochemically active (i.e. oxidizable or reducible) solute present. The potential of the working electrode is now varied with time from a negative value (E_2) through to a positive value (E_1). This potential–time profile is shown in Fig. 5.23(a). The experimental configuration will require that a reference electrode be in electrolytic contact (but not necessarily inserted into) the sample solution. Current flow will be measured between the working electrode and an inert, auxiliary electrode, which is also inserted into the sample solution. Three electrodes are therefore present; the working electrode to which the potential is applied, the auxiliary electrode which measures the current flowing, and the reference electrode.

The current which flows in the above experiment varies with the applied potential and follows the general shape depicted in Fig. 5.23(b), which is called a voltammogram. The current is negligible over most of the potential range because no electroactive solute is present, and appreciable current flow is observed only at the extremes of the potential range. At the most positive potential (E_1), the observed current is due to oxidation of the electrolyte in the solution, or of the working electrode itself, and is therefore represented as an oxidative current. At the most negative potential (E_2), the observed current is due to reduction of the electrolyte or of the working electrode, and is therefore represented as a reductive current.

Clearly, the actual potentials at which these oxidative and reductive currents occur will be dependent on the nature of the working electrode and the type of electrolyte used. Thus, for a given combination of electrode material and supporting electrolyte, there exists a range of potentials over which current flow is minimal when no electroactive solute is present. This range of potentials is known as the potential window for that system. The potential window in Fig. 5.23(b) is indicated by E_W. Many HPLC applications of electrochemical detection are performed at positive (oxidative) potentials because the reduction of oxygen occurs at small, negative potentials and this severely limits the reductive potentials which can be used in solutions that have not been deoxygenated.

When an electroactive solute (i.e. one capable of Faradaic oxidation or reduction at the working electrode) is added to the electrolyte solution, and the potential is again varied from E_2 to E_1, a different potential–current relationship results. Figure 5.23(c) shows the potential–current relationship when the added solute is oxidizable. An S-shaped step (i.e. a wave) appears in the voltammogram. The height of the wave (given by the anodic current, I_a) is proportional to the concentration of the oxidizable solute (C_O), while the position of the middle point of the wave on the potential axis (called the half-wave potential, $E_{1/2}$) is characteristic of the oxidizable solute. Figure 5.23(d) shows the voltammogram which results for a reducible solute and it can be seen that a cathodic, reduction current I_c results, which is proportional to the concentration of the reducible species (C_R). Again, the half-wave potential is characteristic of the reducible solute involved.

Figure 5.23 shows that voltammetry can be used to identify a solute and to determine its concentration. Provided only one electroactive species is present, the same result could be achieved by monitoring the current flow at a fixed potential, that is, by amperometry. The potential used should be one at which the analyte gives maximal current flow, but the residual current due to the electrolyte, working electrode, etc., is minimal. Thus, E_3 would be a suitable potential for amperometric measurement of the oxidizable solute in Fig. 5.23(c), and E_4 would be suitable for the reducible solute in Fig. 5.23(d).

Successful amperometric or coulometric detection can result only if the applied potential is chosen correctly. Extensive compilations of half-wave potentials for many solutes, measured with various working electrodes and under differing experimental conditions, are available in the literature (e.g. [25]). These compilations are valuable, but sometimes do not contain information which is specific to the actual experimental conditions being employed. In these circumstances, voltammograms such as those shown in Fig. 5.23 must be obtained. This is a relatively simple matter when voltammetric instrumentation is available which permits the potential to be varied continuously while the current is monitored. However, amperometric instruments are designed to operate at a fixed applied potential, so true voltammograms cannot be generated. In these cases, a voltammogram can be obtained by measuring the current output for a fixed concentration of analyte, using a series of

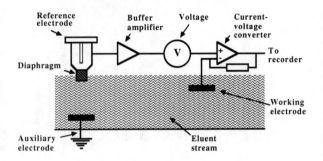

Fig. 5.24.   Basic configuration of an electrochemical detector.

discrete potentials covering the desired range. When the amperometric instrumentation used for this procedure consists of a flow-through cell, the voltammogram is obtained under flowing conditions and is referred to as a hydrodynamic voltammogram [26]. Hydrodynamic voltammograms are used extensively in chromatographic applications of amperometry and coulometry and are especially useful for the comparison of the performance of different electrochemical detectors.

Basic instrumentation for electrochemical detection

An electrochemical detector can be formed from a potential supply, appropriate circuitry for the measurement of current, and a suitable flow-through sample cell. As discussed above, the cell should accommodate three electrodes. Figure 5.24 shows a schematic representation of the basic configuration of an electrochemical detector for flowing sample streams. The recorded analogue signal is commonly generated by conversion of amplified oxidative or reductive currents generated in the cell to voltages, using the current–voltage converter shown in Fig. 5.24. More detailed descriptions of the electronic circuitry for voltammetric and amperometric measurements can be found elsewhere [26–28]. It should be noted that considerable versatility of detection can be achieved by varying the manner in which the potential is applied to the working electrode. It is not necessary that a simple d.c. voltage be used for this purpose. Indeed, it is often very advantageous to use potential pulses, or sequences of pulses, and to measure the resultant current at specific times.

Table 5.4 provides a summary of the more important characteristics of voltammetric, amperometric and coulometric detection.

Usage patterns for electrochemical detection in HPLC

Voltammetry (including polarography), amperometry and coulometry have all found application as detection methods for HPLC; however, it is fair to say that amperometry and coulometry are the most widely applicable methods, and of these, amperometry predominates. It is interesting to speculate on why this usage pattern exists. The difficulty in constructing a low-volume cell containing the dropping mercury electrode, coupled with the limited range of oxidative potentials available with a mercury electrode, combine to restrict the application of polarographic detection in HPLC. Voltammetry at solid electrodes does not offer significant advantages over amperometry using the same electrode types, and requires more

Table 5.4 Basic principles of voltammetric, amperometric and coulometric detection.

Method	Controlled quantity	Measured quantity	Cell design	Electronic unit	Remarks
Amperometry	Potential is held constant	Current	Three electrodes	Potentiostat	Less than 10% conversion of the analyte at the electrode. Electrode materials: carbon, silver, gold, mercury. Well-established technique. Large choice of commercial instrumentation.
Coulometry	Potential is held constant	Current, charge	Working electrodes larger than used for amperometry	Potentiostat	Approximately 100% conversion of the analyte. Well-established technique. Few commercial instruments available.
Voltammetry	Potential is changed as a function of time	Current is evaluated either continuously or by sampling	Three electrodes	Voltammetric instrumentation providing modulated potentials and various current sampling modes	This technique can be used by combining amperometric flow-cells with voltammetric instrumentation.
Pulsed amperometry	Potential is applied as pulses to clean the electrode between measurements	Current is evaluated between two cleaning pulses	Three electrodes	See voltammetry. Microprocessor instrumentation is necessary	Reproducibility of amperometric detection on some solid electrodes is improved with this technique.

sophisticated instrumentation. Amperometry and coulometry therefore remain as the electrochemical methods of choice.

The relative merits of these two techniques have been the subject of some debate [29]. At first, it might appear that coulometry would offer better sensitivity than amperometry because of the greater extent of reaction of the electroactive species which occurs in coulometric detection. However, as the electrode surface area is increased in order to improve the efficiency of the Faradaic reaction, the background current due to breakdown of the solvent electrolyte also increases. Little, if any, gain in sensitivity therefore results. Coulometric cells are often awkward in design and difficult to dismantle and maintain. In addition, these cells are sometimes expensive and can be used with only a limited range of working electrode materials. Considered together, these factors provide some cogent reasons why amperometry is used more frequently than coulometry.

In view of the above, the ensuing discussion of electrochemical detection techniques will focus on amperometry.

Application of the potential in amperometry

In single potential (d.c.) amperometry, a selected potential is applied continuously between the working electrode and the reference electrode. The auxiliary electrode (also called the counter electrode) prevents a deleterious current flow through the reference electrode. The sample is in direct contact with the working electrode. A serious problem encountered commonly with this mode of amperometric detection is a gradual loss of detection sensitivity. To understand the nature of this problem, it must be appreciated that the amperometric detector utilizes a heterogeneous electrochemical reaction that occurs at the interface between the working electrode and the sample solution. In some cases, reaction products can accumulate and adhere to the electrode surface, thereby blocking the surface and hindering further reaction. In other cases, the electrode surface itself can show deterioration. The outcome of either of these processes is a decline in the efficiency of the Faradaic reaction occurring at the working electrode. This leads to a decrease in the current produced and is manifest in a flowing system as a decrease in the peak height recorded for the particular solute under study. In addition, baseline noise and drift may also increase as the electrode becomes coated (or poisoned). The sensitivity must therefore be monitored carefully by frequent injections of a standard and, when the detector response becomes unacceptable, the cell must be reconditioned by replacing or polishing the working electrode.

One elegant method for overcoming the problem of adsorption of reaction products on certain electrode surfaces is to use potential pulses instead of a continuously applied d.c. potential [30]. In the pulsed mode, the detector measures current only during a short sampling interval, so there is less likelihood of electrode fouling. In addition, the potential can be stepped to values more negative or positive than the measuring potential as a means of cleaning the electrode surface or activating it to improve detection response. Figure 5.25 shows a typical pulsed potential waveform which might be applied to a working electrode. The measuring potential (E_1) is applied and the current is determined over a suitable time period. Some reaction products become deposited on the electrode during this process, but

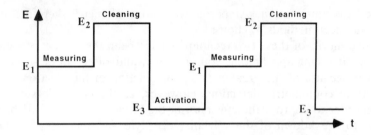

Fig. 5.25. Typical triple-pulse potential waveform used for pulsed amperometric detection.

if the potential E_1 is applied for only a short time, then these deposits can be expected to occur in only very small amounts. After measurement, the potential is raised for a short time (E_2) and then lowered (E_3) for a further short interval. Note that E_2 and E_3 are more positive and more negative, respectively, than E_1. These steps permit the desorption from the working electrode of reaction products which are oxidizable or reducible. At the conclusion of this cycle, the measuring potential can then be applied to a clean electrode surface. A further advantage of pulsed amperometric detection is that activation of the working electrode may occur because of the cycling of potentials.

Electrodes for amperometric detection

Early cell designs included only two electrodes: the working and reference electrodes. We have already noted that this configuration is undesirable because the reference electrode must carry a current and hence its potential does not remain constant. The auxiliary electrode is therefore included to carry the cell current, so that the reference electrode is maintained under conditions of zero current flow.

The most widely utilized reference electrodes are the Ag–AgCl electrode and the saturated calomel electrode (SCE). A palladium–hydrogen electrode and quasi-reference electrodes based on platinum and other metals have also been employed. Potentials of reference electrodes are available in the literature [31] and should be considered when an attempt is made to reproduce a detection method developed originally with a different reference electrode.

Auxiliary electrodes should be constructed of inert material and should ideally be situated as close as possible to the working electrode. This minimizes potential drop owing to the resistance of the sample solution. Typical auxiliary electrode materials are platinum and glassy carbon. In some detector cells, the stainless steel capillary tubing used to connect the chromatographic column to the cell may serve as the auxiliary electrode. This function may also be filled by the cell body itself, provided that it is constructed of a suitably conducting inert material.

Many different materials have been used for the construction of working electrodes for amperometric detection. Several comprehensive reviews on electrochemistry include discussion on this subject [e.g. 32]. For reduction reactions, the electrode material of choice has been, and remains, mercury. The primary reasons for this choice are the high overpotential for reduction of the hydrogen ion, the formation of amalgams with many metals, and the ease of replacement of mercury-drop electrodes. The high hydrogen overpotential on mercury means that mercury electrodes can be used in acid solution without interference by the reduction of hydrogen ions. Mercury electrodes have the widest negative potential range of any electrode material. Conversely, mercury is quite easily oxidized (at about 0.4 V versus SCE) and this prevents its use for the study of most oxidative processes.

Carbon can be used as an electrode material in a number of forms, including carbon paste, carbon impregnated into a suitable binder, glassy carbon, pyrolytic graphite and carbon fibres. Carbon paste electrodes are manufactured from particles of carbon suspended in an oil (such as Nujol) or a wax which is immiscible with the solution phase. These electrodes are relatively simple to manufacture and replace, and give high detection sensitivity owing to the very low residual currents produced. Disadvantages of carbon paste electrodes include some variability in electrode performance, which results even when successive electrodes are prepared from the same constituent materials, and the long equilibration times required for sensitive operation. Composite electrodes, in which the carbon is impregnated into a suitable binder such as polyvinylchloride, neoprene rubber or Kel-F, give more consistent performance but are more difficult to prepare.

Glassy carbon is a very popular electrode material because it can be formed into a variety of shapes and can be polished easily. Glassy carbon is a gas impermeable material formed by the heating of phenol-formaldehyde resins in an inert atmosphere. Electrodes of this type give higher residual currents than carbon-paste electrodes, but can be used over a 2 V working potential range which covers both positive and negative potentials. In addition, carbon is resistant to the formation of oxides at the electrode surface, so the electrode maintains integrity over prolonged periods of usage. Glassy carbon is sometimes also used for the construction of auxiliary electrodes as well as working electrodes.

Working electrodes may also be constructed of pure metals, which are usually inert materials so that the available potential window is not unduly restricted. It should be noted in passing that some reactive metals, such as copper [33], nickel [34] and copperized cadmium [35], have also found limited application as working electrodes in amperometry. Platinum and gold have wide potential windows which extend above +1.0 V in the oxidative region. Silver has a much smaller potential window, but is a valuable electrode material because of the reactions it can undergo with solutes during the detection process.

Some guidelines for selection of the correct material for a given application are required and the following aspects should be considered when making this selection [32]: (i) the potential window for the working electrode in the chromatographic eluent to be used (see Fig. 5.26); (ii) the involvement of the electrode itself in the electrochemical reaction; and (iii) the kinetics of the electron-transfer reaction.

Flow-cells for amperometric detectors

Electrochemical cells for use with flowing streams may be classified as flow-by, in which the eluent flows parallel to the surface of the working electrode; flow-through, in which the eluent follows a tortuous path between surfaces of the working electrode; and flow-at, in which the eluent impinges perpendicularly onto the surface of the working electrode.

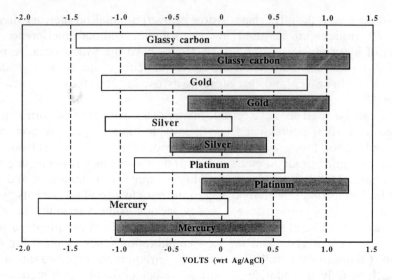

Fig. 5.26. Approximate potential windows accessible with various electrode materials in acidic solution (white boxes) and in alkaline solution (shaded boxes). Data from Rocklin (1984) *LC*, **2**, 588.

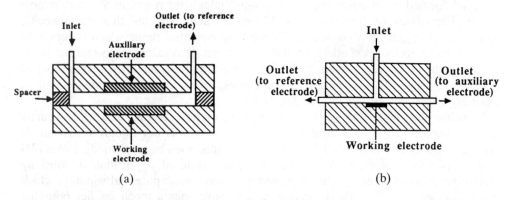

Fig. 5.27. Schematic illustration of (a) thin-layer and (b) wall-jet amperometric flow-cells.

The most common type of amperometric cell is the flow-by, which is illustrated schematically in Fig. 5.27(a). A thin spacer (in the form of a gasket) held between two rigid blocks defines the thickness, width and length of the flow-channel, and thereby the cell volume. In early versions of this type of cell the working electrode was housed in one of the blocks comprising the cell, while the auxiliary and reference electrodes were mounted downstream in another compartment. Newer designs have the auxiliary electrode positioned close to the working electrode and, in some cases, even the reference electrode is positioned in the same cell compartment.

The wall-jet cell (shown schematically in Fig. 5.27b) is the most common example of the flow-at type of electrochemical cell. The sample flow is directed perpendicular onto the surface of the working electrode and may exit from the cell through a narrow flow-tube (as shown in Fig. 5.27b), or may pass into a larger volume compartment surrounding the working electrode. Such cells are called unconstrained wall-jet cells, or large-volume wall-jet cells.

Some important characteristics of wall-jet cells deserve mention:

- The effective cell volume is much smaller than the volume of the electrode compartment, since the effective volume actually consists only of the thin layer of liquid at the electrode surface. After detection, the sample band disperses rapidly in the relatively large electrode compartment, so wall-jet detectors can be used in tandem with other detectors only when placed last in line.
- Wall-jet cells show a greater flow-rate dependence than thin-layer types and are therefore more susceptible to flow-pulsations in the pumping system.
- Wall-jet cells are suitable for miniaturization without loss of concentration detection limits because the ratio of electrode radius to jet diameter is more important than electrode radius alone.
- The relatively small size of the working electrode results in small double-layer capacitance effects and therefore a shorter time constant for the cell. Thus, this cell is suitable for those applications in which the electrode potential is changed rapidly, such as pulsed-amperometric detection.

Thin-layer cells can be modified by inclusion of a second working electrode. This electrode can be arranged in parallel or in series with the first working electrode. The ratio of the currents obtained at each electrode can be used for identification of solutes and as a measure of peak homogeneity, while the difference between the currents can be used to eliminate interferences.

Applications of electrochemical detection

Electrochemical detection is used frequently for the determination of phenols and amines (for example neurotransmitters, such as adrenaline and dopamine), heterocyclic nitrogen compounds (such as morphine alkaloids), sulfur compounds (such as penicillin), ascorbic acid, and some inorganic anions (such as bromide, iodide and thiocyanate). This detection mode is often applied in situations where extreme sensitivity or special selectivity is required and the electrochemical detector is generally operated in tandem with a universal, nonselective detector so that a more general sample analysis can be obtained than is possible with the electrochemical detector alone.

5.5.5 Refractive index detectors

Principles of operation of refractive index detectors

The refractive index (RI) detector is perhaps closest to the ideal of a universal detector than is any other, since in principle the RI of the mobile phase will be altered by the presence of any solute having an RI different to that of the mobile phase. Thus, comparison of the RI of the pure mobile phase with the column effluent will indicate the presence of an eluted solute. Since RI detectors operate in this way they are sometimes referred to as differential refractometers.

There are two major types of RI detector: the deflection type and the reflection type. Deflection RI detectors, as illustrated schematically in Fig. 5.28(a), measure

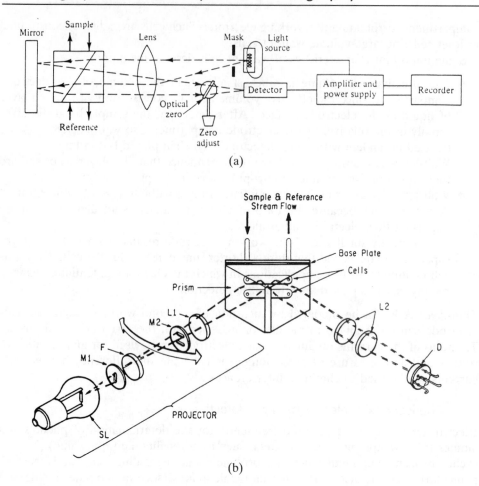

Fig. 5.28. Schematic diagrams of (a) deflection type and (b) Fresnel type refractometers. Reprinted with permission of Waters Chromatography Division of Millipore and Laboratory Data Control.

the deflection of a beam of monochromatic light passing through a double prism created by separating a rectangular cell into two compartments with a diagonal glass divider. The column effluent passes through one compartment, while pure mobile phase either passes through or fills the other compartment. The light beam from a suitable collimated source passes through the prisms onto a beam splitter, which directs the light to twin photomultipliers. The signal to the amplifier is adjusted to zero when the column effluent contains no analyte. When an analyte is eluted, the refractive index of the sample compartment alters, causing refraction of the light beam and a change in the angle at which the beam strikes the beam splitter. The relative amounts of light falling on the two photomultipliers therefore changes, as does their relative output. The change in signal is noted at the recorder and is proportional to the change in RI of the column effluent.

Reflection RI detectors measure the change in percentage of reflected light at a glass–liquid interface as the RI of the liquid changes. This mode of operation is based on Fresnel's law of reflection, which states that the amount of light reflected from a glass–liquid interface is dependent on the angle of incidence and the RI of the liquid. A typical design of a reflection RI detector is illustrated in Fig. 5.28(b), which shows that two collimated beams from a light source pass through a prism into two thin-layer cells, one of which accommodates the column effluent (sample) and the other the mobile phase (reference). The cells are formed by clamping a thin gasket between the prism and a stainless steel plate, which serves as the reflector. The two light beams emerge from the cell and are focused onto dual photomultipliers. As with the deflection-type RI detector, the output signal is zeroed when both the reference and sample compartments contain pure mobile phase. Any change in the RI of the sample solution resulting from the elution of an analyte will therefore cause a change in output signal. This type of RI detector has a relatively limited range and the prism must be changed if the entire useful RI range is to be accessed.

Other types of RI detectors also exist, such as the Christiansen Effect detector and the interferometric refractometer. The former operates in a manner similar to a spectrophotometer using a cell packed with a solid that has the same RI as the mobile phase, so that light is transmitted through the cell. When the RI of the solution flowing through the cell changes the level of transmitted light is reduced, giving rise to a signal. The interferometric design operates by measuring the difference in optical path-length resulting from changes in the RI of the sample solution, using an interferometer.

Characteristics of RI detectors

RI detectors will show response to most solutes, but the magnitude and direction of this response depends on the difference in RI between the mobile phase and the solute(s). Sensitivity reaches a maximum when this difference is greatest. The RI of some common solvents is listed in Table 5.5. The output of an RI detector may show both positive and negative peaks in the same run. For example, a tetrahydrofuran (RI = 1.408) mobile phase will give a large, negative peak for hexane (RI = 1.375), no peak for nonane (RI = 1.408) and a small, positive peak for decane (RI = 1.412). The polarity of the output is usually adjusted to give peaks (or at least the majority of peaks) in the positive direction. The sensitivity of RI detectors is, for

Table 5.5   Refractive indices of some common solvents.

Solvent	RI	Solvent	RI
Methanol	1.329	Nitromethane	1.394
Water	1.330	Cyclopentane	1.406
Acetonitrile	1.344	Tetrahydrofuran	1.408
n-Pentane	1.358	n-Decane	1.412
Ethanol	1.361	Dioxane	1.422
Acetic acid	1.372	Methylene chloride	1.424
Isopropanol	1.380	Ethylene glycol	1.427
n-Propanol	1.380	Chloroform	1.443
Methylethylketone	1.381	Carbon tetrachloride	1.466
Methylisobutylketone	1.394	Toluene	1.496

both the major designs described above, rather moderate, so that RI detection is not usually employed at trace level.

Temperature has a particularly profound effect on RI detectors, with a 0.001°C change causing a change of 10^{-6} RI units. Because of this temperature dependence, most commercial RI detectors have either a heat sink or have active temperature control facilities, but the incorporation of these devices can lead to increased dead volume. Conversely, RI detectors are not sensitive to pressure changes so that pump pulsations do not represent a major problem. The cell size, and hence the sample dispersion produced, depends on the operating principle of the detector, with the deflection type typically using a larger volume cell than the reflection type. Deflection detectors are generally extremely rugged, while reflection detectors offer high sensitivity, compatibility with low mobile phase flow-rates, and ease of cleaning. One of the major drawbacks of all RI detectors is that they are generally unsuitable for use with gradient elution techniques unless the change in RI between the different mobile phases comprising the gradient is very small.

Applications of RI detection

The RI detector was initially one of the most widely used HPLC detectors because of its universal response but finds most use today in applications where high sensitivity is not required, or where the analytes do not possess properties which enable them to be detected by other means. Typical applications include the detection of carbohydrates, alcohols, and polymers (it is common practice to couple RI detection with size-exclusion chromatography).

5.5.6 Conductivity detectors

Principles of operation of conductivity detectors

Conductivity detection is an important example of bulk property detection and is commonly used when the eluted solutes are ionic, for example acids and bases. However, the major use of this form of detection is for inorganic anions and cations after their separation by ion-exchange chromatography. Conductivity detection is universal in response for such solutes, and the detectors themselves are relatively simple to construct and operate. Conductivity detection will be discussed here in terms of the principles of its operation, the modes of detection employed, cell designs, post-column signal enhancement (suppression), performance characteristics of conductivity detectors, and applications.

A solution of an electrolyte will conduct an electrical current if two electrodes are inserted into the solution and a potential is applied across the electrodes. It is relatively straightforward to show that the conductance of a solution, G (having the units of microSiemens, which are represented by the symbol μS), is given by:

$$G = \frac{1000 \, \Lambda \, C}{K} = \frac{\Lambda \, C}{10^{-3} K} \qquad 5.3$$

where Λ_0 is the limiting equivalent conductance of the electrolyte (with units S.cm^2 Eq^{-1}), C is the concentration of the electrolyte, expressed as equivalents per litre of solution, and K is the cell constant (with units of cm^{-1}), determined by the geometry of the electrodes. The conductance can be seen to be proportional to the equivalent conductance of the electrolyte and its concentration. In addition, the lower the cell constant, the higher the conductance. This occurs for cells with large surface area electrodes which are close together.

Since the conductance of the solution results from both the anions and cations of the electrolyte, we must therefore calculate conductance using values for the limiting equivalent ionic conductances (λ) of the individual anions and cations in solution. Equation 5.3 can now be rewritten as:

$$G = \frac{(\lambda_+ + \lambda_-)\, C}{10^{-3}\, K} \qquad\qquad 5.4$$

where λ_+ and λ_- are the limiting equivalent ionic conductances of the cationic and anionic components of the electrolyte, respectively. Limiting equivalent ionic conductances for some common ionic species are listed in Table 5.6.

The operating principles of conductivity detection can be illustrated by considering the conductance of a typical eluent prior to and during the elution of a solute ion. The conductance change, ΔG, produced when an anionic solute S^- is eluted by an anionic eluent E^-, is given by [37]:

$$\Delta G = G_{Elution} - G_{Background} = \frac{(\lambda_{S^-} - \lambda_{E^-})\, C_S I_S}{10^{-3}\, K} \qquad\qquad 5.5$$

Table 5.6 Limiting equivalent ionic conductances of some ions in aqueous solution at 25°C [36].

Anion	λ_-(S.cm^2 Eq^{-1})	Cation	λ_+(S.cm^2 Eq^{-1})
OH$^-$	198	H$_3$O$^+$	350
Fe(CN)$_6^{4-}$	111	Rb$^+$	78
Fe(CN)$_6^{3-}$	101	C$_s^+$	77
CrO$_4^{2-}$	85	K$^+$	74
CN$^-$	82	NH$_4^+$	73
SO$_4^{2-}$	80	Pb^{2+}	71
Br$^-$	78	Fe^{3+}	68
I$^-$	77	Ba^{2+}	64
Cl$^-$	76	Al^{3+}	61
C$_2$O$_4^{2-}$	74	Ca^{2+}	60
CO$_3^{2-}$	72	Sr^{2+}	59
NO$_3^-$	71	CH$_3$NH$_3^+$	58
PO$_4^{3-}$	69	Cu^{2+}	55
ClO$_4^-$	67	Cd^{2+}	54
SCN$^-$	66	Fe^{2+}	54
ClO$_3^-$	65	Mg^{2+}	53
Citrate^{3-}	56	Co^{2+}	53
HCOO$^-$	55	Zn^{2+}	53
F$^-$	54	Na$^+$	50
HCO$_3^-$	45	Phenylethylammonium$^+$	40
CH$_3$COO$^-$	41	Li$^+$	39
Phthalate^{2-}	38	N(C$_2$H$_5$)$_4^+$	33
C$_2$H$_5$COO$^-$	36	Benzylammonium$^+$	32
Benzoate$^-$	32	Methylpyridinium$^+$	30

where C_S is the concentration of the solute, I_S is the fraction of the solute present in the ionic form. Equation 5.5 shows that the detector response depends on solute concentration, the difference in the limiting equivalent ionic conductances of the eluent and solute anions, and the degree of ionization of solute. The last of these parameters is generally governed by the eluent pH.

Sensitive conductivity detection can exist as long as there is a considerable difference in the limiting equivalent ionic conductances of the solute and eluent ions. This difference can be positive or negative, depending on whether the eluent ion is strongly or weakly conducting. If the limiting equivalent ionic conductance of the eluent ion is low, then an increase in conductance occurs when the solute enters the detection cell. This detection mode is direct, since the solute has a higher value of the measured property than does the eluent ion. Alternatively, an eluent ion with a high limiting equivalent ionic conductance can be employed, and a decrease in conductance would occur when the solute enters the detection cell. This type of detection is indirect, where the solute has a lower value of the measured property than does the eluent ion.

Design of conductivity detectors

The measurement of conductivity in liquids is usually performed by the application of an electric potential between two electrodes. Under the influence of this field, anions move towards the anode, while cations move towards the cathode. The current that results is dependent on the applied potential and also on the nature and concentration of ionic species present in the solution. It is usual for the potential to be applied in a pulsed or sinusoidal manner, e.g. as an alternating current. The amplitude of this potential must be such that thermal effects or chemical reactions at the electrode surfaces do not occur to any significant extent.

The simplest alternating current conductance circuit is based on the Wheatstone bridge design. Figure 5.29(a) shows the circuit used [38]. There are several problems involved in the balancing of such a bridge circuit and these arise chiefly from capacitance and resistance phenomena occurring within the ce itself. An alternative experimental arrangement (Fig. 5.29b) is the four-contact m e, where leads and contacts supplying the current are separated from those prob the voltage drop across the sample solution. This approach can be applied to flow solutions using a series of annular electrodes.

Signal enhancement devices for conductivity detection

When the conductivity detector is mounted in the usual position for a chromatographic detector, that is, immediately after the column, the choice of eluent composition must also consider the requirements for sensitive conductimetric detection. In many cases, the requirements of conductivity detection also impose constraints on the characteristics of the column used. One way to diminish this interdependence of column, eluent and detector is to insert a device between the column and detector which can chemically or physically modify the eluent. A commonly used device of this type is a suppressor, which achieves signal enhancement in conductivity detection by reducing the conductance of the eluent and simultaneously increasing the conductance of the sample band.

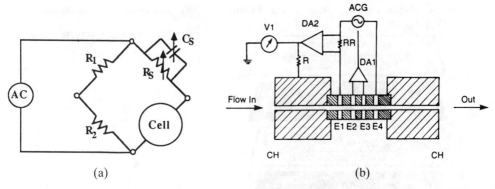

Fig. 5.29. (a) a.c. conductance bridge (b) four-electrode conductimetric cell. ACG, a.c. generator; E1–E4, electrodes; DA, differential amplifier; RR, range resistor; R, resistor; V, voltmeter. Reprinted from Baba and Housako (1984) *US Patent* 4 462 962 with permission.

The operation of a suppressor can be illustrated by considering the elution of choride ion with an eluent of $NaHCO_3$. If the suppressor is capable of exchanging sodium ions in the eluent with hydrogen ions from the suppressor, the first of the reactions below occurs. The HCO_3^- ions are converted into weakly conducting H_2CO_3, and the background conductance of the eluent is said to be suppressed. At the same time, the eluted solute (in this case, chloride) will also undergo the second of the reactions below in the suppressor.

$$\text{Suppressor-}H^+ + Na^+HCO_3^- \rightleftharpoons \text{Suppressor-}Na^+ + H_2CO_3 \qquad 5.5$$

$$\text{Suppressor-}H^+ + Na^+ + Cl^- \rightleftharpoons \text{Suppressor-}Na^+ + H^+ + Cl^- \qquad 5.6$$

The combined result of these processes is that the eluent conductance is decreased greatly, while the conductance of the sample is increased by virtue of the replacement of sodium ions ($\lambda_+ = 50\,S.cm^2.Eq^{-1}$) with hydrogen ions ($\lambda_+ = 350\,S.cm^2.Eq^{-1}$). The detectability of the solute is therefore enhanced.

It is important to note that suppression reactions are not limited to acid–base reactions, such as those shown in the above examples, nor are they limited to the detection of anions. Indeed, any post-column reaction which results in a reduction of the background conductance of the eluent can be classified as a suppression reaction. However, the ensuing discussion of suppressor design and performance will be restricted to those that employ acid–base reactions, since these are the most widely used.

Modern suppressors are based on dialysis reaction occurring through ion-exchange membranes, with the membrane as either a hollow fibre or a flat sheet. Figure 5.30 gives a schematic representation of the design of a typical flat-sheet (or micro-membrane) suppressor. The eluent contacts one side of the membrane while a regenerant solution flows in a countercurrent direction on the opposite side of the membrane. In the case of the example considered earlier, the membrane would be a cation-exchanger and the regenerant would be a solution containing H^+ ions. The eluent passes through a central chamber, which has ion-exchange membrane sheets

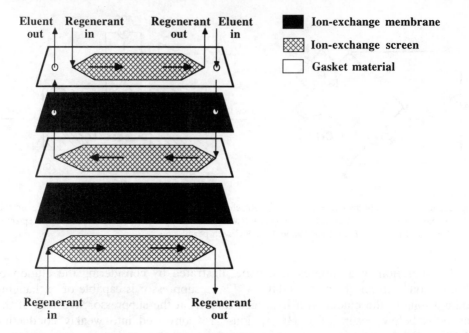

Fig. 5.30. Design of a micromembrane suppressor. Adapted from Stillian (1985) *LC*, **3**, 802.

as the upper and lower surfaces. Regenerant flows in a countercurrent direction over the outer surfaces of both of these membranes. Mesh screens constructed from a polymeric ion-exchange material are inserted into the eluent cavity and also into the cavities which house the flowing regenerant solution. The entire device is constructed in a sandwich-layer configuration, with gaskets being used to define the desired flow-paths. The volume of the eluent chamber is very small ($<50\ \mu$l [39]), so band broadening is minimal.

The mode of operation of the suppressor with an HCO_3^- eluent is illustrated schematically in Fig. 5.31(a). Sodium ions from the eluent diffuse through the cation-exchange membrane and are replaced by H^+ ions from the regenerant, producing the desired suppression reaction. Solute anions (e.g. chloride) are prevented from penetrating the membrane by the repulsion effect of the anionic functional groups of the membrane and therefore remain in the eluent stream. A schematic representation of the operation of this type of suppressor in separations of cations using a hydrochloric acid eluent and a barium hydroxide regenerant solution is shown in Fig. 5.31(b).

The micromembrane suppressor combines the advantages of other suppression devices and, at the same time, eliminates their drawbacks. These advantages can be summarized as:

- Small internal volume, leading to minimal band-broadening effects and hence low detection limits.
- Continuous regeneration.
- High dynamic suppression capacity which can be varied readily by changing the nature, concentration and flow-rate of the regenerant.

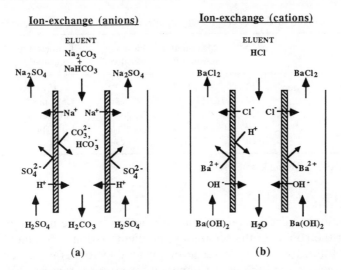

Ion-exchange (anions) **Ion-exchange (cations)**

(a) (b)

Fig. 5.31. Schematic operation of a micromembrane suppressor for eluents used with (a) anion-exchange and (b) cation-exchange chromatography. Hatched bars are cation exchange membranes in (a) and anion-exchange membranes in (b).

- Suitable for gradient elution with appropriate eluents.
- Resistant to many organic solvents and ion-interaction reagents.
- A wider choice of eluent types is possible because of the high dynamic suppression capacities which can be achieved.

Applications of conductivity detection

The most common application of conductivity detection is in the detection of inorganic anions after their separation by ion chromatography, especially using ion-exchange methods. Detector performance in this application is close to optimal because of the high limiting equivalent ionic conductances of many of the solute ions, especially when the conductivity detector is employed in conjunction with a suppressor.

5.5.7 Comparison of detectors

Some of the important characteristics of common HPLC detectors are summarized in Table 5.7.

5.6 Gradient Elution

5.6.1 Introduction

Gradient elution in HPLC refers to the technique of altering the composition of the mobile phase during the course of the chromatographic run. While the majority of

Table 5.7 Characteristics of common high-performance liquid chromatography detectors.

Detector	Type*	Maximum sensitivity[†]	Flow-rate sensitive?	Temperature sensitivity	Compatible with gradient elution?
u.v.–Visible light absorption	SP	2×10^{-10}	No	Low	Yes
Electrochemical	SP	10^{-12}	Yes	1.5% per °C	Yes
Fluorescence	SP	10^{-11}	No	Moderate	Yes
Refractive Index	BP	10^{-7}	No	$\pm 10^{-4}$ per °C	No
Conductivity	BP	10^{-8}	Yes	2% per °C	No

*SP, solute property; BP, bulk property.
[†]sensitivity in g/ml.

HPLC separations utilize isocratic elution conditions, i.e. the mobile phase remains constant over the course of the separation, gradient elution is commonly used when a mixture of solutes with a wide range of capacity factors is to be separated. This technique is analogous to the use of temperature programming in GC, as discussed in Section 3.6.1, except that changes in solvent strength rather than temperature are used to elute the solutes. Regardless of the mode of chromatography, gradient elution involves using a weak eluting solvent at the start of the chromatographic run and adding increasing proportions of a strong eluting solvent over the course of the separation. This allows for improved resolution of the poorly retained solutes at the beginning of the separation, while the more strongly retained solutes are eluted within a shorter time, as shown schematically in Fig. 5.32, for the case of an ionic strength gradient with conductivity detection.

The changes that occur in sample k' during a gradient separation are illustrated in Fig. 5.33, which shows the interrelation of solvent strength and band migration [5]. Three sample components (X, Y and Z) are considered, with X the least strongly retained and Z the most strongly retained. The solid lines mark the fractional distance (r) migrated by each compound between column inlet and outlet. The dashed lines indicate the instantaneous k' value of each band at time t. This value decreases with time because of the increase in solvent strength during the gradient formation. The migration of the bands all follow a similar pattern, as described for compound X. At some point in time, t_x, the k' value (dashed line) starts to decrease as X migrates along the column. As the k' value of X continues to decrease with time, X moves more rapidly through the column, finally eluting when $r = 1$. Similarly, Y begins its migration along the column at time, t_y, eluting from the column when $r = 1$ for Y, and likewise for compound Z. This means that the k' value for each band is very small (≈ 1) at the time the band leaves the column. A small value of k' results in narrow bands and increased sensitivity; hence, gradient elution permits both maximum resolution and sensitivity for every band in the sample [5].

5.6.2 Equipment for gradient elution

Two distinctly different approaches to gradient formation exist in HPLC. Gradient elution can be performed using either low- or high-pressure mixing systems, as

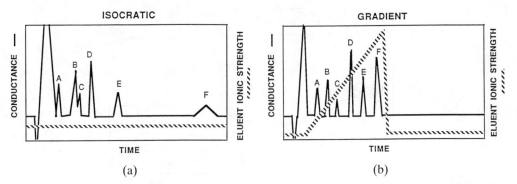

Fig. 5.32. Advantages of a gradient separation. (a) Isocratic separation; (b) gradient separation; solid line is conductance and broken line is eluent ionic strength.

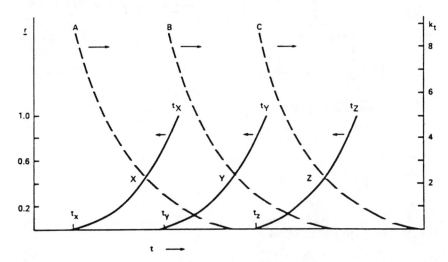

Fig. 5.33. Band migration in gradient elution. The solid lines show the fractional migration (r) of each band along the column, as a function of time. The dashed lines show the instantaneous k' value of each band. Reprinted with permission from Snyder and Kirkland (1979) *Introduction to Modern Liquid Chromatography*, 2nd edn, John Wiley & Sons, Inc., New York.

shown in Fig. 5.34. Low-pressure gradient formation involves blending two (or more) solvents at atmospheric pressure and pumping the mixed solvent to the column with a single high-pressure solvent delivery system. The gradient is formed using a microprocessor-controlled pump which proportions the time that the solenoid valves connected to the various solvent reservoirs are opened and closed. The use of two solvents to form a gradient is referred to as a binary gradient, three solvents creates a ternary gradient, while a quaternary gradient utilizes four different solvents. High-pressure gradient formation involves the high-pressure mixing of two (or more) solvents and requires two (or more) pumps and a microprocessor controlling device. The gradient is formed by controlling the output flow from each of the pumps.

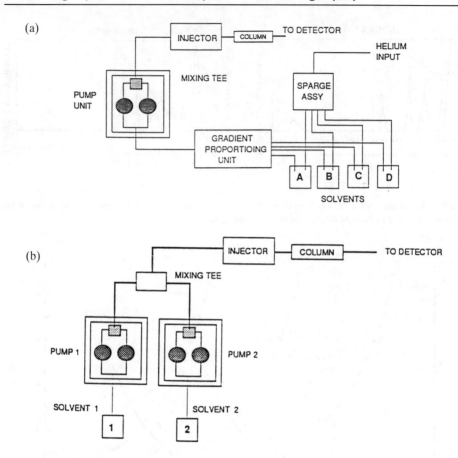

Fig. 5.34. Gradient elution systems. Reprinted with permission from *Developing HPLC Separations, Book Two*, Millipore Corporation, Milford.

Modern instrumentation allows the solvent composition to be varied using either a linear profile or, for more complicated separations, exponential gradient profiles. Figure 5.35 shows examples of typical gradient solvent composition profiles [40]. Both high- and low-pressure gradients similarly allow the generation of linear and exponential curves. Linear profiles (Fig. 5.35, curve 6) are used for the majority of gradient separations; convex gradient profiles (curves 2–5) approach the final conditions more rapidly, while concave gradient profiles (curves 7–10) delay the onset of final conditions. Curves 1 and 11 are used to effect step changes in mobile phase composition, such as to return to initial conditions.

In general, low-pressure gradient systems offer greater flexibility than the high-pressure approach as more than two solvents can be conveniently used and the equipment cost can be kept within reasonable limits. Low-pressure multi-solvent pumps with four solvent lines are commonplace in HPLC today. When comparing the two approaches, perhaps the most significant limitation of high-pressure gradients is that reciprocating pumps often have poor precision at very low flow-

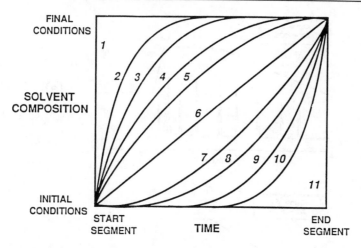

Fig. 5.35. Typical gradient solvent composition profiles. Reprinted with permission from *Developing HPLC Separations, Book Two*, Millipore Corporation, Milford.

rates, such as < 0.1 ml min^{-1}. This creates imprecision in the very early stages of the gradient which, unfortunately, is where precision is most important. Low-pressure systems have an additional advantage in that any solvent volume change that occurs from mixing is completed before the solvent is pressurized, hence flow-rate changes that arise from this effect in high-pressure systems are minimized with low-pressure systems [5].

The major drawback of low-pressure systems is the fact that the solvents must be thoroughly degassed and continually helium sparged prior to mixing in order to prevent outgassing in the pump heads. A further disadvantage of this approach is that the solvents pass through a mixing device, which creates additional delay volume, whereas high-pressure gradients solvents are commonly mixed through a T-piece and have less gradient delay volume. Both approaches to gradients are widely used in HPLC, although there appears to be an increasing trend toward the use of low-pressure gradient systems.

5.6.3 Gradient optimization

The optimization of a gradient separation involves first selecting the appropriate strengths for the solvents A and B which are to be mixed. An isocratic run using solvent B should elute the most strongly retained peaks with a $k' < 5$, while an isocratic run of solvent A should permit retention and resolution of the very early eluting peaks. Obviously, the two solvents must be fully miscible and also the strength of the two solvents should not be too different, as solvent demixing may occur. A simple linear profile from solvents A to B over a moderate timeframe is then generated. The rate of change (or slope) of the gradient profile should then be adjusted by varying the time of the gradient step. The steepest possible profile which allows adequate retention should be used. When the sample contains a mixture of homologues, where retention increases with molecular size, the gradient should be

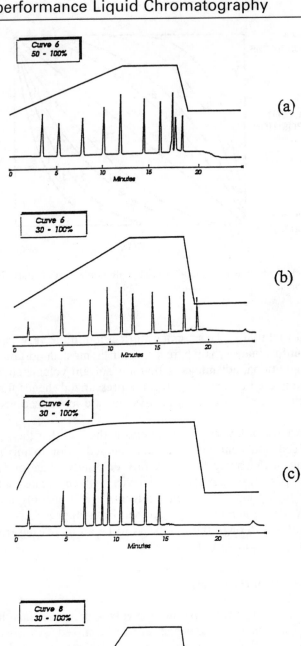

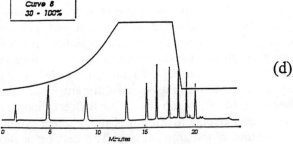

Fig. 5.36. Effect of gradient curve profiles. Reprinted with permission from *Developing HPLC Separations, Book Two*, Millipore Corporation, Milford.

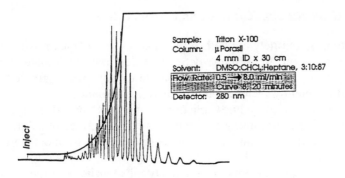

Sample: Triton X-100
Column: µPorasil
           4 mm ID x 30 cm
Solvent: DMSO:CHCl₃:Heptane, 3:10:87
Flow Rate: 0.5 ➔ 8.0 ml/min
           Curve 8, 20 minutes
Detector: 280 nm

Inject

Fig. 5.37. Flow programming in HPLC. Reprinted with permission from *Developing HPLC Separations, Book Two*, Millipore Corporation, Milford.

less steep in the later part of the separation than for samples of similar or random size distribution. Also, final conditions should be maintained for as long as necessary to elute all the peaks, before returning to initial conditions.

Figure 5.36(a) shows the initial linear gradient run for a mixture of alkylphenones on a Waters Resolve C_{18} reversed-phase column from 50–100% gradient of acetonitrile (ACN) over 13 min, with the final conditions being held for 5 min, before a step change returning to initial conditions in order for the column to re-equilibrate. It is apparent from Fig. 5.36(a) that the starting solvent is too strong, as the early peaks are eluting with a $k' < 2$ and probably within the gradient delay volume (that is, the time it takes the gradient mixed in the pump to reach the column). Figure 5.36(b) shows the same gradient, except that the starting solvent is now 30% ACN:water rather than 50% ACN:water. This change permits greater retention of the early eluted peaks. The gradient can now be further improved by selecting a convex curve that will approach the final conditions more rapidly than the linear profile. Figure 5.36(c) shows the same gradient as (b), except using the gradient profile shown in curve 4 of Fig. 5.35. Although a concave gradient profile is inappropriate in this case, the effect of such a profile (in this case, curve 8 in Fig. 5.35) on the separation is shown in Fig. 5.36(d).

If this procedure does not produce an appropriate separation, combinations of different profiles can be used, or the selectivity of the separation can be varied by changing solvent B. If still unsuccessful, a third solvent can be introduced, although this significantly complicates the optimization procedure. The separation of very complex mixtures may require a significant amount of experimentation in order to perfect the gradient conditions, see Snyder *et al.* [41] for further details. An alternative to solvent strength gradients in HPLC is to use flow programming. This technique involves starting the separation at a low flow-rate and programming an increase in flow-rate using one of the curve profiles shown in Fig. 5.35, while keeping the mobile phase composition constant. While not frequently used in HPLC, this approach can sometimes be useful to speed up a separation by pushing the later-eluted peaks through the column more quickly, as shown in Fig. 5.37.

5.6.4   Gradient applications and considerations

Gradient elution is routinely applied to the analysis of complex mixtures. Gradients enable the separation of solutes with a wider polarity range than can be accomplished using isocratic elution. Separations for which gradient elution in HPLC is used frequently include amino acids, polyaromatic hydrocarbons (PAHs), carotenoids, phenols, tetracyclines, oligomers and proteins. In general, a gradient will be required when attempting to separate a large number of solutes of differing polarities in a short time-frame. Gradients are employed most commonly with reversed-phase chromatography since a wide range of compounds can be resolved using this approach, the solvents are relatively inexpensive and the re-equilibration time between gradient runs is short. The gradient is achieved by decreasing the polarity of the mobile phase during the chromatographic run, e.g. from water to methanol. Figure 5.38 shows a typical reversed-phase gradient application — the separation of 6-aminoquinolyl-N-hydroxysuccinimidyl carbamate-(AQC)-derivatized amino acids. Gradients are less commonly used in normal-phase chromatography as the column re-equilibration period is often lengthy and the range of analytes is limited when compared with reversed-phase chromatography. With normal-phase chromatography, the solvent polarity is increased during the course of the separation, e.g. from hexane to chloroform. Proteins are frequently separated using ion-exchange gradients in which solvent A contains a buffer, while solvent B usually contains the same buffer plus a high concentration of salt (e.g. 1.0 M NaCl). The ionic strength of the mobile phase is therefore increased during the separation, eluting the proteins from the column. Gradients are not commonly used with other modes of chromatography, such as gel permeation or size exclusion.

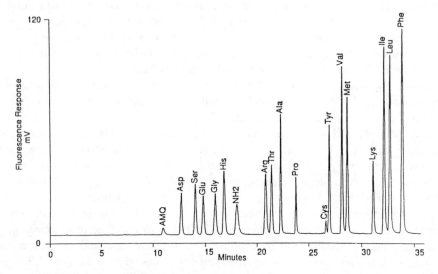

Fig. 5.38. Gradient separation of AQC derivatized amino acids. Solvent A contained acetate buffer, triethylamine and EDTA. Solvent B was acetonitrile and solvent C was water. Fluorescence detection was used. Chromatogram courtesy of Waters.

Successful HPLC gradient chromatography requires considerable attention to experimental procedure. Mobile phases must always be filtered and degassed. The solvents used for low-pressure gradients should always be helium sparged. When performing gradient separations, impurities from the weak solvent can accumulate on the column during the early portion of the run and be eluted later as the solvent strength increases. This can result in additional peaks in the final chromatogram, which reduces the quality of the separation. Hence, the best possible quality solvents should be used. The grade of water is of critical importance, particularly for reversed-phase gradients. Freshly generated 18 megohm, carbon-scrubbed water (e.g. Millipore Milli-Q water) should be used as bottled HPLC-grade water is inadequate for gradient use. Sufficient time must always be allowed between runs to allow the columns to re-equilibrate. Insufficient re-equilibration time leads to nonreproducible chromatography and significant variation in retention times. However, re-equilibrating the column for longer than necessary can allow additional impurities to build up on the column and also wastes time.

The mobile phases used in the gradient must be fully compatible, i.e. all components must be totally miscible and their strengths should not be too dissimilar, as solvent demixing can occur. In addition, baseline shifts may occur during the gradient run if one solvent has an appreciably different background absorbance to the other. Therefore, a blank gradient should always be run to monitor the extent of the baseline drift and also to check for eluting impurity peaks. This baseline drift may severely limit the range of solvents that can be used, although some detectors and many data stations have the capability automatically to subtract blank baselines. Finally, the effect of solvent mixtures on column backpressure must be considered. For instance, the viscosity of a 50:50 water:methanol mixture is significantly higher than for either solvent individually. In practice, this means that column backpressure will approximately double in the middle of a gradient from 100% water to 100% methanol and this may require that lower flow-rates be used during the gradient to avoid overpressuring the column.

5.7 Derivatization Methods

5.7.1 Introduction

Despite the wide variety of separation and detection systems available in modern HPLC, there still arise many circumstances where either the selectivity or sensitivity of a particular analysis may be considered inadequate. This situation occurs commonly in the determination of trace levels of drugs and metabolites in complex biological samples such as blood, plasma, urine, saliva, etc. In such cases, a chemical manipulation, or derivatization, of the analyte is necessary. Some of the aims of derivatization include [42]:

- Improvement in the detectability of the analyte(s).
- Improvement of the resolution of the analyte(s).
- Establishment of solute identity.
- Improvement of the selectivity of the analyte in a complex matrix.

- Improvement of the chromatographic behaviour (e.g. peak shape) of the analyte.
- Improvement of the stability of the analyte during chromatography.
- Achievement of more desirable physical properties (e.g. solubility) of the analyte.
- Introduction of an additional sample clean-up step.

When the aims of the derivatization step are centred around improvement of the chromatographic behaviour of the analyte (e.g. aims (ii), (iv)–(vii) above), the derivatization step must be performed prior to the chromatographic separation. That is, pre-column derivatization is used. However, when enhancement of detectability and selectivity is the prime aim, the derivatization can be performed either before or after the chromatographic separation. That is, either pre-column or post-column derivatization may be used. Both approaches can be performed on-line or off-line; however, the most commonly employed combinations are off-line pre-column derivatization and on-line post-column derivatization. The characteristics of these methods are listed below [42]:

Pre-column derivatization

- No restrictions on reaction kinetics provided that complete reaction can be achieved in a reasonable time.
- Free choice in varying reaction conditions to optimize reaction time and yield.
- The solvent used for the reaction may be incompatible with the chromatographic mobile phase provided that this solvent is removed before injection.
- Excess reagent and byproducts formed during the derivatization may be present after reaction, but there is an opportunity for these to be separated from the derivative either on the chromatographic system or by using a prechromatographic clean-up step.
- Each sample must be handled individually.
- An internal standard should be used in case the derivatization reaction is incomplete.

Post-column derivatization

- The derivatization reaction must be reproducible without the necessity of being quantitative.
- Underivatized compounds can be detected after elution with one detector and the derivatized compounds then monitored with a second detector, thereby providing maximum information on the sample.
- Sample preparation is limited.
- Accuracy and precision of the derivatization can be enhanced through automation of the derivatization procedure.
- The choice of reaction conditions is limited because the derivatization must be carried out in the chromatographic mobile phase.
- Band broadening usually occurs in the post-column reactor, leading to loss of chromatographic resolution.
- Additional equipment is required to perform the post-column reaction.

5.7.2 Hardware and procedures for pre-column derivatization

Pre-column derivatization can be performed either manually or with the aid of specialized reactors. In addition to choosing the correct chemistry for the derivatization, a wide range of additional factors need consideration. The choice of reaction vials is important since many polymer-based vials are incompatible with some solvents, contain plasticizers which may cause interference, and may limit the amount of heating that can be applied. The shape and volume of the vials must also be appropriate for the amount of sample available; for example, vials with a volume of $100 \, \mu l$ are available for separations where sample volume is limited. Heating blocks for reaction at high temperature or solvent evaporation are useful, as are vortex mixers and suitable dispensers for addition of micro-volumes of reagents. Even with such apparatus, manual derivatization is often a time-consuming process and must be performed with great care if reproducible results are to be obtained.

Automated sample processors are available for procedures where the additional expense of these items is warranted, such as when sample numbers are large or when reaction conditions must be standardized in order to attain acceptable reproducibility. These sample processors, which can range in sophistication from a simple device up to a full robotic arm, are able to perform a limited number of sample manipulations, such as addition of solvents, heating, mixing, etc. Automated derivatization can also be achieved using an appropriate autoinjector and this approach is often used in the derivatization of amino acids with o-phthalaldehyde (OPA). Here, the OPA reagent is drawn from a reagent vial, followed by the uptake of analyte from the sample vial. The derivatization takes place while the sample is being transported from the needle tip to the injection loop [43]. A more complex procedure, also performed with the aid of an autoinjector, involves the derivatization of amino acids with a fluorescent probe (9-fluorenylmethylchloroformate) [44]. The steps involved in this process are illustrated in Fig. 5.39.

5.7.3 Hardware and procedures for post-column derivatization

Post-column derivatization is performed by the introduction of suitable reagents into the column effluent. These reagents may be added as a solution which is mixed directly with the column effluent, or through the use of a solid-phase (or packed-bed) reactor of some kind, where the reaction occurs at a solid surface. The first of these possibilities can be defined as solution reaction and the second as packed-bed reaction. Figure 5.40 shows schematic representations of both of the above methods and indicates that a reaction coil is sometimes necessary to provide sufficient time for the reaction to proceed.

The desirable features of reagents for solution reaction can be summarized as [45]:

- The reagent must give a suitable chemical reaction with the analytes to provide suitable detection. Typically, the reagent (or its reaction product with the analyte) contains a strong chromophore or fluorophore.
- The reagent should be stable in order to minimize baseline drift and noise in the detector.
- The reagent must be miscible with the eluent and should not form precipitates

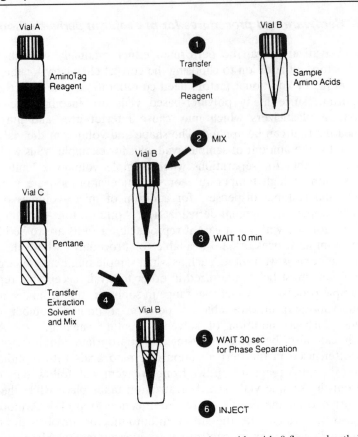

Fig. 5.39. Automatic pre-column derivatization of amino acids with 9-fluorenylmethylchloroformate. Reprinted with permission from Bell *et al.* (1986) *Int. Lab.*, **16**, 62.

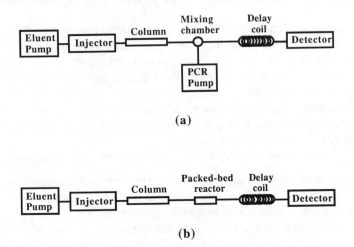

Fig. 5.40. Schematic representation of (a) solution and (b) packed-bed reagent systems.

in the presence of the eluent. Solubility of the reaction products with the analyte is of lesser importance because of the very low concentrations of analyte typically used.

- The reaction time between the analyte and the added reagents should be very short so that the use of reaction coils can be avoided. This, in turn, minimizes the band-broadening effects resulting from the post-column reaction (PCR) procedure.
- The reagent should have similar detection properties to those of the eluent, so that variations in the degree of mixing of the reagent and the eluent do not cause elevated baseline noise.

In packed-bed reaction, the derivatization reagent is immobilized onto a solid support which can be organic or inorganic in nature. The reagent can be physically adsorbed or attached to the support by ionic or covalent bonds. A heterogeneous reaction occurs between the analyte and the immobilized reagent. Packed-bed reaction offers a number of significant advantages when compared with solution reaction. Additional instrumentation, such as a pump, reaction coil and mixing chamber, are not required and the band broadening introduced is generally less than that for solution reaction. Moreover, there is no derivatizing reagent present in the eluent which could lead to an increase in the level of baseline noise. Finally, reactions which are possible in solution only at low concentration because of low solubility of the reagent may be carried out at higher concentrations using a solid support. The requirements of a packed-bed reaction system can be summarized as [46]:

- The immobilized reagent should be stable in the eluent used.
- The solid support should be mechanically stable at the operating pressures used.
- The packed-bed reactor should not contribute excessively to broadening of the analyte band.
- The packed-bed reactor should be capable of extended use before the immobilized reagent is exhausted.
- Reaction products should not be retained irreversibly on the packed-bed reactor, nor should gases or precipitates be produced from reactions with the analyte or eluent.

Reagent delivery systems

When solution reaction is to be used, some form of pump is required to deliver a constant flow of reagent to the mixing chamber. The baseline noise measured at the detector is given by the sum of the detector noise, the cell noise, the mixing noise, and the flow noise. The detector noise arises from the electronics of the detector itself, while the cell noise occurs when the background mobile phase flowing through the detector (in the absence of analyte) produces a finite detector signal. Cell noise usually results from thermal effects and is dependent on the magnitude of the detector signal produced by the background mobile phase. Mixing noise is the result of imperfect mixing of the column effluent and the post-column reagent. The final contribution to baseline noise is flow noise resulting from pulsations in either the mobile phase pump or the pump used to deliver the post-column reagent. Any variations in the rate at which the reagent and column effluent are mixed will cause baseline noise in the detector, especially when the reagent itself gives some detector signal under the conditions used to monitor the reaction product formed between the analyte and the added reagent. Syringe pumps are often preferred for delivery of the post-column reagent. Alternatively, a simple overpressure pneumatic delivery system (of the type illustrated in Fig. 5.8) may be used. In such devices, a regulated gas pressure is maintained over the surface of the reagent, which is housed in a suitable vessel. This gas pressure drives the reagent solution to the mixing chamber without the pressure pulsations that would result from a reciprocating piston pump.

Mixing chambers

The function of the mixing chamber is to accomplish intimate mixing of the column effluent and the reagent in the smallest volume possible. That is, mixing efficiency should be maximized while band broadening of the analyte should be minimized. The simplest type of mixing chamber is a three-way junction which accommodates inlet tubes for the column effluent and reagent and an outlet tube to pass the mixed reagents to a reaction coil or directly to the detector. The angles at which the inlet and outlet tubes are arranged in such a junction can vary, but a 90° T-piece is conventionally used. Examples of such T-pieces are shown in Fig. 5.41. The device shown in Fig. 5.41(a) can be constructed by cutting two lengths of stainless steel tubing at 45° and butting the ends together. An outlet tube is then pressed against the joined inlet tubes. The mixed eluent and PCR reagent flow out and around the small spaces in the tubing joint and pass into the outlet tube. The band broadening of this mixing chamber is minimal. Figure 5.41(b) shows a T-piece constructed from a length of conventional 1/16 inch o.d. stainless steel HPLC tubing (0.007 inch i.d.) joined at its centre with a perpendicularly arranged length of the same tubing. The tubing is joined by machining the components of the joint at 60°. The final T-piece design, depicted in Fig. 5.41(c), uses square-cut tubing and incorporates a 120-mesh stainless steel screen situated between the eluent inlet and detector outlet tubes. The reagent solution flows around the exterior of the eluent inlet tube, before passing

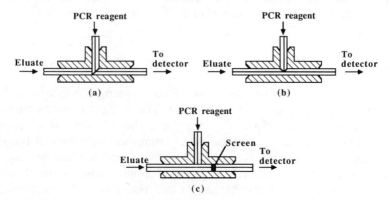

Fig. 5.41. Typical T-pieces used as mixing chambers for solution post-column derivatization applications. All tubing is ¹⁄₁₆ inch o.d. and 0.007 inch i.d. PCR, post-column reaction.

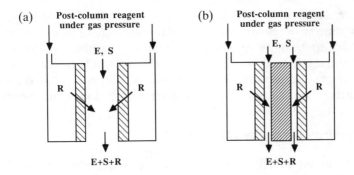

Fig. 5.42. Schematic illustration of (a) membrane reactor and (b) annular membrane reactor used for post-column derivatization applications. E, mobile phase; S, analyte; R, post-column reagent; light hatching, semipermeable hollow-fibre membrane; heavy hatching, monofilament nylon fishing line.

into the screen and from there to the detector. Each of the T-pieces shown in Fig. 5.41 provides very efficient mixing; for example, the screen–T reactor illustrated in Fig. 5.41(c) has been shown to give a mixing homogeneity better than 99.9% of theoretical perfect mixing [47]. It is noteworthy that although the angle of impingement between the column effluent and the reagent is 90° for each of the T-pieces shown in Fig. 5.41, it is also possible for other angles of impingement to be used. A detailed treatment of the flow and mixing characteristics of various tubes and mixing devices can be found elsewhere [48].

The post-column reagent solution can also be introduced into the column effluent by means of a semipermeable membrane in the form of a hollow fibre. The eluent stream passes through the centre of the fibre, while the reagent flows around the exterior. The reagent diffuses through the fibre wall under external pressure and mixes with the eluent. Longitudinal mixing, which causes broadening of the analyte bands, is reduced by coiling the fibre, which also enhances radial mixing. This type of device is illustrated schematically in Fig. 5.42(a). The type of membrane material used in the construction of the hollow fibre should be chosen to permit maximum penetration of the reagent, with minimal loss of the analyte species. The dead volume of a membrane reactor is normally high (up to $100 \mu l$) in comparison with that of a typical T-piece because the reactor must be of sufficient length for diffusion of the post-column reagent to occur. This can result in significant broadening of the analyte band. The elastic nature of the membrane may be of some benefit in reducing pressure pulsations from the eluent pump. The problem of large dead volume can be overcome through the use of an annular membrane reactor in which a monofilament nylon fishing line is inserted into a microporous hollow fibre (Fig. 5.42b). The composite tubing can be heated in water and coiled into a helical configuration. A 5-cm membrane reactor formed in this way has an internal volume of $1.5 \mu l$, with an effective hydraulic radius of $97 \mu m$.

Reactors

After the post-column reagent and the analyte are mixed, it may be necessary for them to be passed to a suitable reactor to provide appropriate conditions for reaction to occur. For example, the kinetics of the reaction might be slow, so that a sufficient delay time is necessary before detection can take place; alternatively, it may be necessary to heat the mixture to stimulate reaction. Reactors can take various forms, ranging from a simple coil designed to provide time for the reaction to occur, to more complex devices such as stitched capillaries or knitted open tubes. The characteristics of the last two types have been described in detail elsewhere [48].

5.7.4 Applications of derivatization

Derivatization reactions in HPLC are used very commonly and entire texts (see Bibliography) have been devoted to discussion of the merits of different reactions. The diversity of reactions is such that it is impossible to summarize them here, but for the purposes of illustration, Table 5.8 gives examples of some typical pre- and post-column derivatization reactions.

Table 5.8 Examples of pre- and post-column derivatization reactions.

Solute	Derivatization reagent	Type of reaction	Detection method
Amino acids	Phenylisothiocyanate	Pre-column	Spectrophotometry
Carboxylic acids	Phenacyl bromide	Pre-column	Spectrophotometry
Ketones	2,4-Dinitrophenylhydrazine	Pre-column	Spectrophotometry
Aflatoxins	Iodine	Post-column	Fluorescence
Carbohydrates	2-Cyanoacetamide	Post-column	Electrochemistry
Formaldehyde	Acetylacetone	Post-column	Spectrophotometry
Barbiturates	Borate buffer	Post-column	Spectrophotometry
Inorganic cations	Pyridylazoresorcinol	Post-column	Spectrophotometry
Catecholamines	Ethylenediamine	Post-column	Fluorescence

5.8 Preparative Chromatography

Preparative chromatography refers to the process of using HPLC to isolate material from an injected sample. In its simplest form, preparative chromatography involves collecting separated peak fractions as they emerge from the detector. In practical terms, preparative chromatography generally means the isolation of significant quantities of material (>0.1 g) using large-dimension columns. The objectives of a preparative separation are often different from those of conventional (analytical scale) HPLC. When developing an analytical separation, factors to be considered include sensitivity, sample complexity, sample throughput, degree of accuracy and/or precision required and, perhaps, ease of use as a routine assay. For preparative separations, factors to be considered include whether the material to be isolated is a major or minor component of the sample, whether biological activity of the solute must be maintained, the degree of purity required for the isolated fraction, and the quantity of material to be isolated.

5.8.1 Instrumentation for preparative chromatography

The instrumentation used for preparative chromatography is essentially similar to that used for analytical chromatography, except that the dimensions are somewhat greater. Reciprocating pumps typically have larger (extended flow) pump heads in order to generate higher (10–200 ml min^{-1}) flow-rates. Conventional injectors may be used for small sample sizes (< 2.0 ml); alternatively, specialized ultraloop sample loaders, or even an HPLC pump may be used to load (or pump) the sample onto the column. Preparative columns generally have greater dimensions and use larger particle sizes than conventional columns. The connectors and tubing are also larger (0.02–0.04 inch i.d.) to reduce the backpressure generated by high flow-rates. Finally, the detector may be fitted with a preparative flow-cell, which allows less sensitivity than an analytical flow-cell, to avoid saturating the detector response at high sample loads. Refractive index is a useful detection approach in preparative chromatography, where its lower sensitivity is an advantage. Several HPLC companies supply accessories (such as extended-flow pump heads or preparative flow-cells) to enable a conventional instrument to be converted to a (semi)preparative liquid chromatograph; specialized preparative HPLC instruments are also available from a number of manufacturers [12].

5.8.2 Preparative separations

The sample load injected during conventional analytical separations is appreciably less than the total column capacity. At such sample loads, peak resolution can be quantitatively predicted using the resolution equation, as discussed in Section 2.5. However, when a column is overloaded, as is typically the case in preparative chromatography, the resolution equation no longer applies quantitatively. Large sample weights drastically affect the capacity factor and efficiency terms of the resolution equation. As sample load increases, the peaks broaden and both the retention and column efficiency decrease, resulting in reduced resolution. Therefore, maintaining adequate resolution as sample load increases requires either greater column efficiency, longer chromatographic runs or, most commonly, using a column with larger dimensions.

When developing a preparative separation it is preferable first to optimize the separation on a small (analytical) scale column. An appropriate packing material is selected and a separation is developed to be consistent with the goals of the purification, in terms of resolution, sample load and speed. The optimum column volume, mobile phase, flow-rate and gradient profile (if applicable) are then identified. Where possible, resolution should be maximized to allow for increased separation between the peak of interest and the other sample components. Once the desired separation is achieved, the maximum amount of sample that can be loaded onto the analytical column, while still maintaining the resolution required to achieve the desired purity, is determined. Finally, a larger preparative column containing the same type of packing material is selected and scale-up factors are used to guide the calculation of both sample size and flow-rate for the larger column. The mobile phase flow-rate and sample load increase proportionally to the column volume as described in equations 5.7 and 5.8 (where D is the column diameter and L is the column length), while Table 5.2 shows sample load and scale-up factors achieved when converting from conventional size analytical columns to both 25 and 40 mm preparative radial compression cartridges [18].

$$\text{Flow-rate} = \text{Flow-rate}_{\text{initial}} \times \frac{(D_{\text{final}})^2}{(D_{\text{initial}})^2} \qquad \text{5.7}$$

$$\text{Load} = \text{Load}_{\text{initial}} \times \frac{(D_{\text{final}})^2 \, L_{\text{final}}}{(D_{\text{initial}})^2 \, L_{\text{initial}}} \qquad \text{5.8}$$

In addition to being physically larger, preparative columns and cartridges are usually packed with larger-diameter particles. As the flow-rate is scaled up, backpressure increases unless particle size increases; larger particle sizes are also easier to pack and are significantly less expensive than smaller diameter particles. While column efficiency decreases as particle size increases, the surface area per theoretical plate is higher for large particle size packings, therefore such columns are less susceptible to overloading than those packed with smaller diameter material [18]. Hence, preparative columns are commonly packed with particle sizes of 20–100 μm. Figure 5.43 shows an example of the scale-up and preparative chromatograms for a carotenoid separation.

If inadequate resolution of the compounds of interest is obtained at high sample loads, the recycle technique can be used. With this approach, the resolved (pure) portions of peaks of interest are collected, while the co-eluting fractions are reinjected, or recycled, through the liquid chromatograph. In the example shown in Fig. 5.44, the well-resolved peak on the left of the triplet is collected as a pure fraction, while the unresolved doublet is recycled through the system. As the

Table 5.9 Sample load scale-up factors achieved when converting from analytical column to preparative cartridge.

Stainless steel and cartridge columns: dimensions and volumes (ml)	Scale-up factor to 25 × 100 mm (49 ml)	Scale-up factor to 40 × 100 mm (125 ml)
Cartridge columns		
8 × 100 mm (5.0 ml)	10	25
25 × 100 mm (49.0 ml)	–	2.5
Stainless steel columns		
3.9 × 150 mm (1.8 ml)	27	70
4.6 × 150 mm (2.5 ml)	20	50
7.8 × 150 mm (7.2 ml)	7	17.5
7.8 × 300 mm (14.3 ml)	3.5	9
9.4 × 300 mm (20.8 ml)	2.5	6
19 × 300 mm (85.0 ml)	–	1.5

To determine scale-up factors for 200 and 300 mm lengths cartridges, multiply the 100 mm scale-up factor by 2 or 3, respectively.

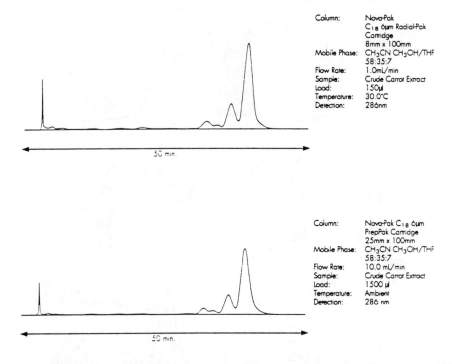

Column:	Nova-Pak C18 6µm Radial-Pak Cartridge 8mm x 100mm
Mobile Phase:	CH3CN CH3OH/THF 58:35:7
Flow Rate:	1.0mL/min
Sample:	Crude Carrot Extract
Load:	150µl
Temperature:	30.0°C
Detection:	286nm

Column:	Nova-Pak C18 6µm PrepPak Cartridge 25mm x 100mm
Mobile Phase:	CH3CN CH3OH/THF 58:35:7
Flow Rate:	10.0 mL/min
Sample:	Crude Carrot Extract
Load:	1500 µl
Temperature:	Ambient
Detection:	286 nm

Fig. 5.43. Scale-up of a carotenoid separation. The upper chromatogram shows the injection of 150 µl of crude carrot extract on an 8 × 100 mm cartridge and the lower chromatogram shows the injection of 1.5 ml of the same extract on a 25 × 100 mm cartridge. Nova-Pak 6 µm C₁₈ packing material was used for both columns, with a mobile phase of 58:35:7 acetonitrile:methanol:tetrahydrofuran. Reprinted with permission from *Waters Sourcebook of Chromatography* (1992), Millipore Corporation, Milford.

doublet is subsequently reinjected at a lower total sample load, column efficiency and resolution increase, hence the separation improves. The edges of the doublet can then be isolated, or shaved, while the centre of the peak can then be reinjected, again at a lower sample loading. The process can be continued until pure fractions of all three peaks are obtained. Preparative instruments are generally fitted with a recycle valve to enable the process to be automated [49]. If only one component is to be isolated when resolution is poor, a pure fraction can be isolated by collecting only the centre portion of the peak. This process is termed a peak heart-cut.

5.8.3 Preparative applications

Chromatography is a relatively expensive separation tool and HPLC purification should only be used when other options, such as open column or solvent extraction, have been considered. Preparative isolation of compounds is possible with any separation mode, although reversed-phase chromatography is most commonly used. This mode is compatible with the widest range of samples, the solvents are relatively inexpensive and the columns can be easily cleaned, although they are somewhat expensive. One drawback is that the solvents used are often of low volatility, hence

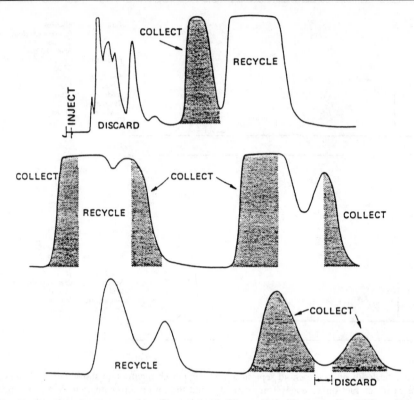

Fig. 5.44. Preparative chromatography with recycle. A Waters Porasil column (7 × 2440 mm) was used with a mobile phase of 160:36:4 trichloromethane:methanol:water. An experimental drug reaction mixture (150 mg) was injected.

their removal from the isolated material may pose a problem. Normal-phase chromatography utilizes inexpensive columns and readily volatile solvents. The columns, however, are more difficult to clean, which is an important consideration as large volumes of crude materials are often injected during preparative applications. A useful advantage of this separation mode is that TLC can conveniently be used as a means for checking fraction purity; however, this mode is limited in terms of the range of solutes for which it can be used. Both ion-exchange and size exclusion chromatography are frequently used for preparative chromatography, particularly for protein purification. Size exclusion is also important as a preparative sample clean-up tool to remove lipids from samples prior to the analysis of pesticides and drug compounds [5].

Perhaps the most important application of preparative chromatography is the isolation of biological materials, such as proteins and peptides. It is also an important technique for the isolation of isomers and chiral mixtures, which are difficult to separate by nonchromatographic means and can be an invaluable tool for the synthetic chemist. Preparative chromatography is also applied widely in the pharmaceutical industry for both purification and production purposes. Bidling-meyer [49] details examples of a significant number of specific preparative HPLC applications.

5.9   Multidimensional Liquid Chromatography

The HPLC methodology discussed so far in this chapter has assumed that an analysis can be developed using an appropriate combination of a single separation and detection method. In some cases, this approach may not be satisfactory and it becomes necessary to utilize a combination of separation methods. This approach is termed multidimensional, or coupled, chromatography and involves the use of column switching valves in conjunction with two (or more) separator columns. Coupled chromatography is distinct from simply joining two different columns in series, as the latter approach rarely offers much advantage in HPLC. Gel permeation chromatography is one notable exception, where columns of different pore sizes linked in series are commonly used to maximize the separation of a sample with a wide molecular weight distribution [5]. The simultaneous separation of anions and cations using tandem anion- and cation-exchange columns in series is perhaps another useful application of simply joining two columns in series [50].

5.9.1   Column switching

Coupled chromatography enables the diversion of various sample components, from a single injection, to different columns in order to achieve maximum resolution in the minimum time. This is usually performed by switching columns with a rotary six-port valve at an appropriate time during the separation, as shown in Fig. 5.45. The sample is injected with the valve in the position shown in Fig. 5.45(a), so that the sample is passed through column A to the detector. Rotation of the valve to the position shown in Fig. 5.45(b), directs some of the effluent from column A to the second column (B), where further separation occurs before the sample components pass to the detector.

   This approach is useful when a sample contains a mixture of strongly and weakly retained solutes, in which case column A is chosen to have significantly less retention capacity than column B. Column A is often just a shorter version of column B, or

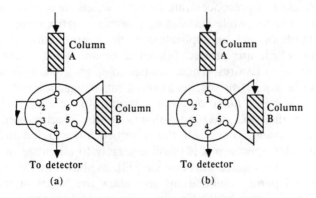

Fig. 5.45.   Basic instrumental configuration for column switching. The circle represents a high-pressure six-port rotary switching valve. Reprinted from Haddad and Jackson (1990) *Journal of Chromatography Library Series No. 46*, Elsevier, Amsterdam, with permission.

even a guard column. The weakly retained solutes are allowed to elute rapidly through the low-capacity column A onto column B. The valve is then switched so that just column A is in line, after which the trapped, more strongly retained solutes are eluted off the lower capacity column to the detector. The valve is rotated again and the weakly retained solutes on column B are then eluted through to the detector. Hence, in the final chromatogram, the strongly retained solutes appear first (from column A only) followed by the weakly retained solutes, which passed through both columns. In practice, however, this technique will only be successful for certain sample types and is somewhat limited when compared with gradient elution. Detailed information of system configurations for different approaches to column switching in HPLC is available elsewhere [5, 10, 51].

5.9.2 Applications of coupled chromatography

There is a variety of uses of coupled chromatography besides the example described above for a sample containing both weakly and strongly retained components. Pre-column preconcentration, a specialized technique for trace analysis discussed in Chapter 8, in addition to both preparative recycle and heart-cut reinjection, discussed in Section 5.8.2, could all be considered examples of multidimensional chromatography. Coupled chromatography is also widely used to perform auto-mated sample clean-up, or on-column matrix elimination. This approach can be used in reversed-phase chromatography when a sample contains high levels of unwanted polar material, and the solutes of interest are more strongly retained. Using the same configuration described above, the sample is injected onto the lower capacity column A. The effluent from this column is initially directed through the detector to waste until most of the band containing the weakly retained, polar solutes has eluted. Column B is then switched in line to capture and separate the band containing the more strongly retained solutes, resulting in reduced interferences in the early portion of the chromatogram. This process is sometimes referred to as pre-column venting [5].

The combination of one pump with two switching valves (or two pumps with one switching valve) permits still greater flexibility for automated sample clean-up. This approach can be used to preconcentrate analytes which show very strong retention on a concentrator column while eliminating unwanted matrix components from the sample. An example of such an application is the determination of aurocyanide in tailings solutions which are produced from a cyanidation process used for the extraction of gold from its ores. These tailings solutions contain very low levels of aurocyanide (10–20 p.p.b.) in the presence of much higher levels of other metal cyano complexes. Matrix elimination on the concentrator column can be achieved by careful control of the strengths and volumes of the eluents used to wash the concentrator column and to transfer the adsorbed aurocyanide to the analytical column. By using this approach it is possible largely to eliminate the interferences without loss of preconcentrated aurocyanide [52], as shown in Fig. 5.46. The same switching valve and pump combinations also allow the heart-cut technique to be used for matrix elimination. With this approach, the bulk of the sample is diverted to waste and only a selected portion of the sample reinjected and completely analysed. This allows for improved resolution and has also been shown to

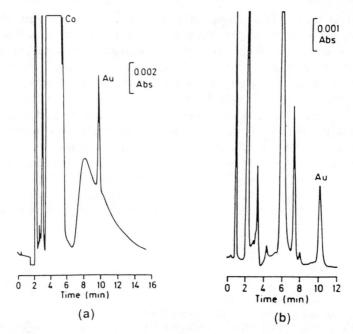

Fig. 5.46. Matrix elimination in the analysis of gold(I) cyanide in mine process liquors. Chromatogram (a) shows the interference from a synthetic liquor on the preconcentration of 2 ml of 50 p.p.b. gold(I) cyanide. Chromatogram (b) shows a process liquor containing 25 p.p.b. gold and large excesses of other metal–cyano complexes after preconcentration using automated matrix elimination. Reprinted from Haddad and Rochester (1988) *J. Chromatogr.*, **439**, 23, with permission.

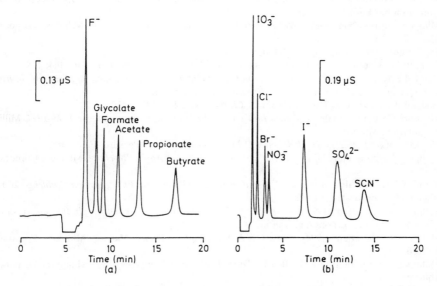

Fig. 5.47. Chromatograms obtained from coupled ion-exclusion and ion-exchange systems. Chromatogram (a) is from the ion-exclusion system (Waters Fast Fruit Juice column with 1.0 mM octanesulfonic acid as eluent, using conductivity detection). Chromatogram (b) is from the anion-exchange system (Waters IC Pak A column with 3.0 mM sodium octanesulfonate as eluent, using conductivity detection). Reprinted from Jones *et al.* (1989) *J. Chromatogr.*, **473**, 171, with permission.

significantly reduce column abuse. A generic approach to matrix elimination in HPLC using heart-cut techniques has recently been described by Villasenor [53].

Finally, perhaps one of the most elegant examples of multidimensional liquid chromatography is the coupling of ion-exclusion and ion-exchange separation modes. The selectivities of the two modes are virtually opposite, in that weakly ionized solutes (such as carboxylic acids and fluoride) are retained much less strongly than fully ionized strong acid anions (such as chloride and sulphate) on an anion-exchanger; whereas the reverse is true for ion-exclusion chromatography. The solute mixture is injected onto an ion-exclusion column and the early part of the chromatogram (containing the fully ionized solutes) can be collected and injected onto an anion-exchange system using a switching valve [54]. Figure 5.47 shows the representative chromatograms obtained from a single injection of a mixture of weak and strong acid solutes, recorded at the conductivity detector on each of the two linked systems.

References

1. *LC.GC, Marketplace Issue*, August (1992).
2. C. & E. News, *Instrumentation '92 issue*, March (1992).
3. Millipore Corporation (1991). *Waters 600E/600-MS Multi Solvent Delivery System, Service Manual*. Millipore Corporation, Milford.
4. Shimadzu Corporation (1992). *Shimadzu LC-10A Series HPLC System Brochure*. Shimadzu Corporation, Tokyo.
5. Snyder L.R. and Kirkland J.J. (1979). *Introduction to Modern Liquid Chromatography*, 2nd edn. John Wiley & Sons, Inc., New York.
6. Millipore Corporation (1989). *Waters Ion Chromatography Issues and Answers*, 5th edn. Millipore Corporation, Milford.
7. Millipore Corporation (1989). *Waters ActION Analyzer, Operators Manual*. Millipore Corporation, Milford.
8. Franklin G. (1985). *Lab. News*, October 50.
9. Sadek P.D., Carr P.W., Bowers L.D. and Haddad L.C. (1985). *Anal. Biochem.*, **144**: 128.
10. Haddad P.R. and Jackson P.E. (1990). *Ion Chromatography: Principles and Applications, Journal of Chromatography Library Series, No. 46*. Elsevier, Amsterdam.
11. Haddad P.R. and Foley R.C.L. (1987). *J. Chromatogr.*, **407**: 133.
12. Millipore Corporation (1993). *Waters Instruments, Parts and Supplies Catalogue 1993–1994*. Millipore Corporation, Milford.
13. Hewlett Packard (1989). *HP 1090 Series II/M Brochure*, Hewlett Packard, Palo Alto.
14. Millipore Corporation (1993). *The ConSep Liquid Chromatography Brochure*. Millipore Corporation, Bedford.
15. Spruce B. and Bakalyar S.R. (1992). *Troubleshooting Guide for HPLC Injection Problems*, 2nd edn. Rheodyne Incorporated, Cotati.
16. Millipore Corporation (1993). *Waters U6K Universal Liquid Chromatograph Injector, Instruction Manual*. Millipore Corporation, Milford.
17. Millipore Corporation (1993). *Waters 717 Autosampler, Service Manual*. Millipore Corporation, Milford.
18. Millipore Corporation (1993). *Waters Sourcebook of Chromatography*, Millipore Corporation, Milford.
19. Scott R.P.W. (1988). *J. Chromatogr.*, **486**: 99.
20. Hirata Y. (1990). *J. Microcol. Sep.*, **2**: 214.
21. Millipore Corporation (1993). *Developing HPLC Separations, Book One*. Millipore Corporation, Milford.
22. Bidlingmeyer B.A. and Warren F.V. (1984). *Anal. Chem.*, **56**: 1583A.

23. Dolan J.W. (1983). *LG.GC*, **11**: 22.
24. Dolan J.W. and Snyder L.R. (1989). *Troubleshooting LC Systems*. Humana Press, Clifton.
25. Meites L., Zuman P. *et al.* (1980–1985). *CRC Handbook Series in Inorganic Electrochemistry*, Vols. I–V. CRC Press Inc., Boca Raton.
26. Kissinger P.T. and Heineman W.R. Editors (1984). *Laboratory Techniques in Electroanalytical Chemistry*. Marcel Dekker Inc., New York.
27. Sawyer D.T. and Roberts J.L. (1974). *Experimental Electrochemistry for Chemists*. John Wiley, New York.
28. Johnson D.C., Weber S.G., Bond A.M., Wightman R.M., Shoup R.E. and Krull I.S. (1986). *Anal. Chim. Acta*, **180**: 187.
29. Roe D.K. (1983). *Anal. Lett.*, **16**: 613.
30. Hughes S. and Johnson D.C. (1983). *Anal. Chim. Acta*, **149**: 1.
31. Serjeant E.P. (1984). *Potentiometry and Potentiometric Titrations*. John Wiley & Sons, New York.
32. Rocklin R.D. (1984). *LC*, **2**: 588.
33. Kok W.Th., Brinkman U.A.Th. and Frei R.W. (1983). *J. Chromatogr.*, **256**: 17.
34. Buchberger W., Winsauer K. and Breitwieser Ch. (1982). *Fres. Z. Anal. Chem.*, **311**: 517.
35. Sherwood G.A. and Johnson D.C. (1981). *Anal. Chim. Acta*, **129**: 101.
36. Shpigun O.A. and Zolotov Yu.A. (1988). *Ion Chromatography in Water Analysis*. Ellis Horwood, Chichester, p. 85.
37. Fritz J.S., Gjerde D.T. and Pohlandt C. (1982). *Ion Chromatography*, Huethig, Heidelberg, p. 119.
38. Baba N. and Housako K. (1984). *U.S. Patent* 4,462,962.
39. Stillian J. (1985). *LC*, **3**: 802.
40. *Developing HPLC Separations, Book Two* (1991). Millipore Corporation, Milford.
41. Snyder L.R., Glach J.L. and Kirkland J.J. (1988). *Practical HPLC Method Development*, John Wiley & Sons, Inc., New York, NY.
42. Lingeman H. and Underberg W.J.M. Editors (1990) *Detection-Oriented Derivatization Techniques in Liquid Chromatography*. Marcel Dekker, New York.
43. Liang Y. (1986). *Chromatogr. Rev.*, **13**: 10.
44. Bell J.P., Simpson R.A. and Cunico R.L. (1986). *Int. Lab.*, **16**: 62.
45. Schlabach T.D. and Weinberger R. (1986). In Krull I.S. Editor, *Reaction Detection in Liquid Chromatography, Chromatographic Science Series*, Vol. 34 Marcel Dekker, New York, Ch. 2.
46. Colgan S.T. and Krull I.S. (1986). In Krull I.S. Editor, *Reaction Detection in Liquid Chromatography, Chromatographic Science Series*, Vol. 34 Marcel Dekker, New York, Ch. 5.
47. Cassidy R.M., Elchuk S. and Dasgupta P.K. (1987). *Anal. Chem.*, **59**: 85.
48. Lillig B. and Engelhardt H. (1986). In Krull I.S. Editor, *Reaction Detection in Liquid Chromatography, Chromatographic Science Series*, Vol. 34 Marcel Dekker, New York, Ch. 1.
49. Bidlingmeyer B.A. Editor (1987). *Preparative Liquid Chromatography, Journal of Chromatography Library Series*, No. 38 Elsevier, Amsterdam.
50. Tarter J.G. (1986). *J. Chromatogr.*, **367**: 191.
51. Harvey M.C. and Stearns S.D. (1984). In Lawrence J.F., Editor, *Liquid Chromatography in Environmental Analysis*. Humana, Clifton, Ch. 7.
52. Haddad P.R. and Rochester N.E. (1988). *J. Chromatogr.*, **439**: 23.
53. Villasenor S.R. (1991). *Anal. Chem.*, **63**: 1362.
54. Jones W.R., Jandik P. and Schwartz M.T. (1989). *J. Chromatogr.*, **473**: 171.

Bibliography

General HPLC

Ahuja S. (1992). *Trace and Ultratrace Analysis by HPLC*. John Wiley & Sons, New York.
Dorsey J.G., Foley J.P., Cooper W.T., Barford R.A. and Barth H.G. (1992). Column liquid chromatography: theory and methodology, *Anal. Chem.*, **64**: 352R.
Harvey M.C. and Stearns S.D. (1984). In Lawrence J.F., Editor, *Liquid Chromatography in Environmental Analysis*. Humana, Clifton.

Lindsay S. (1992). *High Performance Liquid Chromatography*. John Wiley & Sons, Chichester.
Meyer V.R. (1985). High performance liquid chromatographic theory for the practitioner. *J. Chromatogr.*, **334**: 197.
Millipore Corporation (1991). Developing HPLC Separations, *Books One and Two*. Millipore Corporation, Milford.
Poole C.F. and Poole S.K. (1991). *Chromatography Today*. Elsevier, Amsterdam.
Schoenmakers P.J. (1986). Optimization of Chromatographic Selectivity: A Guide to Method Development, *Journal of Chromatography Library Series, No. 35*. Elsevier, Amsterdam.
Snyder L.R. and Kirkland J.J. (1979). *Introduction to Modern Liquid Chromatography, 2nd edn*. John Wiley & Sons, Inc., New York.
Snyder L.R., Glach J.L. and Kirkland J.J. (1988). *Practical HPLC Method Development*. John Wiley & Sons, Inc., New York.

Instrumentation

Am. Lab. (1993). Buyers Guide, February (1993).
Bahowich T.J., Murugaiah V., Sulya A.W., Taylor D.B., Synovec R.E., Bergman R.J., Renn C.N. and Johnson E.L. (1992). Column liquid chromatography: equipment and instrumentation. *Anal. Chem.*, **64**: 255R.
C. & E. News (1992). Instrumentation '92 issue, March (1992).
Ewing G.E. Editor (1990). *Analytical Instrumentation Handbook*. Marcel Dekker, New York.
LC.GC (1992). Marketplace Issue, August (1992).
Millipore Corporation (1993). *Waters Instruments, Parts and Supplies Catalogue 1993–1994*. Millipore Corporation, Milford.

Troubleshooting

Dolan J.W. (1992). Preventative maintenance: just three things. *LC.GC*, **10**: 842.
Dolan J.W. (1992). Problem isolation: three more things. *LC.GC*, **10**: 920.
Dolan J.W. (1993). Column care and feeding. *LC.GC*, **11**: 22.
Dolan J.W. (1993). Avoiding the pitfalls of published methods. *LC.GC*, **11**: 412.
Dolan J.W. and Snyder L.R. (1989). *Troubleshooting LC Systems: A Systematic Approach to Troubleshooting LC Equipment and Separations*. Humana Press, Clifton.
Millipore Corporation (1991). *Guide to Successful Operation of Your LC System*. Millipore Corporation, Milford.
Spruce B. and Bakalyar S.R. (1992). *Troubleshooting Guide for HPLC Injection Problems*, 2nd edn. Rheodyne Incorporated, Cotati.

Columns

Issaq H.J. and Gourley R.E. (1983). High-performance liquid chromatography separations using short columns packed with spherical and irregular shaped ODS particles. *J. Liquid Chromatogr.*, **6**: 1375.
Millipore Corporation (1991–1993). *Waters Sourcebook(s) of Chromatography*. Millipore Corporation, Milford.
Scott R.P.W. (1988). Contemporary liquid chromatography column design. *J. Chromatogr.*, **486**: 99.
Verzele M., Van Dijck J., Mussche P. and Dewaele C. (1982). Spherical versus irregular-shaped silica gel particles in HPLC. *J. Liquid Chromatogr.*, **5**: 1431.

Detection

Frei R.W. and Zech K., Editors (1988). *Selective Sample Handling and Detection in HPLC, Parts A and B*. Elsevier, Amsterdam.
Krull I.S. Editor (1986). *Reaction Detection in Liquid Chromatography, Chromatographic Science Series, Vol. 34*. Marcel Dekker, New York.

Li J.B. (1992). Signal-to-noise optimization in HPLC UV detection, *LC.GC*, **10**: 856.

Lingeman H. and Underberg W.J.M., Editors (1990). *Detection-Oriented Derivatization Techniques in Liquid Chromatography*, Marcel Dekker, New York.

Scott R.P.W. (1987). *Liquid Chromatography Detectors*. 2nd, completely revised edn. Elsevier, Amsterdam.

HPLC techniques

Berridge J.C. (1986). *Techniques for the Automated Optimisation of High Performance Liquid Chromatography Separations*. Wiley-Interscience, Chichester.

Bidlingmeyer B.A., Editor (1987). *Preparative Liquid Chromatography, Journal of Chromatography Library Series, No. 38*. Elsevier Science Publishers B.V.

Cazes J., Editor (1981). *Liquid Chromatography of Polymers and Related Materials, Chromatographic Science Series, No. 19*. Marcel Dekker, Inc., New York.

Haddad P.R. and Jackson P.E. (1990). *Ion Chromatography: Principles and Applications, Journal of Chromatography Library Series, No. 46*. Elsevier Science Publishers B.V., Amsterdam.

Hancock W.S. and Sparrow J.T. (1984). *HPLC Analysis of Biological Compounds: A Laboratory Guide, Chromatographic Science Series, No. 26*. Marcel Dekker, Inc., New York.

Jandera P. and Churacek J. (1985). *Gradient Elution in Column Liquid Chromatography, Journal of Chromatography Library Series, No. 31*. Elsevier, Amsterdam.

Schoenmakers P.J. (1987). *Optimization of Chromatographic Selectivity*. Elsevier, Amsterdam.

Snyder L.R., Stadalius M.A. and Quarry M.A. (1983). Gradient elution in reversed phase HPLC separation of macromolecules. *Anal. Chem.*, **55**: 1412A.

High-performance Liquid Chromatography— Separations

6

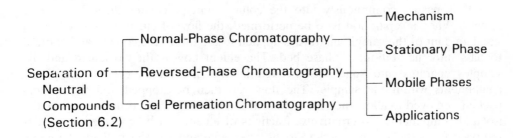

Separation of Neutral Compounds (Section 6.2)
- Normal-Phase Chromatography
- Reversed-Phase Chromatography
- Gel Permeation Chromatography
  - Mechanism
  - Stationary Phase
  - Mobile Phases
  - Applications

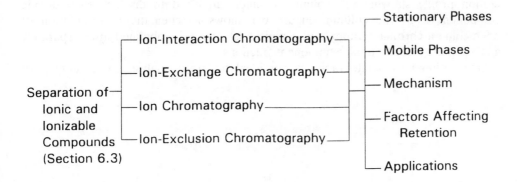

Separation of Ionic and Ionizable Compounds (Section 6.3)
- Ion-Interaction Chromatography
- Ion-Exchange Chromatography
- Ion Chromatography
- Ion-Exclusion Chromatography
  - Stationary Phases
  - Mobile Phases
  - Mechanism
  - Factors Affecting Retention
  - Applications

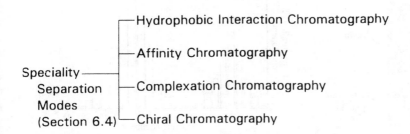

Speciality Separation Modes (Section 6.4)
- Hydrophobic Interaction Chromatography
- Affinity Chromatography
- Complexation Chromatography
- Chiral Chromatography

Choosing a Chromatographic Method
(Section 6.4)

6.1　Introduction

6.1.1　Chromatographic separations

Historically, chromatographic separations were first performed in the open-column mode, as discussed in Section 1.1. In open-column chromatography, the adsorbent, or stationary phase is packed loosely as small particles into a glass column of 1–2 cm diameter. A displacing solution, alternatively termed either the mobile phase or eluent, is passed continuously into the column and percolates through it under gravity. When a separation is to be performed, the flow of eluent is stopped and a small amount of the sample mixture is applied to the top of the column and allowed to pass into the stationary phase bed. The eluent flow is then resumed and the sample separates into zones, as dictated by the distribution coefficients of the components within the sample. The flow can then be stopped and the column packing removed to allow the separated zones to be extracted in order to remove the isolated components. Alternatively, fractions of eluent can be collected at regular intervals from the column outlet, to be later analysed for the individual sample components [1]. The results of these analyses can then be combined to provide an elution profile showing the volume of eluent required to displace each sample component from the column. Figure 6.1 shows a schematic representation of open-column chromatography and a typical set of results obtained after separately analysing the eluted sample component fractions.

Open-column chromatography is not ideal for trace analysis for a number of

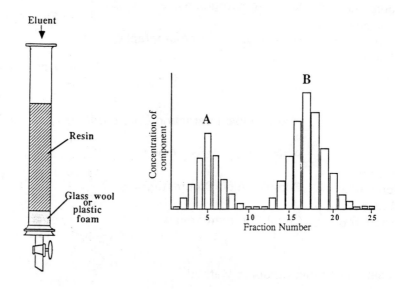

Fig. 6.1. Apparatus for classical open-column chromatography and type of result typically obtained. A and B are two components eluted from the column. Reprinted from Haddad and Jackson (1990) *Journal of Chromatography Library Series No. 46*, Elsevier, Amsterdam, with permission.

reasons. First, the process is very slow because of the low eluent flow-rates employed. Attempts to increase the flow-rate (e.g. by raising the height of the eluent container above the head of the column) invariably result in poorer separation efficiency. This effect is primarily a result of the poor mass-transfer characteristics of the relatively large particles used for the column packing. Second, the chromatographic efficiency attained is rather poor and many important separations simply cannot be achieved using this method. Third, it is inconvenient to either unpack the column bed to extract the separated zones or to collect and separately analyse fractions of eluent [1]. Despite these significant limitations, open-column chromatography still finds use in the modern laboratory, although perhaps more for sample pretreatment rather than for analysis.

Modern liquid chromatographic separations circumvent these difficulties through the use of high-efficiency packing materials combined with continuous flow-through detection. Separations are performed in the column mode with particles of uniform size packed into a column housing constructed of rigid material. The particles of packing material are generally very much smaller than those used for classical open-column chromatography. This requires that mobile phase be pumped through the column, since the small particle size of the stationary phase precludes appreciable flow under gravity. The sample mixture is introduced into the eluent via an injection port and is carried to the column where separation takes place, after which the separated solutes are detected with a flow-through detection device.

6.1.2 Organization of this chapter

In the previous chapter, instrumentation and techniques used to achieve high-performance liquid chromatographic separations were discussed. In this chapter, we deal with the manner by which the separation process occurs, i.e. how the sample components interact with the stationary and mobile phases. One of the major reasons for the widespread success of HPLC as an analytical technique lies in the variety of separation mechanisms (or modes) which may be exploited. This makes possible the separation of a very diverse range of solutes, including, ionic, ionizable, polar, non-polar and polymeric compounds. The actual mechanism by which the separation occurs is perhaps the most common means for classifying chromatographic methods, as discussed in Section 1.4. In this chapter, we take a less conventional approach and divide the separation modes into three broad groups based on both the nature of the solute and the separation mechanism. The three groups are the separation of neutral compounds, the separation of ionogenic compounds and speciality separations. The separation of neutral compounds covers normal-phase, reversed-phase and gel-permeation chromatographic modes. The separation of ionogenic compounds (i.e. those solutes which are ionic or can be ionic under appropriate conditions) includes ion-suppression, ion-interaction, ion-exchange and ion-exclusion modes of chromatography. A number of speciality separation modes, including hydrophobic interaction chromatography, affinity chromatography, ligand-exchange chromatography and chiral separations are then discussed. The final section in the chapter discusses aspects of method selection.

6.2   Separation of Neutral Compounds

In this section, chromatographic modes used primarily for the separation of neutral compounds will be discussed, i.e. normal-phase, reversed-phase and gel-permeation chromatography. While these modes may be modified to effect the separation of ionic compounds, only their application to neutral compounds will be considered in this section. Gel-permeation chromatography (GPC) is the one exception, as the separation of proteins (which are charged macromolecules) will be discussed in this section.

6.2.1   Normal-phase chromatography

Introduction

Normal-phase chromatography, alternatively termed adsorption or liquid–solid chromatography, represents the oldest of the chromatographic separation modes. It was first used in the classical open-column form in 1906 by Tswett to separate plant pigments [2]. In normal-phase chromatography, the sample components are retained on the stationary phase through the interaction of permanent dipoles on the component with permanent dipoles on the stationary phase [3]. This results in an adsorption mechanism and gives rise to the general class of adsorption chromato-graphic methods in which polar stationary phases and non-polar mobile phases are employed. With this separation mode, the more polar the solute, the greater the retention; increasing the polarity of the mobile phase results in decreased solute retention.

Mechanism of normal-phase chromatography

Retention in normal-phase, or adsorption, chromatography is due to interaction of polar functional groups on the solute with discrete sites on the stationary phase surface. The selectivity of the separation depends upon the relative strengths of these polar interactions for different solutes. The extent to which a solute can be accommodated on the stationary phase depends upon its spatial configuration and its ability to form hydrogen bonds with the adsorbent. This chromatographic mode is therefore sensitive to spatial (or steric) differences in solutes and is generally better suited to isomer separations than reversed-phase chromatography [4]. Normal-phase chromatography exhibits a unique ability to distinguish between solutes with different numbers of electronegative atoms, such as oxygen or nitrogen, or molecules with different functional groups. Adsorption chromatography is therefore widely used for class separations [4].

A number of mechanisms have been proposed to account for the nature of the adsorption process [5]. In the case of non-polar and moderately polar mobile phases, which interact with the adsorbent surface largely by dispersive and weak dipole interaction, the competition model assumes that the entire adsorbent surface is covered by a monolayer of mobile phase molecules [4]. Solute retention then occurs by a competitive displacement of a mobile phase molecule from the surface of the adsorbent. The solvent interaction model proposes the formation of mobile phase layers adsorbed onto the stationary phase surface. The composition of the bilayer

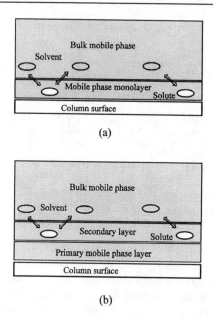

Fig. 6.2. Schematic illustration of the separation mechanism of normal phase chromatography. The competition model is shown in (a), while the solvent interaction model is shown in (b).

formation depends upon the concentration of polar solvent in the mobile phase and solute retention occurs by interaction (either association or displacement) of the solute with the secondary layer of adsorbed mobile phase molecules [6]. Like gel-permeation chromatography, where the basis of the separation is quite simple, retention in adsorption chromatography is easier to understand and predict than other separation modes. Semiquantitative estimates of k' are often possible with knowledge of only the experimental conditions and the structure of the eluted compound and, in some cases, prediction of k' can be made within a few per cent over a wide range of experimental conditions [4]. Figure 6.2 illustrates schematically the separation mechanism of normal-phase chromatography.

Stationary phases for normal-phase chromatography

A wide range of stationary phase materials has been used in adsorption chromatography. These include sucrose, cellulose, starch, silica gel (both 'bare' and functionalized), florisil, charcoal, magnesium oxide, hydroxylapatite and alumina. However, virtually all modern normal-phase separations are performed on either silica or alumina stationary phases. Silica is the preferred stationary phase, for a number of practical reasons [4]. Silica allows higher sample loadings and is less likely to catalyse the decomposition of any sample components while alumina has been known to catalytically decompose many organic compounds. A much wider range of silica columns is available commercially and there is also greater coverage on the application of silica stationary phases for different analyses in the literature. However, it may be occasionally advantageous to use alumina, particularly for basic compounds (i.e. amines), which are very strongly retained on silica. Conversely,

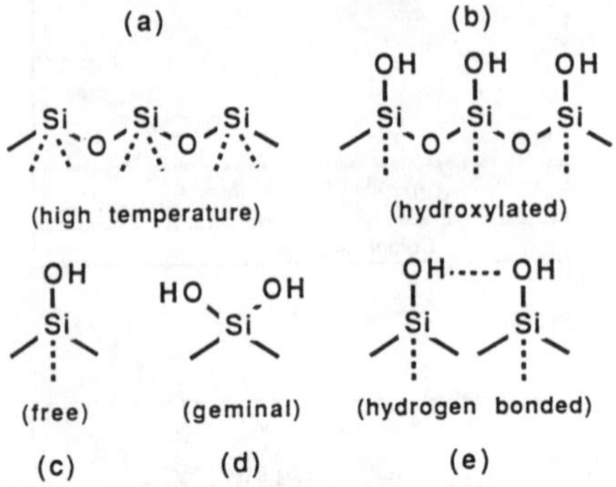

Fig. 6.3. Silanol group configurations on a silica surface. Reprinted with permission from Snyder *et al.*, (1988) *Practical HPLC Method Development*, John Wiley & Sons, New York.

carboxylic acids are very strongly retained on alumina, sometimes irreversibly, and hence are best chromatographed on silica. In general, the retention and separation characteristics on both silica and alumina are similar with the more polar samples being more strongly retained. The usual elution order is: saturated hydrocarbons < olefins < aromatic hydrocarbons ≈ organic halides < sulfides < ethers < nitro compounds < esters ≈ aldehydes ≈ ketones < alcohols ≈ amines < sulfones < sulfoxides < amides < carboxylic acids [4].

In normal-phase chromatography, the adsorbent surface consists of discrete adsorption sites, in a similar fashion to a charged functional group on an ion-exchanger. In the case of silica, the adsorption sites are hydroxyl (–OH) groups and the grouping (–Si–OH) is known as a silanol group. These silanol groups can exist on the silica surface in a number of configurations, depending upon the manner in which the silica has been treated prior to use. This treatment of the silica controls its activity, that is, the number of active sites per unit surface area and also the adsorptive strength (acidity) of these sites. The activity is varied by heating the silica in the presence of controlled quantities of water and acid. Figure 6.3 shows silanol configurations that can exist on a silica surface. Silica heated above 800°C is largely devoid of silanol groups, as shown in Fig. 6.3(a), and is of little practical use in HPLC. Silica used in HPLC is typically heated to 200–300°C, which allows the formation of a fully hydroxylated surface, as shown in Fig. 6.3(b). The individual silanol groups can exist in three general forms, as shown in Figs. 6.3(c–e), with free silanols being the most acidic and geminal silanol groups being the least acidic [7]. Free silanols generally occur in relatively low concentrations on the silica surface. Their highly acidic nature causes strong retention of basic solutes, hence their presence is undesirable for the separation of these compounds.

In addition to the type of material and its activity, other factors that affect the retention of sample molecules on adsorbent packing materials are the surface area and the average pore diameter. To a first approximation, sample retention is directly proportional to the surface area. This requires that the surface area of an adsorbent be held within narrow limits if retention properties are to remain constant from batch to batch [4]. Porous silica materials typically have surface areas in the order of $100–250 \, m^2 \, g^{-1}$ for spherical particles and about $300–400 \, m^2 \, g^{-1}$ for irregular

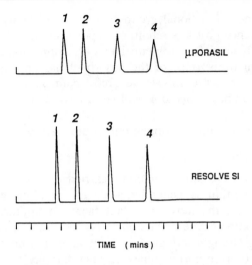

Fig. 6.4. Separation of phthalates on different silica columns. A mobile phase of 7:93 ethyl acetate:trimethyl pentane at a flow-rate of 2.0 ml min^{-1} was used. Detection was by u.v. at 280 nm. Solute identities (phthalate): 1, dioctyl; 2, dibutyl; 3, diethyl; 4, dimethyl. Chromatogram courtesy of Waters.

particles. A practical optimum value of 400 m^2 g^{-1} exists for silica packings. Higher surface areas can be achieved but only at the expense of small pores, which results in poor mass transfer and lower column efficiency values [4]. Table 6.1 shows the base silica characteristics of three normal-phase packings available from a commercial column manufacturer.

Probably the greatest drawback of normal phase chromatography is the lack of separation selectivity between different packing materials: virtually all compounds elute in the same order regardless of the column selected. For instance, Fig. 6.4 shows the separation of phthalates on μPorasil (10 μm) and Resolve (5 μm) silica columns. Despite appreciable differences in particle shape, size and surface area, as shown in Table 6.1, the columns show virtually identical retention and selectivity for the phthalates. Hence, in normal-phase chromatography, changes in selectivity are achieved primarily through changing the mobile phase. Even then, the selectivity changes are somewhat limited when compared with those that may be obtained using reversed-phase chromatography.

A more recent alternative to the use of 'bare' silica for normal-phase chromatography is the use of silica functionalized with either diol, cyanopropyl or amino

Table 6.1 Base silica characteristics of Waters normal-phase columns.

Packing type	Nominal particle size	Particle shape	Nominal pore distribution	Nominal surface area
Nova-Pak	4 μm	Spherical	60 Å (60–100 Å)	120 m^2/g^{-1}
Resolve	5 or 10 μm	Spherical	90 Å (60–125 Å)	200 m^2/g^{-1}
μPorasil	10 μm	Irregular	125 Å (50–300 Å)	300 m^2/g^{-1}

groups [8]. While these functional groups are all significantly less polar than the silanol group, and hence give less retention of polar solutes, the columns are easier to clean and water is less of a problem than with unfunctionalized silica columns. The water bound on the surface of the adsorbent has an important effect on its properties; however, because this is generally controlled by varying the water content of the mobile phase, this topic will be discussed in the following section.

Mobile phases for normal-phase chromatography

Solvent strength

Adsorption separations are most commonly carried out on silica stationary phases and variation in retention is generally achieved by altering the mobile phase composition. Therefore, the most important factor in optimizing a normal-phase separation is the selection of the mobile phase composition. A vast range of solvents with different eluting strengths are available, enabling enormous variation in sample retention and separations in normal-phase chromatography.

The total interaction of a solvent molecule with a sample compound is the result of four interactions, these being dispersion, dipole, hydrogen bonding and dielectric interactions [4]. The ability of a molecule to interact in these four ways is referred to as the polarity of the compound. Hence, polar solvents preferentially attract and dissolve polar compounds. In a similar fashion, the chromatographic strength of a solvent in a normal-phase system is related directly to its polarity. However, it has been shown that a better index of solvent strength for adsorption chromatography is given by the experimental adsorption solvent strength parameter, ε^0. This parameter, which is a measure of the adsorption energy per unit area of solvent, can be used to quantitatively define solvent strength for a given adsorbent [4]. Table 6.2 lists values of ε^0 for a number of different solvents with both silica and alumina adsorbents and such a listing is termed an eluotropic series. An increase in the value of ε^0 means a stronger solvent, leading to lower retention for a given solute. On average, the values of ε^0 for silica are generally 0.8 times less than those for alumina.

An appropriate mobile phase strength can be determined by selecting two solvents whose ε^0 values are too small (A) and too large (B), respectively. The two solvents can then be blended together to allow a continuous variation in strength from pure A up to that of pure B. An increase in solvent ε^0 value by 0.05 units typically decreases retention (k') by a factor of 3–4. The solvent strength does not vary in a linear fashion with %B, rather it follows a convex curve [4]. Care must be taken to ensure that the solvents are miscible, for instance hexane is not totally miscible with all the solvents listed in Table 6.2. It has been suggested that FC-78 or FC-113 are more appropriate weak solvents than hexane, as they are miscible with a wider range of other solvents [8]. Alternatively, a cosolvent (such as methylene chloride) may be added to hexane mobile phases to enhance solubility. Also, methyl *t*-butylether has been recommended as an alternative to either isopropyl or ethyl ether as it is less susceptible to peroxide formation and not as hazardous [9]. It must be noted that many of the solvents listed above are prone to peroxide formation and are often stabilized by the addition of an antioxidant, such as 3–4% ethanol. Any solvent strength considerations must therefore take account of the fact that the solvent may have been stabilized. In practice, a series of mobile phases from hexane, to hexane

Table 6.2   Solvents commonly used with normal-phase chromatography.

Solvent	Solvent strength, ε^0	
	Silica	Alumina
Fluorochemical FC-78	−0.2	−0.25
n-Hexane		
n-Heptane	0.01	0.01
Iso-octane		
1-Chlorobutane	0.20	0.26
Chloroform	0.26	0.40
Methylene chloride	0.32	0.42
Isopropyl ether	0.34	0.28
Ethyl acetate	0.38	0.58
Tetrahydrofuran	0.44	0.57
Acetonitrile	0.50	0.65
Methanol	0.7	0.95

Data from Snyder and Kirkland [4].

containing 0.1–2.0% ether, to hexane containing 5–10% dichloromethane, to dichloromethane containing 5–10% acetonitrile provides a sufficient solvent strength range for most normal-phase separations.

Solvent selectivity

Once a solvent mixture (A/B) of the desired strength (i.e. $2 < k' < 10$) is located for a given separation, the separation selectivity can be varied by holding ε^0 constant and exchanging the strong solvent (B) for another solvent. Such solvent mixtures of equivalent eluting strengths are termed isoeluotropic. For instance, 50% ether in hexane has the same eluting strength (ε^0 of 0.30) as 75% methylene chloride in hexane and 1.7% acetonitrile in isopropyl chloride [4]. While these mobile phases give similar total elution times, they may permit some changes in sample selectivity, i.e. changes in the relative elution order of the peaks. Once the best solvent mixture (in terms of selectivity) has been located, it may then be necessary to fine-tune the solvent strength by minor adjustment of the %B in the mobile phase. Figure 6.5 shows an example of changing the selectivity of a normal phase separation for phthalate plasticizers. In this separation, changing solvent B from ethyl acetate to butyl acetate radically changes the selectivity between the diethyl and diphenyl phthalate peaks.

Tables of solvent strength as a function of binary mobile phase composition are available for normal-phase chromatography to aid in mobile phase selection [10] and general rules have also been devised which predict the variation of selectivity with solvent composition [4]. In practice, it has been shown that three binary mobile phases, each containing hexane with (i) a non-localizing solvent (e.g. methylene chloride), (ii) a localizing basic solvent (e.g. methyl t-butylether) and (iii) a localizing nonbasic solvent (e.g. ethyl acetate) permit the broadest changes possible in normal-phase separation selectivity [7]. More detail on varying the selectivity of normal-phase separations may be found elsewhere [4, 7, 11].

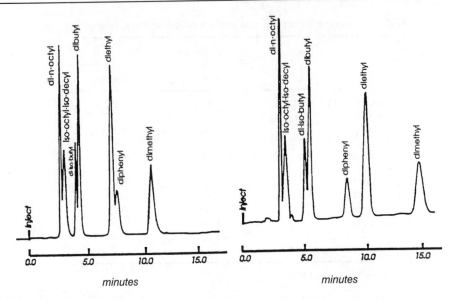

Fig. 6.5. Mobile phase selectivity effects in normal phase chromatography. A mobile phase of 5:95 ethyl acetate:iso-octane was used for the chromatogram on the left and a mobile phase of 5:95 butyl acetate:iso-octane was used for the chromatogram on the right. Reprinted with permission from *Developing HPLC Separations, Book One* (1991), Millipore Corporation, Milford.

Effect of water

Perhaps the greatest practical problem associated with the operation of normal-phase chromatography is the effect that water has on the activity of polar adsorbents. Water is adsorbed onto the strongest adsorption sites, leaving a more uniform distribution of weaker sites to retain the sample. This leads to a decrease in solute retention and the adsorbent is said to be deactivated. This creates problems since, in practical terms, it is very difficult to control the amount of water present in the mobile phase. The water content of the mobile phase varies because [4]: (i) variation exists in the amount of water present in supposedly 'dry' solvents; (ii) water is readily picked up and lost to the atmosphere, depending upon the humidity and water content of the solvent; and (iii) changes occur as a result of contact with the walls of solvent reservoirs and other containers.

These problems are typically overcome by the initial addition of a small amount of water to the mobile phase to give a resulting solution which is 25–50% saturated. The beneficial effects of adding water to the mobile phase are [4]: (i) less variation in sample retention from run to run; (ii) higher sample loadings; (iii) higher column efficiencies and reduced peak tailing for basic compounds; (iv) reduction in batch-to-batch variations of the adsorbent; and (v) reduced catalytic activity of the adsorbent.

However, the addition of water does create some additional problems. Column re-equilibration is slow when changing from one solvent to another which, in turn, may significantly restrict the use of gradient elution. New columns first must be equilibrated with a solvent that has a high water solubility, e.g. 50% saturated ethyl acetate, and once a column has been deactivated, it is very difficult to remove the bound water. Cyanopropyl or amino-functionalized silica normal-phase columns do not require the addition of water to the mobile phase and it has recently been suggested that the process of normal-phase method development should begin with one of these columns, rather than unfunctionalized silica [7].

Applications of normal-phase chromatography

Because adsorption chromatography was the first chromatographic separation mode developed, a great range of applications using this technique has been reported. In addition, this separation mode has found much use resulting from the scale-up of thin layer chromatographic separations. In modern HPLC, however, normal-phase chromatography is by far the least commonly employed of the major separation modes. The main reason for this is the fact that reversed-phase chromatography readily allows for greater variation of separation selectivity for the same types of solutes used in normal-phase chromatography. Normal-phase chromatography is therefore most commonly applied to the analysis of samples that are soluble in non-polar solvents, such as hexane, while solutes which are more polar, water-soluble, and/or ionic tend to be best chromatographed using reversed-phase chromatography. As mentioned previously, adsorption chromatography is generally better suited to isomer separations than reversed-phase chromatography [4]. Adsorption chromatography also demonstrates a unique ability to distinguish solutes with different numbers of electronegative atoms, such as oxygen or nitrogen, or for molecules with different functional groups.

Both fat- and water-soluble vitamins have been analysed by normal-phase chromatography. For instance, vitamin D can be analysed in milk on a silica column,

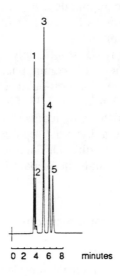

Fig. 6.6. Normal-phase chromatography of pesticides. An Exmere CN-8/5 column was used with a mobile phase of iso-octane at 0.75 ml min^{-1}. Solute identities: 1, aldrin; 2, heptachlor; 3, DDT; 4, endrin; 5, dieldrin. Chromatogram courtesy of SGE.

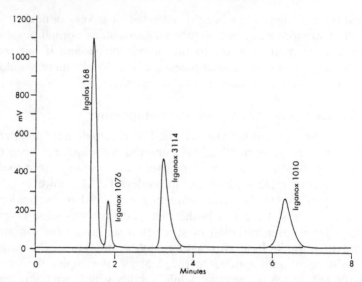

Fig. 6.7. Normal phase chromatography of polymer antioxidants. A Waters Resolve silica column was used with a mobile phase of *n*-butyl chloride containing 30% methylene chloride at 1.5 ml min^{-1}. Chromatogram courtesy of Waters.

using a mobile phase of hexane containing 1% isopropyl alcohol. The less-polar plant pigments, such as carotenoids and porphyrins have been successfully analysed using this approach for over 50 years. Natural oils and flavour extracts are also readily analysed by adsorption chromatography [4]. Hydrocarbons have been analysed on an amino-functionalized silica column, using hexane as the mobile phase, and a wide range of pesticides have also been separated using adsorption chromatography. While these compounds are perhaps better suited to analysis by GC, Fig. 6.6 shows the separation of a number of organochlorine pesticides on a cyano-functionalized silica column.

Normal-phase chromatography is still widely used for the determination of non-polar additives in a variety of commercial products and pharmaceutical formulations, e.g. the separation of non-polar components in Triton X-100, a non-ionic surfactant, as was shown in Fig. 5.37. Figure 6.7 shows an example of the separation of a mixture of polymer antioxidants. Finally, one of the more important uses of normal-phase chromatography is the separation of isomers. A wide range of isomers has been separated using this chromatographic mode, including the six isomers of trinitrotoluene [12] and the separation of 1,2 and 1,6 isomers of 7-hydroxy tetrahydrocannabinol.

6.2.2   Reversed-phase chromatography

Introduction

Reversed-phase chromatography, alternatively termed reverse phase or bonded phase chromatography, is the most widely used of the liquid chromatographic separation modes. It has been reported that over 75% of all HPLC separations are performed using this approach [4]. The term reversed-phase arises from the fact that this separation mode utilizes a non-polar stationary phase with a polar mobile phase, which is the reverse of the situation in normal-phase (adsorption) chromatography. This approach was originally applied in liquid–liquid chromatography in which the stationary phase was a film of liquid held on a solid support, but problems with stationary phase bleeding were often encountered. This led to the use of stationary phases in which the desired non-polar functionality was chemically bound onto a packing substrate, usually silica, to produce a non-polar stationary phase. The mechanism of reversed-phase chromatography is complex and sample components are retained through nonspecific hydrophobic interactions with the stationary phase. With this separation mode, the more polar the solute, the lower the retention. Increasing the polarity of the mobile phase results in increased solute retention.

Mechanism of reversed-phase chromatography

Separation in reversed-phase liquid chromatography (RPLC) is more difficult to explain than the simple polar–polar interactions of normal-phase chromatography. The interactions between solute molecules and the non-polar stationary phase are much too weak to account for the degree of solute retention observed in RPLC. The retention mechanism is complex and could be best described as a combination of partition and adsorption. In a number of instances, RPLC has been referred to as partition chromatography, because the bonded organic surface layer can be regarded as a bound liquid film, although other workers have concluded that the bonded phase acts more like a modified solid than a liquid film [13]. In addition, the heterogeneous nature of silica-based materials means that different mechanisms can operate over different regions of the surface. Furthermore, the mechanism is not necessarily constant over the entire mobile phase composition range as the solvation of binding sites on the stationary phase surface is strongly affected by the mobile phase composition.

It is generally accepted that the solvophobic theory offers the most valid interpretation of RPLC retention [14]. This mechanism, which is perhaps most appropriate when using aqueous mobile phases with low organic modifier content, is based on the assumption that the stationary phase is a uniform layer of a non-polar ligand. The solvophobic theory assumes that the solute binds to the stationary phase which then reduces the surface area of the solute exposed to the mobile phase. The solute is sorbed as a result of this solvent effect; that is, the solute is sorbed because it is solvophobic. Hence, solutes are retained more as a result of (solvophobic) interactions with the mobile phase rather than through specific interactions with the stationary phase. The fact that retention occurs primarily as a result of strong interactions between the mobile phase and the solute molecules means that the mobile phase composition has more influence on separation selectivity than the stationary phase in this mode of chromatography [7]. Figure 6.8 illustrates schematically the separation mechanism of reversed-phase chromatography.

Stationary phases for reversed-phase chromatography

The majority of RPLC separations are carried out on bonded silica stationary phases, although a significant number of polymeric reversed-phase columns is now

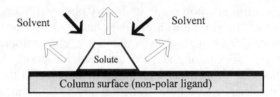

Fig. 6.8. Schematic illustration of the separation mechanism of reversed-phase chromatography. The binding is essentially caused by solvent effects [14]. As indicated by the solid arrows, the association of the two species is facilitated by the decrease in the molecular surface area exposed to the solvent upon complex formation. As indicated by the open arrows, attractive interactions with the solvent have the opposite effect. The magnitude of the non-polar interaction between the solute and the ligand is then given by the difference between the two effects. Modified with permission from Horvath and Melander (1977) *J. Chromatogr. Sci.*, **15**, 393.

also available commercially. While variation in the mobile phase composition is the most common means of altering separation selectivity in RPLC, appreciable differences in selectivity exist between different bonded phase columns. Vast numbers of silica bonded phase columns are available and subtle differences in their selectivity often means that separations on apparently similar column types may show appreciable differences.

Bonded phase silica columns

The performance of a bonded phase silica column is determined by four factors: (i) the base silica material and its pretreatment; (ii) the type of stationary phase bonded onto the silica; (iii) the amount of stationary phase material (or carbon load) bonded onto the silica; and (iv) any secondary bonding (or end-capping) reactions.

Generally, reversed-phase columns are manufactured by functionalizing the same type of silica material used for normal-phase columns. Silica characteristics, such as pore size and distribution, particle size and shape, surface area and degree of activity, will all affect the final performance of the bonded phase column. Table 6.1 detailed base silica characteristics of a number of Waters normal-phase columns. The strength of the interactions in RPLC between the stationary phase and polar solutes has been shown to depend on the surface activity of the silica used as the support material [15]. As previously mentioned, activity depends upon the pretreatment of the silica, in terms of the degree of heating and water content. Small differences in silica pretreatment prior to bonding can result in significant changes in selectivity, particularly for solute mixtures containing polar or basic compounds.

The various methods for attaching the bonded phase to a silica support all rely upon reaction with the surface silanol groups. The most common method involves the reaction of silica with organochlorosilanes to produce siloxane ($Si–O–Si–R_3$) reversed-phase packings. Either mono-, di- or tri-functional organochlorosilane bonding reagents may be used, as shown in Fig. 6.9 for mono- and difunctional reagents.

Reversed-phase columns are identified (and named) by the nature of the bonded R group; C_{18} is the most common functional group on bonded reversed-phase columns. Other functional groups include C_8, phenyl, C_6, C_4, C_2, CN, diol, NH_2 and

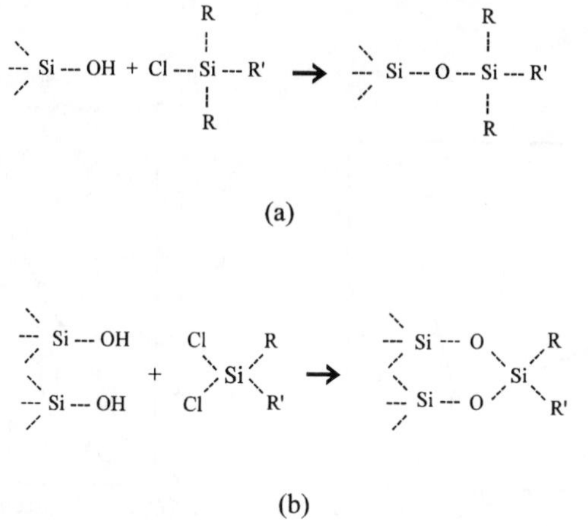

(a)

(b)

Fig. 6.9. Reaction of silica with (a) mono- or (b) difunctional bonding reagents.

NO_2. Retention in RPLC increases exponentially as the alkyl chain length of the bonded functionality increases [4]. The amount of material bonded onto the silica is described by the term carbon load, which is the amount of carbon expressed as a weight percentage of the bulk silica packing. Monofunctional C_{18} bonding reagents typically produce columns with carbon loads in the range of 7–15% (w/w). Monofunctional bonding reactions are generally preferred as they can be carefully controlled, producing monolayer functionalized columns with less batch-to-batch variation than when bi- or tri-functional bonding reagents are used. Monofunctional dimethylsilanes are the most widely used bonding reagents. The use of bi- or tri-functional bonding reagents results in both cross-linking and linear polymerization reactions and produces 'polymeric' C_{18} stationary phases [4]. Columns produced in this fashion can have carbon loads as high as 25%, although they often exhibit low chromatographic efficiency as a result of poor mass-transfer characteristics.

As a general rule, the higher the carbon load of the stationary phase, the more hydrophobic the column and the greater its reversed-phase retention, although the density of the packing must also be considered. Table 6.3 compares the k' for acetophenone using a mobile phase of 50:50 methanol:water on a number of C_{18}

Table 6.3 Acetophenone retention on different C_{18} stationary phases.

Column	Percentage carbon load	k' for acetophenone
DuPont Zorbax ODS*	20	15.7
Merck Lichrosorb PR-18	22	10.3
Beckman Ultrasphere ODS	12	11.3
Whatman Partisil ODS-3	10	12.0
Waters Resolve C_{18}	10	12.4
Waters μBondapak C_{18}	10	7.9
Waters Nova-Pak C_{18}	7	5.5

*ODS, octadecylsilane.

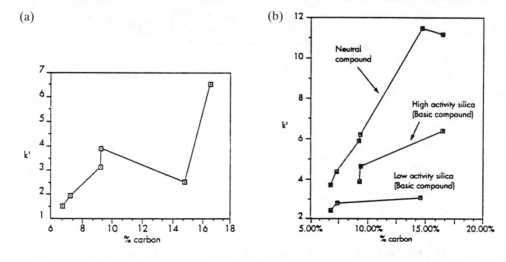

Fig. 6.10. Variation in retention for basic compounds on different C_{18} columns. Data courtesy of Waters.

stationary phases. These data suggest a general increase in k' with increasing carbon load, however it is evident that this is only an approximation. Even taking into account the different density of the packing materials, it can be seen that acetophenone has a somewhat different selectivity on each column.

This variation in retention is particularly evident when basic compounds are chromatographed by RPLC. Figure 6.10 shows the variation in capacity of a base on six different end-capped C_{18} columns using identical mobile phase conditions. It is difficult to draw any meaningful conclusions from Fig. 6.10(a), however, when the stationary phases are subdivided into low- and high-activity silica columns (b), the trend of increasing retention with % carbon becomes evident. Conversely, a similar plot (uppermost curve) for a small neutral hydrocarbon compound shows a smooth increase in retention with increasing percentage carbon for the same six columns.

The type of functional group bonded onto the silica affects both the carbon load and column selectivity in RPLC. The larger the R group, the greater the carbon load and stationary phase hydrophobicity for a given base silica material. Figure 6.11 compares the retention of a mixture of four solutes on Waters Nova-Pak C_{18}, Phenyl and CN columns using a mobile phase of 35:65 acetonitrile:water. The three columns have carbon loads of 7%, 5% and 2%, respectively. However, the major reason for the use of different functional groups in RPLC is to produce changes in selectivity, rather than retention. Columns with different functional groups may show a different separation selectivity for certain solute mixtures. Figure 6.12 compares the separation of a mixture of four solutes on Waters μBondapak CN and Phenyl columns using tetrahydrofuran–water mobile phases of approximately equivalent eluting strength. Peaks 1 and 2 are unresolved on the cyano column, however they are partially separated on the phenyl column. The converse is true of peaks 3 and 4, which are well separated on the cyano column but unresolved on the phenyl column. A range of functionalized (alkyl, phenyl and cyano) columns has been employed in procedures for stationary phase optimization in RPLC [16].

Fig. 6.11. Comparison of retention on Nova-Pak C_{18}, Phenyl and CN columns. Radial-Pak (8×100 mm) columns were used with a mobile phase of of 35:65 acetonitrile : water at 2.0 ml min^{-1}. Solute identities: 1, benzyl alcohol; 2, acetophenone; 3, p-tolualdehyde; 4, anisole. Chromatogram courtesy of Waters.

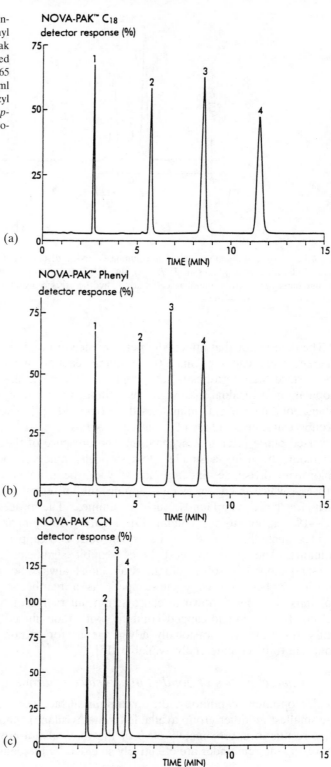

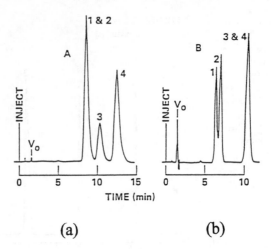

Fig. 6.12.   Effect of functional group on column selectivity. μBondapak CN (A) and Phenyl (B) columns were used with mobile phases of 23:77 (A) and 33:67 (B) tetrahydrofuran:water. Solute identities: 1, dexamethansone; 2, triamcinolone acetonide; 3, hydrocortisone acetate; 4, methyl testosterone. Chromatogram courtesy of Waters.

The final factor that affects the performance of bonded phase silica columns is the degree of end-capping. Since the bonding reaction of the organochlorosilane with the surface silanol groups is not complete, all bonded phase columns have a certain population of residual silanol groups, although the use of monofunctional dimethyl-silane bonding reagents minimizes this population [4]. These residual, acidic silanol groups can cause tailing of basic compounds as a result of a mixed adsorption–reversed-phase retention mechanism. The presence of these residual silanols can be overcome by using a smaller, less sterically hindered bonding reagent, typically chlorotrimethylsilane, in conjunction with the primary bonding procedure. This process, which is known as end-capping, minimizes mixed-mode retention, particu-larly for basic compounds, such as amines. The process usually adds between 0.1—1% carbon to the column [3]. Figure 6.13 compares the separation of five solutes, including two strong bases, on both end-capped and non-end-capped C_{18} columns. The two strongest bases exhibit significant peak distortion on the non-end-capped Resolve column. Care must always be taken to select a column which has been correctly end-capped, as a number of supposedly end-capped columns have been shown to cause significant tailing of basic compounds [3]. The selectivity of non-end-capped columns, with their mixed hydrophobic and silano-philic retention, is occasionally advantageous for certain separations and so they continue to be commercially available [15].

Characteristics of bonded phase silica columns

Under optimum conditions, the maximum surface concentration achievable using the smallest modifier group available (trimethylsilane) is approximately 4.7 μM m^{-2}. This results in approximately 60% of the total silanol groups being functionalized; however, the remainder are effectively shielded by the dense layer of trimethylsilane

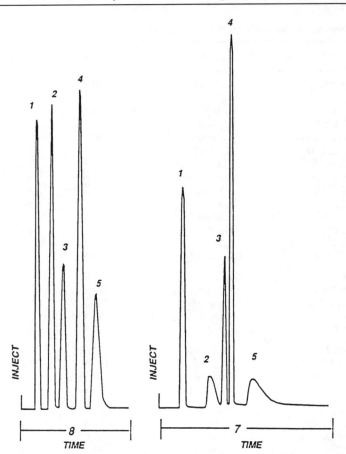

Fig. 6.13. Effect of end-capping on column performance. Waters Nova-Pak C_{18} (left) and Resolve C_{18} (right) columns were used with a mobile phase of 25:75 methanol:water containing PIC-B6. Solute identities: 1, ascorbic acid; 2, niacinamide; 3, pyridoxine HCl; 4, riboflavin; 5, phenol; 6, thiamine HCl. Chromatogram courtesy of Waters.

groups [4]. Bonding the monolayer to porous silica reduces the average pore diameter by about twice the thickness of the layer. The specific surface area and pore volume are decreased accordingly. The longer the chain length of the bonded phase, the more pronounced the effect, although the fact that retention increases with increasing chain length confirms that the percentage carbon (or hydrophobicity), rather than surface area, is the critical factor in determining retention.

The length of the bonded phase chain also affects the column efficiency. In general, for the same support, a decrease in chain length appears to increase column efficiency, particularly for well-retained solutes. The sample capacity of the column is also affected by the chain length, with higher capacities resulting when the chain length is increased. Typically, the maximum sample loading for a C_{18} functionalized column is about 2 mg sample g^{-1} stationary phase, although this varies with the mobile phase organic modifier concentration and temperature [4]. One concern over the performance of bonded phase columns is the reproducibility of some commer-

Table 6.4 Characteristics of some common bonded phase RPLC columns.

Column	Particle size/shape	Pore size	End-capped	% Carbon
Merck Lichrosorb PR-18	5, 10 μm/IRR	100 Å	No	22%
DuPont Zorbax ODS	6 μm/S	70 Å	No	20%
Waters Delta-Pak C_{18}	5 μm/S	100 Å	Yes	17%
Whatman Partisil ODS-2	10 μm/S	80 Å	No	15%
Beckman Ultrasphere ODS	3, 5 μm/S	80 Å	Yes	12%
Supelco Supelcosil LC-18	3, 5 μm/S	100 Å	Yes	11%
Shandon Hypersil ODS	3, 5, 10 μm/S	110 Å	Yes	10%
Waters μBondapak C_{18}	10 μm/IRR	125 Å	Yes	10%
Whatman Partisil ODS-3	5, 10 μm/S	80 Å	Yes	10%
Waters Resolve C_{18}	5, 10 μm/S	90 Å	No	10%
Vydac 201TP C_{18}	5, 10 μm/IRR	330 Å	No	10%
Waters Nova-Pak C_{18}	4 μm/S	60 Å	Yes	7%
Waters Delta-Pak C_{18}	5 μm/S	300 Å	Yes	7%
Whatman Partisil ODS	5, 10 μm/S	80 Å	No	5%
DuPont Zorbax C_8	6 μm/S	70 Å	No	14%
Merck Lichrosorb PR-8	5, 10 μm/IRR	100 Å	No	13%
Supelco Supelcosil LC-8	3, 5 μm/S	100 Å	Yes	7%
Waters Resolve C_8	10 μm/S	90 Å	No	5%
Waters Nova-Pak C_8	4 μm/S	60 Å	Yes	4%
Waters μBondapak Phenyl	10 μm/IRR	125 Å	Yes	8%
Supelco Supelcosil LC-DP	3, 5 μm/S	100 Å	Yes	6%
Waters Nova-Pak Phenyl	4 μm/S	60 Å	Yes	5%
Vydac 209TP	5, 10 μm/IRR	30 Å	Yes	5%
Waters μBondapak CN	10 μm/IRR	125 Å	Yes	6%
Beckman Ultrasphere CN	5 μm/S	80 Å	Yes	5%
Supelco Supelcosil LC-CN	3, 5 μm/S	100 Å	Yes	4%
Waters Resolve CN	10 μm/S	90 Å	No	3%
Waters Nova-Pak CN HP	4 μm/S	60 Å	Yes	2%
DuPont Zorbax-NH_2	6 μm/S	70 Å	No	10%
Waters μBondapak NH_2	10 μm/IRR	80 Å	Yes	9%
Whatman Partisil PAC	10 μm/S	80 Å	No	5%
Shandon Hypersil APS	3, 5, 10 μm/S	110 Å	No	2%
Supelco Supelcosil LC-NH_2	3, 5 μm/S	100 Å	Yes	2%
Merck Lichrosorb PR-2	5, 10 μm/IRR	40 Å	No	5%
Supelco Supelcosil LC-diol	3, 5 μm/S	100 Å	Yes	5%
RSL RSil-NO_2	5, 10 μm/IRR	100 Å	N/A	5%
Waters Delta-Pak C_4	5 μm/S	300 Å	Yes	3%
Supelco Supelcosil LC-1	3, 5 μm/S	100 Å	Yes	3%

IRR, Irregular; ODS, octadecylsilane; S, Spherical; PAC = 2:1 amino:cyano functional groups.

cially available packings. Great variation in manufacturing processes exist and not all columns are created in accordance with current good manufacturing practices (GMP). Care must be taken when developing methods to choose columns from reputable manufacturers to avoid the problem of having to fine tune a separation each time a column is replaced.

An enormous number of bonded phase columns is available. Table 6.4 shows the characteristics of only a limited selection of commonly used bonded phase columns. Catalogues available from column suppliers are perhaps the best source of information on column specifications. It is also worth noting that many column manufacturers sell a wide range of speciality bonded phase columns for particular applications. The use of these columns reduces the need for extensive method development for these particular separations. Examples of speciality bonded phase columns include those for the analysis of amino acids, fatty acids, triglycerides, carbohydrates and catecholamines [15]. For the majority of RPLC separations, a good general-purpose column would be one 15 to 30 cm in length packed with C_{18} functionalized 3–5 μm spherical silica with a pore size of 60–120 Å and a carbon load of 7–10%. The column should be end-capped and have an efficiency of 5000–10 000 theoretical plates.

The stability of bonded phase columns is determined largely by the limitations of the silica support. Generally, mobile phases should be used within the pH range of 2–8. Below pH 2, hydrolysis of the bonded functional groups occurs, resulting in decreased retention. Above pH 8, silica, a weak acid, dissociates and the silica support starts to dissolve. This creates voids in the packing material, resulting in decreased column efficiency. The majority of bonded phase columns can be operated with a wide range of organic solvents at elevated temperatures, up to a maximum temperature of approximately 80°C [4].

Polymeric reversed-phase columns

While most RPLC separations are carried out on bonded silica stationary phases, some polymeric materials are sufficiently hydrophobic to enable reversed-phase retention. The major advantage of these materials is that they can be operated over a pH range of 2–12 and they also have a different selectivity when compared with silica-based columns. For instance, the wide pH range enables weakly basic compounds to be chromatographed at high pH values without the need for ion interaction reagents (see Section 6.3.2 below). The greatest disadvantages of polymeric columns are that they are usually less efficient than silica-based columns and also tend to be less retentive. Despite these drawbacks, several polymeric reversed-phase columns are commercially available, with the Hamilton PRP-1 column manufacturer, while Fig. 6.14 shows the separation of proteins and peptides details the characteristics of a range of polymeric RPLC columns from a commercial column manufacturer, while Fig. 6.14 shows the separation of proteins and peptides on a neutral polystyrene column.

Table 6.5   Characteristics of the Shodex range of polymeric RPLC columns.

Column	Base polymer	Surface	Functional group
DS-613	Polystyrene	Hydrophobic	None
DE-613	Polymethacrylate	Hydrophilic	None
DM-614	Hydroxymethacrylate	Hydrophilic	None
D18-613	Polymethacrylate	Hydrophilic	C_{18}
D8-613	Polymethacrylate	Hydrophilic	C_8
D4-613	Polymethacrylate	Hydrophilic	C_4

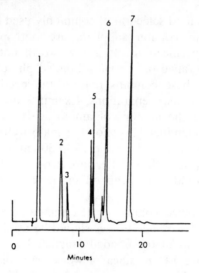

Fig. 6.14. Separation of proteins and peptides on a neutral polystyrene column. A Shodex D18-613 column was used with a mobile phase of 0.05% TFA and a 30-min linear gradient from 80:20 to 30:70 water:acetonitrile. Solute identities: 1, bradykinin; 2, neurotensin; 3, substance P; 4, insulin; 5, insulin B chain; 6, lysozyme; 7, myoglobin. Reprinted with permission from *Waters Sourcebook of Chromatography* (1992), Millipore Corporation, Milford.

Mobile phases for reversed-phase chromatography

While the previous section indicated that different RPLC columns provide some variation in retention and selectivity, control of the separation in RPLC is most commonly achieved by altering the mobile phase composition. Therefore, as was the case for adsorption chromatography, the most important factor in optimizing a reversed-phase separation is the selection of the mobile phase. The great utility of RPLC arises from the fact that the wide range of solvents available provides much scope for easily altering the separation selectivity.

Solvent classification

The development of the solvent classification scheme by Snyder [17] enabled the classification of a large number of potential RPLC solvents into groups comprising those solvents having similar properties. This scheme, depicted as a triangle, was formulated by considering solvent properties, such as proton donor, proton acceptor and dipole characteristics, and is shown in Fig. 6.15, while Table 6.6 lists different solvents classified according to the groups identified in the figure.

Solvent selectivity is controlled by the selectivity of the groups I–VIII. Hence, solvents from group VIII are good proton donors and interact preferentially with basic solutes, such as amines or sulfoxides. Conversely, solvents from group I are good proton acceptors and will interact preferentially with hydroxylated solutes, such as acids or phenols. Solvents from group V interact strongly with solutes that have large dipole moments, such as nitriles or amines. When developing a reversed-phase separation, the first step is to select a solvent mixture which permits

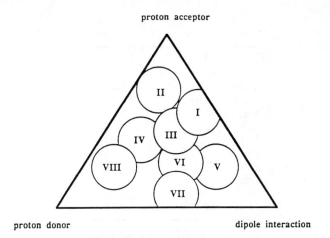

Fig. 6.15. The solvent classification triangle which illustrates the eight selectivity groups determined by polarity and selectivity. Modified with permission from Snyder (1974) *J. Chromatogr.*, **92**, 223.

Table 6.6   Classification of solvents according to their groups.

Group	Solvents
I	Aliphatic ethers, hexamethylphosphoric acid amide, trialkyl amines
II	Aliphatic alcohols, methanol
III	Pyrine derivatives, amides, glycol ethers, sulfoxides, tetrahydrofuran
IV	Glycols, benzyl alcohol, acetic acid, formamide
V	Methylene chloride, ethylene chloride
VI	Tricresyl phosphate, aliphatic ketones and esters, polyethers, dioxane, sulfones, nitriles, polypropylene carbonate, acetonitrile
VII	Aromatic hydrocarbons, nitro compounds, aromatic ethers
VIII	Fluoroalkanols, *m*-cresol, chloroform, water

Data from Snyder [17].

optimal sample retention, i.e. $2 < k' < 10$. The separation selectivity can then be varied by choosing other solvents with similar retention characteristics, but from a different selectivity group [4].

Solvent strength

As was the case for normal-phase chromatography, solvent strength in RPLC can be expressed as a function of polarity, and the eluting strength of a RPLC solvent is generally inversely related to its polarity. Therefore, polar solvents (such as water) are weak eluting solvents in RPLC, while non-polar solvents (such as tetrahydro-furan) are strong eluting solvents. Table 6.7 lists solvents commonly used with RPLC and their solvent polarity parameters (P'), where the lower the value of P', the stronger the solvent.

Table 6.7   Solvents commonly used with reversed-phase chromatography.

Solvent	P'
Water	10.2
Dimethyl sulfoxide	7.2
Ethylene glycol	6.9
Acetonitrile	5.8
Methanol	5.1
Acetone	5.1
Dioxane	4.8
Ethanol	4.3
Tetrahydrofuran	4.0
Isopropanol	3.9

Data from Snyder and Kirkland [4].

Water is generally used as the base solvent for most RPLC separations and the mobile phase strength is determined by mixing water with an appropriate volume of another solvent (termed the organic modifier) from Table 6.7. For instance, a mixture of 50:50 water:methanol would be a weaker mobile phase than a 50:50 mixture of water:tetrahydrofuran. As a general rule, solute retention decreases by a factor of 2 for every 10% volume addition of solvent to water. Plots of the logarithm of solute capacity factor versus the percentage organic modifier are generally linear for binary RPLC mobile phases [14]. This fact allows retention time data to be extrapolated over a wide k' range from only a few chromatographic experiments. In practice, methanol (MeOH) is the most widely used organic modifier as it has a relatively low u.v. cut-off (205 nm), reasonable eluting strength and is inexpensive. Acetonitrile (ACN), which has a lower u.v. cut-off (190 nm), better mass transfer properties, but is more expensive, is the next most commonly used solvent, followed by tetrahydrofuran (THF). Generally, these three solvents, combined with water, permit sufficient retention and selectivity variation for the majority of RPLC separations. When developing a RPLC separation, the optimal mobile phase strength is usually located by starting with a solvent-rich mobile phase and successively decreasing the organic modifier content until the desired retention is obtained, as shown in Fig. 6.16.

Solvent selectivity

Once a binary mixture of the desired strength is located for a given separation, the selectivity can be altered by changing the nature of the organic modifier, while maintaining a constant eluotropic strength. Table 6.8 shows a comparison of the eluting strengths of binary mixtures of water with MeOH, ACN and THF. For instance, mobile phases of 30% methanol:water, 22% acetonitrile:water and 16% tetrahydrofuran:water should all give approximately equal retention in RPLC separations. Each of the different solvent mixtures, however, may permit selectivity changes.

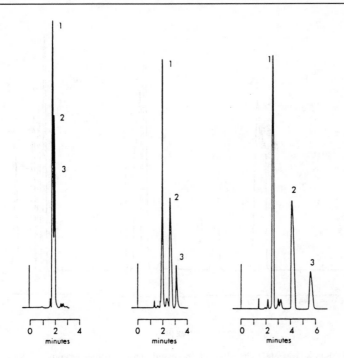

Fig. 6.16.   Effect of organic modifier on a RPLC separation. A Waters μBondapak C_{18} column was used with mobile phases, from left to right, 60:40, 50:50 and 40:60 acetonitrile:water. A flow-rate of 2.0 ml min^{-1} was used with u.v. detection at 254 nm. Solute identities: 1, methyl; 2, propyl; 3, butyl-parabens. Chromatograms courtesy of Waters.

A more accurate method for calculating the organic modifier content of isoeluotropic mobile phases is to use empirically derived equations, known as transfer rules [18]. Equations 6.1 and 6.2 are based on the retention behaviour of a large number of compounds with a wide variety of functional groups. The terms ϕ_{MeOH}, ϕ_{ACN} and ϕ_{THF} represent the mobile phase volume fractions of MeOH, ACN and THF, respectively.

$$\phi_{ACN} = 0.32\phi_{MeOH}^2 + 0.57\phi_{MeOH} \qquad\qquad 6.1$$

$$\phi_{THF} = 0.66\phi_{MeOH} \qquad\qquad 6.2$$

Table 6.8   Relative strengths of binary water:organic modifier mixtures.

Methanol (%)	Acetonitrile (%)	Tetrahydrofuran (%)
30	22	16
40	32	23
50	40	30
60	50	36
70	60	43
80	73	52
90	87	62

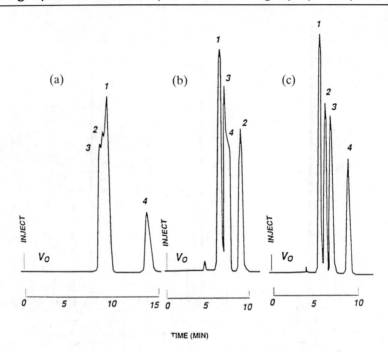

Fig. 6.17. Mobile phase selectivity effects in reversed phase chromatography. A Waters μBondapak C_{18} column was used with mobile phases of 28:72 THF:water (a), 58:42 MeOH/water (b) and 38:62 ACN:water (c) were used. Solute identities: 1, dextramethasone; 2, fluoxymesterone; 3, triamcinolone acetonide; 4, hydrocortisone acetate. Reprinted with permission from *Developing HPLC Separations, Book One* (1991), Millipore Corporation, Milford.

These equations attempt to provide generalized values for the isoeluotropic strength of the three common modifiers used in RPLC, although some variation for particular solutes must be expected. Hence, it is generally necessary to fine-tune the solvent strength by minor adjustment of the percentage organic modifier once the best solvent mixture (in terms of selectivity) has been located. Figure 6.17 shows an example of the differences in selectivity obtained with MeOH, ACN and THF for the separation of four closely related steroids. If satisfactory selectivity is not obtained using binary mixtures, ternary (water and two other solvents) and quaternary (water and three other solvents) mobile phases may be used, although this complicates the selection of an optimal mobile phase. Computer programs which aid in the optimization of mobile phases in RPLC are available. It should also be remembered that water does not form an essential component of the mobile phase and some RPLC separations are carried out using only organic modifiers. More detail on varying the selectivity of reversed-phase separations may be found elsewhere [4, 7, 11].

Other factors

Other factors which affect retention in RPLC are temperature, pH and mobile phase additives. Generally, retention decreases with an increase in temperature and plots of log k' versus $1/T$ are approximately linear [4]. As a rough estimate, a 30°C increase in temperature produces a twofold decrease in retention. Elevating the temperature can also be used to decrease the viscosity of the mobile phase in order to improve chromatographic efficiency and reduce column backpressure. In addition, elevated temperatures can be used to enhance sample (and mobile phase) solubility. Another advantage of controlling the column temperature is that retention time stability often improves; however, temperature generally has little effect on RPLC separation selectivity [19]. The major concern with operating at elevated temperatures is increased deterioration of the bonded phase. Column manufacturers usually provide recommendations on operating limits and these should be adhered to strictly.

The mobile phase pH requires consideration if the solute molecules are affected by pH, i.e. if they are acidic or basic compounds. The retention of such compounds can be manipulated through selective pH control, as will be discussed in Section 6.3. After buffers for pH control, the next most common mobile phase additive used in RPLC are ion-interaction reagents which are used to manipulate the retention of ionic compounds on hydrophobic columns (these will also be discussed in Section 6.3). Finally, various salts may be added to RPLC mobile phases to reduce peak tailing, particularly for basic compounds on non-end-capped, bonded phase silica columns [4]. For example, triethylamine (0.1–1.0%) is often added to reduce tailing of amines on such columns.

Applications of reversed-phase chromatography

As mentioned previously, reversed-phase chromatography is the most widely used separation mode; consequently, an enormous range of applications has been reported using this technique. Neutral solutes that are soluble in water or other relatively polar solvents, and with molecular weights less than 2000–3000, are typically best chromatographed using RPLC. In addition, numerous ionic and ionogenic solutes have been separated using this mode, typically through the use of pH control or the addition of ion-interaction reagents to the mobile phase, as will be discussed in Section 6.3. In fact, so many separations have now been reported that the best start to any methods development procedure is to undertake a comprehensive literature search. It is almost certain that any low molecular weight, water-soluble or moderately polar compound, or at least a structurally similar molecule, will have been chromatographed by RPLC. While it is beyond the scope of a text such as this to provide comprehensive detail on RPLC applications, a brief overview of some important applications is given in the following paragraphs.

The separation of non-polar compounds, such as hydrocarbons, was among the original applications of RPLC [4]. Other non-polar solutes have also been separated using this chromatographic mode, including fatty acids. These acids are resolved on the basis of their carbon chain length and degree of saturation, and typical applications include the analysis of coconut oil, tallow and cosmetics. Figure 6.18 shows the separation of a number of fatty acids. Additionally, nonaqueous reversed-phase chromatography has been used for the analysis of non-polar materials, such as triglycerides and cholesterol, while pigments, such as carotenoids, can be separated in a similar fashion [15].

Perhaps the biggest individual HPLC user group is the pharmaceutical industry. RPLC finds wide application in the area of pharmaceutical and drug analysis, where

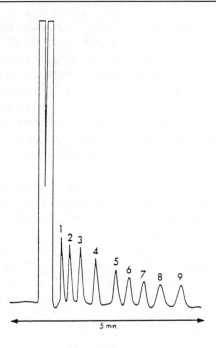

Fig. 6.18. Separation of fatty acids by reversed phase chromatography. A Waters Free Fatty Acid Analysis Column was used with a mobile phase of 45/20/35 ACN/THF/water at 1.5 ml min^{-1}. RI detection was used. Solute identities: 1, capric acid; 2, lauric acid; 3, myristic acid; 4, palmitic acid; 5, stearic acid; 6, nonadecanoic acid; 7, arachidic acid; 8, heneicosanoic acid; 9, behenic acid. Reprinted with permission from *Waters Sourcebook of Chromatography* (1992), Millipore Corporation, Milford.

it is used commonly for quality control assays. Pharmaceutically active compounds including parabens, vitamins, beta-blockers, alkaloids, steroids, tetracyclines, prostaglandins, etc., are all commonly analysed using RPLC. Figure 6.12 shows the separation of steroids by RPLC, while Fig. 6.16 shows the separation of parabens in an acetonitrile extract of lotion. HPLC is also used frequently in clinical laboratories, where perhaps the most common application of RPLC would be the analysis of plasma and urinary catecholamines with electrochemical detection, as shown in Fig. 6.19. Other industrial applications include the analysis of chemical products, such as antioxidants and polymer additives in a diverse variety of sample matrices.

RPLC also finds wide application in the analysis of biologically important molecules. Typical applications include amino acids, proteins and peptides, nucleic acids, nucleosides, nucleotides, and oligosaccharides. These compounds are often analysed in complex matrices and the use of gradient elution is common in this area. The ease with which gradients can be used with RPLC is one of the reasons for its widespread use. A RPLC gradient separation of AQC-derivatized amino acids is shown in Fig. 5.38, while a gradient separation of proteins and peptides on a neutral polystyrene reversed-phase column is shown in Fig. 6.14. RPLC is also commonly used for the analysis of food, beverage and agricultural products. Carbohydrates, food additives, aflatoxins, sweeteners, organic acids, etc., can all be analysed using

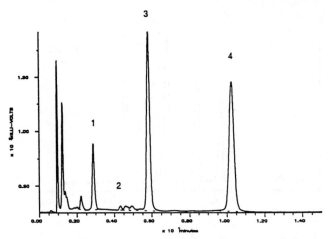

Fig. 6.19.   Separation of catecholamines in urine by RPLC. A Waters Nova-Pak column was used with a mobile phase of 50 mM sodium acetate, 1 mM octane sulfonate, 0.1 mM EDTA in 96:4 water:MeOH. Amperometric detection was used. Solute identities: 1, noradrenaline; 2, adrenaline; 3, dihydroxybenzylamine; 4, dopamine. Chromatogram courtesy of Waters.

this separation mode. A typical example from the food industry is illustrated in Fig. 6.20, which shows the separation of subparts-per-billion levels of the aflatoxins G_2–B_1 in a standard solution and spiked peanut meal.

Finally, the other major application area of RPLC is for analysis of environmental samples. Pesticides and herbicides, including, paraquat and diquat, carbamates, and organophosphorous compounds, are frequently analysed using this technique. RPLC is also used for the measurement of pollutants, such as polyaromatic hydrocarbons (PAHs), phenolics and aldehydes/ketones. Figure 6.21 shows the separation of 16 PAHs using gradient RPLC.

6.2.3 Gel permeation chromatography

Introduction

GPC is alternatively termed size exclusion chromatography or, when applied to the separation of water soluble polymers and proteins, gel filtration chromatography. This chromatographic mode is the simplest of the major HPLC separation modes and is the preferred method for separating high molecular weight (>2000–3000) neutral solutes. In GPC solutes are separated as a result of their permeation into solvent-filled pores within the column packing. Large molecules are excluded from some or all of the pores by virtue of their physical size, while smaller molecules permeate into a greater proportion of the pores. GPC is mainly used to characterize the molecular weight distribution of polymer materials, although it is also useful for the separation high molecular weight proteins.

Mechanism of gel permeation chromatography

Separation in GPC is very straightforward and molecules are separated according to their effective size in solution. Solute molecules which are too large to physically

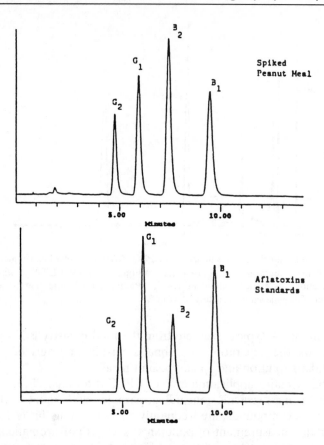

Fig. 6.20. Determination of aflatoxins in spiked peanut meal. A Waters Nova-Pak C_{18} column was used with a mobile phase of 60:40 MeOH:water at a flow-rate of 1.0 ml min^{-1}. Fluorescence detection was used. Chromatogram courtesy of Waters.

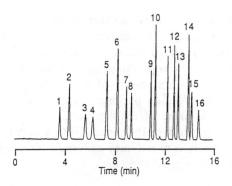

Fig. 6.21. Analysis of PAHs using gradient RPLC. A Vydac 201 TP column was used with a linear gradient from 50/50 ACN/water to 100% ACN over 16 min. Direct u.v. detection at 254 nm was used. Solute identities: NIST Standard # 1647. Chromatogram courtesy of Waters.

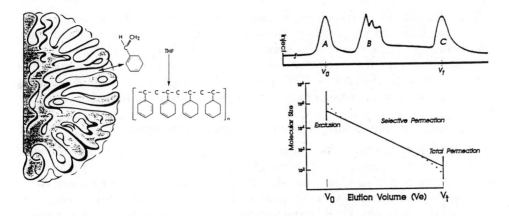

Fig. 6.22. Schematic illustration of the permeation process. In the diagram on the left, the smaller molecule is able to diffuse into solvent within the pores, while the larger solute is excluded. In the diagram on the right, V_o and V_t represent the totally excluded and included volumes, respectively. Reprinted from *Developing HPLC Separations, Book One* (1991), Millipore, Milford, with permission.

permeate into the porous stationary phase show no retention and are eluted at the volume of solvent equivalent to the interstitial column volume. Such large molecules are therefore excluded from the stationary phase pores. Very small molecules, e.g. methanol, are able to permeate fully into the solvent trapped within the stationary phase pores and are eluted at the volume of solvent equivalent to the total of the interstitial column volume plus the pore volume. Solutes of intermediate size are able to permeate partially into the pores and hence exhibit retention between the above extremes. The most significant difference between this separation mode and others is that because solvent molecules are generally small, they are eluted last. The mechanism of GPC and the permeation process are illustrated schematically in Fig. 6.22. A represents solutes too large to permeate into the pores of the stationary phase, B represents solutes that diffuse partially into the pores, while C represents solutes that diffuse totally into the pores. A plot of the logarithm of molecular size (or weight) versus elution volume is termed a calibration curve.

When a strict GPC mechanism is operating, the solute molecules do not interact with the stationary phase. Retention of the solute by other mechanisms, e.g. adsorption or hydrophobic interaction, is undesirable in this mode of chromatography and the column and mobile phase are typically selected to eliminate such interactions. GPC offers a number of advantages over other separation modes. Since the solutes are all eluted within a small retention volume, the peaks are generally narrow. This enhances detection sensitivity, so that a relatively insensitive detection method, such refractive index detection, may be utilized with GPC. RI detection is particularly suitable for GPC because many polymers have no chromophores or other detectable properties. However, the greatest advantage of GPC is that retention is very predictable, consequently retention times can be used to predict the molecular weight (or size) of an unknown molecule. A further advantage is that all injected solutes must be eluted within a limited solvent volume range. Therefore, GPC is a useful alternative to RPLC for samples that contain high levels of triglycerides and other hydrophobic materials, such as creams, ointments or lotions, since these materials are strongly bound onto reversed-phase surfaces and cause column fouling. The major disadvantage of GPC is the lack of resolving power, i.e. its limited peak capacity. Generally, solutes must differ in molecular size by at least 10% in order to be resolved by GPC. Fortunately, complete resolution is often not required for the determination of molecular weight distribution for polymers.

Stationary phases for gel permeation chromatography

GPC differs from other modes of liquid chromatography in that the composition of the mobile phase is generally not the variable used to control retention or selectivity. In this chromatographic mode, the stationary phase is used almost exclusively to vary these parameters. Stationary phases can be divided into two classes: the cross-linked, rigid polystyrene divinylbenzene gels traditionally used for polymer characterization, and those used for the analysis of water soluble solutes. Column packings used for aqueous GPC include polydextrans, polyvinylalcohol gel, hydroxylated methacrylate gel and silica gel [15]. GPC columns are then further classified according to their pore size. The development of a method in GPC involves selecting a solvent which permits sample solubility, then choosing a column (or series of columns) that has an appropriate range of pore sizes. The pore size of the column is matched to the expected range of solute sizes so that they show selective permeation on the gel. Historically, the pore size of a GPC column is defined as the extended chain length (Å) of a molecule of polystyrene which is just large enough to be excluded totally from all the pores of the stationary phase [20]. This value is referred to as the exclusion, or pullulan, limit of the column. Figure 6.23(a) shows a series of calibration curves for several polystyrene gels of differing pore sizes.

Hence, small molecules (molecular weight < 5000) are separated using columns with the smallest pore sizes, typically 60–100 Å. Samples with a narrow molecular weight distribution (< 2.5 decades) can be separated using a single pore-size column, e.g. solutes in the molecular weight range 10^3 to 10^5 can be separated with a 300 Å column [4]. Solutes with a broad molecular weight distribution (e.g. polymers) are commonly analysed using a number of different pore size columns in series. Alternatively, mixed bed columns packed with a blend of different pore size materials are also available. These 'linear' columns are ideal for an initial indication of the molecular weight distribution of an unknown sample. Figure 6.23(b) shows a calibration curve for a linear polystyrene gel column.

Modern, analytical scale GPC columns are typically packed with 7–10 μm particles and have separation efficiencies in the order of 10–15 000 plates per column. They are often available packed in a number of different solvents and care must be taken during solvent changeover in order to avoid excessive shrinking or swelling of the column bed. This will cause voids and channels in the packing material, resulting in loss of efficiency. Columns with tightly controlled, narrow particle size distributions exhibit enhanced lifetimes when exposed to rigorous solvent changeover, and these are commercially available [15]. In addition, many GPC columns, particularly rigid polystyrene divinylbenzene gels used for polymer characterization, may be operated at elevated temperatures (up to 150°C) to assist the separation of complex samples, such as polyolefins, which are not readily soluble at room temperatures. Table 6.9 shows the characteristics of the range of nonaqueous and aqueous GPC packings

(a)   (b)

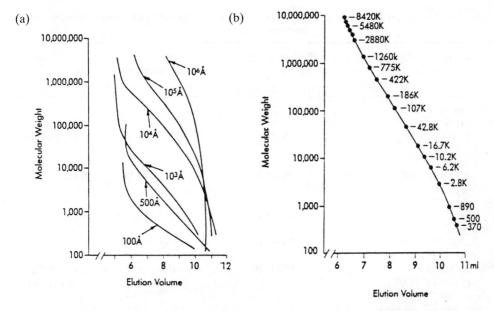

Fig. 6.23.   GPC calibration curves for polystyrene gels. (a) Waters Ultrastyragel columns were used with toluene as the mobile phase at 1.0 ml min^{-1}. (b) A Waters Ultrastyragel linear column was used with tetrahydrofuran as the mobile phase at 1.0 ml min^{-1}. Detection was by refractive index and polystyrene and *n*-hydrocarbon individual molecular weight standards were used for all curves. Reprinted from *Waters Sourcebook of Chromatography* (1987), Millipore, Milford, with permission.

available from a column manufacturer. A wide range of preparative GPC columns is also available and many other column manufacturers also sell extensive ranges of GPC columns.

Mobile phases for gel permeation chromatography

As previously mentioned, the mobile phase in GPC is not normally used to control retention or selectivity. The primary consideration for selection of the mobile phase is sample solubility. Ideally, the mobile phase should have a low viscosity at the separation temperature and it should be compatible with the detection method chosen. The mobile phase must also be selected on the basis of its compatibility with the column. Polar solvents, including methanol, ethanol, isopropanol, and dimethyl sulfoxide should not be used with polystyrene packings because they cause excessive shrinkage, resulting in permanent damage to the column bed. These packings are most commonly used with solvents such as toluene, tetrahydrofuran, chloroform, methylene chloride, dimethyl formamide, *m*-cresol and trichlorobenzene, although the last three must be operated at elevated temperatures in order to reduce viscosity. Silica gel columns can be used with a wide range of solvents, including water, although they are limited to a pH operating range of 2–8.

In addition to sample solubility, viscosity and column compatibility considerations, the mobile phase is also chosen to eliminate (or minimize) the interaction of the solute with the stationary phase. When a strict GPC mechanism is operating, the solute does not interact with the stationary phase. Interactions, such as hydrophobic

Table 6.9   Characteristics of non-aqueous (upper) and aqueous (lower) GPC column ranges available commercially from Waters.

Column range and individual designations	Packing material	Column range exclusion limits (lowest to highest)	Solvent(s)
Nonaqueous columns			
Ultrastyragel 100, 500, 10^3, 10^4, 10^5, 10^6, linear	PS-DVB	500 up to 10^7	THF, toluene
μStyragel 100, 500, 10^3, 10^4, 10^5, 10^6, linear	PS-DVB	500 up to 10^7	Toluene
μStyragel HT 10^3, 10^4, 10^5, 10^6, linear	PS-DVB	30 000 up to 10^7	MEK
Styragel 60, 100, 200, 500, 10^3, 10^4, 10^5, 10^6, 10^7	PS-DVB	500 up to 2×10^7	Toluene
Shodex K series 801, 802, 802.5, 803, 804, 804L, 805, 806, 805, 806, 807, 80-linear	PS-DVB	1.5×10^3 up to 2×10^8	THF, DMF, chloroform
Shodex HFIP series 803, 804, 805, 806, 807, 80-linear	PS-DVB	7×10^4 up to 2×10^8	HFIP
Shodex AT series 803, 804, 805, 806, 807, 80-linear	PS-DVB	7×10^4 up to 2×10^8	Toluene
Aqueous columns			
Shodex OHpak B series 803, 804, 805, 806	Methacrylate	10^5 up to 2×10^7	Water
Shodex OHpak Q series 803, 804, 805, 806	Polyvinylalcohol	10^5 up to 2×10^7	Water
Ultrahydrogel 120, 250, 500, 1000, 2000, linear, DP	Methacrylate	5×10^3 up to 7×10^6	Water
Protein-Pak 60, 125, 300SW	Silica gel	2×10^4 up to 3×10^5	Propanol
Shodex Protein KW 802.5, 803, 804	Diol-bonded silica gel	5×10^4 up to 6×10^5	Water

PS-DVB, polystyrene divinylbenzene; THF, tetrahydrofuran; MEK, methylethyl ketone; DMF, dimethylformamide; HFIP, hexafluoroisopropanone.

association, adsorption, ion-exclusion or ion-exchange, are undesirable because they affect retention times and calibration curves [21]. Hence, solvents which reduce hydrophobic interactions, such as THF or toluene, are used commonly for polymer GPC, while salt solutions, i.e. 0.1–0.5 M sodium sulfate, are used for the aqueous GPC separation of polyelectrolytes. Other agents (such as lithium bromide) are also added to solvents, such as dimethylformamide (DMF) when separating polyether sulfones to reduce unwanted interactions [15]. Table 6.10 shows the properties of solvents commonly used for polymer GPC.

Table 6.10 Properties of solvents commonly used for polymer GPC.

Solvent	u.v. cut-off (nm)	Viscosity at 20°C (cP)	Refractive index at 20°C
Dimethyl formamide*	295	0.924	1.4294
Ethylene chloride	225	0.84	1.4444
Dioxane	220	1.439	1.4221
Trichloroethane	225	1.20	1.4791
Cyclohexane*	220	0.98	1.4262
Carbon tetrachloride	265	0.969	1.4630
Toluene	285	0.59	1.4969
Trifluoroethanol*	190	1.996	1.2910
Benzene	280	0.652	1.5011
Hexafluoroisopropanol*	190	1.021	1.2752
Chloroform	245	0.58	1.4457
Hexane*	210	0.326	1.3749
Xylene	290	0.81	1.4972
Tetrahydrofuran	220	0.55	1.4072
Methylene chloride	220	0.44	1.4237

* Care must be taken when using these solvents with 100–500 Å pore size columns.
Data from Snyder [4].

An additional consideration in GPC when separating polymers is that the injected sample concentration may be limited by its viscosity. As a rough guide, the sample solution injected onto the column should have a viscosity no greater than twice that of the mobile phase [4]. Injecting samples with a higher viscosity than this leads to 'viscous fingering' effects. These effects result from the fact that it takes a finite time for the mobile phase to dilute a highly viscous sample, causing peak broadening and longer elution times, and hence incorrect molecular weight assignment. Generally, if a polymer has an average weight over 1×10^6, a 0.01% solution should be used; between 20 000 and 10^6, a 0.1% solution should be used; between 800 and 20 000, a 0.2% solution should be used; and below 800, a solution as concentrated as 0.5% can be used. The majority of samples are simply dissolved in the mobile phase and filtered through a 0.45 μm filter before injection in GPC.

Applications

The primary application of GPC has traditionally been the characterization of polymers using columns packed with rigid polystyrene divinylbenzene. The molecular weight distribution is responsible for many of the properties which characterize a bulk polymer, including tensile strength, brittleness, hardness, melt viscosity and so on [20]. Subtle differences between materials, and hence final product performance, can be identified by the molecular weight distribution obtained using GPC, which is why this mode of chromatography has found widespread application in the polymer industry.

Columns are typically calibrated using a universal calibration method which first involves injecting a series of standard polymers (usually polystyrene) with narrow molecular weight ranges onto the column. A calibration curve of the logarithm of

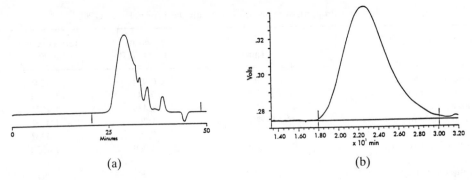

Fig. 6.24. GPC chromatograms showing molecular weight distributions. (a) Waters Ultrastyragel 10^3, 500 and 100 Å columns were used in series with tetrahydrofuran as the mobile phase at 6.0 ml min^{-1}. (b) Shodex KB-80M (2×) and KB-802.5 columns were used with 0.1 M sodium nitrate as the mobile phase at 1.0 ml min^{-1}. Detection was by refractive index. Chromatograms courtesy of Waters.

molecular weight versus elution volume is constructed, after which the unknown polymer sample is injected. Since different types of polymers have varying structures in solution, factors (known as Mark–Houwink constants) which account for these differences must be incorporated into calculations to obtain accurate molecular weight information when polystyrene standards are used [4]. Computer data packages are available which integrate (or slice) the polymer peak and automatically calculate important molecular weight information, such as the number-average molecular weight (Mn), weight-average molecular weight (Mw), z-average molecular weight (Mz) and the polydispersity. Figure 6.24(a) shows the GPC chromatogram of an epoxy resin.

Polymers commonly characterized by GPC include acrylics, alkyds, carboxymethyl cellulose, ethylene-ethyl acrylate, nitrocellulose, nylon, polybutadiene, polycarbonate, polyether sulfone, polyethylene, polyisoprene, polypropylene, polystyrene and silicones [15]. Similarly, water-soluble ionic, non-ionic and amphoteric polymers can be characterized by aqueous GPC. Such polymers include polysaccharides, starches, dextrans, polyacrylamide, hyaluronic acid, carrageenan, sulfonated lignin, collagen and gelatin [21]. Figure 6.24(b) shows the aqueous GPC chromatogram of a polyalginic acid (Na salt), which is a gelling, emulsifying and swelling agent used extensively in the food industry. The weight average molecular weight was found to be 850 000.

While GPC has limited resolving power compared with other separation modes, it is still a useful method for the analysis of matrices which contain high levels of triglycerides and other hydrophobic materials. Samples such as cooking oil, creams, ointments or lotions, typically require extensive clean-up prior to RPLC in order to prevent column fouling. However, small molecules can be analysed in such samples by GPC after simple dilution in the mobile phase. Figure 6.25(a) shows an overlay of the chromatograms of antioxidants obtained from a standard solution and a cooking oil sample, while (b) shows the separation of benzocaine and benzyl alcohol in a benzocaine ointment obtained using small molecule GPC. Neither triglycerides in the cooking oil, nor polyethylene glycols in the ointment interfered with the

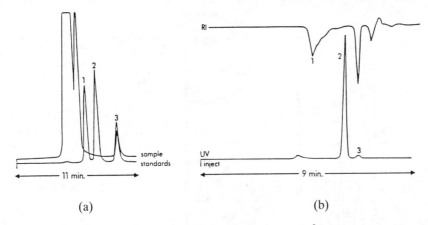

(a)   (b)

Fig. 6.25. Small molecule GPC of complex samples. (a) A Waters 100 Å Ultrastyragel column was used with chloroform as the mobile phase at 1.0 ml min^{-1}. Detection was by direct u.v. at 280 nm. Solute identities: 1, butylated hydroxy toluene; 2, butylated hydroxy anisole; 3, *t*-butylated hydroxyquinone. (b) A Waters 100 Å Ultrastyragel column was used with tetrahydrofuran as the mobile phase at 1.0 ml min^{-1}. Detection was by direct u.v. at 254 nm and refractive index. Solute identities: 1, polyethylene glycols; 2, benzocaine; 3, benzyl alcohol. Reprinted from *Waters Sourcebook of Chromatography* (1987), Millipore, Milford, with permission.

separation, or affected the long-term performance of the column. Sample preparation was dilution in the mobile phase in both cases.

GPC is also applied commonly to the separation of biological molecules, such as proteins, where it provides a rapid, simple method for the separation, purification and characterization of proteins and permits high recoveries of total protein and retention of biological activity. Figure 6.26(a) shows the separation of a complex protein mixture while (b) shows the resulting calibration curve obtained from the injection of a series of standards. Finally, another major application area of GPC is for the preparative clean-up of environmental samples, e.g. EPA method 3640A. In such applications, the sample is passed through a large GPC column, using methylene chloride as the mobile phase. Appropriate fractions are collected after being separated from low volatility, high molecular weight interferences, such as lipids and natural resins. These fractions are subsequently analysed for target compounds, such as pesticides and/or polyaromatic hydrocarbons, using either HPLC or GC [22].

6.3 Techniques for Ionic and Ionizable Species

Solutes which are ionic in nature or which are easily ionized are normally not separated using the chromatographic techniques discussed in Section 6.2 above, at least not without some modification. The reason for this is that retention of such solutes is usually inappropriate; for example, an ionized solute will generally show only very weak retention in a reversed-phase system because it will interact more strongly with the polar mobile phase than it will with the non-polar stationary phase.

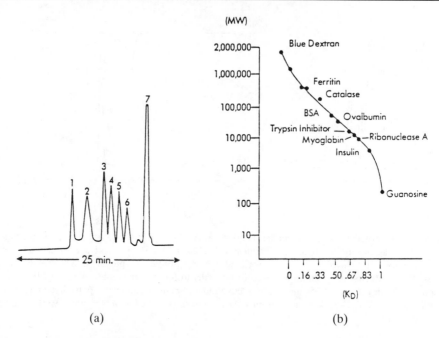

Fig. 6.26. Gel filtration chromatography of proteins. Waters Protein-Pak 300 SW columns (×2) were used with 0.1 M potassium phosphate monobasic as the mobile phase at 1.0 ml min⁻¹. Detection was by direct u.v. at 280 nm. Solute identities: 1, blue dextran; 2, ferritin; 3, bovine serum albumin; 4, ovalbumin; 5, trypsin inhibitor; 6, ribonuclease A; 7, guanosine. Reprinted from *Waters Sourcebook of Chromatography* (1992), Millipore, Milford, with permission.

In this section we will examine the use of a range of chromatographic techniques which are designed specifically to cater for ionized or ionizable solutes. Accordingly, these techniques all have the word "ion" in their descriptive title so that, to a novice, the techniques may sound confusingly similar. The techniques to be discussed embrace such methods as ion-suppression chromatography, ion-interaction chromatography, ion-exclusion chromatography, ion-exchange chromatography, etc. Each method will be discussed in sufficient detail to enable the reader to understand the nature of the stationary and mobile phases used, the mechanism by which separation is effected, the factors which govern solute retention, and some general applications.

6.3.1 Ion-suppression chromatography

This technique is a relatively simple extension of standard reversed-phase chromatography which enables it to be used for ionizable solutes. The approach taken is to suppress the ionization of these solutes by adding a buffer of appropriate pH to the mobile phase. In this way, the solutes are rendered either neutral or only partially charged and their retention on non-polar stationary phases is therefore increased and separation can then be accomplished. Acidic buffers are used for the separation of weak acids, while alkaline buffers are used for the separation of weak bases. Thus, with the exception of the buffer added to the mobile phase, ion-suppression

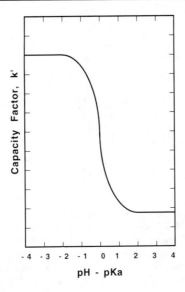

Fig. 6.27. pH versus capacity factor for an organic acid.

chromatography has characteristics which are identical to those of reversed-phase chromatography.

If we assume that the solute to be chromatographed is a weak organic acid, then a plot of capacity factor versus mobile phase pH will be sigmoidal in shape, as illustrated in Fig. 6.27. At pH values substantially less than the pK_a value of the acid, the solute is present in its neutral form and shows a large capacity factor (i.e. a long retention time). Further decreases in mobile phase pH show no effect on capacity factor since there will be no changes in the degree of ionization of the solute. Conversely, mobile phase pH values substantially greater than the pK_a value will result in complete ionization of the solute, leading to a small capacity factor. At intermediate pH values, the solute charge, and hence its retention, will be dependent on the particular pH used and its proximity to the pK_a value. It is clear from Fig. 6.27 that the effective pK_a of the solute acid can be determined from the inflection point in the sigmoidal curve. It should be noted that this pK_a will usually differ from the value published for purely aqueous solution because of the effects exerted by the organic modifiers present in typical reversed-phase mobile phases.

Ion-suppression chromatography is generally considered to be applicable only to those weak acids and bases for which the ionization can be suppressed using buffers with pH values in the range 3–8 [23]. The reason for this is that silica-based C_{18} stationary phases are often unstable outside this pH range. Although this is undoubtedly a major limitation, many organic carboxylic acids may be separated on C_{18} columns. Restrictions in eluent pH do not apply to the use of non-polar polymeric stationary phases, so these materials can therefore be employed for the separation of a wider range of solutes by the ion-suppression technique than is possible with C_{18} stationary phases.

The utility of ion-suppression on polymeric stationary phases may be appreciated by considering the separation of the homologous series of aliphatic carboxylic acids.

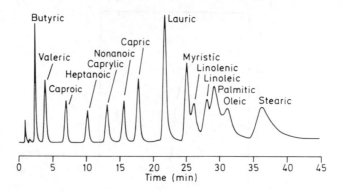

Fig. 6.28. Gradient elution ion-suppression chromatogram of carboxylic acids, obtained on a polymeric reversed-phase column. A Dionex MPIC-NS1 column was used with a gradient of 100% eluent A ($t = 0$) to 100% eluent B ($t = 20$ min), with maintenance of eluent B after this time. Eluent A is 24% acetonitrile and 6% methanol in 0.03 mM HCl. Eluent B is 60% acetonitrile and 24% methanol in 0.05 mM HCl. Detection was by suppressed conductivity. The baseline conductance for a blank gradient has been subtracted in the chromatogram shown. Reprinted from Slingsby (1986) *J. Chromatogr.* **371**, 373, with permission.

Neither ion-exchange nor ion-exclusion chromatography (to be discussed below) yields a complete separation of these species. However, ion-suppression coupled with gradient elution and conductivity detection enables the separation of butyric through to stearic acid, as illustrated in Fig. 6.28 [24]. The gradient used involved an increase in the percentage of organic modifier in the eluent and a decrease in eluent pH. Carboxylic acids more hydrophilic than butyric acid were eluted as a single peak at the column void volume.

6.3.2   Ion-interaction chromatography

Introduction

Strong acids or bases cannot be separated by ion-suppression chromatography because the mobile phase pH values necessary to effect suppression of their ionization are outside the limits tolerable by most reversed-phase columns. A better approach is to chromatograph these solutes in their ionic forms (rather than in their neutral or partly ionized forms, as was done in ion-suppression chromatography). Normally we would expect such ionic solutes to show little or no retention on lipophilic stationary phases when typical reversed-phase eluents are used. However, retention and subsequent separation of these solutes on non-polar stationary phases can be achieved by the addition to the eluent of a lipophilic reagent ion with the opposite charge sign to that of the solute ion. This added reagent ion, and the chromatographic process itself, have been described by a variety of names, some of which are listed in Table 6.11. Most of these names imply some sort of mechanism for the process and may therefore be misleading. Throughout this discussion, the terms ion-interaction chromatography and ion-interaction reagent (IIR) will be used, since these are general terms.

Table 6.11 Alternative names used to describe ion-interaction chromatography and the reagent ion added to the mobile phase.

Chromatographic process	Reagent ion	Reference
Ion-pair chromatography	Pairing ion	25
Paired-ion chromatography	PIC reagent	26
Surfactant chromatography	Surfactant ion	27
Dynamic ion-exchange chromatography	Ion-pairing reagent	28
Ion-interaction chromatography	Ion-interaction reagent	29
Hetaeric chromatography	Hetaeron	30
Mobile phase ion chromatography (MPIC)	Pairing reagent	30

Stationary and mobile phases for ion-interaction chromatography

Ion-interaction chromatography is a reversed-phase technique in which the stationary and mobile phases are the same as those used conventionally, with the exception that the mobile phase contains an additional species, namely the ion-interaction reagent. The technique can be performed successfully on a wide range of stationary phases, including neutral polystyrene divinylbenzene (PS-DVB) polymers and bonded silica materials with C_{18} [e.g. 31], C_8 [32], phenyl [33] and cyano [34] groups as the chemically bound functionality. Each of these stationary phases gives satisfactory retention of ionic solutes, provided the eluent composition is such that an appropriate amount of the IIR is adsorbed. The choice between stationary phases is usually based on such considerations as chromatographic efficiency, pH stability and particle size, rather than on differences in chromatographic selectivity. However, it has been noted [35] that the elution order for solutes can vary when the nature of the stationary phase used to support the IIR is altered. Further factors to be considered in the selection of a stationary phase for ion-interaction chromatography are specific interactions existing between the stationary phase and either the IIR or the solutes, and the role of residual silanol groups on silica-based stationary phases.

The most important component of the eluent in ion-interaction chromatography is the IIR itself. The prime requirements of the IIR are as follows: (i) an appropriate charge, which is unaffected by eluent pH; (ii) suitable lipophilicity to permit adsorption onto non-polar stationary phases; (iii) compatibility with other eluent components; and (iv) compatibility with the desired detection system. Separations of anionic solutes are normally performed using strong base cations, such as tetra-alkylammonium ions (usually tetrabutylammonium), as the IIR. A solution of such an IIR has a pH value of approximately 8 and this ensures that most solute acids are dissociated fully into their anionic form. Cation separations are usually performed using strong acid anions, such as aliphatic sulfonate ions (usually C_4–C_8 alkanesulfonates) as the IIR. These reagents are often supplied in acetic acid solution at a pH of approximately 4 in order to ensure that basic solutes are fully protonated and are therefore present in their ionic form. In each case, the IIR is employed at a constant, specified concentration in the eluent in order to maintain a desired concentration of IIR on the stationary phase. That is, the coating of IIR is considered to be in dynamic equilibrium with the stationary phase and the column can be said to be dynamically coated with IIR.

It should be noted, in passing, that the IIRs mentioned above are only moderately lipophilic and since this lipophilicity governs the degree of adsorption of the IIR onto the stationary phase, we would expect that the stationary phase would be only partially coated with the IIR. A quite distinct alternative to the dynamic coating method can also be used. In this approach, a very lipophilic IIR is used to initially equilibrate the stationary phase and is then removed from the eluent in the actual separation step [36]. The equilibration process establishes a very strongly bound coating of IIR on the stationary phase and this coating persists for long periods of subsequent usage. For this reason, the method is known as permanent coating ion-interaction chromatography and has been used widely in the separation of inorganic anions and cations. For simplicity, we will confine the discussion to the conventional method, in which the IIR is present in the mobile phase at all times.

Mechanism of ion-interaction chromatography

It is instructive to begin this discussion of mechanism by comparing the retention of a solute on a lipophilic stationary phase using an eluent consisting of an IIR dissolved in a mixture of water and one or more organic solvents with the retention of the same solute under the same chromatographic conditions, except using an eluent that does not contain the IIR. When this comparison is made, the following trends are observed:

- The retention of neutral solutes is not altered significantly when the IIR is added to the eluent.
- The retention of solutes having the same charge as the IIR is decreased when the IIR is added to the eluent.
- The retention of solutes having the opposite charge to the IIR is increased when the IIR is added to the eluent.

In addition, the following effects on retention are observed when the composition of the eluent is altered:

- The retention of solutes having the opposite charge to the IIR is increased when the concentration of IIR in the eluent is increased.
- The retention of solutes having the opposite charge to the IIR is increased when the lipophilicity of the IIR is increased.
- Retention of all solutes decreases when the percentage of modifier in the eluent is increased, and vice versa.

Any mechanism suggested for ion-interaction chromatography must necessarily explain these trends in retention behaviour. A large volume of literature has been devoted to the study of such mechanisms for the retention of carboxylic acids and organic bases. Three mechanisms, namely the ion-pair, dynamic ion-exchange, and ion-interaction mechanisms, have been proposed and the important features of each are summarized below.

In the ion-pair model [26], an ion-pair is envisaged to form between the solute ion and the IIR. This occurs in the mobile phase and the resultant neutral ion-pair is

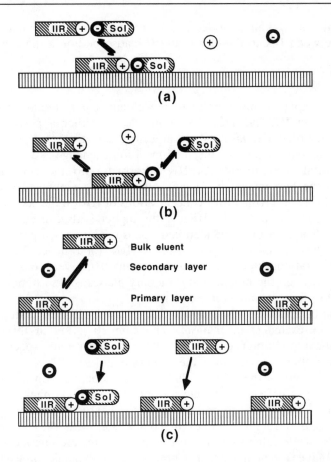

Fig. 6.29. Schematic illustration of (a) the ion-pair, (b) the dynamic ion-exchange and (c) the ion-interaction models for the retention of anionic solutes in the presence of a lipophilic cationic IIR. The solute and the IIR are labelled on the diagram. The large, hatched box represents the lipophilic stationary phase, the black circle with the negative charge represents the counter-anion of the IIR, while the white circle with the positive charge represents the counter-cation of the solute. Adapted from Bidlingmeyer (1980) *J. Chromatogr. Sci.*, **18**, 525.

then considered to partition onto the lipophilic stationary phase in the same manner that any neutral molecule with lipophilic character is retained in reversed-phase chromatography. Retention therefore results solely as a consequence of reactions taking place in the mobile phase. The degree of retention of the ion pair is dependent on its lipophilicity, which in turn depends on the lipophilicity of the IIR itself. Neutral solute molecules are unaffected by the presence of the IIR in the eluent and interact with the stationary phase in the conventional reversed-phase manner. An increase in the percentage of organic solvent in the eluent decreases the interaction of the ion-pairs with the stationary phase and therefore reduces their retention. The ion-pair model is illustrated schematically in Fig. 6.29(a), using a positively charged IIR and a negatively charged solute as an example.

The dynamic ion-exchange model [37, 38] proposes that a dynamic equilibrium is established between IIR in the eluent and IIR adsorbed onto the stationary phase, as follows:

$$\text{IIR}^{\pm}_{(M)} \rightleftharpoons \text{IIR}^{\pm}_{(S)} \qquad\qquad 6.3$$

where the subscripts M and S refer to the eluent and stationary phases and the superscript on the IIR indicates that it may carry either a positive or negative charge. The adsorbed IIR imparts a charge to the stationary phase, causing it to behave as an ion-exchanger (the mechanism of ion-exchange is discussed in Section 6.3.3). The total concentration of IIR adsorbed onto the stationary phase is dependent on the percentage of organic solvent in the mobile phase, with higher percentages of solvent giving lower concentrations of IIR on the stationary phase. In addition, the more lipophilic the IIR, or the higher is its concentration, then the greater is its adsorption onto the stationary phase. Thus, for a given mobile phase composition, the concentration of adsorbed IIR (and hence the ion-exchange capacity of the stationary phase) remains constant. However, constant interchange of IIR occurs between the mobile and stationary phases, so the stationary phase can be considered to be a dynamic ion-exchanger.

Introduction of a solute with opposite charge to the IIR results in retention by a conventional ion-exchange mechanism. The competing ion in this ion-exchange process may be the counter-ion of the IIR, or another ionic species deliberately added to the mobile phase. Since the retention times will be dependent on the ion-exchange capacity of the column, they are also dependent on the lipophilicity of the IIR and the percentage of organic solvent in the mobile phase. Solutes having the same charge as the IIR are repelled from the charged stationary phase surface and show decreased retention times in comparison with those observed in the absence of IIR, while retention times for neutral solutes are unaffected by the IIR. Figure 6.29(b) gives a schematic representation of the dynamic ion-exchange model, again using a positively charged IIR and a negatively charged solute as an example.

The ion-interaction model [23, 29, 39] can be viewed as intermediate between the two previous models in that it incorporates both the electrostatic effects which are the basis of the ion-pair model and the adsorptive effects which form the basis of the dynamic ion-exchange model. The lipophilic IIR ions are considered to form a dynamic equilibrium between the mobile and stationary phases, as depicted in equation 6.3. This results in the formation of an electrical double-layer at the stationary phase surface. The adsorbed IIR ions are expected to be spaced evenly over the stationary phase owing to repulsion effects, which leaves much of the stationary phase surface unaltered by the IIR. The adsorbed IIR ions constitute a primary layer of charge, to which is attracted a diffuse, secondary layer of oppositely charged ions. This secondary layer of charge consists chiefly of the counter-ions of the IIR. The amount of charge in both the primary and secondary charged layers is dependent on the amount of adsorbed IIR, which in turn depends on the lipophilicity of the IIR, its concentration, and the percentage of organic solvent in the mobile phase. The double layer is shown schematically in the top frame of Fig. 6.29(c).

Transfer of solutes through the double-layer to the stationary phase surface is a function of electrostatic effects and of the solvophobic effects responsible for retention in reversed-phase chromatography. Neutral solutes can pass unimpeded through the double layer, so their retention is relatively unaffected by the presence of IIR in the mobile phase. A solute with opposite charge to the IIR can compete for a position in the secondary charged layer, from which it will tend to move into the primary layer as a result of electrostatic attraction and, if applicable, reversed-phase solvophobic effects. The presence of such a solute in the primary layer causes a decrease in the total charge of this layer so, to maintain charge balance, a further IIR ion must enter the primary layer. This means that the adsorption of a solute ion with opposite charge to the IIR will be accompanied by the adsorption of an IIR ion. The overall result is that solute retention involves a pair of ions (that is, the solute and IIR ions), but not necessarily an ion-pair. This process leads to increased retention of the solute compared with the situation in which the IIR is absent from the eluent. The lower frame of Fig. 6.29(c) depicts this process for a positively charged IIR and a negatively charged solute. Solutes with the same charge as the IIR will show decreased retention owing to electrostatic repulsion from the primary charged layer.

Factors affecting retention in ion-interaction chromatography

The preceding discussion has already highlighted most of the parameters used to manipulate solute retention in ion-interaction chromatography. These parameters, all of which relate to the mobile phase composition, include the type and concentration of the IIR, and the percentage of organic modifier in the mobile phase. The effects of varying these parameters can be viewed most simply in terms of the changes produced in the amount of IIR adsorbed to the surface of the stationary phase. Any increase in this amount will result in increased solute retention and, for example, can be produced by increasing the concentration or hydrophobicity of the IIR or decreasing the percentage of organic modifier in the mobile phase.

Figure 6.30 shows the effect of increasing the hydrophobicity of the IIR in the separation of a mixture of acidic, neutral and basic solutes. Here, IIR hydrophobicity is increased by extending the chain length of some alkyl sulfonates over the range C_5 (Fig. 6.30a) to C_7 (Fig. 6.30b). It will be noted that this change causes a marginal decrease in retention of an acidic solute (maleic acid) and virtually no change in retention of a neutral solute (phenacetin). In contrast, the retention times of the remaining, basic solutes increase with increased hydrophobicity of the IIR.

Some more subtle effects on retention can be generated by considering the role of the counter-ion of the IIR. This counter-ion competes with the solute for the charged sites provided by the adsorbed IIR, and manipulation of retention is therefore possible by altering the nature of the counter ion. For example, altering the counter ion from hydroxide to sulfate will cause a decrease in the retention times of solute anions because of the greater competitive effect of sulfate (this will be explained further in Section 6.3.3). Alternatively, a competing ion may be added to the mobile phase as a separate component, usually as an inorganic salt.

Applications of ion-interaction chromatography

Ion-interaction chromatography finds widespread use for the separation of strong acids and bases. It is particularly suitable for the separation of inorganic anions and

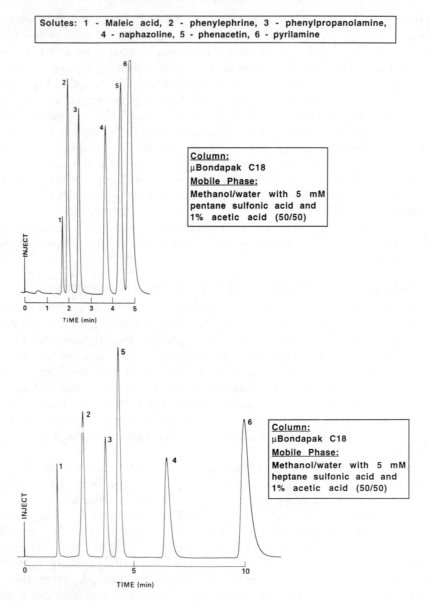

Solutes: 1 - Maleic acid, 2 - phenylephrine, 3 - phenylpropanolamine, 4 - naphazoline, 5 - phenacetin, 6 - pyrilamine

Column:
μBondapak C18
Mobile Phase:
Methanol/water with 5 mM pentane sulfonic acid and 1% acetic acid (50/50)

Column:
μBondapak C18
Mobile Phase:
Methanol/water with 5 mM heptane sulfonic acid and 1% acetic acid (50/50)

Fig. 6.30. Effects of hydrophobicity of the IIR on solute retention in ion-interaction chromatography. Chromatograms courtesy of Waters.

cations and provides a very useful alternative to ion-exchange chromatography. In the case of inorganic cations, it is fair to say that ion-interaction chromatography gives separation performance superior to all other forms of liquid chromatography. This performance is illustrated in Fig. 6.31, which shows the separation of rare earth cations on a reversed-phase column using octanesulfonate as the IIR. In this chromatogram, a complexing agent (α-hydroxyisobutyric acid) has been added to

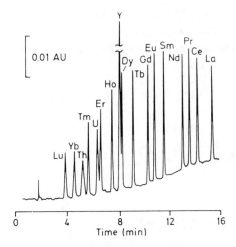

Fig. 6.31. Separation of cations by dynamically coated ion-interaction chromatography. A 5 μm Supelco C$_{18}$ column was used with an eluent formed from a linear gradient of 0.05–0.40 mM α-hydroxyisobutyric acid at pH 4.2, containing 30 mM octanesulfonate and 7.5% methanol. Reprinted from Barkley *et al.* (1986) *Anal. Chem.*, **58**, 2222, with permission. Post-column reaction detection was used.

the mobile phase in order to partially complex the solute cations and thereby decrease their retention.

6.3.3 Ion-exchange chromatography

Introduction

The technique of ion-exchange chromatography is one of the best-established forms of LC, having been used for many years for the separation of ionic solutes, usually in aqueous solution. In this section, we will discuss traditional ion-exchangers and their use in modern chromatography, leaving the more specialized aspects of ion-exchange to be discussed in Section 6.3.4, which deals with ion chromatography.

An ion-exchanger in aqueous solution consists of anions, cations and water, where either the cations or the anions are chemically bound to an insoluble matrix. The chemically bound ions are referred to as the fixed ions and the ions of opposite charge are referred to as the counter ions. The insoluble matrix may be inorganic or may be a polymeric organic resin and, especially in older types of ion-exchangers, is generally porous in nature. These pores contain water from the aqueous solution, together with a sufficient concentration of counter ions to render the whole exchanger electrically neutral. Counter ions may move through the matrix by diffusion and in the ion-exchange process itself, are replaced by ions of the same charge from the external solution. The ion-exchanger is classified as a cation-exchange material when the fixed ion carries a negative charge, and as an anion-exchange material when the fixed ion carries a positive charge.

Mechanism of ion-exchange chromatography

The ion-exchange process can be illustrated by considering an anion-exchange material, for which the counter ion is E^{-}. The exchanger can therefore be represented as M^+E^-, where M^+ denotes the insoluble matrix material containing the fixed (positive) ion. When a solution containing a different anion, A^-, is brought into contact with the ion exchanger, an equilibrium is established between the two mobile ions E^- and A^- as follows:

$$M^+E^- + A^- \rightleftharpoons M^+A^- + E^- \qquad 6.4$$

Since the electroneutrality of the solution must be maintained during the ion-exchange process, the exchange is stoichiometric, such that a single monovalent anion A^- displaces a single monovalent counter-anion E^-. Equation 6.4 can be generalized for y moles of A^{x-} exchanging with x moles (i.e. the stoichiometric amount) of E^{y-} to give:

$$yA_m^{x-} + xE_r^{y-} \rightleftharpoons yA_r^{x-} + xE_m^{y-} \qquad 6.5$$

where the subscript m denotes the mobile (i.e. solution) phase and r denotes the stationary (or resin) phase. It should be noted that equation 6.5 is an equilibrium and, under conditions where the exchanged ions remain in contact with the ion-exchange matrix, complete exchange will not be attained. Furthermore, the solution phase contains a co-ion of the same charge as the fixed ion, but this co-ion plays no part in the ion-exchange process and is therefore not shown.

The equilibrium constant for the reaction shown in equation 6.5 is called the selectivity coefficient, and is given by:

$$K_{A,E} = \frac{(A_r^{x-})^y (E_m^{y-})^x}{(A_m^{x-})^y (E_r^{y-})^x} \qquad 6.6$$

where the parentheses indicate the activity of the designated species. Since the ion activity in the resin phase cannot be determined, $K_{A,E}$ is not a thermodynamically defined equilibrium constant but a coefficient which is defined according to practical requirements. Under conditions where the activity coefficients approximate unity, equation 6.6 can be simplified to:

$$K_{A,E} = \frac{[A_r^{x-}]^y [E_m^{y-}]^x}{[A_m^{x-}]^y [E_r^{y-}]^x} \qquad 6.7$$

where the brackets indicate molar or molal concentrations, or equivalent fraction units. For convenience, concentrations are often expressed in the units moles l^{-1} for the solution phase, and millimoles g^{-1} for the resin phase.

The selectivity coefficient derives its name from the information it provides on the likelihood of exchange between two particular ions. In the above example, if $K=1$, then the ion-exchange matrix shows no selectivity for anion A^{x-} over E^{y-}; that is, the ratios of the concentrations of these ions in the matrix and solution phases are equal. If K is greater than unity, the matrix (or resin) phase will contain a higher concentration of ion A^{x-} than the solution phase, and will select A^{x-} preferentially over E^{y-}. The reverse applies for values of K less than one. Clearly, a competition

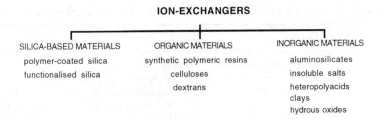

Fig. 6.32. Classification of ion-exchangers.

for the ion-exchange sites exists between the two ions A^{x-} and E^{y-}. It is convenient to designate E^{y-} as the competing anion and to describe A^{x-} as the solute anion.

For cation-exchange equilibria, a similar series of equations can be derived. Thus, for an ion-exchange reaction between the ion A^{x+} and ion E^{y+}, the equilibrium can be written:

$$y A_m^{x+} + x E_r^{y+} \rightleftharpoons y A_r^{x+} + x E_m^{y+} \qquad 6.8$$

and after assuming that the activity coefficients are close to unity, the selectivity coefficient is given by:

$$K_{A,E} = \frac{[A_r^{x+}]^y [E_m^{y+}]^x}{[A_m^{x+}]^y [E_r^{y+}]^x} \qquad 6.9$$

Stationary and mobile phases for ion-exchange chromatography

Ion-exchangers are characterized both by the nature of the ionic species comprising the fixed ion and by the nature of the insoluble ion-exchange matrix itself. The matrix types used for ion-exchange chromatography can be subdivided broadly into silica-based materials, inorganic materials and organic (polymeric) materials, as illustrated in Fig. 6.32. Within these classifications, functionalized silica and synthetic polymeric resins are those that are used most commonly and the ensuing discussion of ion-exchange stationary phases will be confined to these materials.

The silica-based stationary phases are created from the same microparticulate or pellicular silica particles which form the basis of the bonded phases used in reversed-phase chromatography. In these phases a functional group is covalently bound to the surface of the silica and the fixed ion of the ion-exchanger forms part of this functional group. Synthetic polymeric ion-exchangers are formed predominantly from copolymers of styrene and divinylbenzene, or methacrylic acid and divinylbenzene, in which cross-linked, porous beads are formed. These macroporous materials are then treated with suitable reagents, which introduce the desired functional group onto the backbone material. Details of these reactions are beyond the scope of this text, but may be found elsewhere [1]. Table 6.12 shows the types of functional groups commonly encountered in synthetic ion-exchangers.

Cation-exchange resins are classified into strong acid and weak acid types. The former retain the negative charge on the fixed ion over a wide pH range, whereas

Table 6.12   Functional groups found on some typical synthetic ion-exchange materials.

Cation exchangers		Anion exchangers	
Type	Functional group	Type	Functional group
Sulfonic acid	$-SO_3^-$ H^+	Quaternary amine	$-N(CH_3)_3^+$ OH^-
Carboxylic acid	$-COO^-$ H^+	Quaternary amine	$-N(CH_3)_2(EtOH)^+$
Phosphonic acid	$-PO_3H^-$ H^+	Tertiary amine	$-NH(CH_3)_2^+$ OH^-
Phosphinic acid	$-PO_2H^-$ H^+	Secondary amine	$-NH_2(CH_3)^+$ OH^-
Phenolic	$-O^-$ H^+	Primary amine	$-NH_3^+$ OH^-
Arsonic acid	$-AsO_3H^-$ H^+		
Selenonic acid	$-SeO_3^-$ H^+		

the latter type are ionized (and hence act as cation-exchangers) only over a much narrower pH range. Sulfonic acid exchangers are strong acid types, while the remaining cation-exchange functional groups in Table 6.12 are weak. The weak acid types require a sufficiently high pH for use, and this is exemplified by the use of a NaOH eluent with a carboxylic acid cation-exchanger. Similarly, the anion-exchangers are classified as strong base and weak base exchangers. Quaternary amine functional groups form strong base exchangers, while less-substituted amines form weak base exchangers. A weak base material will function only when the pH is sufficiently low to protonate the nitrogen atom in the functional group.

While a diverse range of ion-exchange functionalities exists, most separations with silica and organic ion-exchangers are performed on strong acid cation-exchangers of the sulfonic acid type, and on strong base anion-exchangers of the quaternary ammonium type. These strong cation-exchangers and strong anion-exchangers are often labelled SCX and SAX, respectively.

An important property of an ion-exchanger is its ion-exchange capacity. This is determined by the number of functional groups per unit weight of the resin and may be measured in a variety of units, the most common of which are milliequivalents (of charge) per gram of dry resin, or milliequivalents per millilitre of wet resin. The ion-exchange capacity is often measured by saturating a known weight of resin with a particular ion, followed by washing of the resin and then quantitative displacement of this ion. The number of moles of the displaced ion can then be determined. It should be noted that the capacity measured in this way is often somewhat higher than that applicable when the resin is packed as the stationary phase in a chromatographic column. Polymeric ion-exchange resins, which have been reacted fully to produce the maximum number of functional groups, have capacities which typically fall into the range 3–5 mEq g^{-1}, whereas silica-based resins typically have much lower capacities. It is important to note that high-capacity polymeric resins have functional groups residing within the internal pores and this may lead to a reduction in chromatographic efficiency through the establishment of long solute diffusion paths.

Polymeric resin ion-exchangers consist of cross-linked polymer chains containing ionic functionalities. When such a material comes into contact with water, the outermost functional groups are solvated and the randomly arranged polymer chains unfold to accommodate the larger solvated ions. A very concentrated internal solution of fixed ions and counter-ions therefore exists and the mobile counter-ions tend to diffuse out of the exchanger into the external aqueous solution. The fixed ions cannot diffuse and, as a result, external water molecules are forced into the resin in an attempt to reduce the internal ionic concentration in the resin. The cross-linking of the resin provides mechanical stability that prevents dissolution of the resin, but swelling persists as a result of the equilibrium pressure caused by the differences in concentration between the external and internal ionic solutions. The swelling pressure may be as high as 300 atmospheres for a polymeric resin of high ion-exchange capacity. The degree of swelling of the resin is dependent on the composition of the solution with which it is equilibrated. Thus, changes of eluent are accompanied by changes in the level of swelling and this effect has important ramifications on the types of resins suitable for use as stationary phases in chromatographic columns of fixed volume. Resins of low cross-linking ($<2\%$) exist as soft gels in aqueous solution and exhibit large volume changes when the eluent is altered; for this reason, they are unsuitable as stationary phases for high-performance applications where the eluent is delivered under pressure. Macroporous resins are more rigid owing to their high degree of cross-linking and their resistance to swelling effects renders them more suitable as chromatographic stationary phases for column packing purposes.

The mobile phase (more commonly called the eluent) used in ion-exchange chromatography generally consists of an aqueous solution of a suitable salt or mixture of salts, with a small percentage of an organic solvent being sometimes added. The salt mixture may itself be a buffer, or a separate buffer can be added to the eluent if required. The prime component of the eluent is the competing ion, which has the function of eluting sample components through the column within a reasonable time.

Factors affecting retention in ion-exchange chromatography

The three foremost properties of the eluent affecting the elution characteristics of solute ions are: (i) the eluent pH; (ii) the nature of the competing ion; and (iii) the concentration of the competing ion. The eluent pH can have profound effects on the form in which the functional group on the ion-exchange matrix exists, and also on the forms of both the eluent and solute ions. The selectivity coefficient existing between the competing ion and a particular solute ion will determine the degree to which that competing ion can displace the solute ion from the stationary phase. Since different competing ions will have different selectivity coefficients, it follows that the nature of the competing ion will be a prime factor in determining whether solute ions will be readily eluted. Finally, the concentration of the competing ion can be seen to exert a major effect by influencing the position of the equilibrium point for ion-exchange equilibria, such as those depicted in equations 6.5 and 6.8. The higher the concentration of competing ion in the eluent, the more effectively the eluent displaces solute ions from the stationary phase, and thus the more rapidly is the solute eluted from the column. The ion-exchange capacity of a resin plays a large role in determining the concentration of competing ion used in an eluent to be employed with that resin. Higher capacity resins generally require the use of more concentrated eluents. In addition to the above three factors, elution of the solute is influenced by the eluent flow-rate and the temperature. Faster flow-rates lead to lower elution volumes because the solute ions have less opportunity to interact with the fixed ions. Temperature has a less predictable effect, which is somewhat dependent on the type of ion-exchange material used. An elevated temperature increases the rate of diffusion within the ion-exchange matrix, generally leading to increased interaction with the fixed ions and therefore larger elution volumes. Chromatographic efficiency is usually improved at higher temperatures.

Prior to sample injection, the column must be equilibrated with eluent so that all the exchange sites on the stationary phase contain the same counter ion. A point of terminology arises here, when one considers that equilibration of a column with a solution of competing ion (i.e. the eluent) results in the counter ions associated with the fixed ions being completely replaced with competing ions. When the column is in this condition, the competing ions become the new counter-ions at the ion-exchange sites and the column is said to be in the form of that particular ion. For example, an anion-exchange column which is fully equilibrated with a NaOH eluent is in the hydroxide form. Reproducible elution volumes are obtained only when the column is converted to the same form prior to the injection of each sample and this becomes especially important when the eluent is changed. The time required for a column to equilibrate to a new eluent depends on the selectivity coefficient for the competing ion in that eluent over the previous competing ion, and also on the concentration of the competing ion in the new eluent.

Selectivity coefficients (equations 6.6 and 6.9) provide a means for determining the relative affinities of an ion-exchanger for different ions. It might be considered that a well-defined affinity series for anions and cations could be obtained by simple experiment but, in reality, the relative affinities show considerable variation with the type of ion-exchanger and the conditions under which it is used. In some cases, simple ion-exchange may not be the sole retention mechanism operating; for example, partitioning of solute ions between the eluent and the pores of the stationary phase may occur, or the solute ion could be adsorbed onto the surface of the ion-exchange matrix. In view of these factors, it is possible to provide only approximate guidelines for the relative affinities of ion-exchangers for different ions.

Selectivity coefficients for the uptake of cations by a strong acid cation-exchange resin are generally in the following order [41]: $Pu^{4+} \gg La^{3+} > Ce^{3+} > Pr^{3+} > Eu^{3+} > Y^{3+} > Sc^{3+} > Al^{3+} \gg Ba^{2+} > Pb^{2+} > Sr^{2+} > Ca^{2+} > Ni^{2+} > Cd^{2+} > Cu^{2+} > Co^{2+} > Zn^{2+} > Mg^{2+} > UO_2^{2+} \gg Tl^+ > Ag^+ > Cs^+ > Rb^+ > K^+ > NH_4^+ > Na^+ > H^+ > Li^+$. It follows from this series that a cation-exchange eluent of 0.1 M KCl will be stronger than that containing 0.1 M NaCl, provided other factors are equal. Selectivity coefficients for anions on strong base anion-exchangers follow the general order: citrate > salicylate > $ClO_4^- > SCN^- > I^- > S_2O_3^{2-} > WO_4^{2-} > MoO_4^{2-} > CrO_4^{2-} > C_2O_4^{2-} > SO_4^{2-} > SO_3^{2-} > HPO_4^{2-} > NO_3^- > Br^- > NO_2^- > CN^- > Cl^- > HCO_3^- > H_2PO_4^- > CH_3COO^- > IO_3^- > HCOO^- > BrO_3^- > ClO_3^- > F^- > OH^-$.

Some general rules can be offered to assist in the prediction of the affinity order. These are based on a number of properties of the solute and the ion-exchanger and include:

1. The charge on the solute ion.
2. The solvated size of the solute ion.

3. The degree of cross-linking of the ion-exchange resin.
4. The polarizability of the solute ion.
5. The ion-exchange capacity of the ion-exchanger.
6. The functional group on the ion-exchanger.
7. The degree to which the solute ion interacts with the ion-exchange matrix.

An increase in the charge on the solute ion increases its affinity for an ion-exchanger through increased coulombic interactions. This trend is known as electroselectivity and becomes more pronounced as the external solution in contact with the ion-exchanger becomes more dilute. Electroselectivity may be explained in terms of the Donnan potential, which is the potential difference arising because of the imbalance in the ionic concentrations in the resin bead and in the external solution [42]. An exchange involving the replacement of two bound monovalent ions with a single divalent ion causes this imbalance to be diminished, and is thus a favourable process. Electroselectivity is reflected in the following series of selectivity coefficients: $Pu^{4+} \gg La^{3+} \gg Ba^{2+} \gg Tl^{+}$. For cations, the trend is very strong, such that the selectivity differences between these four ions are large.

The size of the solvated solute ion also exerts a significant effect, with ions of smaller solvated size showing greater binding affinity than larger ions. Thus, the selectivity sequence $Cs^{+} > Rb^{+} > K^{+} > Na^{+} > H^{+} > Li^{+}$ is the exact reverse of the sequence of ionic radii for the hydrated ions, and follows the well-known lyotropic series, with the most strongly hydrated ion, Li^{+}, being held most weakly. This behaviour is related directly to swelling of the resin, since a smaller ion is more easily accommodated in the resin pores. Thus, the higher the degree of cross-linking, the greater the preference of the resin for smaller solute ions. The combination of factors 1 and 2 above suggests that binding affinity should increase with increasing polarizing power, that is, for ions with a high charge and small hydrated radius.

Ion-exchange selectivity coefficients increase with the degree of polarizability of the solute ion. Thus, sulfonic acid fixed ions show greater affinity for the more polarizable Ag^{+} and Tl^{+} than for the alkali metal ions. Similarly, I^{-} is more strongly retained on an anion-exchanger than Br^{-} or Cl^{-}. However, polarization does not explain why ClO_4^{-} has a higher anion-exchange affinity than I^{-}. The strong retention of anions such as ClO_4^{-}, which are large, have low charge and are weak bases, can be attributed to the interaction of these ions with the water structure at the resin surface. Large, polarizable ions with a diffuse charge do not easily form a well-orientated layer of water molecules at their surface, and so tend to disrupt the surrounding water structure. This leads to an increase in free energy, which is the driving force for these ions to bind (that is, to form an ion-pair) with the fixed ion of an ion-exchanger, thereby diminishing both the disruption to the water structure and the free energy. This binding process is called water-structure induced ion-pairing [43].

The trends evident from the remaining factors (5–7) listed above are not as clear-cut. The ion-exchange capacity of the ion-exchanger can affect the selectivity coefficients for some anions and cations, but for most ions, the selectivity coefficient remains essentially constant as the ion-exchange capacity is decreased. Similarly, the nature of the functional group exerts little effect for most ions, but significantly affects the selectivity coefficients for other ions. Large, polarizable anions such as BF_4^{-}, I^{-}, ClO_4^{-} and ClO_3^{-} show changes in selectivity as the alkyl substituents in trialkylammonium strong-base anion-exchangers are varied. This may be related to the water-structure induced ion-pairing mechanism discussed above, with the larger functional groups (i.e. those with the largest alkyl substituents) causing greatest disruption to the water structure, causing them to bind large, polarizable ions more strongly than smaller functional groups [44, 45]. Interactions between the solute ion and the ion-exchange matrix are difficult to predict, and are specific to individual ions.

Applications of ion-exchange chromatography

Applications of high capacity (SAX and SCX) ion-exchangers include the separation of a wide range of organic and inorganic solutes. Examples of separations of almost any desired type of ion can be found in the literature. However, ion-exchange chromatography is particularly important for the separation of amino acids and proteins. This application is illustrated in Fig. 6.33 which shows the separation of physiological amino acids on a fully sulfonated styrene-divinylbenzene polymeric stationary phase.

6.3.4   Ion chromatography

Introduction

In the strict sense, ion chromatography is not a separate chromatographic technique but rather a specialized application of a collection of established techniques. When introduced in 1975 [46] ion chromatography referred only to the separation of inorganic anions and cations using a specific combination of ion-exchange columns coupled to a conductimetric detector. Since that time, the definition of ion chromatography has expanded greatly and it can be best categorized in terms of the type of solutes separated rather than the manner in which the separation is achieved. Thus, ion chromatography can be defined as the use of liquid chromatographic methods for the separation of inorganic anions and cations and low molecular weight water-soluble organic acids and bases. Numerous approaches to this separation can be considered to fall within the definition, for example, any of the chromatographic techniques mentioned in Section 6.3, provided that they were used for the particular solutes which form the area of interest of ion chromatography. However, it is true to say that the majority of ion chromatographic separations are performed on specialized ion-exchangers.

In this section we will look at some of the unique developments that have occurred in the maturation of ion chromatography. In the interests of brevity, the discussion will be confined to ion-exchange methods only. The interested reader seeking a broader coverage of the technique is referred to any of the standard texts listed at the end of this chapter. These ion-exchange methods have been divided somewhat arbitrarily into two main groups, largely on the basis of historical development and commercial marketing influences. These groups of methods are referred to as non-suppressed ion chromatography and suppressed ion chromatography.

Non-suppressed ion chromatography comprises all those methods in which an ion-exchange column is used to separate a mixture of ions, with the separated

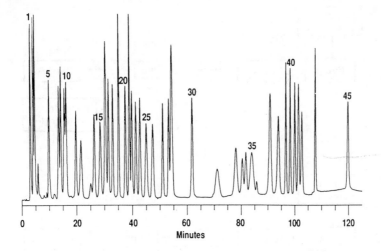

Fig. 6.33. Chromatogram of physiological amino acids. Stationary phase: fully sulfonated polystyrene-divinylbenzene, 5 μm; Mobile phase: lithium citrate buffers. Detection by post-column reaction with ninhydrin; 2.5 nmol of each amino acid was injected. Chromatogram courtesy Dionex Corporation.

solutes being passed directly to the detector. The hardware configuration employed is shown schematically in Fig. 6.34(a), from which it can be seen that this configuration parallels the hardware used in traditional HPLC. Some of the names proposed for this technique are: (i) non-suppressed ion chromatography; (ii) single-column ion chromatography; (iii) electronically suppressed ion chromatography. The first two names indicate that only a single chromatographic column is employed and that the eluent is not chemically modified prior to entering the detector, whereas the last name pertains to the fact that the background conductance of the eluent can be nulled electronically by certain types of conductivity detectors. Non-suppressed IC is the most frequently used term and is recommended.

The second group of ion-exchange methods consists of those in which an additional device, called the suppressor, is inserted between the ion-exchange separator column and the detector, as shown in Fig. 6.34(b). The function of the suppressor is to modify both the eluent and the solute in order to improve the detectability of the solutes with a conductivity detector. The suppressor requires a regenerant (or scavenger) solution to enable it to operate for extended periods. Methods using this hardware configuration are referred to as: (i) suppressed ion chromatography; (ii) chemically suppressed ion chromatography; (iii) eluent-suppressed ion chromatography; and (iv) dual-column ion chromatography. The last of these names is misleading because modern suppressors are not columns, but rather flow-through membrane devices. The term suppressed IC is recommended.

Stationary and mobile phases for ion chromatography

The ion-exchange stationary phases used for ion chromatography fall into the same classes as those discussed earlier in Section 6.3.3. There are two factors which differentiate the ion-exchangers used in ion chromatography. The first is their

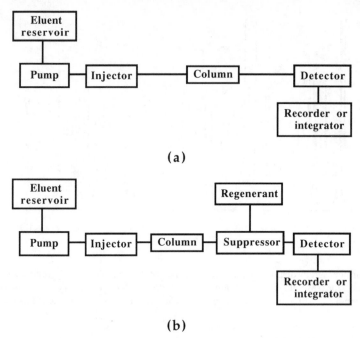

Fig. 6.34. Block diagram showing the instrumental components used in (a) nonsuppressed and (b) suppressed IC.

ion-exchange capacity. Ion chromatography requires ion-exchangers with low ion-exchange capacity, typically in the range 10–100 μEq g^{-1}. This requirement can be attributed chiefly to the fact that ion chromatography was developed originally for use with conductivity detection, which introduces a preference for eluents of low background conductance in order to enhance the detectability of eluted solute ions. The diversity of detection methods currently available now makes it possible to use columns of much higher ion-exchange capacity, but because conductivity detection is still the most commonly employed detection mode, the majority of separations continue to be performed on low-capacity materials. The second characteristic of ion-exchangers for ion chromatography is their higher chromatographic efficiency in comparison with traditional ion-exchangers.

Both of the above-mentioned differentiating characteristics can be achieved by using ion-exchangers in which the functional groups are confined to a thin shell around the surface of the stationary phase particle. This both reduces the number of functional groups (and hence the ion-exchange capacity) and also limits the diffusion path of solute ions, thereby improving mass-transfer characteristics and, hence, chromatographic efficiency. Two main approaches to synthesizing such ion-exchangers can be identified. The first involves the use of only a very short reaction time when the matrix material (i.e. either silica or a polymer) is derivatized in order to introduce the ion-exchange functional group. For example, a macroporous PS-DVB bead immersed in concentrated sulfuric acid for less than 30 s will give a material in which sulfonic acid functional groups are confined to a very shallow depth (of the order of 200 Å) around the outside of the particle. This produces a

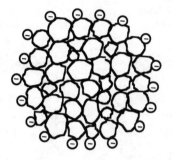

Fig. 6.35. Schematic representation of the cross-section of a surface-sulfonated cation-exchange resin. The negative charges represent sulfonic acid groups which are located on the surface of the resin bead. Note that the interior of the bead is not sulfonated as occurs in high-capacity, fully functionalized materials.

surface-functionalized cation-exchanger, represented schematically in Fig. 6.35, in which the confinement of functional groups to the outer layer has been achieved by chemical means. Surface-functionalized anion-exchangers can be produced in a similar manner. Historically, surface-functionalized ion-exchangers find most use in nonsuppressed ion chromatography.

The second approach to synthesis of ion-exchangers for ion chromatography involves a physical process for confining the functional groups to the outer layer. These ion-exchangers, known as agglomerated materials, consist of a central core particle, to which is attached a monolayer of small-diameter particles which carry the functional groups comprising the fixed ions of the ion-exchanger. Provided the outer layer of functionalized particles is very thin, the agglomerated resin exhibits excellent chromatographic performance owing the very short diffusion paths available to solute ions during the ion-exchange process. Schematic illustrations of agglomerated anion- and cation-exchangers are given in Fig. 6.36. The central core (or support) particle is generally PS-DVB of moderate cross-linking, with a particle size in the range 10–30 μm, which has been functionalized to carry a charge opposite to that of the outer particles. The outer microparticles consist of finely ground resin or monodisperse latex (with diameters in the approximate range 20–100 nm) which has been functionalized to contain the desired ion-exchange functional group. It is this functional group that determines the ion-exchange properties of the composite particle, so that aminated (positively charged) latexes produce agglomerated anion-exchangers (as illustrated in Fig. 6.36a), while sulfonated (negatively charged) latexes produce agglomerated cation-exchangers (Fig. 6.36b). Electrostatic attraction between the oppositely charged core particles and outer microparticles holds the agglomerate together, even over long periods. Figure 6.37 shows details of this electrostatic attraction for an agglomerated anion-exchanger. Agglomerated ion-exchangers are used most frequently in suppressed ion chromatography.

Mobile phases (or eluents) for ion chromatography are similar to those used for regular ion-exchange separations (as discussed in Section 6.3.3), except for the stringent requirement that they should be compatible with conductivity detection. In the case of non-suppressed ion chromatography (in which the eluent is not involved in further reaction before reaching the detector) this means that eluent competing

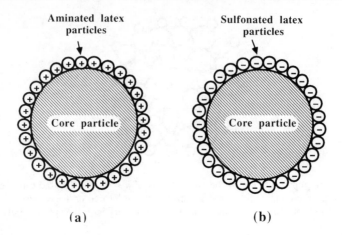

Fig. 6.36.   Schematic representation of agglomerated (a) anion- and (b) cation-exchangers.

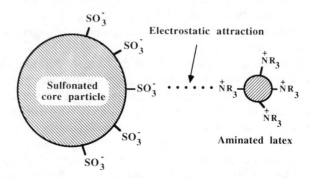

Fig. 6.37.   Formation of an agglomerated anion-exchange resin using electrostatic binding. Note that the core and the latex particles are not drawn to scale.

ions of low limiting equivalent ionic conductance (see Section 5.5.6) are required if direct conductivity detection is to be used. Aromatic carboxylates (such as benzoate and phthalate) aromatic sulfonates (such as toluenesulfonate) and complex ions (such as the anionic complex formed between gluconate and borate) are ideal for anion separations, whereas aromatic bases are useful for cation separations. All of these species are bulky ions with low ionic mobility (and hence low conductance), so that direct detection of more mobile solute ions (such as chloride, sulfate, etc.) is possible using conductivity. Alternatively, indirect conductivity detection is possible using eluent competing ions with very high values of limiting ionic conductance, such as hydronium ions for cation separations and hydroxide ions for anion separations.

Conversely, suppressed ion chromatography offers the opportunity for further reaction of the eluent before detection. The purpose of this reaction is to reduce the conductance of the eluent and in most cases, acid–base reactions are used. The mechanism of eluent suppression will be discussed further, but for the present it can be assumed that the process works best when applied to eluents comprising

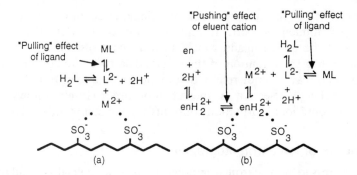

Fig. 6.38. Schematic illustration of the equilibria existing between a solute cation (M^{2+}), ethylene-diamine (en) and an added ligand (H_2L) at the surface of a cation-exchanger. In (a) the eluent contains only the ligand, while in (b), the eluent contains both ligand and ethylenediamine.

competing ions that can be easily neutralized in an acid–base reaction. For example, carbonate and bicarbonate (or mixtures of the two) can be used for anion separations, while dilute solutions of mineral acids can be used for cation separations.

Many cation separations cannot be achieved simply through correct choice of a suitable eluent competing cation. Polyvalent cations show such strong electrostatic attraction to sulfonic acid cation-exchangers that they cannot be displaced except using concentrated eluents. This, in turn, renders detection difficult. Useful alternatives are created by the use of a complexing agent as the eluent, or by the addition of a complexing agent to an eluent that already contains a competing cation. This serves the dual purpose of reducing the effective charge on the solute cation (and hence its affinity for the cation-exchange sites) and also introduces a further dimension of selectivity between solutes which does not exist when ion-exchange is the only retention mechanism in operation. The above approaches are illustrated schematically in Fig. 6.38, which shows the equilibria existing between a divalent metal solute ion M^{2+}, a complexing agent (H_2L), and an ethylenediamine (en) eluent, at the surface of a cation-exchange resin. In Fig. 6.38(a), the eluent contains only the ligand species. Retention of the solute ion on a cation-exchange resin is moderated by the complexation effect of the deprotonated ligand, which can be said to exert a pulling effect on the solute. The eluent pH determines the degree to which the ligand is deprotonated which, in turn, governs the retention of the solute. Retention is also regulated by the type and concentration of the ligand. An example of this same approach applied to ion-interaction chromatography is shown in Fig. 6.31 using α-hydroxyisobutyric acid as the ligand.

Figure 6.38(b) shows the case where the eluent contains both a ligand and a competing cation (enH_2^{2+}). The retention of the solute ion, M^{2+}, is influenced by the competitive effect for the sulfonic acid groups exerted by enH_2^{2+}, and also by the complexation of M^{2+} by the deprotonated ligand L^{2-}. Once again, complexation reduces the effective concentration of M^{2+} and the solute is therefore less successful in competing for the cation-exchange sites. This shows that elution of the solute results from a combination of the pushing, or displacement, effect of the competing cation in the eluent and the complexation, or pulling, effect of the complexing agent. The eluent pH influences both the protonation of ethylenediamine and the deprotonation of the added ligand which, in turn, controls the degree of complex formation and hence the retention of the solute. The type and concentration of the added ligand again play a major role in determining solute retention. For solute ions of similar ion-exchange selectivities, the retention order closely follows the reverse sequence of the conditional formation constants for the solute–ligand complexes.

Mechanism of ion chromatography

Since non-suppressed ion chromatography is just a specialized adaptation of ion-exchange chromatography, there is no need to discuss further the mechanism of this technique. However, suppressed ion chromatography merits further explanation, at least with regard to the suppressor itself. The manner in which the suppressor functions was described in Section 5.5.6 and it is worth remembering that suppressors generally perform the following reactions:

- In the case of anion-exchange separations, cations from the eluent are replaced by hydrogen ions from the suppressor which, in turn, react with the eluent anion (e.g. bicarbonate) to form an undissociated weak acid (e.g. carbonic acid).
- In the case of cation-exchange separations, anions from the eluent are replaced by hydroxide ions from the suppressor, which in turn react with the eluent cation (e.g. hydrogen ion) to form an undissociated weak base (e.g. water).

Solute ions which are the conjugate of strong acids or bases do not react under these conditions, but the exchanged hydrogen ion or hydroxide ion is eluted with the sample band and thereby increases its conductance. The suppressor therefore reduces the background conductance of the eluent and, where possible, simultaneously enhances the detectability of the solute ions. The above reactions of the suppressor are summarized in equations 6.10 and 6.11 for anion separations (here the eluent is represented by E^- and the solute by S^-) and in equations 6.12 for an acidic eluent (HA) used for the separation of cations:

$$\text{Suppressor-H}^+ + \text{Na}^+ + \text{E}^- \rightleftharpoons \text{Suppressor-Na}^+ + \text{HE} \qquad 6.10$$

$$\text{Suppressor-H}^+ + \text{Na}^+ + \text{S}^- \rightleftharpoons \text{Suppressor-Na}^+ + \text{H}^+ + \text{S}^- \qquad 6.11$$

$$\text{Suppressor-OH}^- + \text{H}^+\text{A}^- \rightleftharpoons \text{Suppressor-A}^- + \text{H}_2\text{O} \qquad 6.12$$

Applications of ion chromatography

Ion chromatography is the unrivalled analytical method of choice for the determination of inorganic anions at both trace and ultra-trace levels. Figure 6.39 shows typical separations achievable by both the suppressed and non-suppressed methods. Because of the availability of alternative spectroscopic methods of analysis, ion chromatography of cations is used less frequently than for anions.

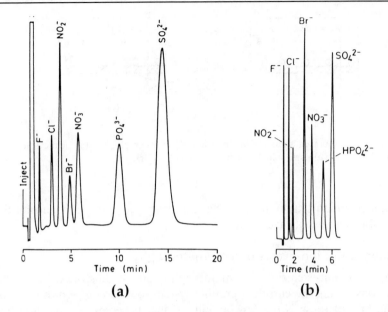

Fig. 6.39. Anion separations obtained using (a) nonsuppressed and (b) suppressed ion chromatography. (a) A Waters IC Pak A column was used with gluconate–borate eluent. (b) A Dionex HPIC-AS4A column was used with a carbonate–bicarbonate eluent. Conductivity detection was employed in both cases. Chromatograms courtesy of Waters and Dionex.

6.3.5 *Ion-exclusion chromatography*

Introduction

Ion-exclusion chromatography involves the use of strong anion- or cation-exchange resins for the separation of ionic solutes from weakly ionized or neutral solutes. In this mode of chromatography, the charge sign on the ion-exchange resin used is the same as that of the weakly ionized solutes. That is, solutes with a partial negative charge (such as carboxylic acids) are separated on a cation-exchange resin having anionic sulfonate functional groups, whereas solutes with a partial positive charge (such as weak bases) are separated on an anion-exchange resin having cationic quaternary ammonium functional groups. This is the opposite to that which occurs in ion-exchange chromatography.

As with other IC separation techniques, ion-exclusion chromatography has been described by a variety of alternative names, some of which are listed below:

- Ion-chromatography exclusion (ICE)
- Ion-exclusion partition chromatography
- Donnan exclusion chromatography
- Ion-moderated partition chromatography

Each of these names implies a mechanism for the separation process and as we will see in the ensuing discussion, the actual mechanism of the process is not clearly defined, but is certainly quite complex. We shall therefore continue to use the term

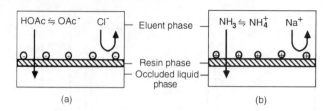

Fig. 6.40. Schematic representation of ion-exclusion chromatography for (a) acidic solutes, such as acetic acid and HCl, and (b) basic solutes, such as NH_3 and NaOH.

ion-exclusion chromatography to describe the technique, while recognizing that this title is probably somewhat inaccurate.

Mechanism of ion-exclusion chromatography

The principles of ion-exclusion chromatography can be illustrated in a schematic manner by considering the chromatographic system to be comprised of three distinct phases. The first of these is the flowing eluent, which passes between the beads of the ion-exchange resin (i.e. through the interstitial volume). The second zone is the polymeric network of the resin material itself, together with its bound ionic functionalities, while the third zone is liquid occluded (i.e. trapped) inside the pores of the resin bead. The polymeric resin can be considered as a semi-permeable, ion-exchange membrane which separates the flowing eluent from the stationary occluded liquid inside the resin [47].

The manner in which solutes are separated in ion-exclusion chromatography is illustrated in Fig. 6.40. We first consider the behaviour of two solutes, hydrochloric acid and acetic acid, on a cation-exchange resin using water as the eluent. From Fig. 6.40(a), we see that Cl^- cannot penetrate into the occluded liquid phase because it is repelled by the anionic functional groups on the resin, in accordance with the Donnan exclusion effect. The Cl^- ions therefore remain in the flowing eluent phase and are not retained by the column. Conversely, the acetic acid is only weakly ionized and exists predominantly as neutral acetic acid molecules, with only a small percentage present as acetate anion. The ionized and neutral acetic acid molecules are in dynamic equilibrium with each other, so that the effective negative charge on the acetic acid is therefore determined by the proportions existing in each form. In a water eluent, this effective charge is quite small and because of this, acetic acid can penetrate the negatively charged resin zone and move into the occluded liquid phase. This results in some degree of retention of acetic acid, so that it is eluted later than hydrochloric acid.

In a similar manner, an anion-exchange resin can be used to separate a weak base (ammonia) from a strong base (NaOH), again using water as eluent. This is illustrated in Fig. 6.40(b), which shows that Na^+ is repelled by the cationic functional groups on the resin phase and is unretained. Ammonia, by virtue of its low degree of ionization and hence its low overall charge, can penetrate into the occluded liquid phase and is therefore retained.

Figure 6.40 suggests that retention of solutes in ion-exclusion chromatography is influenced solely by the charge on the solute. That is, all fully ionized solutes can be expected to be unretained and thus be eluted together at the void volume of the column, and the retention of partially ionized solutes can be expected to increase as the degree of ionization decreases. These predictions are not fully supported in practice and it can be shown that other factors also play a role in solute retention. These factors will be discussed later, but at this stage we will make the assumption that solute charge is the main parameter in determining retention.

Mobile and stationary phases for ion-exclusion chromatography

Ion-exclusion chromatography is usually performed on high capacity, fully func- tionalized PS-DVB polymeric ion-exchange resins. Some of the stationary phase parameters which can exert an influence on solute retention are: (i) particle size; (ii) ion-exchange capacity; (iii) resin structure; (iv) degree of resin cross-linking.

As with other chromatographic techniques, the separation efficiency is strongly influenced by the particle size of the column packing material. Modern ion-exclusion chromatography is generally performed on 5 or 10 μm particles; however, some of the resins used are relatively soft and the use of small-diameter particles means that eluent flow-rates must be kept low to avoid compression of the resin bed. High ion-exchange capacity materials are preferred so that the number of functional groups on the resin is sufficient to exert an appropriate Donnan exclusion effect. The resin structure is also important, with a recent trend towards macroporous materials being evident. The degree of cross-linking of the resin exerts a considerable effect on solute retention, with highly cross-linked materials (e.g. those with 8–12% divinyl- benzene) showing strongest Donnan exclusion effects. Most commercial ion- exclusion resins have approximately 8% cross-linking. Ion-exclusion columns are usually large in comparison with conventional liquid chromatographic columns because a considerable volume of resin material is necessary to provide sufficient occluded liquid phase to permit the separation of solutes of similar size and charge. A typical column would be 30 cm in length, with an internal diameter of 7 mm or more.

The mobile phases (again, more commonly referred to as eluents) used in ion-exclusion chromatography are often very simple in composition. Deionized water may be used and the degree of ionization of the solutes (and hence their retention times) is therefore determined by their pK_a or pK_b values. The limitations of water as an eluent are that stronger acids or bases show too great a degree of ionization to be retained and the peak shape obtained for solutes that are retained is often poor. For this reason, it is common for dilute solutions of strong mineral acids to be employed in the elution of anionic solutes, or dilute solutions of strong bases to be employed in the elution of cationic solutes. In this way, ion-exclusion chromatography can be extended to the separation of relatively strong acids and bases by limiting their degree of ionization.

The most commonly used eluents are formed from strong acids, such as sulfuric acid, hydrochloric acid and aliphatic sulfonic acids. When sulfuric acid is used as the eluent, detection of eluted solutes is generally accomplished by monitoring u.v. absorbance at low wavelengths (200–220 nm). Hydrochloric acid is most often used with conductivity detection, after the eluent is passed through a suitable suppressor.

Aliphatic (and aromatic) sulfonic acids can also be employed for conductivity detection, but because of the relatively low background conductance of these eluents, suppression is not necessary. Weak acids may also be utilized as eluents in ion-exclusion chromatography. Examples include phosphoric acid, tridecafluoro-heptanoic acid (perfluorobutyric acid), and benzoic acid. It is interesting to note that eluents of the same pH, when used on the same stationary phase, produce virtually identical chromatograms, regardless of the nature of the acid used. The choice of eluent acid is therefore governed primarily by the detection method which is to be used.

We can also note, in passing, that organic modifiers, such as methanol, acetonitrile or acetone, are sometimes added to the eluents used in ion-exclusion chromatography. The function of these modifiers is related to the participation of solute adsorption effects in the retention process. This factor is discussed more fully below.

Factors affecting retention in ion-exclusion chromatography

Numerous factors are considered to play a part in the retention process in ion-exclusion chromatography. These factors are listed below in approximate order of importance, at least as far as they influence the retention of carboxylic acids:

1. The degree of ionization of the solute (which is determined by the pK_a of the solute, the eluent pH and the organic modifier content of the eluent).
2. Hydrophobic (reversed-phase) interactions between the solute and the stationary phase (which are determined by the nature of the solute and the organic modifier content of the eluent).
3. The molecular size of the solute.
4. The degree of cross-linking of the stationary phase.
5. The temperature at which the separation is performed.

The degree to which the solute is ionized is the most significant factor that determines solute retention. As the solute becomes more ionized, the Donnan exclusion effect increases in magnitude and this leads to decreased retention. The observed retention time for a particular solute should therefore be dependent on the acid or base dissociation constant of that solute. It should therefore be possible to predict the retention volume of, for example, acidic solutes on the basis of their pK_a values alone. When this is done, the retention behaviour for many solutes is in close accordance with the predictions made above. Figure 6.41 shows a plot of retention volume versus pK_a for a series of solutes [48]. All strong acids, which are fully ionized, are eluted together, while all neutral solutes are eluted together. Solutes having intermediate retention volumes are those which, by virtue of their pK_a values, are partially ionized under the eluent conditions used. It is therefore evident that controlling the level of ionization of a solute by regulating the pH of the eluent will provide a means to manipulate the retention of that solute. For example, we can produce a decrease in solute retention simply by raising the pH of the eluent because this will cause increased ionization of the solutes.

However, some solutes, especially diprotic acids, long-chain aliphatic acids, and aromatic acids show anomalous behaviour which is thought to arise from secondary retention mechanisms operating in conjunction with the ion-exclusion mechanism.

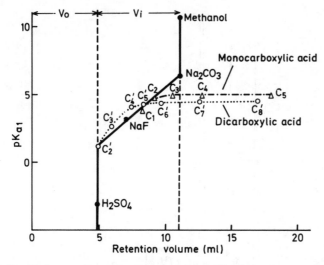

Fig. 6.41. Relationship between retention volume and first dissociation constant (pK_{a1}) for carboxylic acids on a stationary phase with 30% cross-linking, using 1 mM HClO$_4$ as eluent. Reprinted from Kihara *et al.* (1987) *J. Chromatogr.*, **410**, 103, with permission.

These secondary mechanisms include hydrophobic adsorption of the solutes onto the neutral, unfunctionalized regions of the polymeric stationary phase (i.e. a reversed-phase mechanism) and a size-exclusion effect, where certain large solutes are unable to penetrate into the pores of the stationary phase. The former mechanism exerts the greatest effect and opens up the possibility of manipulating the retention of solutes through the addition of organic modifiers to the eluent. Hydrophobic adsorption effects are particularly evident for aromatic solutes.

Temperature can affect retention in ion-exclusion chromatography either by alteration of the chromatographic efficiency in the same manner as observed in most forms of chromatography, or by influencing the degree of ionization of the solute through changes in the dielectric constant of the eluent.

Applications of ion-exclusion chromatography

Ion-exclusion chromatography finds application in the separation of a wide range of small, neutral or partly ionized molecules such as carboxylic acids, inorganic weak acid anions, weak organic bases, and water. It may appear that this restricted group of solutes could diminish the importance of ion-exclusion chromatography in comparison with ion-exchange and ion-interaction chromatography, but in practice ion-exclusion chromatography is of major importance.

The separation of carboxylic acids is the most common application of ion-exclusion chromatography. This mode of chromatography is undoubtedly the method of choice for these solutes. When coupled with direct spectrophotometric detection at low wavelength, ion-exclusion chromatography yields excellent separations and relatively clean chromatograms for a wide variety of very complex sample matrices, such as urine, plasma, foods and beverages, and pharmaceuticals. Figure 6.42 shows a chromatogram for a urine sample, without sample pretreatment, and

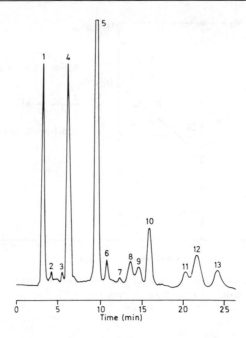

Fig. 6.42. Analysis of human urine using ion-exclusion chromatography. An Interaction ORH-801 column was used with an eluent comprising 10 mN H_2SO_4 containing 10% methanol. Detection was by spectrophotometry at 254 nm. Solute identities: 1, oxalic acid; 2, oxaloacetic acid; 3, α-ketoisovaleric acid; 4, ascorbic acid and α-keto-β-methyl-n-valeric acid; 5, β-phenylpyruvic acid; 6, uric acid; 7, α-ketobutyric acid; 8, homoprotocatechuic acid; 9, unknown; 10, unknown; 11, hydroxyphenylacetic acid; 12, p-hydroxyphenyllactic acid; 13, homovanillic acid. Reprinted from Woo and Benson (1984) *Am. Clin. Prod. Rev.*, Jan, 20, with permission.

illustrates the relatively clean chromatograms which can be achieved for complex samples [49]. Ion-exclusion chromatography has also found increasing usage for the determination of weakly ionized inorganic species. It is especially attractive as an adjunct to ion-exchange chromatography since the selectivities obtained by these two techniques are quite different. Solutes such as fluoride, carbonate, cyanide, borate, sulfite, and ammonium have been determined using this approach. Interference from strongly ionized species is minimal because these solutes are unretained and appear at the column void volume. Ion-exclusion chromatography can therefore readily separate weakly ionized solutes in samples containing high concentrations of ionic species, e.g. seawater and wastewater.

6.4 Speciality Separation Modes

A number of speciality separation modes are also available for specific HPLC applications. These include hydrophobic interaction chromatography, affinity chromatography, complexation chromatography and chiral separations. Each of these separation modes relies upon a very specific interaction between the solute molecules and the stationary phase, with the exception of hydrophobic interaction chromatography which is a specific application of reversed-phase chromatography for the analysis of proteins.

6.4.1   Hydrophobic interaction chromatography

Hydrophobic interaction chromatography (HIC) is a specialized reversed-phase method which allows retention of proteins upon mildly hydrophobic stationary phase surfaces [50]. This mode is commonly used in protein separation schemes because it provides good resolution and recovery of biologically active proteins. In HIC, proteins are bound to a moderately hydrophobic surface using a high ionic strength mobile phase and the proteins are then eluted using a gradient of decreasing ionic strength. Totally aqueous buffered mobile phases are used with HIC and factors which affect retention include the concentration and nature of the buffer salt, buffer pH, presence of surfactants, column temperature and stationary phase functionality [22].

The major advantage of HIC when compared with conventional reversed-phase chromatography is that the use of aqueous buffers avoids any denaturation of biopolymers (proteins) which would otherwise result from the presence of organic modifiers in the mobile phase. Since proteins typically interact with each other under conditions of high ionic strength, better resolution in HIC is achieved when the starting material is partly purified. Also, it is often inconvenient to load large sample volumes at high ionic strength, hence HIC is most commonly used as a second or subsequent step in a protein purification scheme. For instance, this mode of chromatography is particularly appropriate when an ammonium sulfate precipitation has been performed prior to the chromatographic step.

Generally, conventional RPLC columns are simply too hydrophobic (i.e. give too much retention) to be used with purely aqueous mobile phases, hence new materials have been designed specifically for HIC of biopolymers. Stationary phase materials for HIC have larger (e.g. 1000 Å) pore sizes than RPLC columns and are prepared by low density bonding of C_{18}, C_8, phenyl, C_5 or diol functional groups to polyamide-coated silica or gels, such as Sepharose or polymethacrylate [22]. The low density bonding produces stationary phases which are only lightly functionalized, resulting in low to moderate hydrophobicity, while the large pore diameter permits rapid protein diffusion. Figure 6.43 shows the purification of proteins from wheat germ extract using a phenyl functionalized polymethacrylate HIC column. HIC offers the advantages of high sample capacity and excellent recovery of protein activity. The use of resin-based HIC columns is preferable owing to the fact that they can be cleaned with sodium hydroxide. The main limitations of HIC are related to the moderate resolution of the technique and that protein fractions typically have to be desalted prior to the next step in a purification scheme.

6.4.2   Affinity chromatography

This chromatographic separation mode is a more recent development in HPLC which utilizes the unique biological specificity of an interaction between a protein molecule and an immobilized ligand [2]. The ligand, which exhibits a specific binding

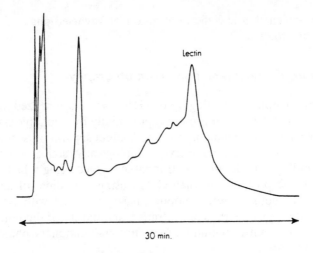

Fig. 6.43. Purification of proteins from wheat germ extract using hydrophobic interaction chromatography. The column was a Waters Protein HIC PH-814 used with a linear gradient from 0.1 M sodium phosphate:1.7 M ammonium sulfate, pH 7.0 to 0.1 M sodium phosphate, pH 7.0. Detection was by direct u.v. at 280 nm. Reprinted from *The Waters Chromatography Handbook* (1993), Millipore, Milford, with permission.

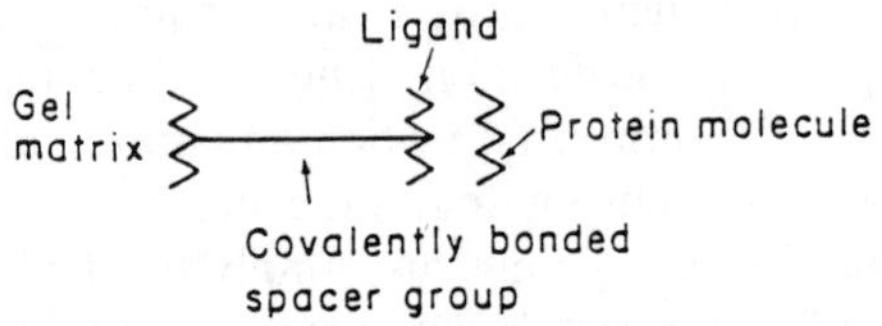

Fig. 6.44. Schematic representation of affinity chromatography. Reprinted from Sewell and Clarke (1987) *Chromatographic Separations*, John Wiley & Sons, Chichester, with permission.

affinity for a given protein or functional group, is covalently bonded to a gel matrix to form the stationary phase [51]. A buffered mobile phase is used and those molecules with a specific affinity for the ligand become bound and hence are retained, while unbound material is eluted from the column. The composition or pH of the mobile phase can then be altered, via a gradient, to weaken the specific interaction between the bound molecules and the ligand, so that the protein is then released from the stationary phase and eluted from the column. This separation mode allows proteins to be recovered with high yields of bioactivity and good purity, hence it is often used for protein purification in the microgram to gram scale. The affinity process can be represented schematically, as shown in Fig. 6.44. The specific interaction between the ligand and the protein molecule is often referred to as a key and lock mechanism [2].

The gel–ligand bond must be stable under the experimental conditions, while the ligand–sample bond reaction must be specific but reversible. This ligand–sample bond forms from a biospecific interaction, such as an antibody–antigen reaction, a

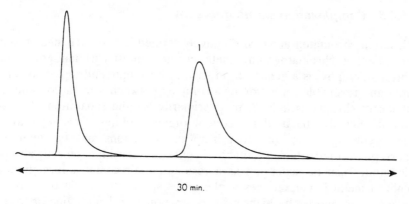

Fig. 6.45. Purification of carbonic anhydrase using affinity chromatography. The column was a Waters Protein-Pak epoxy-activated affinity column with a coupled sulfanilamide ligand. A gradient from 100 mM Tris sulfate:200 mM sodium sulfate, pH 8.7 to 200 mM potassium thiocyanate:50 mM Tris sulfate, pH 6.5 was used. Reprinted from *The Waters Chromatography Handbook* (1993), Millipore, Milford, with permission.

chemical reaction, such as the binding of *cis*–diol groups to boronate, or other types of interactions, such as the attraction of albumin to Cibaron Blue F3G-A dye. An enormous range of affinity columns is available for use with HPLC. These columns typically have a biospecific ligand covalently bonded to a rigid, highly porous, hydrophilic polymer. Examples of bonded ligands include avidin, bovine serum albumin, biotin, dextran sulfate, gelatin, heparin, lysine, protein A, protein G or phenylalanine [15]. In addition, a vast range of affinity chromatography media is also available for low-pressure chromatography of proteins. These materials typically utilize the cyanogen bromide method to attach the ligand to a gel, such as agarose, sepharose, methylcellulose or acrylamide [52].

In some instances, where the gel matrix would interfere with the specific bonding properties of the ligand, the matrix and the ligand are separated by the inclusion of a spacer group [53]. HPLC columns of this type are packed typically with silica gel which has been epoxy activated using, for instance, a glycidoxypropyl functionality. This particular epoxy-activating reaction results in a hydrophilic bonding layer on the silica with a seven-atom spacer arm. The epoxy-activated surface can then be used to immobilize a wide range of ligands via a covalent linkage with amino, hydroxyl or sulfhydryl groups on the protein, using simple coupling procedures. Figure 6.45 shows the isolation of carbonic anhydrase from bovine haemolysate using an epoxy-activated silica gel with a coupled sulfanilamide ligand. The bovine haemolysate sample (24 mg) yielded 0.6 mg of carbonic anhydrase, with a 100% recovery of the esterase activity. The major limitations of affinity chromatography are the high price of the stationary phases, potential ligand leakage from the column and the fact that the columns cannot be cleaned using aggressive chemicals, e.g. sodium hydroxide. Despite these limitations, this mode of chromatography is widely used for protein purification because of its very specific separation selectivity. Additional benefits include high sample capacity and good recovery of protein activity.

6.4.3 Complexation chromatography

Complexation chromatography, alternatively termed ligand exchange chromatography, coordination chromatography, metal chelate affinity chromatography or chelation chromatography, is a generic term used to encompass all separations involving the rapid and reversible formation of a complex between a metal ion and a ligand. This mode of chromatography is used primarily for the separation of metal ions, however it may also be used for certain organic solutes. This separation mode operates according to one of two principles. Metal ions may be separated on a stationary phase onto which a suitable ligand has been immobilized; alternatively, organic solutes can be separated as a result of their complex formation with an immobilized metal ion. Examples of the latter approach include the separation of mono- and disaccharides by hydroxyl coordination to calcium, zinc or lead cations immobilized on ion exchangers [22] and the concentration of tetracycline residues from milk on metal chelating Sepharose [54].

Numerous complexing stationary phases have been synthesized using styrene-divinylbenzene polymers or silica as the support material for the separation of metal ions. The ligand is bonded chemically to the support using an appropriate reaction, such as the siloxane reaction with silica as discussed in Section 6.2.2. Examples of the ligands which can be bound in this way include iminodiacetate (referred to as Chelex 100), propylene-diaminetetraacetate, diketones (e.g. trifluoroacetoacetate), 8-hydroxyquinoline, hydroxamic acids, maleic acid, dithiocarbamates, phenylhydrazones and dithiozone [1]. Solute retention can be manipulated through the addition of a competing ligand to the mobile phase or by varying the eluent pH. The key factor in the success of such materials as chromatographic stationary phases is the

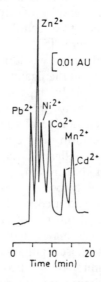

Fig. 6.46. Separation of metal ions on a stationary phase formed by bonding dithiozone functionalities to silica gel. The eluent used was 15 mM tartrate at pH 4.0. Detection was by spectrophotometry after post-column reaction with 4-(2-pyridylazo)-resorcinol. Reprinted from Faltynski and Jezorek (1986) *Chromatographia*, **22**, 5, with permission.

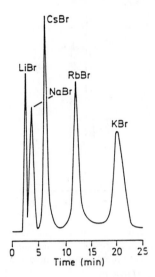

Fig. 6.47. Separation of alkali metal cations on a crown ether stationary phase. A poly(benzo-15-crown-5)-modified silica was used as stationary phase with water as eluent. Detection was by conductivity. Reprinted from Nakajima *et al.* (1983) *Anal. Chem.*, **55**, 463, with permission.

rate at which the metal–ligand complex is formed and dissociated. Slow complexation rates lead to poor peak shape and a recent evaluation of the literature suggests that most ligands give unacceptably slow rates of reaction, hence chromatograms obtained using this approach are typically characterized by very broad peaks [1]. Figure 6.46 shows the separation of metal ions on dithizone functionalized silica [55].

An interesting alternative to the above approach is to use a stationary phase to which a crown ether has been chemically bound. Crown ethers are cyclic compounds which possess an inner cavity, generally consisting of oxygen atoms linked by ethylene bridges. The chromatographic utility of such stationary phases rests in their ability to complex cations of a specified size. Hence, altering the size of the cavity in the crown ether alters the selectivity of the stationary phase. The binding of metal ions on a particular crown ether stationary phase depends on the size of the cation, the nature of the associated anion and the organic modifier content of the mobile phase [1]. Water alone can be used as a mobile phase in this form of chromatography, typically in conjunction with conductivity detection. Figure 6.47 shows the separation of alkali metal cations on a crown ether functionalized silica stationary phase [56].

6.4.4 Chiral chromatography

Chiral chromatography refers to the separation of optical isomers, or enantiomers, by HPLC. There are a number of different approaches which enable the separation of such compounds [4]. Enantiomers may be separated with a chiral bonded stationary phase, through the use of chiral additives in the mobile phase, or by derivatization of the sample to form diastereometric products of the two enantio-

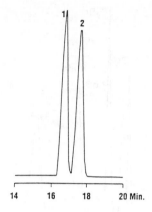

Fig. 6.48. Separation of D- and L-tryptophan on a chiral bonded silica stationary phase. An Astec Cyclobond III column was used with an eluent of 15% methanol and 0.85% triethylamine at pH 5.0. Solute identities: 1, L-tryptophan; 2, D-tryptophan. Reprinted from *Alltech Chromatography Catalogue 300* (1993), Alltech, Deerfield, with permission.

mers, which are then separated by conventional HPLC. An example of the latter approach used for the separation of D- and L- amino acids is to derivatize the sample first with camphorsulfonyl chloride to form the diastereomers and then with *p*-nitrobenzylbromide to introduce the chromophore group. The resulting derivatives can then be separated using normal-phase chromatography on an amino-bonded silica column with u.v. detection at 254 nm [57].

While the above approach was often used for enantiomer separations in the past, most recent interest in chiral separations lies in the use of chiral bonded stationary phases. A wide range of chiral stationary phases is now available commercially. These columns are typically silica gel with bonded optically active functionalities, such as (either D or L) phenyl urea, naphthyl urea, phenyl glycine or leucine. These stationary phases can be used for the separation of enantiomers of compounds including mandelic acid analogues, alkyaryl sulfoxides, propranolol analogues, aryl-substituted phthalides, aryl-acetamides, etc. In addition, cyclodextrins bonded to silica gel also permit the separation of optical isomers, via the formation of inclusion complexes within the cyclodextrin cavity. The D and L enantiomers of a number of compounds form inclusion complexes of different strengths within the cavity and are retained to varying degrees, enabling their separation. Enantiomers of pharmaceutically active compounds such as tryptophan, ibuprofen, atenolol and norgestol can be chromatographed with cyclodextrin-functionalized silica stationary phases. Figure 6.48 shows the separation of D- and L-tryptophan on an alpha-cyclodextrin bonded silica stationary phase [58].

The major drawback of chiral bonded stationary phases is that a different stationary phase is required for each new class of optical isomers if good separation is to be achieved. The addition of optically active compounds to the mobile phase appears to be a more flexible option and additives such as cyclodextrins, copper complexes of L-proline and nickel complexes of L-proline-*n*-octylamide have been used for the separation of chiral compounds [59]. The disadvantage of this approach

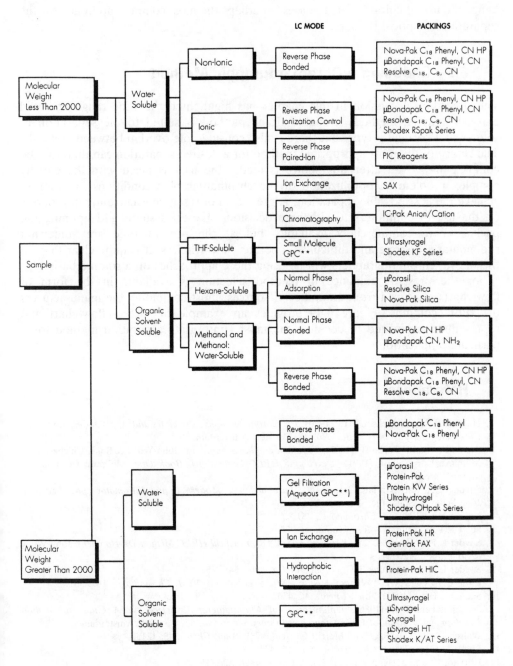

Fig. 6.49. Guide for selection of chromatographic separation modes and stationary phases. Courtesy of Waters.

is that the necessary resolution of particular enantiomeric pairs is rarely achieved [4], hence the use of chiral bonded phases is perhaps the most common approach for the separation of optical isomers.

6.5 Choosing a Chromatographic Method

The preceding discussion in this chapter has highlighted the wide range of liquid chromatographic methods available. Each method is suited to the separation of solutes of a particular type, but there is also considerable overlap between methods. The choice of the most appropriate method for a desired separation can therefore be complex and, even after this choice is made, the user is faced with the equally complex problem of optimization of the chromatographic conditions in order to achieve the best possible separation. A great deal of research is currently in progress on the use of computers to aid these decisions. Expert systems and optimization packages are commercially available, but at the present time are somewhat rudimentary in scope and difficult to operate. Nevertheless, it is quite clear that, in the future, chromatographers will be using these approaches on a routine basis.

For the present, guidance in method selection is available in the form of flow-charts produced from a number of sources, most commonly the manufacturers of HPLC columns. Figure 6.49 provides an example of such a flow-chart and represents a useful aid in deciding the most appropriate separation method for a particular application.

References

1. Haddad P.R. and Jackson P.E. (1990). *Ion Chromatography: Principles and Applications, Journal of Chromatography Library Series, No. 46*. Elsevier, Amsterdam.
2. Sewell P.A. and Clarke B. (1987). *Chromatographic Separations*. John Wiley & Sons, Chichester.
3. Millipore Corporation (1991). *Developing HPLC Separations, Book One*. Millipore Corporation, Milford.
4. Snyder L.R. and Kirkland J.J. (1979). *Introduction to Modern Liquid Chromatography*, 2nd edn. John Wiley & Sons, Inc., New York.
5. Yashin Y.I. (1982). *J. Chromatogr.*, **251**: 269.
6. Scott R.P.W. (1982). *Adv. Chromatogr.*, **20**: 167.
7. Snyder L.R., Glach J.L. and Kirkland J.J. (1988). *Practical HPLC Method Development*. John Wiley & Sons, Inc., New York.
8. Snyder L.R. and Schunk T.C. (1982). *Anal. Chem.*, **54**: 1764.
9. Little C.J., Dale A.D., Whatley J.A. and Wickings J.A. (1979). *J. Chromatogr.*, **169**: 381.
10. Saunders D.L. (1974). *Anal. Chem.*, **46**: 470.
11. Schoenmakers P.J. (1986). *Optimization of Chromatographic Selectivity: A Guide to Method Development, Journal of Chromatography Library Series, No. 35*. Elsevier, Amsterdam.
12. Walsch J.T., Chalk R.C. and Merritt C., Jr. (1973). *Anal. Chem.*, **45**: 1215.
13. Locke D.C. (1974). *J. Chromatogr. Sci.*, **12**: 433.
14. Horvath C. and Melander W. (1977). *J. Chromatogr. Sci.*, **15**: 393.
15. Millipore Corporation (1992). *Waters Sourcebook of Chromatography*. Millipore Corporation, Milford.
16. Glach J.L. and Kirkland J.J. (1983). *Anal. Chem.*, **55**: 319A.
17. Snyder L.R. (1974). *J. Chromatogr.*, **92**: 223.
18. Schoenmakers P.J., Billiet H.A.H. and de Galan L. (1981). *J. Chromatogr.*, **205**: 13.

19. Poppe H. and Kraak J.C. (1983). *J. Chromatogr.*, **282**: 399.
20. Millipore Corporation (1987). *Waters Sourcebook of Chromatography*. Millipore Corporation, Milford.
21. Millipore Corporation (1991). *Waters Column*. Millipore Corporation, Milford.
22. Millipore Corporation (1993). *The Waters Chromatography Handbook*. Millipore Corporation, Milford.
23. Bidlingmeyer B.A. (1980). *J. Chromatogr. Sci.*, **18**: 525.
24. Slingsby R.W. (1986). *J. Chromatogr.*, **371**: 373.
25. Billiet H.A.H., Drouen A.C.J.H. and De Galan L. (1984). *J. Chromatogr.*, **316**: 231.
26. Paired Ion Chromatography, an Alternative to Ion-Exchange, *Waters Bulletin F61*, May (1976).
27. Tomlinson E., Jefferies T.M. and Riley C.M. (1978). *J. Chromatogr.*, **159**: 315.
28. Cassidy R.M. and Elchuk S. (1982). *Anal. Chem.*, **54**: 1558.
29. Bidlingmeyer B.A., Deming S.N., Price W.P., Jr., Sachok B. and Petrusek M. (1979). *J. Chromatogr. Sci.*, **186**: 419.
30. Horvath C., Melander W., Molnar I. and Molnar P. (1977). *Anal. Chem.*, **49**: 2295.
31. Wheals B.B. (1983). *J. Chromatogr.*, **262**: 61.
32. Schwedt G. (1979). *Chromatographia*, **12**: 613.
33. Crommen J., Schill G., Westerlund D. and Hackzell L. (1987). *Chromatographia*, **24**: 252.
34. Cassidy R.M. and Elchuk S. (1983). *J. Chromatogr. Sci.*, **21**: 454.
35. Hilton D.F. and Haddad P.R. (1986). *J. Chromatogr.*, **361**: 141.
36. Cassidy R.M. and Elchuk S. (1983). *J. Chromatogr.*, **262**: 311.
37. Kraak J.C., Jonker K.M. and Huber J.F.K. (1977). *J. Chromatogr.*, **142**: 671.
38. Hoffman N.E. and Liao J.C. (1977). *Anal. Chem.*, **49**: 2231.
39. Cantwell F.F. and Puon S. (1979). *Anal. Chem.*, **51**: 623.
40. Barkley D.J., Blanchette M., Cassidy R.M. and Elchuk S. (1986). *Anal. Chem.*, **58**: 2222.
41. Samuelson O. (1953). *Ion Exchangers in Analytical Chemistry*. Wiley, New York.
42. Paterson R. (1970). *An Introduction to Ion Exchange*. Heyden, London. p. 30.
43. Diamond R.M. (1963). *J. Phys. Chem.*, **67**: 2513.
44. Barron R.E. and Fritz J.S. (1984). *J. Chromatogr.*, **316**: 201.
45. Barron R.E. and Fritz J.S. (1984). *J. Chromatogr.*, **284**: 13.
46. Small H., Stevens T.S. and Bauman W.C. (1975). *Anal. Chem.*, **47**: 1801.
47. Wheaton R.M. and Bauman W.C. (1953). *Ind. Eng. Chem.*, **45**: 228.
48. Kihara K., Rokushika S. and Hatano H. (1987). *J. Chromatogr.*, **410**: 103.
49. Woo D.J. and Benson J.R. (1984). *Am. Clin. Prod. Rev.*, January: 20.
50. Hertjen S. (1981). *Adv. Chromatogr.*, **19**: 59.
51. Walters R.R. (1985). *Anal. Chem.*, **57**: 1099A.
52. Sigma (1989). *Affinity Chromatography Media*. Sigma Chemical Company, St. Louis.
53. Parikh I. and Cuatrecasas P. (1985). *C. & E. News*, August: 17.
54. Carson M.C. (1993). *J. Assoc. of Anal. Chem.*, **76**: 329.
55. Faltynski K.H. and Jezorek J.R. (1986). *Chromatographia*, **22**: 5.
56. Nakajima M., Kimura K. and Shono T. (1983). *Anal. Chem.*, **55**: 463.
57. Furukawa H., Mori Y., Takeuchi Y. and Ito K. (1977). *J. Chromatogr.*, **136**: 428.
58. Alltech Associates (1993). *Alltech Chromatography Catalogue 300*. Alltech Associates, Deerfield.
59. Hancock W.S. and Sparrow J.T. (1984). *HPLC Analysis of Biological Compounds, A Laboratory Guide*. Marcel Dekker, Inc., New York.

Bibliography

General HPLC

Ahuja S. (1992). *Trace and Ultratrace Analysis by HPLC*. John Wiley & Sons, New York.
Dorsey J.G., Foley J.P., Cooper W.T., Barford R.A. and Barth H.G. (1992). Column liquid chromatography: theory and methodology. *Anal. Chem.*, **64**: 352R.
Harvey M.C. and Stearns S.D. (1984). In *Liquid Chromatography in Environmental Analysis*, Lawrence J.F. (Ed.). Humana, Clifton.

Lindsay S. (1992). *High Performance Liquid Chromatography*, 2nd edn. John Wiley & Sons, Chichester.

Meyer V.R. (1985). High performance liquid chromatographic theory for the practitioner, *J. Chromatogr.*, **334**: 197.

Millipore Corporation (1991). *Developing HPLC Separations, Books One and Two*, Millipore Corporation, Milford.

Poole C.F. and Poole S.K. (1991). *Chromatography Today*. Elsevier, Amsterdam.

Snyder L.R. and Kirkland J.J. (1979). *Introduction to Modern Liquid Chromatography*, 2nd edn. John Wiley & Sons, Inc., New York.

Snyder L.R., Glach J.L. and Kirkland J.J. (1988). *Practical HPLC Method Development*. John Wiley & Sons, Inc., New York.

Schoenmakers P.J. (1986). *Optimization of Chromatographic Selectivity: A Guide to Method Development*, *Journal of Chromatography Library Series, No. 35*. Elsevier, Amsterdam.

Separation modes

Berensden G.E., Pikaart K.A. and De Galan L. (1980). Preparation of various bonded phases for HPLC using monochlorosilanes. *J. Liquid Chromatogr.*, **3**: 1437.

Cazes J., Editor (1981). *Liquid Chromatography of Polymers and Related Materials, Chromatographic Science Series, No. 19*. Marcel Dekker, Inc., New York.

Cooke N.H.C. and Olsen K. (1980). Some modern concepts in reversed-phase liquid chromatography on bonded alkyl stationary phases. *J. Chromatogr. Sci.*, **18**: 512.

Haddad P.R. and Jackson P.E. (1990). *Ion Chromatography: Principles and Applications, Journal of Chromatography Library Series, No. 46*. Elsevier, Amsterdam.

Hancock W.S. and Sparrow J.T. (1984). *HPLC Analysis of Biological Compounds: A Laboratory Guide, Chromatographic Science Series, No. 26*. Marcel Dekker, Inc., New York.

Heftmann E., Editor (1992). *Chromatography, 5th Edition*, Parts A and B. Elsevier, Amsterdam.

Horvath C. and Melander W. (1977). Liquid chromatography with hydrocarbonaceous bonded phases; theory and practice of reversed phase chromatography. *J. Chromatogr. Sci.*, **15**: 393.

Issaq H.J. (1980). Effect of alkyl chain length of bonded silica phases on separation, resolution and efficiency in high performance liquid chromatography. *J. Liquid Chromatogr.*, **4**: 1917

Krstulovic A.M. and Brown P.R. (1982). *Reversed-Phase High Performance Liquid Chromatography*. John Wiley & Sons, Inc., New York.

Scott R.P.W. (1980). The silica gel surface and its interactions with solvent and solute in liquid chromatography. *J. Chromatogr. Sci.*, **18**: 297.

Sewell P.A. and Clarke B. (1987). *Chromatographic Separations*. John Wiley & Sons, Chichester.

Soczewinski E. and Waksmundzka-Hajnos M. (1980). Effect of type and concentration of organic modifier in aqueous eluents on retention in reversed phase systems. *J. Liquid Chromatogr.*, **3**: 1625.

Szepesi G. (1992). *How to Use Reverse-Phase HPLC*. VCH Publishers, New York.

Yau W.W., Kirkland J.J. and Bly D.D. (1979). *Modern Size-Exclusion Liquid Chromatography*. John Wiley & Sons, Inc., New York.

Supercritical Fluid Chromatography 7

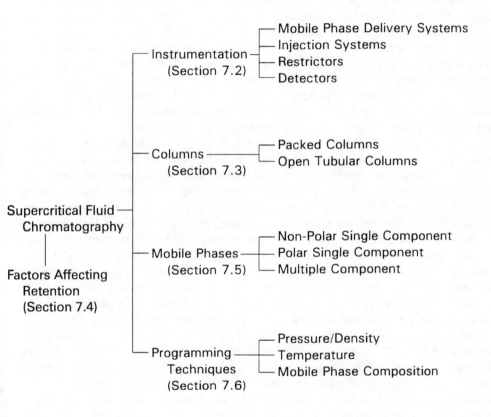

Supercritical Fluid Chromatography

Factors Affecting Retention (Section 7.4)

Instrumentation (Section 7.2)
- Mobile Phase Delivery Systems
- Injection Systems
- Restrictors
- Detectors

Columns (Section 7.3)
- Packed Columns
- Open Tubular Columns

Mobile Phases (Section 7.5)
- Non-Polar Single Component
- Polar Single Component
- Multiple Component

Programming Techniques (Section 7.6)
- Pressure/Density
- Temperature
- Mobile Phase Composition

7.1   Introduction

Supercritical fluid chromatography (SFC) refers to a mode of chromatography in which a fluid above its critical temperature and critical pressure is used as the mobile phase. There has been some debate over the suitability of this description of SFC [1], but it probably represents the only workable definition. The fact that it includes some GC separations and possibly some high-temperature LC separations is not a shortcoming but emphasizes the complementary nature of SFC with GC and HPLC. Indeed, SFC was termed hyperpressure GC in some of the earlier literature, and various aspects of SFC were treated in both the GC and LC sections of *Fundamental Reviews of Analytical Chemistry* prior to 1990 [2]. This method of chromatography was first reported in 1962 [3] but it was not until the 1980s that much developmental work and commercial instrumentation appeared. SFC is now a well-established if not, as yet, a common analytical technique [2, 4–8]. Developments are also occurring in the parallel field of supercritical fluid extraction (SFE) for both analytical scale (see Chapter 8) and preparative scale extractions.

What is a supercritical fluid and what properties make it attractive as a mobile phase? This can be understood by reference to the phase diagram of a substance near, and beyond, the region of gas:liquid equilibrium. For a pure compound, the phase diagram has the general appearance shown in Fig. 7.1. It is commonly known that such diagrams indicate the phase (solid, S; liquid, L; gas, G) present for any combination of pressure and temperature. Along the solid lines, equilibrium exists between the adjacent phases, that is the phases coexist, while at the triple point (T) all three phases solid, liquid and gaseous coexist. Crossing one of the solid lines results in a phase change during which the physical properties (density, viscosity, diffusivity, etc.) change abruptly. The point C, the critical point, which is defined by the critical temperature (T_c) and critical pressure (P_c) for each substance is of particular interest in this discussion since, at this point and beyond, there is no distinct liquid phase and the fluid is said to be in a supercritical state, that is, it is a supercritical fluid. The reason for this nondistinction is that above the critical point the kinetic energy of the molecules is so great compared with the condensing influence of intermolecular forces that the molecules can no longer cling together to form a distinct liquid phase. A supercritical fluid, however, does not represent a distinct phase since its properties, e.g., density, viscosity, refractive index, etc. vary smoothly with temperature and pressure. It can be of a low density indistinguishable from the gaseous state, or of a density comparable to that of liquids, or anywhere in between. Notice that in Fig. 7.1, there is no complete barrier between the gas and liquid phases. It is therefore possible to move from the gaseous state to the liquid state and vice versa without traversing the G–L phase boundary. This is accomplished by following a path which passes above the critical point.

The critical parameters $(T_c$ and $P_c)$ for compounds depend on van der Waals forces characteristic for molecules of the compound. The critical temperature defines the temperature above which a liquid cannot be liquefied regardless of the applied pressure. The minimum pressure required to liquefy a substance, at its critical temperature, is the critical pressure. The values of such parameters for some common substances are given in Table 7.1. Substances most suitable as mobile phases for SFC are those that exhibit low critical parameters, that is, have a

Table 7.1   Properties of some substances that may be suitable for supercritical fluid chromatography [6].

Compound	Normal boiling point (°C)	Critical data		
		T_c (°C)	P_c (MPa)	ρ_c (g cm^{-3})
Carbon dioxide	−78.5	31.3	7.38	0.448
Nitrous oxide	−89	36.5	7.23	0.457
Sulfur hexafluoride		45.5	3.76	0.23
Ammonia	−33.4	132.3	11.27	0.24
Methanol	64.7	240.5	7.99	0.272
Ethanol	78.4	243.4	6.38	0.276
Isopropanol	82.5	235.3	4.76	0.273
Ethane	−88	32.4	4.89	0.203
n-Propane	−44.5	96.8	4.25	0.220
n-Butane	−0.05	152.0	3.80	0.228
n-Pentane	36.3	196.6	3.37	0.232
Ethoxyethane	34.6	193.6	3.68	0.267
Xenon	−107.1	16.6	5.89	1.105

1 atm = 1.01×10^5 Pa.

supercritical region which can easily be accessed by contemporary chromatographic pumps and ovens (up to 100 MPa and 400°C). Such fluids must also be inert and otherwise compatible with the chromatographic system. Indeed, most work currently carried out in SFC uses carbon dioxide (with and without polar additives) as mobile phase, with other fluids being used to a much lesser extent.

It is not immediately apparent from Fig. 7.1, why supercritical fluids should have any distinct advantages over gases or liquids as mobile phases. The answer lies in the solubility properties, physical properties (viscosity and diffusion coefficients) and detector compatibility of the supercritical fluid, which may combine to make SFC more suitable for a given application than either GC or HPLC. In GC, separation

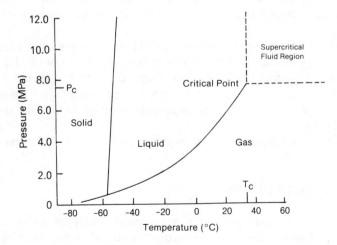

Fig. 7.1.   Phase diagram for carbon dioxide.

depends first on having species that are volatile, and second, on differences in the affinity of analyte species for the stationary phase versus volatility. If an analyte does not exhibit sufficient vapour pressure within the temperature range of GC, or if the compound is thermally unstable, then GC may not be suitable, although it may be possible to resort to chemical derivatization to overcome this problem. In LC and SFC, volatility is not a prerequisite and separation is influenced by differences in analyte affinity for the stationary and mobile phases. Thus, for a given column, it is possible to vary the mobile phase composition to affect a separation, a feature which is not available in GC. A unique feature of SFC is that mobile phase affinity for an analyte is also a function of fluid density; this is a distinctive property of supercritical fluids. It is thus possible to vary the density (in practice, this is accomplished by varying the pressure) to elute compounds with widely differing capacity factors. A disadvantage of SFC is that highly polar systems and particularly aqueous systems are difficult to handle, partly because of the high critical parameters required but also because of the extreme chemical reactivity, for example, of water toward analyte and chromatographic systems alike.

The diffusion coefficients and viscosities of supercritical fluids, which are intermediate between that of gases and liquids (see Table 2.3), have an important bearing on the type and dimensions of suitable chromatographic columns, and on the optimum mobile phase velocity that can be used in SFC. Efficient HPLC and SFC columns tend to have closely similar H (height equivalent to a theoretical plate) values of about $0.01\,mm$; however, the greater solute diffusion coefficients of supercritical fluids means that the resistance to mass transfer in the Van Deemter equation is lower in SFC than HPLC. This, together with the lower viscosities of supercritical fluids, means that greater mobile phase velocities may be used in SFC than in HPLC, and separations may be effected in a shorter time. The consequence is that more practical open tubular columns may be used in SFC than in HPLC. Furthermore, because supercritical fluids are considerably less viscous than liquids, pressure drops across a column are small, especially in open tubular columns, and practical SFC columns of $10–20\,m$ length can be used. Thus, complex mixtures which cannot be suitably analysed by either GC (because of involatility or instability) or HPLC (because of complexity and long retention) may be more effectively analysed by SFC.

Another advantage of SFC over GC and HPLC concerns the greater range of detectors that may be used in SFC, which includes most, if not all, HPLC detectors and also most GC detectors. Hence, with supercritical CO_2 as mobile phase, it is possible to use either an FID or u.v. detector to advantage, depending on the application, whereas only the FID is commonly available in GC and only the u.v. is available in HPLC.

A more detailed discussion of these aspects is given in the following sections.

7.2   Instrumentation

The modules in SFC instrumentation are basically adaptations of those used in HPLC and GC. Current commercial instruments are offered as either dedicated SFC systems or add-on modules for existing GC or HPLC systems. A schematic diagram

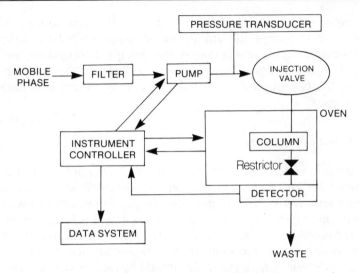

Fig. 7.2.   Schematic diagram of a supercritical fluid chromatograph.

of a basic SFC system is shown in Fig. 7.2. This comprises a mobile phase delivery system, an injector, an oven, a restrictor, a detector and a control/data system. The various modules are discussed in detail in this section. The type, arrangement and characteristics of such modules depend on whether an open tubular or a packed column is used and on whether a liquid phase (i.e., HPLC) or a gas phase (i.e., GC) detector is used.

7.2.1   Mobile phase delivery systems

The most important component in this system is the pump, which functions not only in maintaining a suitable, precise mobile phase flow but also, in conjunction with the restrictor and control system, to apply the necessary pressure to keep the mobile phase in the supercritical state under precisely controlled conditions. This requires not only accurate and precisely controlled pressure but also accurate and precisely controlled temperature. For this reason, the oven temperature must be controlled to within ±1°C with minimal temperature gradients throughout the oven. Mobile phase, e.g. CO_2, N_2O, *n*-pentane, etc., usually enters the pump as a liquid from a cylinder (which may be pressurized under helium) and is pumped to the column, where it is heated to the supercritical state. On-line filters and activated carbon or alumina adsorption cartridges are usually employed to purify the mobile phase prior to entering the pump. Pump design and seals must be selected to tolerate the very high solvent strengths and pressures (up to 60 MPa or more) encountered in such systems. Either syringe pumps (1–10 μl min^{-1}) or reciprocating pumps (1–10 ml min^{-1}) are used, depending on whether open tubular or packed columns, respectively, are utilized. Dual pump systems may also be used for producing composition gradients, as in HPLC. Syringe pumps deliver pulseless fluid flow, an important requirement in SFC, but have a fixed delivery volume (typically 10–100 ml) and require a refilling cycle. Reciprocating pumps, however, do not require a filling cycle but require pressure-dampening devices, and probably also cooling, to prevent vapour locking. These pumps also are not suited to open tubular columns and are less suited to pressure programming than syringe pumps. In operation, mobile phase enters the pumps as a liquid and is pumped through the injector at room temperature to the column at an elevated temperature, where it is usually in the supercritical state.

7.2.2 Injection systems

Sample introduction in SFC [9] is based on the high-pressure rotary valve system commonly found on HPLC instrumentation. As indicated above, the injector is usually at room temperature and samples are commonly introduced dissolved in an organic solvent. In a typical operation, the sample solution is loaded in the sampling loop at ambient conditions then, when the valve is actuated to the inject position, the sample is swept onto the column in a high-pressure mobile phase. Methods involving direct injection from a supercritical fluid extraction system have also been developed. On conventional or narrow-bore packed columns (i.e., 1 mm i.d. or larger) direct valve injection is usually a straightforward method of sample introduction, and problems are encountered with solute or solvent incompatibility with the mobile phase only. However, open tubular columns, as in GC and microcolumn HPLC, are problematic when it comes to sample injection. The difficulty stems from the very small volumes of such columns. A typical 20 m column with an internal diameter of 0.05 mm has an internal volume of only about 100 μl, and so the sample volume and dead volume of fittings has to be quite small to prevent serious loss of column efficiency. It is possible to calculate the injector and detector volumes which would produce a 1% reduction in resolution. For the 20 m column this corresponds to 0.05 μl. Thus, the volume of the injector would have to be of this magnitude and injected sample volumes would have to be smaller than this, say 0.01 μl or less. Commercial rotary injection valves with a sample loop of 60 nl capacity are available; however, this presents problems in column plugging and requires large sweep volumes to clear the loop.

The main problems associated with injection devices for open tubular columns are: (i) peak distortion and loss of resolution; (ii) discrimination and memory effects; (iii) poor reproducibility; and (iv) lack of concentration sensitivity, even with high absolute detector sensitivity owing to the small quantity of sample analysed. Much research has been carried out to overcome these limitations. The main approaches have involved the use of: (i) splitting devices, and (ii) solvent venting, frequently with the use of solute focusing techniques.

Common splitting devices are the split/splitless valve injector and the timed-split injector. The principle of operation is similar to that found in open tubular GC and is illustrated in Fig. 7.3. In the split/splitless system (also called dynamic split system) the valve is turned to the sampling position and sample solution is injected under ambient conditions into the sample loop. When the valve is switched to the load position, sample is forced into the splitter by the flow of the high-pressure mobile phase stream. At this time, the split vent is opened and the sample stream is split, with usually only a small fraction entering the column, the major part exiting from

(a)

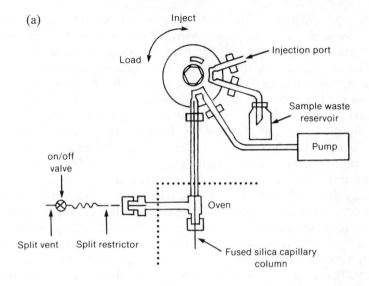

(b)

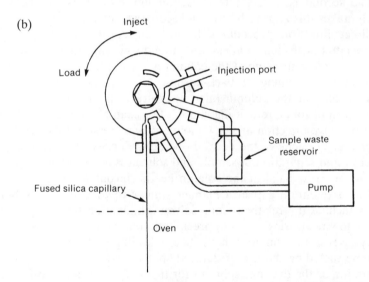

Fig. 7.3. Schematic diagram of a split/splitless valve injector (a) and a timed-split injector (b) for open tubular column supercritical fluid chromatography.

the split vent via the split restrictor. The vent valve is then shut and mobile phase then continues to flow to the column, sweeping the subsample through. The split ratio is determined by the relative flow impedance of the vent and column and is adjusted by adjusting the split vent valve. If the vent valve is kept shut then a splitless mode of operation is used (hence the term split/splitless). With a relatively large sample volume, however, a splitless mode of operation will flood the column and give rise to a poor chromatogram.

The other type of split injector is the timed-split injector. In this method, the splitting of the sample plug is accomplished by rapid switching of the sampling valve from the load to the inject position and then back again sufficiently fast so that only a fraction of the sample enters the column. Electronic or pneumatic actuators capable of moving the valve as fast as 10 ms are suitable for this purpose. The proportion of sample injected can then be adjusted by changing the valve actuation time. For example, sample volumes corresponding to 1–2 nl can be delivered to the column by setting the rotor travel time of an electronically actuated 0.2 μl sample loop valve for 70 ms. Reproducibility in this method will depend on the precision available from the electromechanical system involved (4% has been reported). An advantage of this procedure is that, since only a fraction of the sample loop volume is directed to the column, the part closer to the wall which suffers viscous drag (and is not totally displaced for about 5–10 bed volumes of mobile phase) does not enter the column and therefore does not contribute to band broadening. A similar effect can be obtained by the previous method if the split vent is also shut electronically after one bed volume of sample is displaced into the splitter. Both methods suffer from two potentially undesirable effects. One is the perturbation effect of solvent on chromatographic elution or detection (e.g., the solvent may give a larger detector response). The second is high solute detection limits (as most of the sample is vented) which makes this approach unsuitable for trace analysis.

Various direct injection procedures have been developed to overcome the problems associated with split/splitless and timed-split injection. These methods invariably involve solute focusing at the column head and frequently have provision for solvent venting, skimming or back-flushing [10–12] and manipulation of the mobile phase density at the column inlet. Pressure trapping [13, 14] is the most popular mechanism of solute refocusing in SFC. For example, direct injection can be accomplished by valve injection at a low mobile phase pressure followed by a rapid increase of pressure to achieve the desired operating density. This approach can be effective where rapid separation of sample and solvent is achieved and the sample is precipitated as a narrow band at the column head. Uncoated fused silica tubing is usually used as a retention gap in a manner similar to that in open tubular GC. Disadvantages include the relatively long solvent elimination times (up to 10 min.) and potentially low recoveries owing to precipitation in the injector of solutes with a low solubility. Several techniques have been developed to reduce the solvent elimination time including the use of delayed split injection. This method uses the same configuration as the dynamic split but for the addition of an on/off valve in the vent line.

7.2.3 Restrictors

In order to maintain fluid flow and supercritical fluid conditions along the length of the column, a restrictor is placed in the flow path. Its position in the system depends on the type of detector in use. For those detectors requiring a dense mobile phase (e.g., optical detectors), the restrictor is placed after the detector. Otherwise, it is placed between the column and detector (e.g., with flame-type detectors) in order to achieve flow/pressure restriction and rapid decompression to atmospheric pressure immediately before detection [15]. The manner in which decompression is effected is

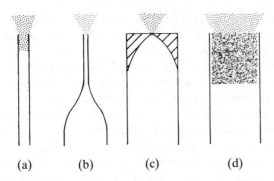

Fig. 7.4. Different designs of flow restrictor: (a), linear; (b), tapered; (c), integral; and (d), frit.

more critical for the latter detectors. In this case, the restrictor must not only transfer the eluent from the supercritical phase to the gas phase but it must achieve this without compromising column efficiency by introducing excessive dead volume or causing the sample components to condense.

The ideal restrictor does not exist but it would be inert, immune from plugging, adjustable, easily replaceable, and effective for all sample types. Initially, short lengths of narrow-bore fused silica restrictors (5–15 cm long, 5–15 mm i.d.) were connected to the column end using zero dead volume connectors. These linear restrictors (Fig. 7.4) generate a linear pressure gradient over the length of the restrictor. They are satisfactory for low molecular mass analytes only and are commonly associated with gas phase detectors. With the pressure gradient generated by linear restrictors, polar and high molecular mass analytes cause detector spiking as a result of molecular association and condensation of the analyte during decompression to yield fog particles that enter the detector and create short ion bursts in the flame [4]. Other restrictor designs overcome problems of detector spiking by ensuring that rapid decompression occurs over a very short length, with the restrictor tip heated to avoid condensation of analyte. Tapered and integral restrictors fit into this category where the abrupt pressure drop prevents any precipitation of analyte.

7.2.4 Detectors

Nearly every detector that has been used in either GC or HPLC has been, or is being, investigated for SFC. Indeed, one of the strengths of SFC is the range of detectors available. Currently favoured detectors for SFC are u.v. absorption and the FID. The latter is constrained to mobile phases which yield no significant ionization during detection, whereas the limitations of u.v. detection, are related to sensitivity. In general, the FID is as easy to use and troublefree in SFC as it is in GC [16]. FID provides sensitive, universal detection of most organic compounds, with almost no response to carbon dioxide, nitrous oxide and sulphur hexafluoride. It is generally incompatible with other mobile phases and mixed mobile phases containing organic modifiers, except for formic acid, but has been used with methanol in concentrations up to 1% [17]. Other detectors, such as the electron capture, flame photometric and alkali flame ionization detector can provide additional specificity if necessary.

The u.v. absorption detector is the second most popular detector for SFC, particularly for those conditions incompatible with flame ionization detection, that is, with wide-bore columns and for mobile phases containing organic solvent modifiers. For open tubular columns, detector cell volumes below 50 nl are required to avoid extra-column band broadening. This can only be achieved with direct on-column detection, in which a section of the polymeric coating is removed from the fused silica tubing, or pseudo on-column detection, where the column is inserted into a slightly wider piece of fused silica tubing which acts as the detection cell. Improved sensitivity is achieved with pseudo on-column detection on account of the greater path length. Dead volume effects are of less concern with packed columns, and standard u.v. detector cells modified, if necessary, to withstand the high operating pressures of SFC are suitable.

SFC has been coupled to mass spectrometric and FTIR detectors. Several commercial SFC–MS interfaces are now available and with columns of 1 mm i.d. or less, the total column effluent can be introduced directly into a chemical ionization (CI) source mass spectrometer. CI has produced detection limits that are superior to those of electron ionization (EI). The latter requires high dilution of the decompressed fluid to the requisite low pressures for EI. The SFC restrictor largely dictates the performance of the combined system and defines the range of applications [18]. In applications involving FTIR, carbon dioxide is especially useful as a mobile phase since it can be condensed in a cold trap and is transparent to infrared in the important C-H stretch region. As an alternative mobile phase, xenon has demonstrated usefulness as it is practically transparent from the vacuum u.v. up to the nuclear magnetic resonance (NMR) region.

7.3 Columns

Two different types of column (Table 7.2) have been used for SFC: open tubular and packed columns. Column technology has been largely borrowed from GC (for the open tubular format) or from HPLC (for the packed format), although

Table 7.2 Columns used for supercritical fluid chromatography.

Parameter	Packed column			Open tubular column
	Conventional	Narrow-bore	Micro	
Internal diameter (mm)	4.6	1.0	<0.5	0.025–0.1
Length (cm)	10–30	10–30	50–150	100–5000
Column material	Stainless steel	Stainless steel	Fused silica	Fused silica
Film thickness (μm)	–	–	–	0.05–0.5
Particle size (μm)	3–10	3–10	3–10	–
Stationary phase	Chemically bonded silica, alumina, polymeric resins			Polysiloxanes
Pressure drop (MPa)	7	–	–	0.01

stationary phases must be more stable than in either HPLC or GC because of the higher operating temperatures than in HPLC and the enhanced solvating ability of supercritical fluids relative to gases. Columns have been compared in a number of papers [19, 20] with the results summarized by Schoenmakers and Uunk [1] as follows:

1. Contemporary packed columns provide a much higher speed of analysis than contemporary open tubular columns.
2. Open tubular columns can provide a much larger number of theoretical plates than packed columns for the same pressure drop. The difference is a factor of almost 500 for the 50 μm i.d. open tubular column and the 5 μm particle size packed column.
3. Volumetric flow-rates are much higher in packed column SFC than in open tubular column SFC, which makes injection and flow control less problematic.
4. The sample capacity of packed columns is much higher.

Many of the packed column versus open tubular column issues are common to GC, HPLC and SFC. Nevertheless, open tubular columns are favoured for SFC as for GC although for different reasons. Thus, pressure (and hence density) control, which has been the most effective means of controlling retention in SFC, is more easily accomplished with open tubular columns owing to the low pressure drop across an open tubular column. Moreover, the small drop in column pressure also minimizes changes in density along the column, which can affect the separation efficiency.

7.3.1 Open tubular columns

Open tubular columns for SFC differ from those in GC only in that column diameters are generally much smaller. Practical open tubular column dimensions of 50 μm i.d. and 10–20 m lengths are a compromise between performance and ease of preparation. In all other respects, open tubular column preparation procedures and stationary phase chemistries are identical to those used in GC.

Polysiloxanes are the most popular stationary phases for open tubular columns. This can be attributed to their versatility, high thermal stability and low viscosity, resulting in high solute diffusivity. Equal efficiencies are predicted [19] for equal values of the reduced film thickness, δ_f, which is defined by equation 7.1:

$$\delta_f = \frac{d_f}{d_c}\left[\frac{D_m}{D_s}\right]^{1/2} \qquad\qquad 7.1$$

where d_f is the thickness of the stationary film, d_c is the column internal diameter, and D_m and D_s are the diffusion coefficients of the solute in the mobile and stationary phase, respectively. Hence, if D_s is increased by using a low viscosity phase, a thicker film can be used resulting in a higher sample capacity without a loss in efficiency.

The full range of polysiloxanes encountered in GC are available. Polarity is varied by incorporating different functional groups in the polymer. Nevertheless, methyl-polysiloxanes are the most popular phases in both open tubular SFC and GC, where selectivity is a secondary consideration compared with the efficiency and ability of the column to elute the solute molecules as sharp, symmetrical peaks. More polar phases are usually used to take advantage of improved peak shape with polar solutes rather than to enhance selectivity. An example where full benefit is taken from the selectivity of the stationary phase is the separation of isomeric naphthylcarbazolyl-ethanes on a liquid–crystal phase [8].

As with other forms of chromatography, column stability in SFC depends on column preparation procedures, operating conditions and treatment of the column during use and storage. Cross-linked stationary phases are essential to ensure solvolytic stability of the polymeric film. Repeated cross-linking is often performed on SFC stationary phases to promote their stability. However, such phases should be distinguished from those which are both immobilized by cross-linking and attached by covalent chemical bonding to the inner wall of the column. This distinction is not always clear in manufacturers' catalogues. The effect of cross-linking on the solute diffusion coefficient in the stationary phase has not been established for SFC but, intuitively, a reduction in the efficiency of mass transfer can be expected for a cross-linked phase. Moreover, cross-linking usually leads to an increase in the density of the polymer and an increase in the glass transition temperature. Excessive cross-linking often turns a flexible polymer, with good solute diffusion characteristics and high chromatographic efficiency, into a hard resin with impaired efficiency. A compromise must therefore be reached between maximum column stability and life (high cross-linking) and maximum efficiency (low cross-linking).

Excessively high operating temperatures reduce stationary phase film stability in SFC and GC. In the case of SFC, the supercritical mobile phase also has a significant effect on column stability. The stability of polysiloxane stationary phase columns decreases with supercritical fluid in the order [4] carbon dioxide, nitrous oxide, carbon dioxide + 1%methanol, n-pentane and ammonia. The decrease in column stability with supercritical n-pentane can be attributed to higher SFC operating temperatures as a result of the higher critical temperature, and to the greater mutual solubility of supercritical n-pentane and the stationary phases. Repeated swelling and compression of the stationary phase on exposure to supercritical n-alkanes may also contribute to reduced film stability [4, 21]. Conversely, although still speculative, swelling may promote efficiency by enhancing diffusion coefficients in the stationary phase and the sample capacity. A slow decompression cycle at the end of an analysis is a wise preventive measure to minimize any detrimental effects of sudden compression/decompression. Polysiloxane stationary phases are even less stable in polar supercritical fluids such as ammonia, which attacks the polysiloxane backbone, leading to chemical degradation.

7.3.2 Packed columns

Packed columns are generally similar to HPLC columns with either 10 μm, 5 μm or 3 μm porous particles. Consequently, the most common stationary phases (Table 7.3) are silica-based chemically bonded phases. These packings are much more retentive than typical open tubular SFC columns. This retention is a function of the

Table 7.3   Characteristics of silica-based chemically bonded stationary phases.

Particle size	3–10 μm
Surface area	50–500 m^2 g^{-1}
Pore size	5–50 nm
Porosity	20–70%
Functionality	Alkyl (e.g., C$_{18}$, C$_8$, C$_2$), phenyl, amino*, cyano, diol.
Residual silanols	Considerable variation between materials in the extent of end-capping.

*Amino phases form carbamates with CO$_2$-based mobile phases.

pore structure and specific surface area, with strong specific interactions at uncapped silanol groups on the surface of the stationary phase [22–24]. The number of silanol groups will be the same irrespective of whether the phase is used for SFC or HPLC but the accessibility of the groups to solute molecules will differ greatly. If the mobile phase used in SFC is pure carbon dioxide, then the surface silanol groups will be essentially uncovered, allowing ready access of solute molecules. The action of low concentrations of organic modifiers added to carbon dioxide mobile phase is attributed to deactivation of adsorptive sites, including residual silanol groups on the stationary phase [25]. There is evidence that chemical reaction between the silanol groups and some polar modifiers [26] and solutes [23, 27] may occur under typical SFC conditions. Problems associated with adsorptive and chemical activity of silica- and alumina-based materials can be avoided by using a different substrate such as graphitized carbon or polymer beads.

To date, few systematic studies have been undertaken on the effects of different stationary phases and, although chemically bonded phases containing amino, diol or cyano groups have been employed, the most often used stationary phase for packed column SFC is octadecylsilyl (ODS or C$_{18}$) modified silica. The retentive nature of chemically bonded phases when used in conjunction with supercritical CO$_2$ limits their use to the elution of non-polar, low molecular mass solutes such as the simpler polycyclic aromatic hydrocarbons [28]. Non-polar phases with short alkyl groups (e.g., C$_2$) or polar phases where the functional group is polar (e.g., cyanopropyl) are usually less retentive and are suitable for a wider range of solutes, including larger hydrocarbons. Nevertheless, chemically bonded phases are generally less suited to the elution of polar hydrogen-bonding solutes, although polar modifiers in the mobile phase can improve the chromatographic performance of these solutes [1].

In addition to the chemically bonded phases, other materials investigated for packed column SFC are the polar adsorbents such as silica and alumina and polymeric materials. At this stage, the data on the use of adsorbents is not particularly encouraging. These materials exhibit good retention and selectivity but, as in LC, their disadvantages outweigh their advantages. Reproducibility of the adsorbent surface is a major concern, as is their low sample capacity and sensitivity to small changes in mobile phase composition. Organic polymers and copolymers based on polystyrene exhibit very high retention, necessitating a high mobile phase density, even for small molecules.

7.4 Factors Affecting Retention

Retention in SFC is not as easily rationalized as in GC or LC. It is dependent upon temperature, pressure, mobile phase density and composition of the stationary and mobile phases. Many of these experimental variables are interactive and do not change in a simple or easily predicted manner. Nevertheless, some general observations can usually be applied. For a given column, solute retention is affected by four major factors: choice of supercritical fluid, pressure, temperature, and gradient profile.

7.4.1 Solubilizing power

An understanding of SFC begins with a knowledge of fluid properties and their dependence on pressure and temperature. Supercritical fluids may exhibit gas-like or liquid-like properties at the extremes of the range represented by the dotted lines in Fig. 7.1. The properties of a fluid most relevant to its use as a mobile phase in SFC are the viscosity, diffusivity and solvating power, which depend on the density of the fluid. Density, in turn, depends on the applied pressure and temperature in a nonlinear manner described by an equation of state. This can be shown diagrammatically, as in a phase diagram (Fig. 7.5). The data in Table 7.4 represent conditions close to the critical pressure (7.3 MPa) and towards the upper operating pressure (40 MPa) of many SFC systems. Computer software is available to compute pressure/density data which can be used to construct phase diagrams as in Fig. 7.5. Close to the critical point there is a rapid sigmoidal variation in density with small changes in pressure. The changes in density with pressure are less dramatic at temperatures further removed from the critical point. Figure 7.5 also shows that a decrease in density accompanies an increase in temperature at constant pressure.

The solvating ability or solubilizing power and, hence, eluting power of a supercritical fluid is related to its capacity for specific intermolecular interactions and its density. Solvating power of substances can be expressed in terms of the Hildebrand solubility parameter, δ, which is defined as the square root of the cohesive energy density, i.e., the energy content of the fluid per unit volume relative to the ideal gas state. The solubility parameter is a much better indicator of the overall polarity of a solvent than the dipole moment or the dielectric constant. There is no correlation between the dipole moment of a fluid and its solubility parameter under SFC conditions (see Table 7.5).

Table 7.4 Temperature and pressure dependence of density for supercritical carbon dioxide.

Pressure (MPa)	Temperature (°C)	Density (g cm^{-3})	Pressure (MPa)	Temperature (°C)	Density (g cm^{-3})
7.3	40	0.22	40	40	0.96
7.3	60	0.17	40	60	0.90
7.3	80	0.14	40	80	0.82
7.3	100	0.13	40	100	0.76
7.3	120	0.12	40	120	0.70
7.3	140	0.11	40	140	0.64

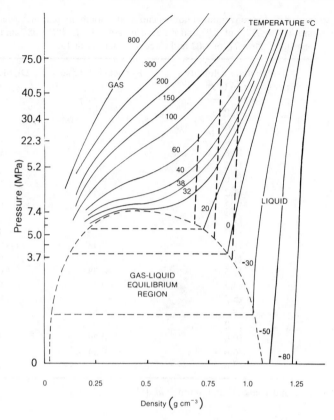

Fig. 7.5. Phase diagram showing pressure–density behaviour of carbon dioxide at various temperatures. Broken lines indicate isotherm crossing at either constant density or pressure.

The density of supercritical fluids is intermediate between that of gases and liquids but closer to that of liquids. Supercritical fluids therefore have a much greater solubilizing power than gases. They do not, however, provide any advantages in solubilizing ability over liquids, given a similar temperature constraint, and should not be regarded as 'super solvents'. They do, however, have the advantage that solvent strength can be varied by controlling density through changes in applied pressure and temperature, as shown in Fig. 7.6 by the dependence of the Hildebrand solubility parameter on temperature and pressure for supercritical carbon dioxide. In contrast, liquids are virtually incompressible and their solvent strength is essentially constant, regardless of experimental conditions; gases require inconveniently high pressures to approach liquid densities.

The π^* Kamlet–Taft scale of solvent polarity/polarizability based on spectroscopic measurements is an alternative method for expressing solvent strength. The normal range of values for liquids is 0 to 1, with values close to 1 representing strong dipole-type interactions. This scale measures only orientation and induction-type interactions and ignores solute-specific interactions such as hydrogen bonding. With the possible exception of ammonia, this approximation is reasonable since commonly

Table 7.5 Solvent strength data for conventional liquids and fluids. Solubility parameter data are quoted for a reduced temperature T_R (= T/T_c) of 1.02 and reduced pressure P_R (=P/P_c) of 2 and Kamlet–Taft constants for a T_R of 1.02 and a reduced density of 1.02.

Substance	Hildebrand solubility parameter, δ (J cm^{-3})$^{1/2}$	Kamlet–Taft scale π^*	Dipole moment (D)
n-Hexane	10.0	−0.03	–
n-Pentane	10.4	–	–
but-1-ene	10.6	–	0.34
n-Butane	10.8	–	0.05
Ethoxyethane	11.0	–	1.15
Freon 13	11.0	−0.46	–
Sulphur hexafluoride	11.2	−0.54	–
Propane	11.2	–	0.08
Dichloromethane	11.5	–	1.60
Ethyl acetate	11.7	–	1.78
Ethane	11.9	−0.48	0
Ethene	11.9	–	0
Xenon	12.5	−0.38	0
Nitrous oxide	14.7	−0.27	0.17
Carbon dioxide	15.3	−0.19	0
Benzene	–	0.62	0
Methanol	18.2	–	1.70
Ammonia	19.0	0.20	1.47
Ethanol	26.0	0.54	1.69
Water	27.6	–	1.85

Data from Schoenmakers and Uunk [1] and Smith *et al.* [8].

used supercritical fluids are unlikely to exhibit significant proton donor–acceptor interactions. Solvent strengths of several fluids compared with conventional liquids are shown in Table 7.5 using this scale. Kamlet–Taft constants increase with fluid density, confirming our expectation that fluid polarity or polarizability will increase as fluid density increases [8]. The largest effect is noted in the region of the critical point where relatively small changes in pressure cause large changes in the fluid polarity–polarizability. Furthermore, more polar solvents such as ammonia show a large change in solvent strength as density increases, whereas less polar solvents show smaller effects.

7.4.2 Temperature/pressure effects on solute solubility and retention

The relationship between solute solubility in the mobile phase and applied pressure and temperature is complex. At lower pressures, nearer the critical point, solubility decreases significantly as the temperature is increased, whereas at higher pressures, further removed from the critical point, there is an increase in solubility with increasing temperature [29]. This can be explained by the occurrence of two competing phenomena. As temperature increases, the vapour pressure of the solute tends to increase solute solubility but, at the same time, the lower fluid density decreases its solubilizing power. In the situation of lower pressures, there are large

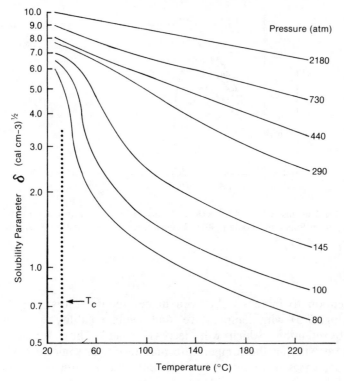

Fig. 7.6. Effect of temperature and pressure on the Hildebrand solubility parameter for supercritical carbon dioxide.

changes in fluid density and this dominates solubility, with the result that solubility decreases with increasing temperature. At higher pressures, the changes in fluid density with temperature are less marked and solute solubility depends largely on volatility. Using CO_2 as an example of density effects on solubilizing power, and therefore retention, at 6 MPa and 22°C (density < 0.1 g cm^{-3}) CO_2 has an elution strength comparable to that of hexane whereas at 35 MPa and 100°C (0.68 g cm^{-3}), supercritical carbon dioxide has an elution strength comparable to dichloromethane.

Varying temperature can have multiple effects on an SFC separation. In the case of a mobile phase at constant density, the logarithm of the capacity factor is proportional to the reciprocal of the column temperature [1, 29], that is, at constant mobile phase density, increasing the temperature will reduce retention.

A rather more complicated relationship is observed if retention (k or log k) is plotted as a function of temperature over a wide range (e.g. from 20°C to 230°C) at constant pressure (Fig. 7.7). The shape of the curves in Fig. 7.7 is dependent on competing effects related to solute vapour pressure and solubility in the mobile phase, as determined by mobile phase density [1]. At low temperatures where the mobile phase is a liquid, retention decreases slowly (not shown in Fig. 7.7) with increasing temperature. At some point, however, the mobile phase becomes supercritical and the decrease in mobile phase density with increasing temperature

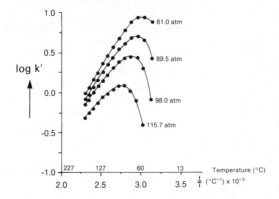

Fig. 7.7. Plot of log k against $1/T$ for hexadecane at constant pressure with carbon dioxide as mobile phase. Reproduced from Poole and Poole (1991) *Chromatography Today*, Elsevier, Amsterdam, p. 624 with permission.

will cause retention to increase. At even higher temperatures, the fluid density changes less (Fig. 7.5) with temperature, and solute volatility is the main effect contributing to retention, which will decrease again. Thus, for separations at constant pressure, varying the temperature may result, in some instances, in both increases and decreases in retention for different solutes, with possible changes in elution order. The overriding consideration in selecting temperature is always the thermal stability of the sample and stationary phase.

Pressure has a major impact on retention in SFC. Data showing the effect of pressure and temperature on the retention of naphthalene on bonded phase silica with carbon dioxide as mobile phase (Fig. 7.8) show the most marked effect of pressure at the lowest temperature, i.e. closest to the critical point. The variation is less at the other temperatures. Another important feature of the curves is the effect of temperature at high pressures (after the three lines have intersected). Here, retention is highest at the highest temperature because of the lower mobile phase density and associated reduction in its solubilizing power. It is a characteristic of SFC that retention increases with increasing temperature at constant pressure.

For a given column and temperature, retention generally decreases as operating pressure is increased (Fig. 7.8) because fluid density (and hence solvating ability) increases at higher pressures. For this reason, pressure programming or density programming has become a popular means of programmed elution in SFC.

7.5 Mobile Phases

The properties of non-polar or low-polarity compounds with moderate critical properties (e.g., CO_2, N_2O, SF_6, xenon, ethane, propane, pentane) have been well explored for use as mobile phases in SFC. On the Kamlet–Taft scale, these fluids provide only very weak polar interactions. The combination of low polarity mobile phase with a relatively polar stationary phase means that most SFC separations

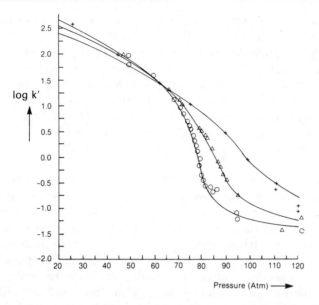

Fig. 7.8. Variation in retention of naphthalene on a Perisorb RP-8 column with carbon dioxide mobile phase as a function of pressure and temperature (35°C, $\bigcirc$; 40°C, Δ; and 50°C, $+$). Reprinted from Schoenmakers and Uunk (1989) in Giddings J.C. (editor) *Advances in Chromatography*, Vol. 30, by courtesy of Marcel Dekker Inc, New York, with permission.

involve normal-phase partition or adsorption. In such systems and, in the absence of specific interactions, elution time is a function of molecular mass and polarity. The larger or the more polar the analyte molecules, the stronger the retention.

7.5.1 Single-component phases

For a number of reasons, carbon dioxide is the fluid of choice for many applications. Its critical properties are favourable (Table 7.1). A low critical temperature means that operating temperatures can be as low as 40°C which is an advantage with thermally labile samples. The critical pressure of 7.3 MPa is moderate and, combined with a critical density of 0.448 g cm^{-3}, this implies that high densities can be achieved at reasonable pressures of less than 40 MPa. In addition to its desirable critical properties, carbon dioxide is nontoxic, nonflammable, noncorrosive (although mixtures with water are corrosive), inert to most substances and does not interfere with most detection methods (particularly the FID). Carbon dioxide also has a wide density range under normal operating conditions, which provides maximum versatility for optimizing solubilizing power. Finally, carbon dioxide is readily available in high purity at low cost. The type of detector in use has a major influence on the required level of purity. For example, with an FID any hydrocarbon impurities in carbon dioxide will lead to an increased noise level and a drifting baseline in programmed elution. Under these circumstances, high-purity carbon dioxide (e.g., 99.9995% CO_2) is required, whereas regular-grade CO_2 (99.8%) is suitable in combination with u.v. detection. Carbon dioxide does have some limitations; it reacts with some amines, particularly primary and secondary amines with a pK_b > 9, to form ureas and carbamates. Nevertheless, most aromatic amines and N-heterocyclic compounds can be separated without problems. The safety and environmental aspects of carbon dioxide are drawing special attention.

The critical properties and chromatographic behaviour of nitrous oxide are very similar to those of carbon dioxide. The solvent strength of the two fluids is similar and only minor selectivity differences are observed [1]. It offers no particular advantages with respect to detection and its only advantage seems to be the ability to chromatograph some specific solutes, such as primary amines, which react with carbon dioxide.

Alkanes are less attractive than carbon dioxide with respect to safety and detector compatibility but show different solvent characteristics. For non-polar solutes, they are often stronger solvents and in some instances show considerable selectivity differences from carbon dioxide. Of the homologous n-alkanes, four members (ethane through pentane) have critical properties which make them suitable for use as mobile phases. Selectivity can be varied by choosing different members of the homologous series and, thus, different operating temperatures and pressures. For example, at the temperatures required for supercritical operation with n-pentane, solute volatility may be a more significant factor than with supercritical ethane as mobile phase.

Other mobile phases such as halocarbons, xenon and sulphur hexafluoride find speciality applications. For example, group separations of hydrocarbons into aliphatic, olefinic and aromatic fractions have been achieved [30] with supercritical sulphur hexafluoride. However, SF_6 is not particularly useful if detection involves FID. Xenon has favourable supercritical properties, performs well as a supercritical solvent [31] and is transparent in the infrared region; however, its cost is prohibitive.

Unfortunately, highly polar and high molecular mass solutes have limited solubilities in CO_2 and other similar low-polarity fluids. Schoenmakers and Uunk [32] considered several substances as potential mobile phases and from this prepared an inventory of possible solvents with suitable critical properties, and with a polarity exceeding that of carbon dioxide. The list, in order of increasing polarity (δ_{SFC} value given), comprised hydrogen bromide (7.54), hydrogen chloride (7.61), methylamine (7.67), hydrogen sulphide (7.83), sulphur dioxide (7.87), methyl bromide (8.24), nitrosyl chloride (8.62), ammonia (9.34) and nitrogen dioxide (11.04). Polar fluids may have the desired solvating power but safety and toxicity aspects associated with their use are major concerns. Ammonia appears to be the only realistic choice as a polar single-component fluid for SFC, but its use has been limited. This can be attributed to the fact that ammonia is highly toxic, corrosive and reactive, and is flammable and explosive under certain conditions. The future growth and wider acceptance of SFC will depend on its ability to analyse polar analytes because many thermally labile, high molecular mass, non-volatile molecules, for which SFC should be the method of choice, have polar functional groups.

7.5.2 Modifiers

Because of the complications of using single-component polar solvents, organic modifiers (entrainers) have been used to enhance the solvent strength of low-polarity fluids such as carbon dioxide. At concentrations below about 2% (v/v) modifiers probably have minimal influence on the solubility of the sample in the supercritical fluid mobile phase, and are used primarily to mask interactions between sample molecules and accessible reactive silanol groups on the surface of the stationary phase. Consistent with this explanation is the observation that there are significant changes in retention when modifiers are used in conjunction with packed columns [1] but only minimal changes in chromatographic behaviour when the same modifiers are used with deactivated fused silica open tubular columns [29].

At higher modifier concentrations, it is possible to adjust the solvent strength and selectivity of the mobile phase by tailoring the nature and extent of solute–solvent intermolecular interactions. A good starting-point in selecting a solvent modifier is to choose a substance that is a good solvent for the analytes. Miscibility limitations and unfavourable changes in the critical point parameters for the binary mobile phase may limit the range over which properties can be varied. Few experimental data are available on the critical properties of mixtures, which may be calculated, however, using various numerical techniques based on an appropriate equation of state and interaction parameters [1].

Modified fluids have included: carbon dioxide doped with methanol, isopropanol, dichloromethane, tetrahydrofuran, dimethylsulphoxide and acetonitrile; trifluoromethane modified with ammonia; and pentane doped with benzene, isopropanol and dichloromethane. Aliphatic alcohols are the most popular modifiers. With reversed-phase systems and supercritical carbon dioxide as the main component of the mobile phase, modifier strength increases with carbon chain length [17, 33]. Methanol has been extensively studied as a polar modifier and, in mixtures with carbon dioxide, behaves particularly well [1]. One of the disadvantages of methanol is that it eliminates the possibility of using the FID. Water has been examined as a modifier with a supercritical carbon dioxide mobile phase [34]. It has the advantage that it does not contribute a background current to an FID, but mixtures of carbon dioxide and water are corrosive.

The scope of SFC can be extended in other ways as, for example, in the addition of ion pairing reagents to supercritical carbon dioxide–modifier mobile phases which extends SFC separations [35] to include ionic and ionizable analytes. Selectivity and solubilizing power can also be enhanced by introducing secondary chemical equilibria into the fluid system. The use of secondary equilibria has not been fully investigated but examples are organized molecular assemblies, such as reverse micelles and microemulsions [8], which greatly increase the solubility of polar analytes in the otherwise non-polar mobile phases. It is difficult to predict the future of micellar SFC, although it must be noted that micellar LC has not become a major analytical technique despite significant research effort.

There are limited data on the stability and safety of many solvents under supercritical conditions. Under these circumstances a cautious and conservative approach is essential. This applies particularly if harmful products can be formed in the detector as, for example, with sulphur hexafluoride and the FID [30].

Methods for adding modifiers

Methods for adding modifiers are largely borrowed from HPLC where the use of mixed mobile phases is extremely common. The following methods can be and have been used: premixing the solvents in the reservoir flask; low-pressure mixing in which the solvents are mixed in, or just before, the pump head or the cylinder of a syringe pump; and high-pressure mixing in which the effluent streams of two pumps are mixed. There are benefits to each method; the first, for example, places no additional demands on the instrumentation. Detailed discussion of the problems associated with accurate and precise mixing, where the main solvent is a gas under ambient conditions, is given elsewhere [1].

7.6 Programming Techniques

Programming techniques in SFC [36] are extremely versatile (Table 7.6). Solvent strength and therefore solute retention can be controlled by gradients in pressure (or density), temperature and mobile phase composition, or any combination of these. Of these possibilities, pressure or density gradients [37, 38] have been the most popular. Both techniques aim to decrease retention during the course of the analysis by increasing the density. Pressure is the physical property that is measured by SFC instruments but density is the important parameter that regulates retention. Microcomputer-controlled systems can provide reproducible programs for multi-step ramping of pressure or density in any combination of linear pressure, linear density, or asymptotic density. Computerized conversion of pressure to density for linear and asymptotic density ramping is essential. A number of software packages are available which generate density/solubility isotherms for binary fluids to assist with supercritical fluid selection. Density programming can be achieved in several ways [29] such as, for example, by programming the column inlet pressure and fixing the column outlet pressure by a restrictor or backpressure regulator.

A similar effect to pressure or density programming can be achieved by programming the temperature downward (negative temperature programming). Combined temperature and pressure programming exploits density, solubility and diffusion effects in order to enhance separation efficiency [34, 40]. Versatility in optimizing a separation can be achieved by either negative, positive or synchronized temperature/density programming.

Solvent strength is conveniently changed with density programming but selectivity is easier to change with a composition gradient. With a composition gradient, the

Table 7.6 Elution programmes used in different forms of chromatography.

Parameter	Constant conditions	Programmed elution	Use
Temperature	Isothermal	Temperature programming	GC, SFC
Composition	Isocratic	Solvent programming, gradient elution	LC, SFC
Pressure	Isobaric	Pressure programming	SFC
Density	Isoconfertic, isopycnic	Density programming	SFC
Flow*	–	Flow programming	GC, LC?

*Flow programming has not been widely used because of the problems of increased pressure drop across the column and loss in separation efficiency as mobile phase flow rate is increased.

percentage of organic modifier added to the supercritical fluid is increased during the separation. Some care is necessary, however, because the critical temperature of the supercritical mobile phase varies with the concentration of the organic modifier. For this, and other mainly practical reasons (e.g., more complex instrument arrangement), composition gradients have not been widely used in SFC until comparatively recently. Composition gradients invariably involve changes in other parameters; for example, viscosity of the mobile phase is likely to change with composition and hence also the flow-rate and/or the pressure.

An important advantage of SFC is the rapid column re-equilibration with both normal- and bonded-phase systems following a programme run. With a normal-phase silica column, 300–1000 column volumes of mobile phase are required to stabilize the column after a change in the composition of a liquid mobile phase. However, with a supercritical carbon dioxide mobile phase, columns re-equilibrate after only two column volumes of flow following a pressure programme and after 10–30 volumes following a modifier change [41].

7.7   Areas of Application

A search of the literature shows the following application areas for SFC:

- Industrial
  dyes, isocyanates, oligosaccharides, polysaccharides, sucrose polyesters, pesticides, surfactants (polyglycols), synthetic oligomers, polymers/additives, waxes
- Biochemical
  antibiotics, drugs of abuse, fatty acids/lipids, prostaglandins, steroids
- Fossil fuels
  fractionation of petroleum and coal-derived fluids, hydrocarbon group analysis, simulated distillation.

A close look at these applications shows that they are, in general, samples that are poorly handled by current technology in either GC or HPLC. In some cases, SFC is favoured by the absence of u.v. absorbing chromophores in the analyte, which complicates detection in HPLC. In many instances, however, the analytes are too low in thermal stability and volatility for ready elution in GC, while they are so complex that it may require considerable effort to develop a suitable HPLC method [7]. The analysis of petrochemicals, characterized as they are by complex mixtures of non-polar compounds, has long been a successful application area for SFC with supercritical carbon dioxide.

Despite this impressive array of applications, the major future of this methodology is probably as an extraction technique for sample preparation (see Chapter 8).

References

1. Schoenmakers P.J. and Uunk L.G.M. (1989). In Giddings J.C., Grushka E. and Brown P.R., Editors, *Advances in Chromatography, Vol 30*. Marcel Dekker, New York, pp. 1–80.
2. Chester T.L. and Pinkerton J.D. (1990). *Anal. Chem.*, **62**: 394R.

3. Klesper E., Corwin A.H. and Turner D.A. (1962). *J. Org. Chem.*, **27**: 700.
4. Lee M.L. and Markides K.E. (1986). *J. High Resol. Chromatogr., Chromatogr. Comm.*, **9**: 652.
5. Chester T.L., Pinkerton J.D. and Raynie D.E. (1992). *Anal. Chem.*, **64**: 153R.
6. Palmieri M.D. (1988). *J. Chem. Educ.*, **65**: A254.
7. Later D.W., Richter B.E., Felix W.D., Andersen M.R. and Knowles D.E. (1986). *Int. Lab.*, **16**: 84.
8. Smith R.D., Wright B.W. and Yonker C.R. (1988). *Anal Chem.*, **60**: 1323A.
9. Smith R.M., Editor (1988). *Supercritical Fluid Chromatography*. The Royal Society of Chemistry, London.
10. Lee M.L., Xu B., Huang E.C., Djordevic N.M., Chang H.-C.K. and Markides K.E. (1989). *J. Microcol. Sepns.*, **1**: 7.
11. Kohler J., Rose A. and Schomburg G. (1988). *J. High Resol. Chromatogr., Chromatogr. Comm.*, **11**: 191.
12. Richter B.E., Knowles D.E., Anderson M.R., Porter N.L., Campbell E.R. and Later D.W. (1988). *J. High Resol. Chromatogr., Chromatogr. Comm.*, **11**: 29.
13. Hirata Y., Tanaka M. and Inomata K. (1989). *J. Chromatogr. Sci.*, **27**: 395.
14. Hawthorne S.B. and Miller D.J. (1989). *J. Chromatogr. Sci.*, **27**: 197.
15. Fjelsted J.C., Kong R.C. and Lee M.L. (1983). *J. Chromatogr.*, **279**: 449.
16. Richter B.E., Bornhop D.J., Swanson J.T., Wangsgaard J.G. and Andersen M.R. (1989). *J. Chromatogr. Sci.*, **27**: 303.
17. Levy J.M. and Ritchey W.M. (1986). *J. Chromatogr. Sci.*, **24**: 242.
18. Smith R.D., Kalinoski H.T. and Udseth H.R. (1987). *Mass Spectrom. Rev.*, **6**: 445.
19. Schoenmakers P.J. (1988). *J. High Resol. Chromatogr., Chromatogr. Comm.*, **11**: 278.
20. Schwartz H.E., Barthel P.J., Moring S.E. and Lauer H.H. (1987). *LC-GC*, **5**: 490.
21. Springston S.R., David P., Steger J. and Novotny M. (1986). *Anal. Chem.*, **58**: 977.
22. Engelhardt H., Gross A., Mertens R. and Petersen M. (1989). *J. Chromatogr.*, **477**: 169.
23. Dean T.A. and Poole C.F. (1989). *J. Chromatogr.*, **468**: 127.
24. Ashraf-Khorassani M., Taylor L.T. and Henry R.A. (1988). *Anal. Chem.*, **60**: 1529.
25. Janssen J.G.M., Schoenmakers P.J. and Cramers C.A. (1989). *J. High Resol. Chromatogr. Chromatogr. Comm.*, **12**: 645.
26. Hirata Y. (1984). *J. Chromatogr.*, **315**: 31.
27. Evans M.B., Smith M.S. and Oxford J.M. (1989). *J. Chromatogr.*, **479**: 170.
28. Gere D.R., Board R. and McManigill D. (1982). *Anal. Chem.*, **54**: 736.
29. Poole C.F. and Poole S.K. (1991). *Chromatography Today*. Elsevier, Amsterdam, pp. 601–648.
30. Schwartz H.E. and Brownlee R.G. (1985). *J. Chromatogr.*, **353**: 77.
31. French S.B. and Novotny M. (1986). *Anal. Chem.*, **58**: 164.
32. Schoenmakers P.J. and Uunk L.G.M. (1987). *Eur. Chromatogr. News*, **1**: 14.
33. Blilie A.L. and Greibrokk T. (1985). *Anal. Chem.*, **57**: 2239.
34. Geiser F.O., Yocklovich S.G., Lurcott S.M., Guthrie J.W. and Levy E. (1988). *J. Chromatogr.*, **459**: 173.
35. Steuer W., Baumann J. and Erni F. (1990). *J. Chromatogr.*, **500**: 469.
36. Klesper E. and Schmitz F.P. (1987). *J. Chromatogr.*, **402**: 1.
37. Wilsch A. and Schneider G.M. (1986). *J. Chromatogr.*, **357**: 239.
38. Linnemann K.H., Wilsch A. and Schneider G.M. (1986). *J. Chromatogr.*, **369**: 39.
39. Peaden P.A. and Lee M.L. (1982). *J. Liq. Chromatogr.*, **5** (Suppl. 2): 179.
40. Klesper E. (1988). *Fresenius Z. Anal. Chem.*, **330**: 200.
41. Steuer W., Schindler M. and Erni F. (1988). *J. Chromatogr.*, **454**: 253.

Bibliography

Camel V., Thiebaut D. and Caude M. (1992). Supercritical fluid chromatography and extraction. *Analusis*, **20**(8): M18.

Charpentier B.A. and Sevenants M.R. (1988). *Supercritical Fluid Extraction and Chromatography: Techniques and Applications*. ACS Symposium Series 366, American Chemical Society, Washington.

Janssen H.G. and Cramers C.A. (1993). Supercritical fluid chromatography: recent developments and new directions. *Anal. Proc.*, **30**: 89.

Smith R.M., Editor (1988). *Supercritical Fluid Chromatography*. The Royal Society of Chemistry, London.
White C.M. (1988). *Modern Supercritical Fluid Chromatography*. Huethig, Heidelberg.

Separation efficiency

Ashraf-Khorassani M., Shah S. and Taylor L.T. (1990). Column efficiency comparison with supercritical fluid carbon dioxide versus methanol-modified carbon dioxide as the mobile phase. *Anal. Chem.*, **62**: 1173.
Poe D.P. and Martire D.E. (1990). Plate height theory for compressible mobile phase fluids and its application to gas liquid and supercritical fluid chromatography. *J. Chromatogr.*, **517**: 3.
Schoenmakers P.J. (1988). Supercritical fluid chromatography: open columns vs packed columns. *J. High Resol. Chromatogr., Chromatogr. Comm.*, **11**: 278.

Injectors

Richter B.E., Knowles D.E., Andersen M.R., Porter N.L., Campbell E.R. and Later D.W. (1988). Reproducibility in capillary supercritical fluid chromatography. Comparison of injection techniques. *J. High Resol. Chromatogr. Chromatogr. Comm.*, **11**: 29.

Restrictors

Green S. and Bertsch W. (1988). Simple restrictors for capillary supercritical fluid chromatography. *J. High Resol. Chromatogr. Chromatogr. Comm.*, **11**: 414.
Raynor M.W., Bartle K.D., Davies I.L., Clifford A.A. and Williams A. (1988). Preparation of robust tapered restrictors for capillary supercritical fluid chromatography. *J. High Resol. Chromatogr. Chromatogr. Comm.*, **11**: 289.

Detectors

Carey J.M. and Caruso J.A. (1992). Plasma spectrometric detection for supercritical fluid chromatography. *Trends Anal. Chem.*, **11**: 287.
Davidson G. and Jenkins T.J. (1992). SFC/FT-IR – a powerful analytical tool. *Spectrosc. Eur.*, **4**: 32.
Hill H.H. Jr. and Gallagher M.M. (1990). Fundamentals of detection after supercritical fluid chromatography. *J. Microcolumn Sep.*, **2**: 114.
Jinno K. (1987). Interfacing between supercritical fluid chromatography and infrared spectroscopy – a review. *Chromatographia*, **23**: 55.

Columns

Schoenmakers P.J., Uunk L.G.M. and Janssen H.G. (1990). *J. Chromatogr.*, **506**: 563.

Sample Handling in Chromatography 8

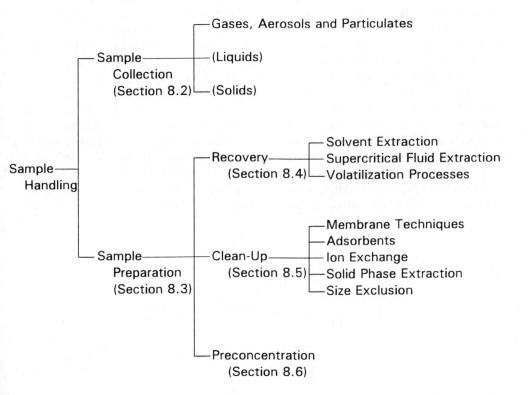

8.1   Introduction

In far too many instances, the chromatogram is automatically assumed to be an accurate representation of the composition of the starting sample. This assumption makes no allowance for the effects of sample preparation, which can have a profound influence on the chromatogram. This is illustrated in Fig. 8.1 which compares the organic material extracted [1] from 0.5 l of surface water by three solvents. Hence, the processes involved in sample preparation may alter the composition of the sample prior to injection. Moreover, discrimination effects may occur during injection and, even in those instances where the sample injected is a true representation of the original sample, there may be on-column decomposition of some components. Further problems may arise in the form of unresolved peaks of which the chromatographer may be unaware and there may be inaccuracies in the peak identification and integration. The use of modern high-efficiency systems and improved sample preparation procedures rectifies some, but not all, of these problems. Sample preparation remains complex, with no single technique being uniformly satisfactory. The chromatographer must understand all possible proce- dures to enable the most appropriate method to be chosen to suit the needs of a particular system and problem. The most important criterion for sample preparation is the success of the method in quantitatively recovering the analyte from the sample matrix. Reference standards, spiked samples and blanks should be processed

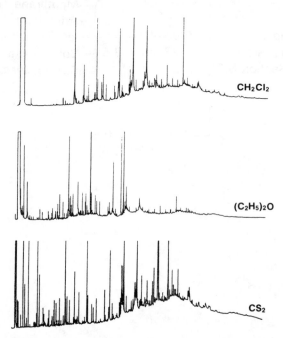

Fig. 8.1. Comparison of the organic components extracted from surface water by different solvents. Analysis performed on a 25 m × 0.23 mm SE-30 glass WCOT column with flame ionization detection. Extracts were recovered from water using the method of Mieure and Dietrich [1]. Reprinted with permission from Freeman R.R., editor (1981) *High Resolution Gas Chromatography*, 2nd edn, Hewlett Packard, p. 142.

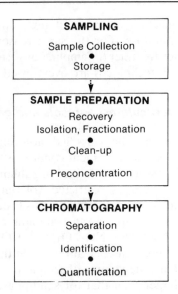

Fig. 8.2. Steps involved in a chromatographic analysis.

together with samples in deciding the suitability of a preparation scheme. Procedures are also chosen in terms of time, reagent use and availability, practicality, cost involved and, increasingly, on the procedure's environmental impact. The last of these accounts for the increasing popularity of procedures involving supercritical fluid extraction. The fewer the steps involved, the simpler, more convenient and cost effective and less time-consuming, the method is. More importantly, simple procedures lead to improved accuracy, reliability and reproducibility.

The steps involved in a chromatographic analysis may be summarized as in Fig. 8.2. In general, there are three stages which may be identified as sampling, sample preparation and the actual chromatography, although the distinction between the three may not always be clear. Since the inception of chromatography, sample preparation has been an important component of chromatographic analysis. Methods of preparation may change but the need and *raison d'être* remain the same; the majority of samples are too complex, too dilute or incompatible with some component of the chromatographic system to permit direct introduction. It may seem somewhat ironic that the needs of sample preparation remain unchanged despite tremendous improvements in resolving power and sensitivity achieved with modern chromatographic systems. This is related to the cycle of advances in methodology revealing more information about a sample, creating the regulatory and scientific desire to know more by characterizing more components, or by determining target components at ever decreasing limits of detection.

8.2 Sample Collection Procedures

The usual concerns of a contamination-free, representative sample apply equally in all methods of analysis. There are numerous treatments of statistically based procedures for sample acquisition in specialist texts and the interested reader is referred to this extensive literature. For the same reasons, sampling of solids and liquids will receive no further treatment. Sampling of gases, however, presents some unique problems and relevant techniques are overviewed in this section.

8.2.1   Sampling of particulates, aerosols and gases

Compounds in ambient air can be classified according to their volatility as: substances whose natural state under ambient conditions is that of a gas; substances that have sufficient vapour pressure to volatilize or sublime into the atmosphere and exist as vapours; and substances of restricted volatility attached to particulate matter. No single sampling method is suitable for this broad spectrum of substances. Aerosol formation may further complicate the sampling requirements.

With few exceptions (e.g. N_2O, NO, NO_2, O_3, CO, CO_2) most of the atmospheric substances are organic and present at concentrations in the mg m^{-3} range or below. Owing to the low density of air, the volume of sample which has to be collected for a trace analysis is generally larger than that for solid-phase or liquid-phase samples. For the same reason, contamination effects or wall losses have a greater impact on the concentration in gas phase samples. For example, contamination of a 1 dm^3 sample by 1 mg of substance changes the concentration of a liquid sample by 10^{-9} g g^{-1} but, for an air sample, by nearly 10^{-6} g g^{-1} [2]. The problems of gas sampling are circumvented by high-pressure sampling and preconcentration. Nonetheless, there is no general procedure for the collection of representative and stable samples for the analysis of organic substances in air. Each procedure has to be adapted to the kind of substance, expected concentration range and type of interfering substances present.

Whole air sampling

Whole air or grab sampling is useful for measurements of low and medium molecular mass hydrocarbons and light halocarbons. Where suitable, it is preferred because it eliminates uncertainties related to sampling efficiency and sample volume. Laboratory analysis of grab samples is usually achieved by GC with a preconcentration step and, occasionally, by direct injection without preconcentration. Whole air samples are taken by drawing air into an evacuated glass or stainless steel container, plastic bag (teflon, Tedlar or aluminized Tedlar) or syringe. The use of polymer bags is limited to measurements in highly polluted air, or for otherwise relatively high trace gas concentrations [3]. Stainless steel canisters are now more popular and are, in general, equipped with metal bellows valves which avoid all polymeric seals. Thorough cleaning and minimization of the active surface area are essential [2]. Moreover, the containers are fairly expensive and the preparation of the containers for sample collection is elaborate. The volume of sample containers is limited, for practical reasons, to a few litres but larger volumes of air can be sampled by pressurizing the samples using pumping or cryogenic collection [2].

Collection devices

Equipment suitable for gas sampling can be classified as filter media, impingers, solid adsorbents and diffusion scrubbers.

Filter media

Filter media, such as PTFE membranes or glass fibre filters, are suitable for collection of gases, aerosols and particulates. The sampling efficiency can be improved by impregnating the filter medium with some form of absorbing solution. For example, acidic and alkaline solutions are used in the collection of basic and acidic gases, respectively. A passive sampling method may be used in which the sampling process is controlled by diffusion. This approach is often used for 'badge' samplers used in personal monitoring where a time-integrated sample is required. Alternatively, a known volume of air may be drawn through the filter medium, often using a dichotomous sampler, which allows fractionation of the sample into aerodynamic size ranges. In a typical application, 2000 m^3 of air can be sampled in a 24-h period, yielding approximately 0.1 g of particulates. The result of an analysis on a sample obtained with impregnated filter media usually represents the total analyte concentration (i.e. the sum of the gaseous, aerosol and particulate components).

Impingers

Impingers (or bubblers) have been used widely for sample collection. They consist of a suitable vessel containing an absorbing solution, through which is drawn a measured quantity of air. Spray and evaporative losses must be controlled and this is conveniently achieved with a second container packed with glass wool and placed downstream from the impinger itself, which serves to collect any aerosol particles from the impinger absorbing solution. The absorbing solution is chosen to provide quantitative retention of the sample components of interest.

In addition to the problems of aerosol formation, impingers suffer from a number of disadvantages. They are clumsy, there is a large dilution factor in the absorbing solution and air sampling rates must be carefully controlled. The last requirement comes from the need for very small bubbles which remain in contact with the absorbing solution long enough to trap the sample gases.

Solid adsorbents

Gas samples can be collected on a solid sorbent packed into suitable sampling tubes in two ways. In passive sampling, the adsorbent is exposed in a defined fashion to the atmosphere, with the sorptive process being controlled by diffusion processes and the adsorption properties of the sorbent. Passive sampling is widely used for monitoring of workplace atmospheres and dosimetry for relatively high levels of pollutants. Active sampling is more suited to measurement of components present at low levels and involves pumping a defined volume of air through a tube packed with an appropriate sorbent. Porous polymers, various forms of carbon and, less commonly, silica gel and alumina are used. Tenax is the most popular material for trapping compounds at room temperature. It is hydrophobic and thus retains little water during sampling. Artefacts can be formed by reaction of ozone, chlorine and nitrating agents with both Tenax and porous polymers. The efficiency of the collection, and the maximum possible sampling volume, are characterized by the breakthrough volume of the analyte with the lowest breakthrough volume which, in turn, depends mainly on the affinity of the analyte for the sorbent, analyte concentration and matrix coadsorption. Ideally, the sample collection should be selective for the compounds of interest but, in reality, it is considered adequate if there is no significant adsorption of the main atmospheric components (nitrogen, oxygen, water vapour and carbon dioxide). The retained components can be recovered for analysis by either thermal desorption or liquid–liquid extraction of the adsorbent.

Diffusion scrubbers

Diffusion scrubbers (or diffusion denuders) are devices for the selective removal of a gas from a gas–aerosol or a gas–particulate mixture. The operating principle is based on the diffusion of gaseous components of an inlet stream at right angles to the direction of flow, whereas aerosols and particulates are carried in the direction of flow.

Further details of ambient air sampling are available in a number of sources [2, 4, 5].

8.3   Sample Preparation—An Overview

After a representative sample has been obtained, it must be prepared in a form suited to the chromatographic technique being used. In some instances, this may involve simple dissolution of the sample. At the other extreme are sample preparation procedures involving complex multi-step extractions, followed by a preliminary chromatographic clean-up and preconcentration of the sample extract. Irrespective of the complexity of the sample treatment, preparation procedures always require the utmost care and meticulous attention to detail. The following sections provide a comprehensive coverage of those sample preparation techniques commonly used and those techniques starting to emerge as important advances over existing methods.

A variety of physical, chemical and chromatographic methods have been used (Table 8.1) to process samples for analysis, with the aim of providing a sample extract uniformly enriched in all components of interest and free from interfering matrix components. This range of sample handling procedures is not unexpected in the context of the diversity of separation and detection methods, and sample types. In addition to the techniques described in this chapter, a number of the procedures described elsewhere in this text are useful for the preliminary processing of samples. Some of these are specialized techniques with restricted application, such as pyrolysis GC (Section 3.9) and headspace analysis (Section 3.9), whereas others such as chemical derivatization (Sections 3.9 and 5.7), multidimensional chromatography (Sections 3.9 and 5.9) and planar chromatography (Chapter 4) have more general application. A comprehensive treatment of sample handling procedures for ion chromatography is available [6].

In general terms, the steps used to overcome the problems associated with the complexity and concentration of the sample components and system incompatibility, may be classified as recovery of analyte(s) (which, in turn, may involve sample fractionation and isolation of analyte), clean-up and preconcentration of the isolated analyte(s). Recovery procedures have traditionally involved some form of extraction to remove the analyte from the bulk sample matrix. In rare instances, the recovery

Table 8.1   Sample preparation procedures.

Technique	Analyte	Sample/application
Ashing	Inorganics	Organic samples
Chromatography	Diverse	Diverse
Conventional filtration	General	Turbid samples; sample clarification
Dialysis	Ions, solutes	Biological materials
Diffusion	Gases	Gas sampling
Fractional distillation	Volatiles	Limited
Headspace analysis	Volatiles such as flavours and odorous compounds	Wide range of sample types including foods and waters
Microfiltration	General	Clean-up procedure preliminary to HPLC or ion chromatography
Precipitation	Diverse	Used to remove proteins from biological samples
Reverse osmosis	?	Biological samples
Saponification	Pesticides stable to strong acid and base	Used as an adjunct to chromatographic clean-up of fatty samples
Solid phase extraction	Diverse	Usually aqueous, preferably free of particulate matter
Solvent extraction	Diverse, e.g., drugs pesticides	Diverse
Sublimation	Compounds with suitable thermal stability and volatility	Limited
Supercritical fluid extraction	Diverse	Diverse
Steam distillation	Low molecular mass polar compounds	Aqueous samples
Sweep codistillation	Pesticides	Fatty samples
Ultrafiltration	Low molecular mass solutes	Used to remove proteins from biological samples

of analyte from the sample matrix is so efficient that unwanted matrix components are left behind. Usually, however, clean-up is necessary to eliminate the unwanted co-extracted matrix components from the sample extract. Samples of high fat content are of particular concern. The first step in sample preparation with such samples is, invariably, extraction of the fatty material from the homogenized sample matrix. This is essential for adequate recovery of most analytes because of their high solubility in lipoidal materials. Isolation of the lipid-soluble analytes from the fatty extract for subsequent determination can be achieved by a variety of techniques, as described later in this chapter. Finally, some form of preconcentration is normally required following recovery because of the low levels of analytes normally encountered in chromatographic analysis and because of the dilution effects of the recovery and clean-up steps. The distinction between the three steps is becoming blurred and, in some instances, it is possible to combine these procedures in a single step.

In some analyses, sample preparation may be undertaken for reasons unrelated to the chromatography. For example, protein is removed from biological fluids, such as serum and urine, prior to the determination of organic acids in order to inactivate metabolic activities and to displace the organic acids from their binding protein [7]. This can be achieved in either of two ways: ultrafiltration or precipitation. In serum, organic acids are bound, to various extents, to proteins. Hence, if the total concentration of organic acids (bound and nonbound) is desired, the removal of protein by precipitation is an essential step in sample treatment. Precipitation is a simple and rapid procedure, but there is the risk that some analyte may be lost by adsorption on the precipitate. Conversely, ultrafiltration is a milder treatment and is used in those instances where only the free fraction of organic acids is required. Where the subsequent analysis involves GC, the additional objectives are to avoid indirect interferences through artefact formation by pyrolytic processes, and to avoid deterioration of the chromatographic system by reducing the build-up of deposits in the injector.

Sample preparation procedures may be performed off-line prior to the chromato-graphic analysis, but an increasing number are being incorporated as on-line processes linked directly to the chromatographic hardware. The increased popularity of on-line operation can be attributed to the greater accuracy which can be expected by minimizing sample losses, sample manipulation and contamination. Furthermore, the coupled system can be automated for unattended operation, with subsequent increase in sample throughput.

8.4   Recovery Procedures

A sample in solution is either mandatory or highly desirable for most chromato-graphic techniques. For this reason, simple dissolution of all or part of a sample matrix is one of the most widely used sample preparation techniques for gases, vapours, liquids or solids. In GC, volatile liquids and gases can be examined directly but, more often, both liquids and solids are dissolved in a volatile solvent to give a dilute solution. The solvent must be compatible with the detector and, preferably, dissolve the entire sample. Any insoluble materials are removed by filtration or centrifugation in order to avoid their being trapped at the head of the column as charred residues, causing column deterioration and/or a high background signal. This simple approach to sample preparation for GC is suitable for samples which contain no non-volatile material, such as lipids and proteins, and where the detector sensitivity is adequate to measure the concentration of analytes present.

There are instances in LC and SFC also, where simple dissolution of the sample is all the preparation that is required. Samples must be free from any insoluble material, which would be trapped at the column head and increase column operating pressure in HPLC and SFC, or would remain at the point of application and interfere with layer development in planar techniques. In LC, care must also be taken that sample components will not precipitate on mixing with the mobile phase. This can often occur with aqueous samples containing salts and proteins, which are frequently insoluble in aqueous-organic solvents used as mobile phases in reversed-phase chromatography. The ideal solvent has the same composition as the mobile phase and this will cause minimum disturbance to the chromatographic system.

8.4.1   Solvent extraction

Partitioning of the sample (or sample extract) between two immiscible liquids, in which the analyte(s) and its matrix have different solubilities, provides additional selectivity and the opportunity to increase the analyte concentration by reduction in the volume of extract. For example, trace organic compounds present in water samples can be isolated and concentrated by extraction of large volumes of water with relatively small volumes of non-polar solvents (e.g. *n*-hexane). This type of extraction, termed liquid–liquid extraction is also described as solvent extraction. The latter term is generic, and also includes extraction of a solid with a liquid solvent. Liquid–liquid extraction is widely used and its advantages and disadvantages are well known. It has the advantages of low equipment costs, the sample extract is compatible with the requirements of chromatographic systems and a wide variety of selectivities can be exploited by varying the solvent. There are, however, a number of disadvantages associated with the relatively large volumes of frequently toxic and flammable solvents used in the extraction process, such as disposal and the resulting environmental concerns. Moreover, solvent extraction is a labour-intensive procedure. In practice, emulsion formation can be a problem particularly if the emulsion is persistent and cannot be broken by techniques such as refrigeration, salting out, filtration through a glass wool plug, centrifugation or addition of a small volume of organic solvent. Water droplets entrapped in the organic extract can be removed by filtration through dry filter paper, followed by washing the paper with fresh organic solvent. Other problems encountered in trace analysis are contamination from glassware and solvents, and analyte losses owing to sorption on laboratory ware. Solvents must be of high purity in order to minimize problems of contamination and background interferences in subsequent chromatographic procedures.

Selection of extracting solvent

The extracting solvent is usually selected for its extraction efficiency and selectivity, its inertness and its boiling point. Other factors which should be considered are the toxicity of the extracting solvent, relative densities of the two phases and their tendency to form emulsions. In general, solvents with low boiling points are favoured in order to facilitate subsequent recovery of the analyte by distillation or evaporation of the solvent. Hence, low molecular mass hydrocarbons such as pentane, hexane and toluene, ethoxyethane and dichloromethane are widely used. Benzene and chloroform have been used but are best avoided because of their toxicity. These relatively non-polar solvents are nonselective and therefore extract lipophilic materials, such as lipids, together with the analyte of interest. Various approaches have been used to improve the selectivity of extractions. *n*-Butyl chloride, for example, provides some selectivity in the recovery of drugs from biological fluids. Methods based on salting-out procedures are also used for this purpose. As an illustration, salt-solvent extractions, such as ammonium carbonate–ethyl acetate [8], yield reproducibly high recoveries for basic drugs. Partitioning between dimethyl sulphoxide and an organic solvent can be very useful because the water solubility of dimethyl sulphoxide can be exploited to alter its selectivity.

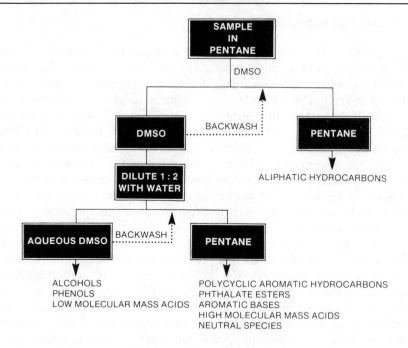

Fig. 8.3. Schematic outline of a fractionation scheme using liquid–liquid extraction.

Dimethyl sulphoxide provides a high extraction efficiency for compounds that contain hydrogen bonding or π-electron-rich sites [4]. As an illustration of an extraction procedure, polar and non-polar solutes can be separated (Fig. 8.3) by partitioning between pentane and dimethyl sulphoxide. The polar solutes recovered in the dimethyl sulphoxide may be further fractionated into hydrogen bonding and neutral polar solutes by addition of water and back-extraction. Dilution of the dimethyl sulphoxide with water reduces the extent of π-electron attraction with the solute but does not appreciably reduce hydrogen bonding interactions.

Further general observations may be made. Liquid organic salts (e.g. ethylammonium nitrate) offer some unique properties as extracting solvents [9]. Carbon disulphide has a very low flame ionization response and is useful in GC for the determination of volatile compounds that are otherwise obscured by the solvent peak. Ethoxyethane is sometimes favoured as an extracting solvent where pentane or isopentane are more suitable. For example, in the extraction of fermentation products [10], lower molecular mass alcohols dominate in an ether extract but are largely unextracted by hydrocarbons, which are the preferred extractants, despite the lower efficiency of hydrocarbon extraction. Ethoxyethane is also prone to peroxide formation which can lead to artefacts.

Extraction efficiency and selectivity

Selectivity refers to the ability of a solvent to extract one component of a solution (represented by A) in preference to another. The selectivity of a solvent for a given component can be determined from phase diagrams or distribution data, in the form of a partition coefficient:

$$D_c = \frac{[A_{organic}]}{[A_{aqueous}]} \qquad\qquad 8.1$$

This strategy is little used in analytical chemistry principally because of the lack of distribution data. This situation exists because of the enormity of the task of compilation and because distribution data depend critically on experimental design. As a result, the choice of an extracting solvent is based on either experience or semi-empirical considerations. In order to avoid a completely empirical approach, the usual rule governing solubility, that is, like dissolves like, is invoked. Thus, polar solvents are used to extract polar substances from non-polar media and vice versa. The ideal solvent has a large partition coefficient for the analyte, thus ensuring quantitative recovery, together with a low tendency to dissolve and co-extract interfering matrix components (i.e. high selectivity). Because even matrices which are seemingly similar are, in fact, highly variable, it is not possible to find a solvent system that satisfies all needs, and compromises have to be made.

The efficiency of extraction depends on the affinity of the solute, A, for the solvent, the relative volume of the two phases (sample versus extracting solvent) and the number of extractions. The per cent extraction (E) may be calculated from equation 8.2.

$$E = 100\left[1 - \frac{1}{(D_c V_r + 1)^n}\right] \qquad\qquad 8.2$$

where V_r is the volume ratio given by $V_{organic}/V_{aqueous}$. It is informative to compare the extraction of a substance under different circumstances, as illustrated by the extraction of a 10^{-2} M aqueous solution (100 ml) of compound Z (assume D_c is 3.0) with: (i) 100 ml of ethoxyethane, (ii) 500 ml of ethoxyethane, and (iii) multiple extraction with 2×100 ml portions of ethoxyethane. In (i) the per cent extraction is given by equation 8.2 as:

$$E = 100\left[1 - \frac{1}{(3.0 \times 1 + 1)^1}\right]$$
$$= 100(1 - 0.25)$$
$$= 75\%$$

Similarly, the per cent extraction in (ii) and (iii) is calculated as 93.8% and 93.8%, respectively. The efficiency of using several extractions with a smaller volume of extracting solvent, rather than one large extraction, is clearly demonstrated by these examples. The improved efficiency brought about by multiple extractions falls off rapidly, however, as the number of extractions increases. For the above example, increasing the number of extractions in (iii) to 3, 4, or 5 increases the extraction efficiency to 98.4%, 99.6% or 99.9%, respectively. In general, little is gained by dividing the extracting solvent into more than five portions.

Allowance must be made for the effect of the sample matrix on the extraction efficiency. In some instances, for example in the determination of total versus bound drug in a serum sample, this may be advantageous. Nevertheless, possible differences in the rate and extent of extraction between idealized test systems and real systems must always be considered in determining recoveries. Furthermore, solvents which are immiscible always have a small, but finite, solubility in each other. This mutual miscibility is such that volumes of liquids after equilibration may differ from those of the original volumes, and some allowance for this effect in calculations may be necessary if the effect is severe. A more important consequence of the miscibility is its effect on the solubility of analyte and matrix in the solvents, and on the selectivity of the extracting solvent.

Types of extraction

Continuous versus discontinuous liquid–liquid extraction

Extractions can be performed either continuously, where equilibrium between the two immiscible phases is not necessarily attained, or discontinuously with the attainment of equilibrium. Continuous extraction procedures include the lighter-than- or heavier-than-water, liquid–liquid extractors and droplet, centrifugal or rotating disk countercurrent separators generally used for isolation of analytes. Discontinuous extractions are liquid–liquid partition (e.g. batch extraction in a separating funnel) and countercurrent distribution, which may be used for either isolation of analyte or clean-up of sample extracts.

Discontinuous batch extraction

In cases where a suitable extracting solvent is available, in which the desired component has a favourable partition coefficient, batch extraction in a separating funnel is favoured for its simplicity. In practical terms, batch extraction consists of simply adding the two phases to a separating funnel, shaking (with regular venting to prevent build-up of pressure), allowing the two liquids to settle and separate, and recovering the desired layer. The extraction can be repeated several times, if necessary, to improve recovery. Multiple extraction is usually used to improve recovery rather than to improve selectivity. Conversely, backwashing of the desired phase is frequently employed to reduce the level of co-extracted material and hence to improve selectivity. With batch extraction, the partition coefficient must be large because of the practical limit to the number of extractions and the volume of extracting solvent that can be used, otherwise extraction becomes a very tedious procedure and results in a very dilute extract of the sample. In some instances, the partition can be controlled by, for example, formation of ion pairs or pH control. As an illustration of this effect, the extraction of organic compounds with acidic or basic properties varies greatly with the pH of the aqueous solution. This fact can be exploited during solvent extraction [11] to effect a fractionation into acidic, basic and neutral extracts as illustrated in Fig. 8.4. Chemical reactivity can be exploited to further fractionate the various extracts. For instance, in the scheme of Fig. 8.4,

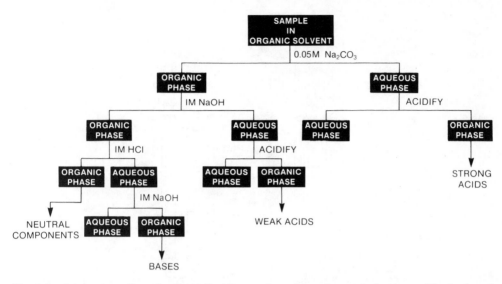

Fig. 8.4. Schematic outline of a liquid–liquid extraction scheme for separation into acidic, basic and neutral fractions.

aldehydes and ketones can be selectively recovered from the neutral organic fraction by the formation of water-soluble complexes [12].

Continuous extraction

Continuous extraction is used when the partition coefficient of the analyte–solvent combination is so small that multiple extraction becomes ineffective in achieving efficient recovery, the sample volume is large or the rate of extraction slow. Numerous continuous liquid–liquid extractors (see Fig. 8.5) using heavier-than- and lighter-than-water solvents have been described [13–15]. In most systems, the extracting solvent is allowed to percolate repetitively through an immiscible solvent in which the sample is dissolved. Where the extracting solvent is more dense, it can simply be allowed to drop through the liquid to be extracted or, where the extracting solvent is less dense, it can rise through the liquid buoyantly. The extracting solvent is continuously recycled and purified by a distillation process. The process can be continued indefinitely with a relatively small volume of extracting solvent. Concentration factors up to about 10^5 can be achieved, but require several hours. This is not as severe a restriction as it sounds because unattended operation is possible and several units can be operated in series.

Countercurrent extraction

A countercurrent extraction is one in which two immiscible liquids are contacted as they flow through each other in opposite directions. The Craig extraction apparatus [16] is probably the most sophisticated and efficient system available for the performance of countercurrent fractionation. It consists of a large number of

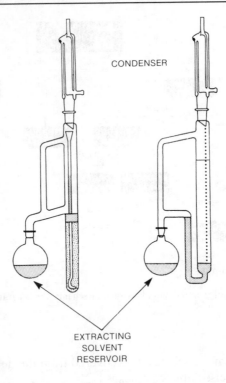

CONDENSER

EXTRACTING
SOLVENT
RESERVOIR

Fig. 8.5.   Continuous liquid–liquid extraction apparatus for use with heavier than water, and lighter than water extracting solvents.

connected glass tubes which are used to perform a series of extractions between fresh portions of the two phases in a series of discrete steps. Countercurrent extraction differs from the continuous technique in that fresh portions of only one phase are introduced in the latter. The process of countercurrent fractionation is rather tedious and it has been replaced by countercurrent chromatography, which is a particularly attractive technique for the preliminary fractionation of complex mixtures. One of the features of countercurrent chromatography is the absence of an adsorbent phase on which irreversible sorption of some polar compounds, or denaturation of polypeptides, may occur. Droplet countercurrent chromatography (Fig. 8.6) is simple and reproducible; solvent consumption is low but typical separation times extend up to several hours.

Applications

The applications of liquid–liquid extraction of both organic and inorganic species are too numerous to present here. Various aspects relating to the extraction of environmental samples [17], fire debris [18] and biological fluids [7, 8, 19, 20] are reviewed elsewhere. The principles of sample recovery are well illustrated by 'A Master Analytical Scheme' which was developed [21] for the analysis of volatile organic compounds in water. The scheme involves recovery of organic compounds from water and their fractionation into five groups. It represents the first attempt at

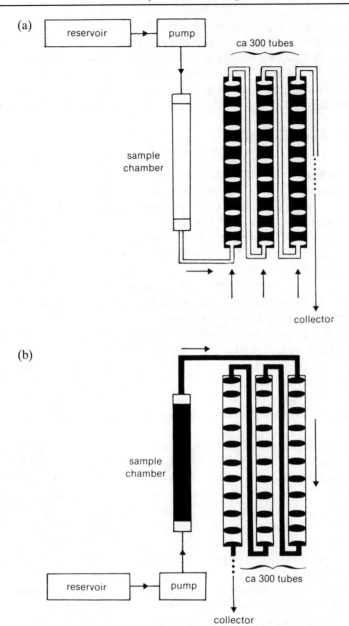

Fig. 8.6. Droplet countercurrent chromatograph. (a) Ascending mode, (b) descending mode.

providing a comprehensive analysis scheme for volatiles in water. Liquid–liquid extraction was an integral part of the scheme for the recovery of extractable organics.

An application of liquid–liquid extraction that is likely to see further development is with inorganic materials. Metal ions can be separated from each other, and recovered from sample matrices, using extraction with an organic solution of complexing agent. There are many complexing agents with broad-spectrum reactivity. Alternatively, complexing agents can be designed with group selectivity, or to exhibit selectivity for an individual cation. The latter approach has considerable potential for preparation of solid phase extraction systems (including membrane systems) (Section 8.5.3) by immobilizing the complexing agent on an inert support.

Extraction of solid samples

The extraction or leaching of a solid sample such as a soil, plant material or tissue with a solvent is related to liquid–liquid extraction. Techniques used for this purpose are Soxhlet extraction, shake-flask methods, sonication and various homogenization procedures [22–24]. Sonic treatment with a probe or in a bath is useful for extracting coarse granular materials, such as soil, sediments and air particulates. Shake-flask methods also work well when the matrix is a porous solid and the analyte is very soluble in the extracting solvent. The solvent may be warmed or heated under reflux to improve extraction efficiency. However, homogenization procedures usually provide better recovery of analyte because of more efficient contact between the sample and solvent. Plant materials, for example, are usually dried, ground and milled prior to extraction to improve efficiency.

Tissue samples present one of the most difficult matrices. Traditionally, the tissue is homogenized with a water-miscible organic solvent to promote efficient extraction. Most methods involve lengthy acid or base hydrolysis steps, and numerous volumetric transfers of the homogenate [25, 26]. In many cases, enzyme digestion provides a rapid alternative to these methods for tissue samples. The liquefied sample produced by hydrolysis can be extracted using procedures developed for liquid samples.

Soxhlet extraction can be used for samples that are difficult to extract efficiently by other means. The sample is finely ground to ensure the greatest possible surface area for liquid contact, and is then placed into an extraction thimble. The thimble is placed in an extraction chamber which allows passage of extracting solvent while it retains the sample. The solvent is vaporized, condensed and collected in the extraction chamber where it percolates through the sample. When the extraction chamber is full, the solvent is returned to the boiling flask via a siphon arrangement. The process is a sequence of fillings and siphonings which is characterized as a discontinuous extraction.

The main disadvantages of Soxhlet extraction are that the extraction times are frequently long (10 h is not uncommon), contact of sample and solvent is a 'static' process, except for the brief period when siphoning occurs, and the analytes must be stable at the boiling point of the extractant as they eventually accumulate in the boiling flask. Samples containing moisture can create problems, particularly when non-polar extracting solvents are used. Water has a small, but finite, solubility in such solvents and its presence may substantially alter the selectivity of the solvent. There are some practical concerns, particularly if mixed solvents are used without proper care. Depending on relative volumes and size of extraction chamber, a low boiling temperature fraction may collect in the chamber and, by the time it is returned to the boiling flask, a substantially higher boiling temperature mixture may be present. Under these circumstances, the low boiling temperature fraction may become superheated and the condenser will be unable to cope. A second problem arises when the cool solvent returning to the boiling flask causes cessation of boiling. In this case, severe bumping may arise during the reheating. In this situation, boiling chips are of little help because they retain their effectiveness for only a few hours.

Reduction in volume of sample extract

Conventional extraction procedures result in dilution of the analyte with a large volume of solvent, which must be reduced by preconcentration prior to chromatographic analysis. Various volatilization procedures are used to achieve preconcentration (see Section 8.6). Alternatively, micro-extraction methods provide an attractive alternative with economy of solvent consumption, generally without the need for further solvent reduction [4]. A number of devices have been described for this purpose [27, 28].

8.4.2   Supercritical fluid extraction

Supercritical fluid extraction (SFE) has a long history as a physicochemical phenomenon and industrial and engineering applications, such as the decaffeination of coffee and extraction of crude oil from rocks, are standard procedures. Supercritical fluids have also been used in a number of analytical techniques [29] such as nuclear magnetic resonance spectroscopy. However, supercritical fluids are now normally associated by the analytical chemist with either SFC (Chapter 7) or SFE. Despite this long history, SFE has developed as an analytical technique only in the last decade. A computer search of *Chemical Abstracts* listings for relevant journals revealed only two papers on SFE prior to 1986, compared with 26 articles on this technique published from 1986 to mid-1989 [30]. As an analytical technique, SFE is currently in an evolutionary stage of development. It is rapidly gaining interest as a sample preparation technique because of its time and cost savings, relative to other methods, for extracting components of interest from solid sample matrices. A review of the literature [31–36] indicates that SFE is much faster, more efficient, safer and less labour intensive than conventional solvent extraction.

Supercritical fluids offer several advantages over liquids for selective extraction [4]. Extraction rates in SFE are greatly enhanced compared with liquid extractions. Extraction times in SFE are typically less than 30 min compared with the several hours using Soxhlet extraction with liquids. This difference is attributed to two factors. First, the low viscosity (Table 8.2) and absence of surface tension in

Table 8.2 Typical properties of common fluids for selected operating conditions.

Property	He	CO_2	CO_2	H_2O
Operating temperature (°C)	200	100	35	20
Operating pressure (MPa)	0.15	8	20	7
Aggregation	Gas	Supercritical fluids		Liquid
Density (g cm^{-3})	10^{-4}	0.1	0.8	1.0
Viscosity (10^{-3} Pa s)	0.02	0.02	0.1	1.0
Diffusivity (cm^2 s^{-1})	10^{-1}	10^{-3}	10^{-4}	10^{-5}

supercritical fluids increase the speed of percolation of the fluid into the interstices of the sample matrix. Second, the transfer of solute out of the sample matrix, as a result of the high diffusion coefficient of solutes in supercritical fluids, is typically 1–2 orders of magnitude higher than in comparable liquid solvents. The efficiency of SFE for extraction of many non-polar to moderately polar compounds has been demonstrated by recovery data using spiked samples [37]. This conclusion is also supported by recent results on samples containing incurred residues [38]. The efficient recovery of more polar analytes still presents problems.

Variables which must be optimized for successful SFE include the choice of supercritical fluid, pressure and temperature conditions, extraction time and sample size. Much of the information presented in Chapter 7 on SFC is relevant to the discussion of supercritical fluid extraction.

Equipment

Instrumentation for SFE consists of a suitable pump, a heating block or oven, a pressure chamber (typical volumes 0.5 ml to 50 ml) to contain the sample for extraction, control valves, and provision for collecting extracted components. A suitable system can be assembled in the laboratory but care must be taken to ensure that all components are rated for the working pressures. Alternatively, commercial units of varying sophistication, including fully automated systems, are available.

A suitable high-pressure pump is the key to successful SFE. The pump must deliver a known pressure of the extraction fluid to the extraction vessel, which is controlled at a temperature above the critical temperature of the supercritical fluid. Diaphragm and reciprocating piston pumps have had limited application but continuous displacement or syringe pumps are superior for SFE. Among the advantages of syringe pumps are better pressure control and convenience, since no cooling is required during operation. Syringe pumps are also more appropriate for dynamic SFE where the range of flow-rates extends from the low ml min^{-1} range up to 10 ml min^{-1} or more. Ideally, the pump should have both pressure and flow control and readouts for both parameters. Most extractions are performed isobarically (at constant pressure/density), although pressure programming is sometimes used to alter selectivity for fractionation procedures.

Current instrumentation is capable of operating up to 100 MPa but theory predicts that many useful extractions can be performed at higher pressures and temperatures. These conditions will enhance the application of SFE to include higher molecular mass solutes. Most instruments offer limited pressure control at the lower extraction pressures required for recovery of volatile materials. Improvements in instrument design at this end of the pressure scale will also enhance the viability of SFE, as an alternative to headspace analysis. In comparison with headspace analysis, SFE has the potential to encompass a wide range of volatile to non-volatile components.

SFE can be performed in either a static or a dynamic mode. In static mode, the sample and fluid are sealed in the extraction vessel, which is heated to the required temperature and, after a set period of time, a valve is opened to allow the analytes to be swept into a collection device. Dynamic SFE is performed by allowing the supercritical fluid to flow through the extraction vessel. Pressure is maintained by a flow restrictor that also allows fluid to depressurize into the collection device. Various flow restrictors have been used. For example, the original flow restrictors consisted of a short length of fused silica tubing with an i.d. of 10–50 μm. More recently, crimped stainless steel and micrometering valves have been examined. The major advantage of dynamic SFE is that the supercritical fluid is constantly renewed during the extraction. However, static SFE requires less fluid. Nevertheless, criteria for the selection of one mode or the other have not been established.

As currently practised, SFE is performed either on-line [39–41] or off-line [42]. In off-line mode, the analytes are collected after depressurization for subsequent analysis, whereas on-line or coupled SFE involves direct transfer of the extract to another system. The off-line mode offers greater flexibility in choice of operating parameters for method development, and is inherently simpler because there is no need to simultaneously optimize operating conditions for the two techniques. For these reasons, the market has been more aligned to off-line SFE. On-line methods are usually combinations of SFE with ancillary techniques such as SFC, GC or, less commonly, LC. The principal advantages of on-line SFE are the elimination of sample handling between extraction and measurement and the ability to more easily automate the entire system.

Most fluids used for SFE are gases at ambient conditions and, in off-line SFE, are therefore vented from the collection vessel through the restrictor to atmosphere. Quantitative trapping of the volatile analytes can be problematic because it depends on the recovery of analytes from the expanded gas flow upon depressurization (e.g. a supercritical CO_2 flow of 1 ml min^{-1} corresponds to a gas flow of 500 ml min^{-1}). In general, quantitative recovery of the extracted analytes is facilitated at lower extraction flow-rates. Nevertheless, with trace analytes, trapping in empty vessels, even with cooling, has been problematic because of losses from aerosol formation [43]. In such cases, collection of the analytes in a small volume of an appropriate solvent, such as methanol, or on a sorbent cartridge (e.g. silica or a bonded phase) packing is essential. Evaporation of the collecting solvent owing to high gas flows is minimized by the cooling of the solvent caused by the expanding supercritical fluid.

Selecting a supercritical fluid and extraction conditions

Theoretical models that are available for predicting supercritical fluid behaviour with a range of analytes and matrices are severely restricted [29] in practise, and experimental optimization of extraction conditions is the most viable alternative, at least for the present. This can be attributed to two reasons. First, analytical SFE rarely involves recovery of a single target compound, but rather the recovery of a complex variety of analytes. In such cases the prediction of optimal conditions is complicated by the need to optimize the extraction for groups of compounds. Second, the matrix may modify the solubility behaviour of the analytes because of

Table 8.3　Characteristics of selected compounds used in supercritical fluid extraction.

Compound	Critical temperature (°C)	Critical pressure (MPa)	Maximum solubility parameter (J cm^{-3})$^{1/2}$
Ethylene	10	5.1	13.5
Ethane	32	4.8	13.5
CClF$_3$	29	3.8	16.0
N$_2$O	36	7.2	21.7
CO$_2$	31	7.3	21.9
NH$_3$	132	11.2	27.0
Methanol	240	7.8	29.4

Data from Monin *et al.* [46].

competitive interactions between the analyte–matrix and analyte–extracting fluid, making the prediction of optimal extraction conditions very difficult [44]. Neverthe-less, programs are available (Isco, Inc.) for generating density/solubility isotherms and Hildebrand solubility parameters that facilitate selection of extraction conditions to replace existing liquid–solvent extraction methods.

Various fluids have been used for SFE (Table 8.3). The solubility parameters listed in Table 8.3 are maximum values (compare Table 7.5) that are approached only at very high pressures. For practical reasons, related to hardware considera-tions, SFE is usually performed at lower pressures that yield lower solvent strengths. Moreover, practical considerations frequently limit the choice of extracting fluid, despite the availability of fluids with a range of polarities that could be used to optimize an extraction. It is because of its attractive practical properties that carbon dioxide has been the supercritical fluid of choice in most analytical applications. These properties include a low critical pressure and temperature, low reactivity and toxicity, high purity at reasonable cost and widespread use as a mobile phase for SFC. The search for alternatives to organic solvents, because of environmental concerns over their disposal, has also enhanced the popularity of supercritical carbon dioxide as an extractant for both Soxhlet-type extractions of lipid material [45] and for recovery of trace level analytes [42].

The solvating ability of supercritical CO$_2$ mimics that of non-polar to moderately polar solvents. Some generalizations [44] regarding the extraction possibilities and limitations of supercritical CO$_2$ are:

- Non-polar lipophilic solutes including hydrocarbons, terpenes, ethers, esters and ketones with molecular masses up to 400 Da can be easily extracted in the pressure range up to 30 MPa. Supercritical CO$_2$ is also a reasonable extraction medium for moderately polar species, including polycyclic aromatic hydrocar-bons, polychlorinated biphenyls, organochlorine pesticides, aldehydes, esters, alcohols and fats.
- The presence of several polar functional groups (e.g. hydroxyl or carboxyl groups) in a solute makes extraction more difficult, or even impossible. Nevertheless, benzene derivatives containing three phenolic groups are still extractable, as are compounds with one carboxylic group.

- Polar solutes such as sugars, glycosides, amino acids, proteins, cellulose and synthetic polymers are not extractable in the range up to 40 MPa. Non-polar oligomers are only slightly soluble.
- Fractionated extraction is possible for solutes differing in molar mass, vapour pressure and polarity.

Unfortunately, the solvent strength of supercritical CO_2 is insufficient to quantitatively extract very polar analytes at typical working pressures (8–60 MPa). Fluids with higher solvent strength are available (see Table 8.3) but their use is limited by practical considerations. Supercritical methanol is an ideal solvent but has a high critical temperature and it is a liquid at ambient conditions, which makes sample recovery and concentration following the extraction more difficult. Supercritical ammonia also has excellent solvent properties and is very attractive for this reason. However, ammonia is chemically reactive and difficult to pump as it tends to dissolve pump seals. Because of the problems of finding a suitable polar fluid, extraction of highly polar analytes is most often achieved by addition of a small amount of an organic modifier to CO_2. The modifier can either be added directly as a liquid to the sample before commencing the extraction or it can be introduced via the pumping system. Selection of modifiers has been largely empirical and a variety of organic modifiers has been used, including alcohols, organic acids, carbon disulphide and dichloromethane, but methanol is most popular. The use of organic modifiers greatly enhances the range of analytes that can be quantitatively extracted at reasonable pressures [30].

Selective extractions

Supercritical fluids approach the densities of liquids and, consequently, the solvation power of supercritical fluids also approaches that of liquids. In contrast to liquids, however, the solvating power of a supercritical fluid is exponentially proportional to its fluid density, and both density and solvation power can be varied by changing pressure and/or temperature. Hence, the efficiency and selectivity of a supercritical fluid extraction can be varied over a wide range by manipulating these variables, or by the addition of small amounts of organic solvent modifiers to the supercritical fluid. Rapid exhaustive extractions are best performed at high pressures where solvation power is a maximum. Conversely, selectivity can be achieved by operating at a lower extraction pressure, but usually at the expense of solute solubility. The application of pressure control to the selective extraction of polycyclic aromatic hydrocarbons is illustrated in Fig. 8.7. This approach allows an extraction to be optimized for a particular compound class by simply changing the pressure and, to a lesser extent, the temperature of the extraction. For example, alkanes can be extracted from urban air particulates with CO_2 at 7.5 MPa (and 45°C), whereas polycyclic aromatic hydrocarbons are not extracted from air particulates below 30 MPa. Thus, selective extraction is possible by sequentially extracting the air particulates at these two pressures [31].

Selective fractionation of, for example, analytes from unwanted co-extracted material can also be achieved by the use of selective sorbents such as reversed-phase materials, silica or speciality packings. These can be either packed in the extraction vessel or incorporated in-line but downstream from the extraction vessel [29].

Fig. 8.7.   Open tubular (15 m × 0.25 mm fused silica) gas chromatograms of polycyclic aromatic hydrocarbons obtained by supercritical carbon dioxide extraction of a complex matrix at progressively higher pressures. The extractions were performed sequentially at 50°C and at pressures of (a) 7.5 MPa, (b) 9.0 MPa, (c) 11 MPa and (d) 15.0 MPa corresponding to CO_2 densities of 0.20, 0.30, 0.52 and 0.71 g cm^{-3}, respectively. The final chromatogram (e) illustrates exhaustive extraction at 25 MPa (density = 0.83 g cm^{-3}) where all components are soluble. Reprinted with permission from Wright *et al.* (1987) *Anal. Chem.*, **59**, 640, Copyright (1987) American Chemical Society.

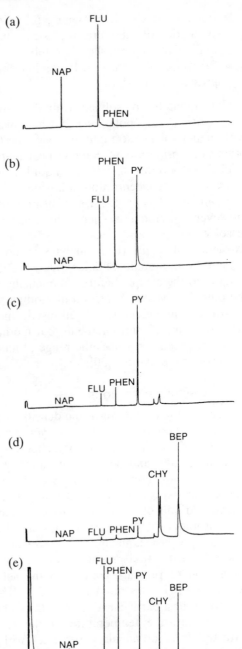

Concluding remarks

The development of analytical SFE is still in its infancy but its continued expansion seems assured as new applications continue to be discovered. For example, SFE has shown potential as an alternative to thermal desorption for the recovery of samples adsorbed on Tenax [40]. Compared with thermal desorption, SFE shows an enhanced recovery of high molecular mass and thermally labile analytes. Furthermore, regulatory agencies may mandate its use for routine analyses. The United States Environmental Protection Agency began a 4-year program in 1989 to develop SFE methods, with the ultimate goal of reducing the amount of dichloromethane used for extractions in EPA methods by 95% from levels current in 1989 [30].

8.4.3 Volatilization processes

Volatilization implies the conversion of a solid or liquid into the gaseous state. Related processes, including distillation and sublimation, can be the basis of a separation and are therefore useful sample recovery procedures. However, a simple change in physical state is not sufficient to effect a resolution. For separation, there must also be a mass transfer of at least one of the sample components. The versatility of volatilization techniques is related to the ease with which a mass transfer can be effected when a substance is in the gas phase. Many volumes have been written which describe the physical basis of volatilization processes and only a brief summary will be provided here.

Sublimation

Sublimation involves direct vaporization and condensation of a solid without formation of an intermediate liquid phase. Sublimation is used most extensively for separation of volatile components from non-volatile components. Hence, it is a useful preparative technique for recovery of compounds that can be sublimed at reasonable temperatures, at either atmospheric or reduced pressure. The temperature at which a sublimate first becomes noticeable in a given system is often referred to as the sublimation temperature, but it is useless for comparative purposes unless the variables such as pressure, the size and geometry of the sublimation apparatus are standardized. The effect of reduced pressure on the sublimation temperature of common substances is indicated in Table 8.4. The list of substances amenable to sublimation includes polycyclic aromatic hydrocarbons, benzoic acid, salicylic acid,

Table 8.4 Effect of reduced pressure on the sublimation temperature of some common substances [47].

Compound	Melting point (°C)	Sublimation temperature (°C)	
		at 101 kPa	at 65 to 130 Pa
Naphthalene	79	36–38	25
Benzoic acid	120	43–45	25
β-Naphthol	122	43–45	33–35
Urea	132	59–61	49–52
Anthracene	215	77–79	28–31

camphor, saccharin, quinine, cholesterol, palmitic and stearic acids, acetylsalicylic acid and atropine. Despite its successes, sublimation has limited application and it is difficult to optimize for quantitative trace analysis because of the dependence of extraction on the vapour pressure of analytes at a given temperature and pressure, and on the nature of the matrix.

Freeze drying

The removal of water from a sample by vacuum sublimation is termed lyophilization or, more commonly, freeze drying. In freeze drying, the material to be dried is customarily frozen and placed in a vacuum chamber and the water passes from the solid to the vapour state directly. It is widely used for preserving biological samples. Freeze-dried biological samples are conveniently ground and sieved to a fine powder, providing a homogeneous sample for analysis. The loss of some volatile organic components may occur if the temperature is allowed to rise while the sample remains under a high vacuum. Freeze drying has also been used to concentrate non-volatile organic compounds in aqueous solution prior to liquid–liquid extraction. The advantage of preliminary freeze drying is the reduction in the volume of solvent required in subsequent extraction steps.

Distillation

Distillation is defined as the partial vaporization of a liquid in a still or retort by the application of heat and the subsequent collection of the condensed vapour in a separate vessel. It may be performed in either a simple or fractional mode of operation. Pure liquids can be distilled but, inasmuch as no separation is involved, this discussion is limited to liquid mixtures. In the broadest sense, distillation is a separation process based on the difference in composition between a liquid and the vapour in equilibrium with it. The more volatile components become concentrated in the vapour phase. The separation efficiency in simple distillation is determined mainly by the relative volatilities of the individual components of the mixture. In fractional distillation, however, because it involves repeated mass and heat exchanges, the length of the fractionating column and the reflux ratio are of decisive importance. Fractional distillation enables components with differences in boiling points of about 0.5°C to be separated; this figure can be as low as 0.05°C if extremely efficient fractionating columns are employed. Nonetheless, complete resolution of all but the simplest mixtures is not possible, and distillation is widely used for the crude fractionation of complex mixtures of hydrocarbons, petroleum products, etc., into broad fractions roughly according to molecular mass. Nowadays, distillation is more commonly used for the isolation of volatile organic compounds from liquid samples or the soluble portion of solid samples. Equipment is available to enable distillations to be performed on the microscale but, more commonly, distillation is used in cases where macroscale fractions are required for spectrometric and other methods of identification.

Fundamental concepts of distillation are defined [48] in a number of ways in the literature. There is even confusion over basic terminology. For sample preparation, four distillation techniques are widely used. These variants are known by an assortment of names but those most recognized by analytical chemists are simple distillation, fractional distillation, steam distillation and assisted distillation.

Simple distillation

In simple distillation, the vapours emerging from the evaporating surface move uniformly without contacting condensed liquid until they reach the condensing surface. Thus, the composition of the vapour leaving the liquid does not change as it moves from the surface of the liquid to the condenser. The apparatus used for simple distillation usually consists of a flask fitted with a condenser and a product receiver. Simple distillation is always a one-stage batch process and, as a separation process, it is not very effective. Nevertheless, it is suitable for the separation of simple mixtures where the components differ widely in boiling point, for crude fractionations into boiling point cuts with defined boiling point ranges, and for stripping a liquid from a soluble involatile solid.

Distillation apparatus is commonly constructed of glass with ground-glass joints. These joints must be sealed to prevent escape of sample vapour or loss of vacuum in reduced pressure distillation. The use of PTFE sleeves or wetting with an appropriate solvent is usually adequate. The application of lubricants is unsatisfactory in analyses involving low levels of analyte owing to the possibility of contamination.

Fractional distillation

Fractional distillation produces more efficient separations than simple distillation. It can be performed in a continuous mode, involving continuous sample feed to the column or in a batch mode, where a discrete batch of sample is placed in the boiling flask. Most laboratory fractional distillations are performed in batch mode. In fractional distillation, part of the condensed vapour, the reflux, returns to the still, meeting the rising vapour in its passage. This type of distillation is an equilibrium process in which the composition of the boiling liquid and the distillate is constantly changing as the distillation proceeds. This equilibrium is achieved by a countercurrent flow process occurring through contact of vapour and condensed vapour at 'plates' or 'trays' in a fractionating column. Hence, a fractionating column and a reflux-ratio controlling device are equipment requirements in addition to those used for simple distillation. The exchange of material and heat is a physical process taking place at the interface between the two phases, and the surface area for exchange should therefore be as large as possible. This surface can be provided by an empty column but, in most cases, surface area is increased by using columns packed with spirals, helices, etc., or plate columns. Plate columns differ considerably in design although the basic requirement is a series of practical plates fixed perpendicular to the axis of the column, to allow the liquid and vapour phases to be mixed thoroughly. Packed columns also have a number of advantages and are widely used in the laboratory. Their efficiency depends on the packing.

Prior to the development of chromatography, fractional distillation was one of the most important techniques for the characterization of complex multicomponent mixtures. The approach usually involved fractional distillation of the mixture into a series of narrow boiling point fractions which were then characterized by their physical properties. Since chromatography is more efficient than distillation, the combination of simple distillation and chromatography is generally more convenient and faster than methods involving fractional distillation. Nevertheless, fractional distillation still warrants consideration when devising a sample preparation program.

Steam distillation

Steam distillation [49] is a special kind of simple distillation in which the total distillation pressure is the sum of the vapour pressures of the steam and of the component to be distilled. Thus, the temperature needed to vaporize the sample component is reduced.

This procedure has been used in many fields for separating components on the principle of differences of their vapour pressures over water. Any inert carrier gas can be used, but steam is normally the most economical because of its low molecular mass compared with the compounds being distilled. Steam distillation involves vaporization of a mixture by continuously blowing steam from an external steam generator through the sample, or by boiling the sample and water in admixture. The volatile organic compounds are entrained by the steam and recovered after condensation in a receiving flask. Various types of apparatus have been described [50–55].

Steam distillation is most suited to the isolation of low molecular mass polar compounds from an aqueous matrix. In many cases, the boiling point of substances can be lowered by azeotrope formation with water and this expands the range of sample types that can be steam distilled. The recovery of some analytes can be improved by pH adjustment, salting out or the addition of a codistillation solvent.

With large sample sizes, phase separation occurs after condensation, but in most cases adequate recovery of components from the distillate requires additional treatment. In general, this has been achieved by liquid–liquid extraction with a non-polar organic solvent. This off-line approach has been used for a variety of samples, despite the low efficiency for recovery of polar organic compounds. One of the advantages of steam distillation is that the solvent extract is, in many instances, suitable for direct examination by chromatography without further treatment. More recently, steam distillation has been combined [56] with solid phase extraction for the recovery, from aqueous samples, of a wide range of solutes having relatively low volatilities.

Assisted codistillation

Assisted or sweep codistillation is a technique originally used for the recovery and clean-up of organophosphate pesticides from fruit and vegetable extracts [57] but is now widely used for the isolation of trace levels of pesticides from a range of fatty samples [58]. It has achieved greatest popularity in Australia for pesticide recovery from meat samples. Following improvements in apparatus design, a commercial instrument with the acronym UNITREX (universal trace residue extractor) was developed by Scientific Glass Engineering (Melbourne, Australia). This apparatus is described in detail by Luke *et al.* [59] in a method reported for recovery of organochlorine pesticides in beef fat. The UNITREX system consists of a flow controller and temperature controller with an oven containing 10 sample fractionating tubes. Simultaneous clean-up of up to 10 samples is therefore possible.

The technique is based on forced volatilization [60]. The oil or liquefied fat sample is injected by syringe through a septum into a heated all-glass fractionating tube packed with silanized glass beads. The sample is swept through the tube with a flow of nitrogen. The high temperature and high velocity flow of nitrogen disperses the fatty material as a thin film on the surface of the beads, while the pesticides are volatilized or stripped and collected in a sodium sulphate/Florisil trap. The pesticides are then eluted from the trap with a suitable solvent for subsequent analysis by chromatography. Tubes are cleaned between samples by washing with hexane. More thorough cleaning is necessary after about 50 samples have been processed. Typical recoveries of pesticides are 80–100%, although individual recoveries depend on the volatility of the pesticide and its thermal and catalytic stability. Optimum recovery requires calibration of flow and temperature. The technique is rapid and economical in comparison with conventional methods of sample preparation. In one programme in Australia, sweep codistillation has been used to recover pesticides from 600 000 samples of beef at a cost of AU\$18 M. The estimated cost using the method prescribed in the Official Methods of the Association of Official Analytical Chemists was AU\$50 M.

8.5   Sample Clean-up Methods

After the analyte has been recovered (by dissolution, extraction, distillation, etc.), some modification of the sample extract is usually necessary prior to chromatographic analysis. This modification may involve a simple filtration, or it may be more extensive and involve a preliminary fractionation if this was not included in the recovery step. Various derivatization processes may also be included where it is necessary to change the chemical form of the analyte to improve its chromatographic performance (separation, detection or identification). The steps involved in sample clean-up often determine the total analysis time and may contribute significantly to the final cost of the analysis, both in terms of labour and the consumption of materials. The ultimate success of the analysis is often critically dependent on the success of the clean-up step. Indeed, manipulation of the sample is frequently a major contributor to the overall analytical precision which may greatly outweigh any variables in the chromatographic process itself.

The goals of clean-up are to achieve removal of matrix interferences, concentration or dilution of the analyte, and preparation of the analyte in a form most suited to the needs of the final chromatographic technique.

8.5.1   Membrane techniques

A number of processes based on membrane technology may be exploited in sample preparation but the most useful are dialysis, microfiltration, ultrafiltration and reverse osmosis. These processes all involve the migration of molecules across membranes by molecular diffusion. Commonly used separation processes based on membrane technology, and their range of application in terms of particle size, are shown in Fig. 8.8.

The range of particles encountered in different sample types is surprisingly large (Fig. 8.8). At the upper extreme, some very coarse particles found in an aqueous suspension may exceed 100–1000 μm in diameter. Some biological materials such as pollen grains exceed 100 μm in diameter, whereas the diameter of the largest microbial cells is greater than 10 μm. Certain bacteria are less than 0.3 μm in

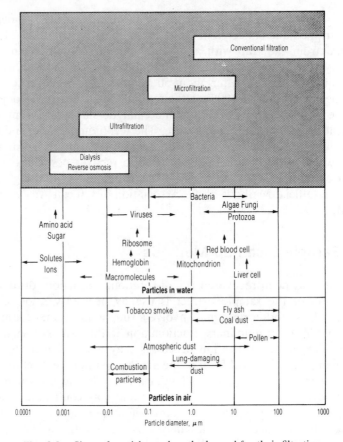

Fig. 8.8. Sizes of particles and methods used for their filtration.

diameter and viruses can be as small as 10 nm. These ranges also encompass the size of particulates encountered in air samples. In water samples, a distinction is often drawn between soluble and particulate matter, with the classification based on filtration through a 0.45 μm filter. Macromolecules range in size from 5 nm to 5 μm while solutes and ions are even smaller. However, most particles are not spherical and their filtration behaviour depends on the direction in which the particle encounters the filter. We can thus speak of a nominal particle size, which is the size of a sphere which behaves in a manner similar to that of the particle in question.

Filtration

Filtration might appear to be a simple process but it is actually very complex [61]. Filters function not only by a sieving mechanism and physically trapping particles, but also various sorption phenomena may be involved. Capillarity may also play a significant role in filter function in some instances. Various filter types are used in separating different sized particles from liquids. The simplest is the net or sieve filter (e.g. nylon bolting cloth) used to separate algae and small animals from ocean or lake water. This is suitable for particles down to about 60 μm. The most common filter used for clarification of samples is the depth filter, consisting of a fibrous sheet usually made from paper or glass fibres. During filtration, particles are trapped within the matrix of the filter. A filter aid such as diatomaceous earth may be deposited on the surface of the filter base to assist in the filtration. Depth filters are frequently used as prefilters to remove larger particles prior to membrane or ultrafiltration. Membrane filters made of a colloidal polymer film are used in general particle filtration. They have a complex, open, colloidal type structure, and in contrast to depth filters, mainly retain caught particles on the surface of the filter. Asymmetrical membranes are used primarily for pressure-driven processes such as ultrafiltration and reverse osmosis. Their structure consists of a thin ($0.1–2\,\mu m$) polymeric film on a highly porous sublayer support $100–200\,\mu m$ thick. Pore sizes in these membranes are of molecular dimensions. They are mechanically very strong in order to support high hydrostatic pressures without deformation. Membranes for ultrafiltration are not as mechanically strong as the reverse osmosis membranes but they have a much higher permeability to pure water.

Pore size is one of the most commonly discussed characteristics of a filter. With the exception of nucleation-track Nuclepore membrane filters, pore sizes are determined by indirect measurements involving calculation using a mathematical model rather than by direct measurement. The pore size may be expressed as a nominal or absolute value indicating that 98% or 100%, respectively, of the particles of that size are retained by the filter. To a first approximation, retention by depth filters can be considered nominal and retention by membrane filters as absolute. In either case, the pore size does not refer to the size of the filter pores, but to the size of the particles retained by the filter. Membrane filters with pore sizes ranging from $0.1–10\,\mu m$ are available.

In addition to the filtration of liquids, microporous membranes are widely used in air filtration both for air pollution analysis and for general air filtration in the determination of the weight distribution of particles in the atmosphere. Membranes made from PTFE are particularly useful. The hydrophobic nature of the membrane makes it virtually impermeable for aqueous solutions in filtrations utilizing hydrostatic pressure up to $1\,MPa$, although vapours can pass the membrane quite freely. Both national and international agencies have established standard methods of analysis and set threshold limits for a large number of airborne contaminants. Two of these agencies, NIOSH (National Institute of Occupational Safety and Health Administration) and OSHA (Occupational Safety and Health Administration) have issued several methods which specify the use of membrane technology for sample collection followed by chromatographic analysis. Airborne contaminants which can be collected on membrane filters include benzidine, dinitrobenzene, picric acid and hydroquinone.

Microfiltration, ultrafiltration and reverse osmosis are basically similar processes differing only in the size of the particles to be separated and the membranes used. In all three processes, the sample is exposed to the surface of a semi-permeable membrane and separation occurs under the influence of a hydrostatic driving force. Microfiltration describes the separation of particles with diameters in the range of $0.1–10\,\mu m$ from low molecular mass solutes or solvent. When the components to be separated are true molecules or particles not larger than $0.3\,\mu m$, the process is termed ultrafiltration. The osmotic pressure of the sample is negligible in both

Table 8.5 Membrane separation processes.

Separation process	Membrane type (pore diameter in μm)	Driving force	Method of separation	Application
Dialysis	Microporous (0.0001–0.01)	Concentration gradient	Diffusion in convection-free layer	Separation of salts and solutes from macro-molecular solutions
Reverse osmosis	Asymmetrical skin-type	Hydrostatic pressure difference, 2–10 MPa	Solution diffusion mechanism	Separation of salts and solutes from solutions
Micro-filtration	Symmetric, microporous (0.1–10)	Hydrostatic pressure difference, 0.01–0.1 MPa	Sieving mechanism due to pore radius, plus sorption	Clarification, sterile filtration
Ultra-filtration	Asymmetric, microporous (0.001–1)	Hydrostatic pressure difference, 0.05–0.5 MPa	Sieving mechanism	Separation of macro-molecular solutions

microfiltration and ultrafiltration and low hydrostatic pressures are used. When low molecular mass (<2000) solutes are to be separated, osmotic pressure becomes significant in comparison with the hydrostatic pressure acting as the driving force. The separation is then referred to as reverse osmosis. Symmetrical, microporous membranes are used for microfiltration. The membranes used in ultrafiltration and reverse osmosis generally have asymmetrical structures. In ultrafiltration, the chemical nature of the membrane has minimal effect on the separation characteristics, in contrast to reverse osmosis. The main areas of operation and other relevant data for some membrane separation processes are summarized in Table 8.5.

Microfiltration

All chromatographic methods require that the sample be free from particulate matter to prevent fouling of syringes, capillary tubing, column end frits and other hardware components. Many samples, such as water samples, appear suitable for direct analysis. Despite appearances, filtration through a membrane filter of porosity 0.45 μm or less is mandatory before processing samples by either HPLC or ion chromatography. Column life will invariably be reduced by failure to perform this simple step. For other chromatographic techniques filtration is a wise precaution. Sample filtration is very straightforward if disposable filter units are employed.

Dialysis

During dialysis, sample components are separated by selective permeation through a semi-permeable membrane as a result of a concentration gradient. Two dialytic techniques may be differentiated; passive dialysis which involves diffusion of particles of a specified molecular mass range through a neutral membrane, and active (or Donnan) dialysis, which is the transfer of ions of a specified charge sign through an ion-exchange membrane. Both approaches have been applied to the clean-up of samples for chromatography. Donnan dialysis is particularly useful in the clean-up of samples for ion chromatography and detailed discussion of this role is available elsewhere [6].

The permeation selectivity of a neutral membrane (i.e. for passive dialysis) is largely determined by the relative sizes of different molecules, although molecular shape and electrical charge have a secondary influence. For this reason dialysis has been most effective in separating small molecules from large molecules and thus finds widest application in areas such as biochemistry or clinical chemistry. The sample extract and a dialyser or acceptor solution are pumped separately through a chamber, where they are separated by a membrane. Static, pulsed or continuous operation is possible. In the conventional static mode, the donor channel is filled with the sample and then the flow is stopped. The sample is subsequently dialysed by using a continuously moving acceptor phase. The continuous flow mode, where both the donor and acceptor phases are moving, gives relatively low recoveries but is advantageous if speed is the main goal. In all three modes, low molecular mass substances pass from the sample through the membrane and into the acceptor solution, while macromolecules are retained. Any ions in the acceptor solution will pass in the opposite direction and equilibrate with the sample extract. Hence, by careful choice of the acceptor solution, the final composition of both phases can be controlled. For example, protein extracts can be desalted by removing low molecular mass solutes, and ionic species and drugs or other small molecules can be recovered from biological fluids. Because solvent flow due to osmosis is in the opposite direction to solute flow, membrane clogging is not a problem. Nevertheless, periodic replacement of membranes exposed to protein is necessary because of ageing effects caused by protein deposition, which reduces the overall efficiency of the process. The quantity of solute that passes through the membrane is controlled by the ratio of the flow rates of the sample and dialysate solutions. The thickness and porosity of the membrane, the fluid channel geometry and area of contact of the solutions, the concentration gradient and composition of fluids on either side of the membrane [62] are additional factors that determine the overall efficiency of the process.

Dialysers are commonly used in many automated analysers, such as the Skalar Flow Analyser, which combines on-line dialysis and solvent extraction with HPLC. This automated sample preparation procedure greatly simplifies analysis and permits rapid and reliable quantitative analysis.

Ultrafiltration

Ultrafiltration is a separation process in which molecules, primarily macromolecules, are separated from solution by filtration through membranes. Membranes were originally constructed of cellulose nitrate but now they are also made of polyvinylidene difluoride, nylon, PTFE, cellulose acetate, polyacrylonitrile and polysulphone. The selection of the right membrane type depends on the nature of the sample and its solvent (see Table 8.6). Conceptually, ultrafiltration is not different from particle filtration processes except that the particles involved are of smaller dimensions. Membranes are characterized by their nominal molecular mass cut-off. This is an indication of the retention characteristics of the membrane in terms of molecules of known size. The word nominal reflects the fact that the shape and size of the

Table 8.6 Properties of membrane filters.

Membrane type	Nature	Characteristics
Nylon	Hydrophobic	Resistant to most organic solvents but not to acids
Polytetrafluoroethylene	Hydrophobic	Resistant to almost anything
Cellulose acetate	Hydrophilic	Least resistant of common filter types but lowest protein-binding membrane
Polysulphone	Hydrophilic	Resistant to most alcohols, ethers, acids and bases; low protein-binding, able to withstand higher temperatures
Polyvinylidene difluoride	Hydrophilic	Resistant to most solvents but not dimethylsulphoxide or dimethylformamide
Polyvinylchloride	Hydrophilic	Resistant to most solvents but sensitive to acids, ketones and dioxane
Acrylic	Hydrophilic	Resistant to most materials but sensitive to acids and ketones

molecule influence its rate of migration through the membrane. For example, linear molecules with a molecular mass exceeding the molecular mass cut-off may pass through the membrane, whereas charged molecules smaller than the molecular mass cut-off may be retained. Membrane filters are commercially available for separating molecules in the approximate molecular mass range of 10^3–10^6.

Ultrafiltration is commonly applied either as pressure ultrafiltration or centrifugal ultrafiltration. The pressures needed [63, 64] for ultrafiltration are not high (<0.5 MPa), so that peristaltic pumps or gas pressure can be used. Simple installations may suffice for many applications but more elaborate equipment is probably desirable in most instances. A number of manufacturers (e.g. Millipore and Nuclepore) provide specially designed equipment for use with ultrafilters. Ultrafilter membranes can be constructed as flat sheets in plate and frame installations, as hollow fibre cartridges in which bundles of hollow fibres are fused to a cylindrical housing and as spiral-wound cartridges in which a flat membrane is wound around a central core. Cartridge units are available (Millipore) that fit into a laboratory centrifuge [65] and consist of a polypropylene sample tube and receiver tube separated by a membrane filter. The procedure for using these tubes is very simple and involves loading the sample into the sample tube which is then centrifuged (1000–2000 g) following which the filtrate is collected from the receiver tube [66, 67].

General applications of ultrafiltration are:

- Concentration of high molecular mass analytes by removal of solvent.
- Fractionation of molecules by separating molecules on the basis of size.
- Desalting of organic solutes.
- Isolation of low or high molecular mass analytes in the presence of each other [68].
- Efficient removal of proteins from samples [69].

In many applications, membrane filters are used to filter samples containing liquids other than water. There is a wide variation in the compatibility of different types of membrane filters to organic solvents, acids and bases (Table 8.6). Manufacturers conduct extensive compatibility tests on the filters they market and provide detailed information in their technical literature. For any new application, compatibility of the sample and membrane can be tested by soaking the membrane for a test period of 24 h at the required temperature.

Reverse osmosis

Osmosis is a molecular diffusion process in which solvent molecules migrate through a membrane from a dilute solution to a more concentrated one. The most common solvent involved in osmosis is water. In reverse osmosis, the movement of solvent is in a direction opposite to that in which it would be in normal osmosis. This requires the application of a hydrostatic pressure to the system which exceeds the osmotic pressure that exists across the membrane. Reverse osmosis is most widely used to reduce the solute concentration of aqueous solutions, as in the desalination of salt water. Analytically, it is suitable for preconcentrating relatively large volumes of dilute solutions such as natural waters. The membranes used in reverse osmosis have lower molecular mass cut-off values than those used in ultrafiltration so that low molecular mass solutes cannot pass. The pressures involved (typically 10 MPa) are, as a consequence, much greater than in ultrafiltration. The equipment is also more substantial (Fig. 8.9) because of the necessity for backpressure regulators and pressure relief valves. The required operating pressure is a function of solute concentration: higher pressures are required with higher solute concentration. The efficiency of the purification is controlled by the flow-rate of the feed and the pressure differential across the membrane. Plate-and-frame, hollow fibre and spiral-wound membrane assemblies are used as in ultrafiltration. Filters must be replaced when backpressure becomes excessive owing to clogging of the filter.

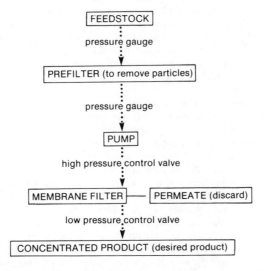

Fig. 8.9. Schematic arrangement of a single-stage system for reverse osmosis to be used for the concentration of organic solutes from natural waters.

8.5.2   Chromatography

Any of the chromatographic techniques described in this book can be used for sample fractionation or clean-up, including planar chromatography [70] and HPLC, especially when used in a semi-preparative mode [71]. Column chromatographic techniques have been favoured for purely practical reasons related to the ease of sample recovery. The application of these techniques to sample preparation follows directly from the principles described in other chapters and will not be further expanded. The separations described in this section for the clean-up of samples vary widely, but all involve column separation with a liquid mobile phase. In general, they are simple, being similar to the original work by Tswett, although some form of pumping and fraction collector may be used. In this format, LC is often described as classical LC or classical column chromatography and involves relatively large particle sizes in order to keep the pressure drop low and facilitate easy dry packing of the column. The techniques exploit polarity differences and differences in molecular size or ion-exchange capacity as the basis of fractionation. Materials used as sorbents are summarized in Table 8.7. Ion-exchange is particularly attractive for the recovery of ionizable substances since neutral molecules are easily washed from the ion-exchanger without affecting the recovery of the ionized components.

Table 8.7   Materials used for sample clean-up.

Type	Comments
Silica gel	Most widely used general purpose adsorbent; may irreversibly bind strongly basic compounds; usually slightly acidic but this may vary; activated by heating at 180°C for 8–12 h.
Alumina	Commercially available in neutral (pH 6.9–7.1), acidic (pH 3.5–4.5) and basic (pH 10–10.5) forms; neutral alumina is a good all-purpose adsorbent; basic alumina has strong cation-exchange properties in aqueous solution and is used to separate acid-labile substances; acidic alumina acts as an anion-exchanger and is used to separate organic acids; activated by heating at 400°C for 8–12 h.
Florisil	Widely used for clean-up of pesticide extracts; irreversible adsorption of basic compounds may occur; there may be some batch-to-batch variation in properties; activated by heating at 130°C for 8–12 h.
Diatomaceous earth (e.g. celite)	Occasional use as an adsorbent for polar molecules or labile samples.
Carbon	General purpose adsorbent for removal of organic species from aqueous solution; large surface area; surface properties vary because of the variable number of acidic functional groups formed during preparation.
Porous polymers	Widely used for the recovery of trace organic compounds from waters and aqueous biological samples; characterized by high affinity for neutral organic compounds, chemical inertness and low water retention; analytes recovered by elution with organic solvents or thermal desorption.
Ion-exchangers	Used for the efficient recovery of acidic and basic substances such as phenols, amides, steroids, organic acids and amino acids for which recovery is often variable and incomplete from adsorbents.
Controlled porosity gels	These materials are not adsorbents; they are used for size exclusion; organic solvent compatible gels (e.g. Sephadex LH-20 and Bio-Beads SX) are used for sample clean-up to separate small molecules from a high molecular mass matrix; water-compatible gels are used to recover large biochemical molecules.

Adsorption

Traditionally, liquid–solid chromatography has been used as an adjunct to solvent extraction for the further fractionation of sample extracts. Early method developments involving adsorbents for clean-up of pesticide extracts were centred on Florisil and magnesia. Following method development and collaborative study, a procedure involving adsorptive clean-up was afforded official status by the Association of Official Analytical Chemists (AOAC). The principles of this method are now 30 years old, but it continues to enjoy official status in the AOAC official methods book [72]. There is a possibility of artefact formation in any adsorptive process but several workers have failed to demonstrate any significant degree of analyte rearrangement or degradation [10].

In a typical application involving adsorptive clean-up, sample is applied to the column in a small volume of a weak solvent and separation is achieved by sequential elution with a series of solvents of increasing solvent strength (increasing polarity for normal-phase systems such as silica or alumina). The components of interest are eluted and collected in a small number of fractions following which solvent is evaporated to preconcentrate the extract. Blank determinations and recovery experiments employing spiked samples should be performed routinely. The extract quality and analyte recoveries are significantly affected by the choice of eluting solvent, its amount, flow-rate and, in the case of ionizable analytes, pH [75]. Optimum recovery of all components cannot be achieved on a single system.

Columns and packings

The columns used for adsorptive clean-up are usually made of glass and consequently can be used only at low pressures. Typical column sizes are 20 to 100 cm in length with diameters of 1 to 4 cm. The volume of the column, and hence the amount of stationary phase required for packing, increases with the square of the column diameter. A typical column contains between 50 and 200 g of packing, which means that low-cost packings are essential. Sample capacity for a column of this size is of the order of 0.1–10 g (or 0.5–40 ml). The sample is usually added directly to the top of the column and flow occurs under gravity. Alternatively, a low-pressure pump or pressurized gas can be used to pressurize a solvent reservoir. The eluent is usually collected manually in separate tubes or an automatic fraction collector can be employed. Cleaning of the column packing can be achieved with polar solvents (for normal-phase systems) or the packing can be simply discarded after a single use. The latter is the preferable option and the one usually adopted.

Silica gel is the most popular general purpose adsorbent. Other adsorbents include alumina, Florisil, carbon and diatomaceous earth (Table 8.7). Silica, alumina and Florisil function as polar sorbents which strongly retain polar components when eluted with organic solvents of relatively low polarity. They are used to fractionate extracts into analytes with the same type and number of polar functional groups. The activity of the adsorbent is an important consideration, as this determines the fraction in which the components of interest elute. Careful control of the adsorbent activity is therefore required to obtain reproducible separations into well-defined fractions. In this connection, the water content of the sample must be considered as this can significantly alter the adsorbent activity and mobile phase polarity, particularly when a non-polar solvent is being used. As a precaution, sample extracts are dried with anhydrous sodium sulphate or molecular sieves prior to passage through the adsorbent column. A 1–2 cm plug of sodium sulphate is often added to the top of the column as a further precaution. Before using a new batch of adsorbent, it must be standardized to establish the elution order of analytes. Various approaches are used for this purpose but the most reliable is to examine a real sample spiked with the analytes of interest.

The surface of silica and alumina can be modified [73, 74] by coating with chemical reagents such as silver nitrate, sulphuric acid, sodium hydroxide or alkaline potassium permanaganate to improve the selectivity of the separation. These chemically modified adsorbents are used to retain selectively chemically active co-extractants. For example, the separation of alkanes from olefins is enhanced on silver-modified silica gel, which selectively retains the olefins by the formation of charge-transfer complexes.

Mobile phase

Mobile phase requirements are identical with those already discussed for liquid–solid chromatography (Sections 4.6.1 and 6.2.1), but the need for high volatility, to facilitate sample recovery, and low cost assume a new importance. Nonetheless, high purity solvents are essential to minimize the effects of any contaminants when the solvent is evaporated. Solvent strength and selectivity are the most important chromatographic characteristics and various measures of these properties have been developed (e.g., see Tables 4.10 and 6.2). The solvent polarity index, for example, is a measure of the intermolecular attraction between a solute and a solvent, but it provides no indication of the selectivity of the solvent. Snyder [76] proposed a solvent classification scheme in which solvents were classified into eight groups (Table 6.6) showing significantly different selectivities. The selectivity parameters, χ_c, χ_d and χ_n (Table 8.8) to a first approximation reflect the relative ability of a solvent to act as a proton acceptor, a proton donor and a dipole interactor, respectively. Comparing the data in Table 4.10 for tetrahydrofuran, ethyl acetate

Table 8.8. Properties of selected solvents.

Solvent	Viscosity (10^{-3} Pa.s, 25°C)*	Boiling point (°C)	Solvent selectivity parameters			Selectivity group
			χ_c	χ_d	χ_n	
n-Hexane	0.30	69				
Chlorobenzene	0.75	132	0.24	0.34	0.42	VII
Ethoxyethane	0.22	35	0.55	0.11	0.34	I
Dichloromethane	0.41	40	0.34	0.17	0.49	V
Tetrahydrofuran	0.46	66	0.41	0.19	0.40	III
Ethyl acetate	0.43	77	0.34	0.25	0.42	VI
Chloroform	0.53	61	0.28	0.39	0.33	VIII
Acetic acid	1.10	118	0.41	0.29	0.30	IV
Methanol	0.54	65	0.51	0.19	0.30	II

*1×10^{-3} Pa. s $\equiv 1$ cP.

and chloroform, for example, shows a similar polarity for the three solvents (P' approximately 4) while their selectivities (Table 8.8) are very different. As a further illustration, ethyl acetate, dichloromethane, ethoxyethane and 1-chlorobutane were compared [75] for the recovery of drugs from blood using diatomaceous earth sorbents. Dichloromethane yielded the cleanest extracts and the highest recoveries for acidic neutral compounds but the lowest recoveries for basic compounds, consistent with the weak proton donor properties (low affinity for basic compounds) of dichloromethane. Ethyl acetate provided high recoveries for the largest number of compounds and was favoured as an extracting solvent for this reason. However, the low selectivity of this solvent yielded lower quality extracts with numerous co-extractants. In general, the highest recoveries were achieved with ethoxyethane although it also co-extracted many interferents. This study highlights the complexity of selecting a solvent for use as an elution solvent.

Ion-exchange

The potential of ion-exchange was realized in the 1940s during the Manhattan Project with the first separations on organic ion-exchangers involving mixtures of uranium(VI), plutonium(IV) and their fission products. Synthetic ion-exchange materials have been available for four decades as relatively large beads used at low pressures, as distinct from the materials described in Section 6.3.3 which are used for HPLC. The most commonly used resins consist of a copolymer of polystyrene with divinylbenzene with functional groups attached to the phenyl rings to provide the ionic exchange sites. Divinylbenzene provides cross-linking that is usually between 4% and 16%. These materials are produced in the form of flexible spherical beads which are capable of swelling upon solvent uptake. Since most of the exchange sites are within the resin beads, swelling is essential for efficient operation. Excessive swelling is a problem as it changes the volume of the phase. In this regard, the extent of cross-linking is critical since it regulates the degree of swelling. Increased cross-linking, while increasing the retention and mechanical strength of the resin, decreases both the degree of swelling and the permeability or porosity. A compromise around 8% is usually necessary to provide sufficient porosity without excessive swelling.

Polystyrene-based resins are unsuitable for the chromatography of labile bio-polymers, which may be denatured by the hydrophobic regions of the resins. Moreover, large molecules cannot penetrate the pores of such matrices. Such difficulties have been overcome by introducing suitable exchange sites into hydrophilic substrates such as cellulose, agarose or dextran which are often cross-linked for added rigidity. A low density of exchange sites is introduced to permit elution of solutes under mild conditions. Like polystyrenes, these materials are limited to low-pressure operation.

Ion-exchange resins can be characterized as strong or weak according to the acid/base strength of the ion-exchange sites (acids for the cation resins and bases for the anion resins). A functional group can only function as an ion-exchanger when it is ionized, and this depends on its pK_a and the pH of the mobile phase. Materials containing sulphonate or quaternary amine functionalities are described as strong ion-exchangers because the extent of ionization of these groups is unaffected by pH

Table 8.9   Some commercially available ion-exchange resins used for sample preparation.

Trade name	Cation exchangers		Anion exchangers	
	Strong ($-SO_3H$)	Weak ($-COOH$)	Strong ($-N(CH_3)_3Cl$)	Weak ($NH(CH_3)_2Cl$)
Dowex*	50 W	–	1	3
Amberlite	IR-120*	IRC-50†	IR-400*	IR-45*
Permutit	Q-100	Q-210	S-100	S-300
Sephadex‡	SP	CM	QAE	DEAE

The matrix is identified as *polystyrene-divinylbenzene, †polymethacrylic acid-divinylbenzene, or ‡cross-linked dextran.

over a wide range. With resins containing weaker acidic or basic groups (e.g. carboxylate or tertiary/secondary amines) the extent of ionization depends greatly on pH, and these materials are called weak ion-exchangers because they are active in a limited pH range. Some common ion-exchange resins are listed in Table 8.9. In purchasing a commercial resin, the counter ion should be noted as it may be possible to avoid the need for lengthy conversions. Cation-exchange resins are usually sold in the hydrogen ion form and anion resins in the chloride form.

Liquid ion-exchangers are a special type of ion-exchange material. These materials include dialkyl esters of phosphoric acid and the monoalkyl esters of alkanephosphonic acids for cation-exchange chromatography and long chain amines or quaternary ammonium compounds such as trioctylmethylammonium chloride as anion-exchangers. The advantages of liquid ion-exchangers include low solubility in water, small surface activity and large capacities.

Prior to use, resins must be swollen by exposure to solvent, and buffered. The resin must first be converted into the form identical to the type of ions present in the mobile phase buffer. This can be achieved by suspending the resin in an excess of a 1–2 M solution of the salt of the new counter ion. Three resin bed volumes are sufficient to equilibrate most columns. After equilibrium has been reached, the supernatant is decanted and, after several washings with the salt solution, the resin is equilibrated in the appropriate buffer, de-aerated and the slurry is packed in the chromatography column. The mobile phase can involve either the starting buffer alone or a gradient in which the pH of the buffer is gradually increased (cation exchange) or decreased (anion exchange). Ionic strength is usually increased simultaneously.

After use resins may be regenerated, that is, reconverted into the form in which they are required for use. If an ion is strongly sorbed to the resin, it may be necessary to form a strong complex with the ion using a ligand such as diammonium citrate before regeneration. For removing organic compounds that strongly sorb on the resin, it may be necessary to use a mixture such as 1 M hydrochloric acid in ethanol. Resins should not be stored for long periods in aqueous suspension because there is the risk of microbial contamination. Cation resins can be stored in the hydrogen ion or salt (e.g. Na^+) form while anion resins should be stored in the salt form (e.g. Cl^-). Further details on regeneration are available in trade literature available from resin suppliers.

Gel permeation

Extracts can be fractionated on the basis of molecular size using controlled porosity gels with aqueous or organic mobile phases. They are particularly useful in the analysis of small molecules in a matrix of high molecular mass polymers or oils. By using gel permeation, the larger molecules are eluted ahead of the smaller analyte molecules thereby providing an efficient clean-up. Automation is easily achieved because total elution of the sample occurs, leaving a clean column which can be used repetitively.

Since separations based on gel permeation are controlled by the pore dimensions of the stationary phase, the pore sizes must be capable of differentiating between the analytes in a sample. A wide range of stationary phases is therefore currently available and these range from synthetic polymers to glasses. The earliest materials were soft nonrigid gels based, for example, on cross-linked polystyrene (e.g. Styragel) or polyvinylacetate (e.g. Merckogel) which covered a wide range of pore sizes suitable for the separation of compounds of differing molecular mass using organic mobile phases. In order to obtain compatibility with aqueous mobile phases, these materials are now available with hydrophilic surfaces, obtained by sulfonation of the polystyrene gels. Alternatively, semirigid hydrophilic gels are now available as cross-linked dextran (a polysaccharide, e.g. Sephadex) or polyacrylamide (e.g. Bio-Gel-P) and nonrigid materials as agarose (a polygalactopyranose, e.g. Sepharose and Bio-Gel-A).

The mobile phase in gel permeation is chosen primarily for its ability to dissolve the sample, although it must also satisfy requirements of compatibility with the stationary phase and the detector, if appropriate. Most gels can tolerate a wide range of organic solvents but there are exceptions and the trade literature should be consulted for up-to-date information.

8.5.3 Solid phase extraction

Solid phase extraction (SPE) has rapidly established itself as one of the premier sample preparation techniques for either trace enrichment or matrix simplification [77–81]. Compared with conventional liquid–liquid extraction, solid phase extraction is convenient, easy to use and less time consuming, requires much smaller amounts of solvents and is capable of producing cleaner extracts [19]. An added advantage is that solid phase extraction procedures are easily automated using flow processing manifolds or robotics [82, 83]. The technique has been approved by the US Environmental Protection Agency, for example, for the determination of organic contaminants in drinking water (Method 525 [84]). The various materials available for solid phase extraction have been compared by Majors [85]. A list of commonly used phases and a discussion of phase-selection criteria, retention mechanisms and conditioning procedures is included in a review by McDowall [86]. Trade names used for these materials include Sep-Pak (Waters), Bond Elut (Analytichem International), TechElut (HPLC Technology), Extra-Sep (Activon), Accubond (J & W) and Extrelut (Merck). Many of the studies of solid phase extraction have focused on recovery experiments and comparison with established solvent extraction methods [75, 87–90]. Although solid phase extraction works well with 'clean' samples, its applicability is limited by particulate matter. This often requires preliminary filtration of the sample and analytes may be lost in this step owing to adsorption on the solid particulate matter. With some sample types (e.g. some soils and effluents), filtration may become impossible and solid phase extraction may not be the method of choice for sample preparation in such cases. Alternatively, it may be possible to combine solid phase extraction with a membrane clean-up or steam distillation [56].

The principles and practice of solid phase extraction are drawn from both low-pressure column chromatography and HPLC. The packing materials for solid phase extraction are contained in low volume cartridges formed from highly purified polyethylene or polypropylene. Cartridges are available in several sizes, containing 35 mg–5.0 g of sorbent with the 100 mg and 500 mg cartridges being the most commonly used. For a given cartridge size, the sample volume that can be processed depends primarily on the breakthrough volume of the analyte and, hence, the elution strength of the mobile phase, the concentration of the analyte(s) and matrix and the flow-rate. The packing materials have particle sizes in the range 30–60 μm and are sandwiched between two polyethylene frits with a pore of about 20 μm. The materials used to construct the cartridges are compatible with most solvents, but there are differences between materials from different suppliers. Junk *et al.* [91] demonstrated that the contaminants leached from the cartridge limit the practical detection limits obtainable for the compounds of interest. For this reason, resistance to a particular reagent or solvent should be verified before use by performing a blank extraction on a cartridge.

Most cartridges have a Luer fitting at one end to connect a syringe needle to direct the effluent into a connection vial, or for easy connection to a sampling manifold. Solid phase extraction procedures are easily automated using robotics or special-purpose flow-processing units [82, 83, 92]. A number of companies supply purpose-designed cartridges for compatibility with automated sample preparation systems. Flow of solution through the cartridge is usually induced by vacuum suction using a vacuum box manifold that can process several samples simultaneously or by forcing liquid through the cartridge by connecting a syringe to one end. With low viscosity organic solvents and adsorbent phases, gravity is usually sufficient to cause flow to occur without any applied pressure. In these situations, a receiving vessel must be in place before adding sample to the cartridge. Typical flow-rates are 1–10 ml min^{-1} for normal and reversed-phase cartridges and 0.5–2 ml min^{-1} for ion exchange. A separation can be optimized through using more than one type of phase by connecting cartridges in series.

The packing materials used for solid phase extraction are similar to those currently used in HPLC, except for the larger particle size (30–60 μm versus 3, 5 or 10 μm). Hence, the whole range of chemistries exploited in HPLC phase technology can be applied to solid phase extraction, although silica and bonded phases are most common. Packing materials commercially available for solid phase extraction include:

- The common inorganic adsorbents (silica, alumina, Florisil).
- Bonded phase silanized silica materials (octadecyl, octyl, hexyl, ethyl, cyclo-hexyl, phenyl, cyanopropyl, diol, aminopropyl, N-propylethylenediamine, sulfonylpropyl, carboxymethyl, diethylaminopropyl, etc.),

- Cation and anion-exchange resins.
- Polymers (e.g., styrene divinylbenzene or polyvinylpyrrolidine).
- Chelating resins.
- Speciality phases.

An example of a speciality phase is provided by the trifunctional C_{18} phase developed by Waters to meet the requirements of US Environmental Protection Agency Method 525 [84] for trace organics in drinking water. Method 525 requires that the water sample be acidified to pH < 2 and passed through a C_{18} bonded silica phase cartridge. However, the large volumes of acidic water hydrolyse some of the bonds attaching the C_{18} functional groups to the silica support in the usual monofunctional C_{18} phases. The hydrolysis product, dimethyloctadecylsilanol, is eluted from the cartridge along with the compounds of interest, and may interfere in the analysis when using sensitive nonselective detectors. The newer trifunctional phase can withstand prolonged exposure to acidic solutions owing to the multiple siloxane bonds attaching the C_{18} functionality to the silica support and to other C_{18} functional groups.

Modes of operation

Cartridge columns can be employed in one of two ways. The first method is the selective removal of the analyte from the sample matrix and, in this approach, chromatographic conditions (notably eluting solvent) are selected to ensure strong retention of the analyte (that is, a very large capacity factor). The alternative operational mode is to selectively retain matrix components under conditions where the analyte is unretained (that is, capacity factor approaches zero). In general, it is inadvisable to attempt separation of solutes with capacity factors intermediate between the above-mentioned extremes. The reason for this is the difficulty of obtaining reproducible separations under these conditions because of the variable experimental factors (e.g., column efficiency, flow-rate, and packing reproducibility).

The general procedure for using solid phase extraction cartridges whether used off-line or incorporated on-line is as follows:

1. For maximum performance and reproducibility, solid phase extraction columns almost invariably require pretreatment in order to remove very fine particles of the packing material, to elute any contaminants resulting from the manufacturing process, or to condition the stationary phase in order to improve the efficiency of sample binding. Pretreatment involves passing a small volume (typically 3–5 ml) of an appropriate solvent through the cartridge. The solvent chosen for pretreatment depends on the nature of the phase: (i) for adsorbents (e.g., silica, Florisil) and normal-phase packings (e.g., bonded phase silica materials with diol or cyano functionality), a non-polar conditioning solvent such as hexane or dichloromethane is chosen. Prewetting of the cartridge may not be necessary for non-bonded phases; (ii) for reversed-phase packings (e.g., alkyl functionality), a polar conditioning solvent such as methanol, acetonitrile or tetrahydrofuran is chosen followed by rinsing with water or a buffer; (iii) for ion-exchange packings, water and/or an appropriate low ionic strength buffer (0.01 M) are chosen.

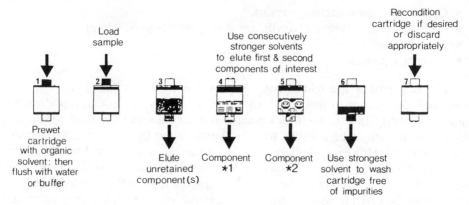

Fig. 8.10. General elution procedure for using reversed-phase Sep-Pak cartridges. Reproduced with permission from Waters Chromatography.

2. After the packing has been conditioned, sample is applied to the cartridge.
3. Depending on the operational mode, the sorbent bed is washed with a solvent or solvent mixture in order to either: (i) elute the analyte(s) and leave the unwanted matrix components retained on the sorbent bed (this approach is adopted in instances where the analyte(s) are present in high concentrations); or (ii) selectively retain the analyte(s) while eluting the interfering matrix components (this is the most common mode used for samples where the analyte(s) is present at low levels, and is called trace enrichment—enrichment factors of 1000 are easily achieved.
4. In cases of trace enrichment, the sorbent bed is washed with a stronger solvent (see Table 8.10) to elute the analyte(s) in as little volume as possible. Subsequent steps in the analysis should be borne in mind so that the solvent is compatible with the system being used.
5. The used cartridges are disposed of after a single use.

The general procedure for reversed phase cartridges is illustrated in Fig. 8.10.

In some instances, it may be desirable [93] to vary the manner in which the sample is applied to, and eluted from, the cartridge column by using frontal development in place of elution. In frontal development, the sample is applied continuously to the cartridge, discarding the first two or three column volumes and then collecting sufficient eluent for analysis. This method appears to have advantages for toxicity reduction evaluations, which are defined by the U.S. Environmental Protection Agency as 'an investigation conducted within a plant or municipal system to isolate the sources of effluent toxicity, specific causative pollutants if possible and determine the effectiveness of pollution control options in reducing the effluent toxicity'. The search for sources of effluent toxicity combines three elements: fractionation of wastewater effluents, testing of the various fractions for aquatic toxicity and analysing for specific chemicals. Solid phase extraction using frontal development is well suited to the fractionation step of this process.

Table 8.10 Guidelines for using solid phase extraction.

Sorbent	Silica, Florisil, alumina, amino or diol functionality	Octadecyl or cyanopropyl functionality	Ion-exchange
Typical sample loading solvent	Hexane, toluene dichloromethane	Water, buffers	Water, buffers
Typical elution solvent	Ethyl acetate, acetone, acetonitrile	Aqueous methanol, aqueous acetonitrile	Buffers, salt solutions
Sample elution order	Least polar sample components first	Most polar sample components first	Most weakly ionized sample components first
Solvent change needed to elute retained solutes	Increase solvent polarity	Decrease solvent polarity	Increase ionic strength or, for anion-exchange increase pH, or for cation exchange, decrease pH

Applications

Applications of solid phase extraction are too numerous to provide a comprehensive survey. An excellent discussion is available [8] on the use of solid phase extraction for the recovery of drugs and pesticides from biological materials. In addition to the more customary scientific literature, catalogues of most chromatography suppliers contain extensive listings of applications for different types of packings. An excellent source of information is the applications bibliography produced by Waters [94] which contains approximately 4000 applications of SPE.

Some examples, taken from four different fields, will illustrate typical applications. In the first application, aldehydes in air can be collected on cartridges impregnated with dinitrophenylhydrazine. The collection of air samples using solid phase extraction is convenient and simple, with many advantages over the use of bubblers and impingers. In the second application, food analysis, sugars, dyes and flavours can be conveniently determined on a single sample following solid phase extraction. A sample of soft drink (2 ml) is applied to an acetonitrile-conditioned C_{18} cartridge. The eluent is discarded and an additional 2 ml of soft drink is applied to the column and the eluent is collected and retained for sugar determination. Further development with aqueous acetonitrile recovers the dyes, and finally flavour oils can be eluted with straight acetonitrile. The third application involves determination of carbamates and pigments in biological materials such as green leaf vegetables after homogenization with dichloromethane. The extract is applied to a silica cartridge and the carotenoids collected in the eluent. Changing the solvent to dichloromethane:methanol elutes the carbamate pesticides followed by chlorophylls.

A fourth and important application area is the trace enrichment of very dilute solutions such as natural waters, where large sample volumes of two or more litres may have to be processed in order to obtain concentrations of analyte sufficient for convenient detection. Semi-volatile and non-volatile neutral and ionic species are conveniently recovered from waters in high yield by solid phase extraction. The recovery of volatile species is generally low and these are more readily isolated by other techniques, such as headspace analysis. With the exception of ion-exchange materials, the phases of most general use for this application are non-polar materials, such as bonded phase silanized silicas, in order to maximize the attractive forces between the analyte(s) and stationary phase. As an example, PCBs can be determined in water following trace enrichment on a C_{18} phase. The water sample (1 l) is drawn through an in-line filter and preconditioned C_{18} cartridge, where both PCBs and organics are retained. Interfering organic substances are removed by elution with aqueous methanol, and the PCBs recovered by elution with either hexane (1 ml) or methanol (1 ml) for determination by GC or HPLC, respectively. An added advantage of this procedure is the ability to collect the sample on the cartridge while in the field. This eliminates the need to transport large sample volumes to the laboratory and reduces the possibility of sample contamination and losses of analyte by adsorption to sample container surfaces. Furthermore, humic acids and organic particulates found in surface waters can shorten the life of the analytical column, cause ghost peaks, and disrupt the baseline if injected directly into a HPLC system. Solid phase extraction and ultrafiltration can be combined to remove this problem; membrane filters can be used to remove the particulates and the humic acids can be retained on a sorbent cartridge.

The versatility of solid phase extraction has been enhanced [95] by the preparation of membranes from efficient sorbent materials enmeshed in a network of inert PTFE fibrils. Membranes can be considered to be large diameter (25 mm and 47 mm membranes are available), short length (0.5 mm) extraction cartridges. The membranes are used with standard filtration equipment and can be made to have a variety of sorptive or reactive properties by varying the composition of the sorbent. Efficiencies equivalent to that of conventional SPE cartridges are achieved by using smaller particles (8 μm) and uniform packing in the membranes.

8.5.4 Chemical modification of the sample by pre-column reaction

For some samples, clean-up can be best achieved by using a chemical reaction to eliminate a matrix component. For example, in the determination of pesticides in high-fat samples, the traditional approach involved saponification of the sample as an adjunct to chromatographic clean-up. Other aspects of derivatization are treated elsewhere in this text, and these include the need to derivatize an analyte in order to enhance its separation or detection.

8.6 Preconcentration Techniques

For many samples, preconcentration is essential either because of the low levels of analytes present in the sample or as a result of the manipulative procedures involved in their recovery. Large solvent volumes resulting from extraction procedures can be concentrated using rotary evaporation, gas blow-down, a Kuderna–Danish evaporative condenser (Fig. 8.11) or an automated evaporative concentrator (EVACS). There are a variety of vaporizers which can be adapted to suit the needs of

SNYDER
COLUMN

Fig. 8.11.   Kuderna–Danish evaporative condenser.

different problems. Flash and film evaporators are available; the latter permit vacuum distillation. The best known types of film evaporators are the rotary-film, climbing-film or falling-film evaporators, operated at either atmospheric or reduced pressure. Film evaporators are suitable for isolating organic volatiles, but are more commonly used for reducing the volume of a sample or extract by removing some or all of the solvent from the sample. Nearly all rotary evaporators can be used to evaporate to dryness. They offer several advantages and are less likely than other distillation procedures to produce artefacts because the sample contact time with the heated surface is much shorter. Nevertheless, some loss of volatile components, proportional to their concentrations and vapour pressures, is inevitable during concentration steps, and the recovery of less volatile compounds may be lower than expected owing to entrainment of the material in the solvent vapour. Such losses of lower boiling temperature volatiles are usually less serious with extracting solvents of lower boiling point.

Gas blow-down is the simplest method of preconcentration. A gentle stream of pure gas is passed over the surface of the sample extract contained in a conical-tipped vessel. Several factors influence the rate of solvent evaporation including the gas flow-rate and the surface area of the solvent. In addition to the dangers of analyte loss by nebulization, heat losses during evaporation cause cooling of the sample and impurities in the gas may become concentrated in the extract. With some solvents, particularly diethyl ether, heat loss is so pronounced that moisture from the atmosphere will condense in the extract. For this reason, the sample tube is usually partly immersed in a water bath. In some instances it is necessary to reduce the extract to dryness (e.g., to ensure compatibility of solvent with chromatographic mobile phase) and, in these circumstances, subsequent dissolution of recovered analytes may be incomplete.

The Kuderna-Danish technique generally provides higher recovery of trace organic compounds but is slow.

Preconcentration of aqueous samples and samples for ion chromatography may be achieved in the manner described earlier for trace enrichment. A more detailed discussion is provided by Haddad and Jackson [6].

8.7　Contamination Effects

It is appropriate to conclude this chapter with a cautionary note. Analyte losses and contamination are important considerations in all sample handling procedures. Contamination may arise from various sources such as collection and storage vessels, volumetric glassware employed, the manipulative procedures used, filtration or clean-up devices, or the chromatographic hardware itself. Such contamination may affect the true level of analyte in two ways: directly, by contributing detectable levels of the analyte to the final solution or, indirectly, by promoting chemical reactions that cause levels of analytes to alter. Further discussion of this topic is beyond the scope of this text but may be found in most analytical chemistry texts.

References

1. Mieure J.P. and Dietrich M.W. (1973). *J. Chromatogr. Sci.*, **11**: 559.
2. Rudolph J., Muller K.P. and Koppmann R. (1990). *Anal. Chim. Acta*, **236**: 197.
3. Seila R.L., Lonneman W.A. and Meeks S.A. (1976). *J. Environ. Sci. Health Environ. Sci. Eng.*, **A11**: 121.
4. Poole S.K., Dean T.A., Oudsema J.W. and Poole C.F. (1990). *Anal. Chim. Acta*, **236**: 3.
5. Keith L.H., Editor (1984). *Identification and Analysis of Organic Pollutants in Air*. Butterworth, Boston.
6. Haddad P.R. and Jackson P.E. (1990). *Ion Chromatography: Principles and Applications, Journal of Chromatography Library Series, No. 46*. Elsevier, Amsterdam, pp. 409–462.
7. Liebich H.M. (1990). *Anal. Chim. Acta*, **236**: 121.
8. Furton K.G. and Rein J. (1990). *Anal. Chim. Acta*, **236**: 99.
9. Shetty P.H., Poole S.K. and Poole C.F. (1990). *Anal. Chim. Acta*, **236**: 51.
10. Jennings W. (1978). *Gas Chromatography with Glass Capillary Columns*. Academic Press, New York, pp. 112, 115.
11. Colgrove S.G. and Svec J.H. (1981). *Anal. Chem.*, **53**: 1737.
12. Wheeler D.H. (1962). *Chem. Revs.*, **62**: 205.
13. Jennings W.G. and Rapp A. (1983). *Sample Preparation for Gas Chromatographic Analysis*. Huethig, Heidelberg.
14. Korte F., Editor (1974). *Methodicum Chimicum*. Academic Press, New York, pp. 54–56.
15. Cziczwa J., Leuenberger C., Tremp J., Giger W. and Ahel M. (1987). *J. Chromatogr.*, **403**: 233.
16. Craig L.C. (1950). *Anal. Chem.*, **22**: 1346.
17. Hites R.A. (1977). In Giddings J.C., Editor, *Advances in Chromatography, Vol. 15*. Marcel Dekker, New York, p. 69.
18. Bertsch W. and Zhang Q.-W. (1990). *Anal. Chim. Acta*, **236**: 183.
19. Dean Rood H. (1990). *Anal. Chim. Acta*, **236**: 115.
20. Ried E., Editor (1981). *Trace-Organic Sample Handling*. Horwood, Chichester.
21. Garrison A.W. and Pellizzari E.D. (1982). In Albaiges J., Editor, *Analytical Techniques in Environmental Chemistry. 2*, Pergamon Press, Oxford, pp. 87–99.
22. Coover M.P., Sims R.C. and Doucette W. (1987). *J. Assoc. Off. Anal. Chem.*, **70**: 1018.
23. Junk G.A. and Richard J.J. (1987). *Anal. Chem.*, **59**: 1228.
24. Alford-Stevens A.L., Budde W.L. and Bellar T.A. (1985). *Anal. Chem.*, **57**: 2452.
25. Nakamura G.R., Liu Y. and Noguchi T.T. (1981). *J. Anal. Toxicol.*, **5**: 162.

26. Ozretich R.J. and Schroeder W.P. (1986). *Anal. Chem.*, **58**: 2041.
27. van Rensberg J.F.J. and Hasset A.J. (1982). *J. High Resol. Chromatogr., Chromatogr. Comm.*, **5**: 574.
28. Attygale A.B. and Morgan E.D. (1986). *Anal. Chem.*, **58**: 3054.
29. King J.W. and Hopper M.L. (1992). *J. Assoc. Off. Anal. Chem. Int.*, **75**: 375.
30. Hawthorne S.B. (1990). *Anal. Chem.*, **62**: 633A.
31. Hawthorne S.B. and Miller D.J. (1986). *J. Chromatogr. Sci.*, **24**: 258.
32. Hawthorne S.B. and Miller D.J. (1987). *J. Chromatogr.*, **403**: 63.
33. Hawthorne S.B. and Miller D.J. (1987). *Anal. Chem.*, **59**: 1705.
34. Hawthorne S.B., Krieger M.S. and Miller D.J. (1988). *Anal. Chem.*, **60**: 472.
35. Hedrick J. and Taylor L.T. (1989). *Anal. Chem.*, **61**: 1986.
36. Wright B.W., Frye S.R., McMinn D.G. and Smith R.D. (1987). *Anal. Chem.*, **59**: 640.
37. Murphy B.J. and Richter B.E. (1991). *J. Microcolumn Sep.*, **3**: 59.
38. Ramsey E.D., Perkins J.R., Games D.E. and Startin J.R. (1989). *J. Chromatogr.*, **464**: 353.
39. Davies I.L., Raynor M.W., Kathinji J.P., Bartle K.D., Williams P.T. and Andrews G.E. (1988). *Anal. Chem.*, **60**: 683A.
40. Hawthorne S.B., Miller D.J. and Krieger M.S. (1989). *J. Chromatogr. Sci.*, **27**: 347.
41. Anderson M.R., Swanson J.T., Porter N.L. and Richter B.E. (1989). *J. Chromatogr. Sci.*, **27**: 371.
42. McNally M.E. and Wheeler J.R. (1988). *J. Chromatogr.*, **435**: 63.
43. Wright B.W., Wright C.W., Gale R.W. and Smith R.D. (1987). *Anal. Chem.*, **59**: 38.
44. Engelhardt H. and Gross A. (1991). *Trends Anal. Chem.*, **10**: 64.
45. King J.W., Johnson J.H. and Friedrich J.P. (1989). *J. Agric. Food Chem.*, **37**: 951.
46. Monin J.C., Barth D., Perrut M., Espitalie M. and Durand B. (1988). *Adv. Org. Geochem.*, **13**: 1079.
47. Tipson R.S. (1951). In Weissberger A., Editor, *Technique of Organic Chemistry, Vol. IV.* Interscience Publishers, Inc., New York, p. 611.
48. Krell E. (1982). *Handbook of Laboratory Distillation*, 2nd edn. Elsevier, Amsterdam.
49. Dix K.D. and Fritz J.S. (1987). *J. Chromatogr.*, **408**: 201.
50. Peters T.L. (1980). *Anal. Chem.*, **52**: 211.
51. Godefroot M., Stechele M., Sandra P. and Verzele M. (1982). *J. High Resol. Chromatogr., Chromatogr. Comm.*, **5**: 75.
52. Bicchi C., Amato A.D., Nano G.M. and Frattini C. (1983). *J. Chromatogr.*, **279**: 409.
53. Donkin P. and Evans S.V. (1984). *Anal. Chim. Acta*, **156**: 207.
54. Curvers J., Noiji T., Cramers C. and Rijks J. (1985). *Chromatographia*, **19**: 225.
55. Page D., Newsome H.W. and MacDonald S.B. (1987). *J. Assoc. Off. Anal. Chem.*, **70**: 446.
56. Dix K.D. and Fritz J.S. (1990). *Anal. Chim. Acta*, **236**: 43.
57. Storherr R.W. and Watts R.R. (1965). *J. Assoc. Off. Anal. Chem.*, **48**: 1154.
58. Brown R.L., Farmer C.L. and Millar R.G. (1987). *J. Assoc. Off. Anal. Chem.*, **70**: 442.
59. Luke B.G., Richards J.C. and Dawes E.F. (1984). *J. Assoc. Off. Anal. Chem.*, **67**: 295.
60. Ott D.E. and Gunther F.A. (1964). *J. Agric. Food Chem.*, **12**: 239.
61. Lukaszewicz R.C. and Meltzer T.H. (1979). *J. Parenteral Drug Assoc.*, **33**: 187.
62. Babson A.L. and Kleinman N.M. (1967). *Clin. Chem.*, **13**: 163.
63. Liebich H.M., Bubeck J.I., Pickert A., Wahl G. and Scheiter A. (1990). *J. Chromatogr.*, **500**: 615.
64. Issachar D., Holland J.F. and Sweeley C.C. (1982). *Anal. Chem.*, **54**: 29.
65. Bock J.L. and Ben-czra J. (1985). *Clin. Chem.*, **31**: 1884.
66. Schoots A.C., Gerlag P.G.G., Mulder A.W., Peeters J.A.G. and Cramers C.A.M.G. (1988). *Clin. Chem.*, **34**: 91.
67. Schoots A.C., Peeters J.A.G. and Gerlag P.G.G. (1989). *Nephron*, **53**: 208.
68. Mehta A.C. (1989). *Trends Anal. Chem.*, **8**: 107.
69. Blanchard J. (1981). *J. Chromatogr.*, **226**: 455.
70. Nyiredy S. (1990). *Anal. Chim. Acta*, **236**: 83.
71. Walters S.M. (1990). *Anal. Chim. Acta*, **236**: 77.
72. Williams S., Editor (1990). *Official Methods of Analysis of the Association of Official Analytical Chemists.* AOAC, Arlington, p. 278.
73. Nestrick T.J. and Lamparski L.L. (1982). *Anal. Chem.*, **54**: 2292.
74. Lienne M., Gareil P., Rosset R., Husson J.F., Emmelin M. and Neff B. (1987). *J. Chromatogr.*, **395**: 255.

75. Logan B.K. and Stafford D.T. (1989). *J. Forensic Sci.*, **34**: 553.
76. Snyder L.S. (1974). *J. Chromatogr.*, **92**: 223.
77. Schuette S.A., Smith R.G., Holden L.R. and Graham J.A. (1990). *Anal. Chim. Acta*, **236**: 141.
78. Wilson I.D., Morgan E.D. and Murphy S.J. (1990). *Anal. Chim. Acta*, **236**: 145.
79. Needham L., Paschal D., Rollen Z.J., Liddle J. and Bayse D. (1979). *J. Chromatogr. Sci.*, **17**: 87.
80. Van Horne K.C. (1985). *Sorbent Extraction Technology*, Analytichem International, Harbor City.
81. Junk G.A. and Richard J.J. (1988). *Anal. Chem.*, **60**: 451.
82. Forbes S. (1987). *Anal. Chim. Acta*, **196**: 75.
83. McDowall R.D., Pearce J.C. and Murkitt G.S. (1989). *Trends Anal. Chem.*, **8**: 124.
84. Method 525, *Determination of Organic Compounds in Drinking Water by Liquid–Solid Extraction and Capillary Column Gas Chromatography–Mass Spectrometry* (Revision 2.1). Environmental Monitoring Systems Laboratory, USEPA, Cincinnati.
85. Majors R.E. (1986). *LC/GC.*, **4**: 972.
86. McDowall R.D. (1989). *J. Chromatogr.*, **492**: 3.
87. Musch G. and Massart D.L. (1988). *J. Chromatogr.*, **432**: 209.
88. Marble L.K. and Delfino J.J. (1988). *Am. Lab.*, **11**: 23.
89. Matsubara K., Maseda C. and Fukui Y. (1984). *Forensic Sci. Int.*, **26**: 181.
90. Steward J.T., Reeves T.S. and Honigberg I.L. (1984). *Anal. Lett.*, **17**: 1811.
91. Junk G.A., Avery M.J. and Richard J.J. (1988). *Anal. Chem.*, **60**: 1347.
92. Juergens U. (1986). *J. Chromatogr.*, **371**: 307.
93. Wells M.J.M., Rossano A.J. and Roberts E.C. (1990). *Anal. Chim. Acta*, **236**: 131.
94. McDonald P.D. (1991). *Waters Sep-Pak Cartridge Applications Bibliography*, 5th edn, Waters, Milford.
95. Hagen D.F., Markell C.G. and Schmitt G.A. (1990). *Anal. Chim. Acta*, **236**: 157.

Bibliography

Frei R.W., Editor (1986). *Handling of Samples in Chromatography*. Gordon and Breach Science Publishers.

Frei R.W. and Zech K., Editors (1988). *Selective Sample Handling and Detection in High-Performance Liquid Chromatography, Part A, Journal of Chromatography Library Series, No. 39A*. Elsevier, Amsterdam.

Jennings W.G. and Rapp A. (1983). *Sample Preparation for Gas Chromatographic Analysis*. Alfred Huethig, Heidelberg.

Karasek F.W., Clement R.E. and Sweetman J.A. (1981). Preconcentration for trace analysis of organic compounds. *Anal. Chem.*, **53**: 1050A.

Poole S.K., Dean T.A., Oudsema J.W. and Poole C.F. (1990). Sample preparation for chromatographic separations: an overview. *Anal. Chim. Acta*, **236**: 3.

McKenzie H.A. and Smythe L.E., Editors (1988). *Quantitative Trace Analysis of Biological Materials, Part 2*. Elsevier, Amsterdam.

Analytica Chimica Acta (1990). **236**: No. 1, is a special issue on sample preparation for chromatographic analysis.

Distillation and sublimation

McCormick R.H. (1987). In Gruenwedel D.W. and Whitaker J.R., Editors, *Food Analysis: Principles and Techniques, Vol. 4, Separation Techniques*. Marcel Dekker, New York, pp. 1–54.

Membrane techniques

Strathmann H. (1987). In Gruenwedel D.W. and Whitaker J.R., Editors, *Food Analysis: Principles and Techniques, Vol. 4, Separation Techniques*. Marcel Dekker, New York, pp. 135–217.

Supercritical fluid extraction

Bright F.V. and McNally M.E.P., Editors (1992). *Supercritical Fluid Technology: Theoretical and Applied Approaches in Analytical Chemistry, American Chemical Society Symposium Series No. 488.* ACS, Washington.

Engelhardt H. and Gross A. (1991). Supercritical fluid extraction and chromatography: potential and limitations. *Trends Anal. Chem.*, **10**: 64.

Hawthorne S.B. (1990). Analytical-scale supercritical fluid extraction. *Anal. Chem.*, **62**: 633A.

Levy J.M. (1991). Advances in analytical SFE. *Am. Lab.*, August.

Wenclawiak B. (1992). *Analysis with Supercritical Fluids: Extraction and Chromatography.* Springer, Berlin.

Westwood S.A., Editor (1992). *Supercritical Fluid Extraction and its Use in Chromatographic Sample Preparation.* Blackie.

Wright B.W., Frye S.R., McMinn D.G. and Smith R.D. (1987). On-line supercritical fluid extraction— Capillary gas chromatography. *Anal. Chem.*, **59**: 640.

Chromatography

Grob K. (1991). Liquid chromatography for sample preparation in coupled liquid chromatography–gas chromatography. A review. *Chimia*, **45**: 109.

Imai H., Masujima T., Morita-Wada I. and Tamai G. (1989). On-line sample enrichment and clean-up for high performance liquid chromatography with column switching technique: A review. *Anal. Sci.*, **5**: 389.

Jackson P.E. and Haddad P.R. (1988). Studies on sample preconcentration in ion chromatography. VII. Review of methodology and applications of anion preconcentration. *J. Chromatogr.*, **439**: 37.

Simpson N. (1992). Solid phase extraction: disposable chromatography. *Am. Lab.*, **24**(12): 37.

Qualitative and Quantitative Analysis 9

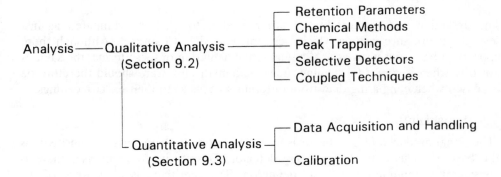

9.1 Introduction

Techniques used to interpret the chromatograms in order to provide both qualitative and quantitative data are described in this chapter. These techniques are essentially the same for all instrumental methods in which the separation occurs in a column. Both qualitative and quantitative analysis impose high demands on instrumentation and methods including those for evaluation of the chromatogram. Various authors [1–3] have studied the relationship between the properties of the detection and recording systems and the distortion of the chromatographic record.

This chapter begins with a general discussion of detectors since they are integral to both qualitative and quantitative analysis.

9.1.1 Detector specifications

The operation and applicability of different detectors can be compared against several performance criteria which were introduced in Chapter 3. Although there appears to be a consensus as to the qualitative definition of these terms, the same is not true when it comes to defining them mathematically. Care should therefore be exercised when comparing data from different sources as to their exact meanings.

Sensitivity

The signal produced by a detector is its most important characteristic. Sensitivity is the basic term used to describe any detector. Despite its frequent use, there is considerable nonuniformity in its definition. The sensitivity is a measure of the magnitude of the signal generated by the detector for a given amount of analyte. Since the signal can have a variety of units (volts, amps, coulombs, absorbance units, etc.) the sensitivity of a concentration-type detector is expressed as signal/concentration in units of, for example, mV/(concentration) or $mV \ ml \ g^{-1}$ and that of a mass flow-rate-type detector as $mV \ s \ g^{-1}$.

Ideally, sensitivity should be a constant independent of analyte concentration in which case a plot of detector signal as a function of analyte concentration is linear. However, in practice, sensitivity falls off at some value dependent on the particular detector and linearity is a measure of the extent to which detector response is directly proportional to sample size. The linear dynamic range of a detector is defined as the range of sample amount over which the response of the detector is constant to within 5%. It is usually expressed as the ratio of the maximum to minimum sample size for which the detector response is linear. The linear range can be obtained from a graph of signal versus concentration but such graphs are usually presented on a log–log basis because the range of analyte sizes usually extends over several orders of magnitude. Such graphs have a tendency to obscure some of the deviations from linearity and give an overly optimistic value of the linear dynamic range. A better presentation for obtaining this value is a graph of sensitivity versus analyte concentration (Fig. 9.1).

The linear dynamic range is the range in which the instrument operates with its greatest precision. The measure of precision is the standard deviation of a series of

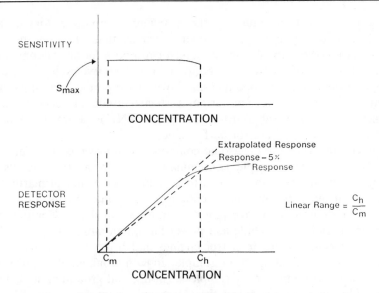

Fig. 9.1. The determination of the linear dynamic range of a detector using two approaches. C_m is the minimum detectable concentration; S_{max} is the maximum sensitivity and C_h is defined from the figure.

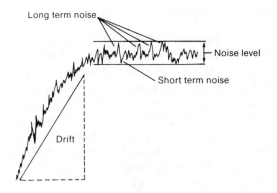

Fig. 9.2. Methods for calculating short- and long-term noise and drift of a chromatographic detector.

detector responses to a set of identical samples. It is essential for accurate quantification that a detector has high precision over the duration of an analysis.

Noise

Noise is the random perturbation in signal produced by a detector in the absence of any sample. Noise must be kept to a minimum because it restricts the minimum signal that can be detected. Figure 9.2 illustrates the typical appearance of noise on a physical measurement. The plot distinguishes between short-term and long-term noise and drift. Short-term noise is the maximum amplitude for all random variations of the detector signal of a frequency greater than one cycle per minute, whereas long-term noise is the maximum detector response for all random variations of the detector signal of frequencies between 6 and 60 cycles h^{-1}. Drift is the average slope of the noise envelope measured over a 1-h period. It is apparent from Fig. 9.2 that noise becomes more important as its magnitude approaches that of the analyte signal. Thus, the signal to noise ratio (S/N) is a much more useful characteristic for describing detector performance.

Noise is associated with the electronic components from which the detector and instrument is made, from sample contamination and from stray signals in the environment. Environmental noise arises from many sources but primarily because each conductor in an instrument is potentially an antenna capable of picking up electromagnetic radiation and converting it to an electrical signal. Common sources of electromagnetic radiation include a.c. power lines, lightning, radio and television transmissions, ignition systems in petrol engines and brushes in electrical motors. Nevertheless, noise originating from sample and environmental sources can be eliminated or at least minimized by sample clean-up, and grounding and shielding, respectively. However, noise from the electronic components is a result of fundamental, intrinsic properties of the system and cannot be eliminated, although it can be minimized by proper circuit design.

Minimum detectable quantity

The minimum detectable quantity (MDQ) or detection limit is the minimum quantity of sample for which the detector will give a visible response. The latter is usually defined as two or three times the noise signal (i.e., S/N = 2 or 3). Commonly, the MDQ is expressed in concentration units for a concentration-type detector and in g s^{-1} for a mass flow-rate detector. Other terms used more or less interchangeably with MDQ are the limit of detection, minimum detectability and detectivity. More recently sensitivity has been used incorrectly in place of the MDQ. The limit of determination is usually defined as the amount of analyte that gives a peak with a height 10 times the background noise level. It represents the smallest peak that can be confidently quantified with accuracy.

Time constant and response time

The detector signal in chromatography reflects fluctuations in analyte concentration as a function of time. An important factor in evaluating a detector is therefore its time behaviour. The time constant and response time are measures of the speed of response of a detector. They are defined as the time (usually in milliseconds) a detector takes to respond to 63.2% and 98%, respectively, of true value following a sudden change of signal. A large time constant will reduce short-term noise (called damping) but increases the retention time and chromatographic peak width. Response time in modern systems should not be a problem but older equipment designed for broad peaks from packed columns may deliberately dampen the signal to reduce noise. In this case, the equipment may not be able to respond to the sharp fast-eluting peaks from high-performance systems.

Selectivity of response

A detector is considered to be selective if its response to a certain type of compound differs markedly from that to another type of compound; the usual meaning attached to markedly is that response differs by a factor higher than 10 [4]. A detector that does not exhibit selectivity can be termed a universal detector. For purposes of screening a sample of unknown composition, a universal detector has definite advantages, whereas a selective detector may aid in the identification of an unknown. Selective detectors are particularly useful for the analysis of complex mixtures, where the selectivity may greatly simplify the chromatogram. It is well to remember that the lack of response to a class of compounds does not mean that the compounds will not have some influence on detector performance. For example, the FID does not respond appreciably to carbon disulfide but the presence of large quantities of carbon disulfide (e.g., used as a solvent) in a sample can result in distorted analyte peaks.

9.1.2 Classification of detectors

Detectors can be usefully classified in a number of ways although Novak [5] has criticized the fundamental principles on which these classifications are based. One classification [6, 7] distinguishes concentration-sensitive from mass-sensitive detectors. The response of concentration-type detectors is proportional to the relative concentration (mass of solute per unit volume of carrier gas), while mass-sensitive detectors produce a signal proportional to the absolute mass of solute reaching the detector per unit time. In contrast to concentration detectors it is independent of detector volume. The response of the two detector types to mobile phase flow changes is very different. The signal (peak height) from the mass detector increases proportionately with carrier gas flow-rate, because as the flow-rate increases, so does the number of nanograms of sample inside the detector over the now shorter period it takes for the peak to pass through the detector. However, the area under the peak will remain the same because the peak width (elution time through the detector) will decrease. Contrast this with using the concentration detector. The peak width will decrease with any increase of flow-rate, but if a concentration-type detector is being used, there will not be any compensating increase in signal (peak height) because there is no change in the concentration of sample. Thus, the area of the peak will be reduced with any increase in flow-rate when a concentration detector is being used.

A second classification is that of destructive and nondestructive detectors [8]. With nondestructive detectors, the original chemical form of the analyte persists throughout the detection process. This is an obvious advantage when the analyte is required for further analysis. In contrast, the process of detection with destructive detectors involves an irreversible chemical change in the analyte. One way to utilize a destructive detector, when recovery of analyte is desired, is to split the effluent stream and send only part of it to the detector, collecting the rest for further analysis.

In his book on detectors for LC, Scott [9] classified detectors on the basis of the property detected. Although less useful for gas chromatographic detectors, it does

Table 9.1 Performance criteria for chromatographic detectors.

	Selective/ Universal (S/U)	Specific/ Bulk (S/B)	Concentration/ Mass (C/M)	Destructive/ Nondestructive (D/N)	Linear range	Sensitivity*
FID	U	S	M	D	10^7	10^{-12} g s^{-1}
TCD	U	B	C	N	10^4	10^{-7} g ml^{-1}
ECD	S	B	C	N	10^4	10^{-15} g ml^{-1}
AFID	S	S	M	D	10^3–10^5	N:10^{-14} g s^{-1} P:10^{-15} g s^{-1}
FPD	S	S	M	D	S:$10^{3\dagger}$ P:10^5	10^{-10} g s^{-1} 10^{-12} g s^{-1}
u.v.	S	S	C	N	10^5	10^{-10} g ml^{-1}
RI	U	B	C	N	10^4	10^{-7} g ml^{-1}
EC	S	S	C	N	10^6	10^{-12} g ml^{-1}
Fluorimeter	S	S	C	N	10^3	10^{-11} g ml^{-1}
Conductivity	S	S	C	N	10^4	10^{-8} g ml^{-1}

*Sensitivity to a favourable compound.
†Following linearization of the response.

Detectors are identified (from top to bottom) as flame ionization, thermal conductivity, electron capture, alkali flame ionization, flame photometric, ultraviolet, refractive index and electrochemical.

have some merits. According to this classification, bulk property detectors measure a property exhibited by both the mobile phase and analyte (e.g. refractive index) and are inherently insensitive (the ECD is an exception) because they only measure changes in a given property. In contrast, a specific property detector produces little or no signal when there is no analyte present but produces a relatively large signal (compared with zero) when the analyte appears.

The most appropriate detector for a given problem will be determined by a consideration of the characteristics discussed above (see Table 9.1) in relation to the needs of the particular analysis. For instance, a detector with a wide linear dynamic range and low detection limit will be adopted for the determination of trace components in addition to main components in a sample. Conversely, the use of a selective detector is convenient if the trace components belong to a particular class of substance or possess some common functional group.

9.2 Qualitative Analysis

Chromatography is primarily a separation process and, as such, the best presently available; it does not provide any structural information about the components of a sample. Indeed, chromatography cannot provide any information on the identity of a totally unknown sample. Some information on polarity and the presence of specific elements may be inferred but otherwise, at the conclusion of a separation, the only data are retention times which are of very little use until you know what to compare them with.

Errors in assigning peak identity can be costly. Consider the case of the woman wrongly imprisoned for the murder of her child because of a mistaken peak identity. The woman's daughter died unexpectedly and blood analysis revealed what was identified as ethylene glycol in the child's blood. The mother was tried and convicted of murder. The child was subsequently identified as suffering from a rare genetic disorder; the peak was a normal metabolite produced in this disorder, and which co-eluted with ethylene glycol. The woman spent two years in jail as a result of this error.

Virtually every conceivable technique has been examined in an attempt to improve the ability of chromatography to identify solutes. Many of the techniques are specific for one sample type or chromatographic procedure. Techniques for enhancing the qualitative content of chromatographic data may be divided into two types as: procedures designed to improve the separation and hence assist in identification, and procedures designed to assist directly in the identification.

9.2.1 Retention parameters

The most common method of peak identification is the correspondence between the retention time of an unknown and a standard. It is important to remember that the retention time is characteristic for a substance but not uniquely so; it is quite possible for several compounds to have the same retention time under identical chromatographic conditions. Nonetheless, the chromatographer usually has some knowledge of the sample and what it is likely to contain and, under these circumstances, it is possible to assign peak identities using retention data. However, this method is only applicable when the chromatogram contains relatively few peaks and there is additional supportive information based on the sample type etc. A retention time is only a single piece of information and coincidence cannot be ruled out. Obviously, the purpose of the analysis should be considered in determining the certainty with which peak identities must be assigned.

One method of increasing the probability of a correct identification is to repeat the analysis on a second column with a stationary phase of markedly different polarity. If the retention of the standard and unknown coincide on both columns, then the presumption of identity is much more certain. This procedure is based on information theory which requires that the two pieces of information are obtained by fundamentally different processes. Many early suggestions that the probability of a correct identification is increased to in excess of 95% are probably misplaced because the physicochemical processes involved in obtaining the two retention times are too similar, despite the differences in phase polarity.

There is always some error in measurement of retention data. To overcome this uncertainty, spiking can be used to assist in confirming the identity of a peak. When the identity of a peak has been assigned, a small amount of the substance is added to the sample and the separation is repeated. If the identification was correct, the peak simply increases in size. If it was incorrect, then a new peak, most likely only partially resolved from the analyte (recall that the retention times were very close when injected separately), appears in the chromatogram. Note that this procedure cannot prove the identity of a peak, although it can provide evidence of an incorrect identification.

Relative retention

Retention time (or volume) is a characteristic parameter for qualitative analysis, but it depends on several experimental variables that must be closely controlled to make comparisons useful, except when samples and standards are run in close proximity on the same system. However, most of these variables affect the phase ratio only and thus the distribution constants, but selectivity should be unaltered. It is likely that a relative retention that is ratioed to a standard would be more reproducible, since the adjusted retention times of the sample and standard would both have been affected proportionally by any differences in the phase ratio. The separation or selectivity factor defined in Chapter 2 is a relative retention. The selectivity factor will vary only with the temperature and the stationary phase, but should otherwise be independent of other experimental variables. Provided the standard, temperature and stationary phase are quoted, selectivity factor is a simple and useful method of standardizing gas chromatographic retention data to facilitate day-to-day comparisons within a laboratory.

Retention data for the standard and sample can be collected in separate injections. If a standard can be chosen so that it does not co-elute with other sample components, then it can be added as an internal standard to the sample before injection. This will also aid accurate quantification (see Section 9.3). The selection of a suitable standard is very important and can be time consuming. The ideal standard will be closely related structurally to the analyte(s) of interest so that it will respond in a similar manner to any changes in the separation conditions. For this reason, structural isomers or homologues of the analyte are usually chosen. The retention time of the standard should be similar to that of the analyte (usually longer) in order to minimize errors associated with taking a ratio of numbers differing greatly in magnitude. Care must be taken to ensure that the chosen standard will not occur naturally in the sample. In instances where there are several analytes with a range of retention times, it may become necessary to select more than one standard.

Although recording retention data as relative retention time overcomes many of the problems associated with small changes in experimental conditions, it does not entirely solve the problem. Within a single laboratory, where the range of interests is relatively small, it may be possible to choose a single reference standard. However, it is desirable to have available a databank of retention values that can be used to identify unknown samples. This clearly requires the use of a single reference standard, which is not possible with the multiplicity of interests of different laboratories. The retention index was proposed to solve this dilemma. The most useful system of retention indices was developed by Kovats (see Chapter 3) for use in GC.

Retention indices for gas chromatography

Kovats proposed the use of n-alkanes, which are readily available and chemically stable as a series of universal standard compounds that could be used to define a retention index scale. Alternative sets of standards have been proposed [10] but none has gained wide acceptance. For example, n-bromoalkanes have been proposed for use with the electron capture detector because of the poor response of n-alkanes. The retention indices of the n-alkanes on any phase are defined as the

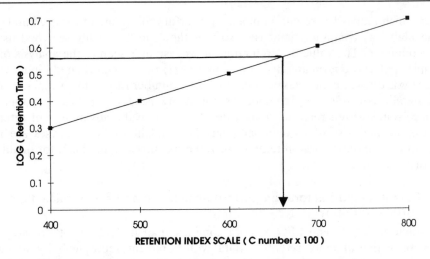

Fig. 9.3. Graphical determination of the Kovats retention index. The logarithm of the bracketing *n*-alkanes are plotted as functions of their retention indices. The retention index of the analyte can be determined from the graph as shown.

number of carbon atoms $\times 100$ (i.e., retention index for butane is 400). The retention indices (I) of analytes are determined by interpolation and are calculated using the equation:

$$I = n \times 100 + 100 \times \left[\frac{\log t_{r.u} - \log t_{r.n}}{\log t_{r.n+1} - \log t_{r.n}} \right] \qquad 9.1$$

where $t_{r.u}$, $t_{r.n}$ and $t_{r.n + 1}$, respectively, are the adjusted retention times of the analyte, and of the *n*-alkanes with n and $n + 1$ carbons, which are eluted immediately before and after the analyte. The data are obtained under isothermal conditions. Alternatively, the value of I can be determined graphically by interpolation, as illustrated in Fig. 9.3.

Programmed temperature operation is very common with complex mixtures analysed on open tubular columns. Under these conditions, the log terms in the Kovats equation can be replaced with the adjusted retention times to give an approximately linear index scale. However, retention index values obtained with programmed temperature operation are best used in conjunction with analyses in which the programmed conditions are exactly duplicated.

Retention indices vary slightly with temperature and markedly with stationary phase but are otherwise independent of conditions. Kovats indicated that with packed columns, a reproducibility of 2 units was within experimental error. With high-resolution columns errors can be held [11] to less than 0.02%. However, a reproducibility of 2–5 units is probably more realistic in routine operation. These small differences can be attributed to a number of causes, such as changes in the properties of the phase as a result of oxidation by traces of oxygen in the carrier gas. Active sites on the column wall or support will also contribute to variation, particularly for polar compounds. There is an extensive literature of published values which makes the retention index a powerful tool. Some of these compilations are available as computer databases, so that they can be readily searched using a microcomputer. The value of a retention index system is seen in the analysis of, for example, polycyclic aromatic hydrocarbons (PAH) or polychlorinated biphenyls. In the analysis of such complex mixtures, the large number of possible isomers and the unavailability of reference compounds for many isomers, makes availability of reliable, standardized retention data essential. Despite the advantages of retention indices, the majority of chromatographers involved in routine analyses are more likely to use simple relative retentions or retention times as the basis for qualitative analysis.

Retention indices for high-performance liquid and supercritical fluid chromatography

The situation in LC and SFC is more complex than in GC, and there are many problems with obtaining retention data with sufficient reproducibility to enable comparisons between laboratories. Hence, despite the plethora of published data on capacity factors for HPLC, they are of little use in identifications other than to suggest possible separation conditions. This is, in part, due to the dependence of retention on the exact composition of the mobile phase, which is often difficult to reproduce exactly [12]. The measurement of capacity factor is a further source of inconsistency. The calculation of the capacity factor requires measurement of the void volume. In theory, this is relatively straightforward as it simply requires the measurement of the retention time of an unretained solute but, in practice, different methods are used for the measurement which frequently give different values on a single system. The column is a major source of variation. There are large differences in the chromatographic behaviour (retention and selectivity) of different brands of nominally equivalent column packing materials (Table 9.2) with smaller batch-to-batch variations.

Many of the problems related to measurement of retention data can be overcome by measuring relative retentions as in GC. With reversed-phase systems, retention indices compared with a homologous series of standards are also used. A number of alternative series of compounds have been described but alkan-2-ones and alkyl aryl ketones are most popular. The interest in developing and using a retention index system for LC and SFC, however, is much less than in GC, probably because the modes of analysis are much more variable and complex so that the data are not as widely usable.

9.2.2   Chemical methods

Chemical methods or reaction chromatography can be performed either pre-column, on-column or post-column. In the pre-column category, are pyrolysis GC (see Chapter 3) and many derivatizations. Derivatization is usually performed to improve the chromatographic behaviour of a solute but, in some cases, the derivatives may provide qualitative information. An example is the use of deuterated reagents to form derivatives that can be easily distinguished by their higher molecular mass when analysed by GC–MS or LC–MS. Peak shift techniques based on pre-column

Table 9.2   Comparison of chromatographic behaviour of a test mixture containing dimethyl phthalate (peak 1), di-*n*-butyl phthalate (peak 2) and pyrene (peak 3) on different chemically bonded ODS-silica packings using a methanol-water (90:10) mobile phase [12, p. 278].

Commercial name	Carbon load (%)	Relative retention		
		Peak 1	Peak 2	Peak 3
Partisil 10 ODS	5	1.0	8	22
Hypersil ODS	9	1.4	13	26
Spherisorb ODS	7	2.0	13	28
Partisil 10 ODS 3	10	0.9	16	31
μBondapak C_{18}	10	13	31	45
Zorbax ODS	15	1.9	16	50
Spherisorb S5 ODS 2	10	2.0	20	49
LiChrosorb RP18	–	1.5	18	65
Partisil 10 ODS 2	15	0.7	25	66
Nucleosil 5 C_{18}	–	1.5	17	50

derivatization can be used to aid in the identification of analytes with reactive functional groups. The formation of a derivative invariably alters the chromatographic properties of a compound, resulting in a characteristic shift in its position in the chromatogram. Abstraction is also a pre-column technique in which the sample is chemically modified to eliminate the analyte. Subtraction techniques usually involve passing the sample through a trap or pre-column containing a reagent that will react with specific functional groups to form insoluble derivatives. Reaction loops for removing alcohols, carboxylic acids, aldehydes, ketones and epoxides have been described [13]. A detailed listing of subtraction reagents for use in GC is available [14]. The absence of the peak in the chromatogram is then used to assist in confirming identity.

Unfortunately, most chemical techniques have now been relegated to a position of historical interest only, as a result of the development and greater access to coupled techniques. For the most part, however, many chemical tests are simple to perform and provide elegant solutions to otherwise complex problems. Of course, derivatization techniques still form an important and integral part of chromatography for purposes other than compound identification. For example, post-column derivatizations are used widely in HPLC for analyte detection and quantification (see Section 5.7).

9.2.3   Peak trapping and off-line spectrometric methods

Prior to the advent of coupled techniques, trapping followed by off-line spectrometry was the most viable method of improving qualitative aspects of chromatography. In GC a common arrangement is for a stream splitter to be attached to the outlet end of the column. A small portion of gas flow is allowed to pass to the detector, while the major part is directed to a series of traps which are designed to retain the separated components. The most obvious approach is to condense the sample components in a cold trap and simple U-tubes, melting point tubes, coils and combinations of bulbs and coils surrounded by a coolant (ice mixtures, solid carbon dioxide, liquid nitrogen) have all been used. Careful design of the trap is important to avoid sample loss owing to mist formation. When a hot gas stream containing a condensable component is cooled rapidly, the molecules tend to agglomerate to form an aerosol rather than liquefy on the walls of the containing vessel. In addition to trap design, the use of a coolant of moderately low temperature minimizes the problem by providing less rapid cooling. Most efficient trapping is achieved with a packed trap. Conventional chromatographic packings are ideal for this purpose.

Fraction collection has also been used in LC, where the nature of the mobile phase means that recovery of separated components is easier and more convenient. The column effluent is collected in a fraction collector which, in the simplest case, may be a series of micro test tubes.

9.2.4 On-line instrumental techniques

On-line instrumental techniques for peak identification are focused on the detector. Detectors are available for both GC and LC (and hence SFC) that show a selectivity for particular groups of compounds or functional groups. Such detectors are unable to identify a completely unknown substance but can be used either alone or in combination with a second detector to assist in confirming the identity of a peak. Dual detection can be performed with the detectors connected in series or in parallel, with the column effluent being split and run through both of them simultaneously. One of the detectors is usually chosen for its selectivity, the other being a universal detector. An example is the analysis of petrol using dual detection with an FID and an ECD. The hydrocarbons are detected by the FID but the alkyl-lead additives are selectively detected by the ECD without interference from the hydrocarbons.

A similar arrangement in LC (and SFC) is the simultaneous, parallel detection at two wavelengths with a u.v./visible light detector. The two signals are ratioed against each other and by comparing the ratio at different points of the peak (on the up-slope, at the peak maximum, and on the down slope) it is possible to draw conclusions about the purity or integrity of the peak. Differences in the ratio between the front and tailing edges of a peak would indicate the presence of an unresolved impurity. Alternatively, the ratio taken at the peak maximum can be used as a characteristic property for peak identification. The availability of multiple-wavelength and photodiode-array detectors makes analyte recognition possible in LC by increasing the number of independent informational degrees of freedom. These detectors simultaneously measure the full absorption spectrum in less than 0.1 s, thus providing a continuous measurement of the full spectral region of the eluent with time. The chromatogram can be displayed in a variety of representations; for example, absorbance at a specific wavelength or at a summed group of wavelengths, or as a three-dimensional plot showing changes in the full spectra against time.

An example of an unusual but very specific detector is the live moth used with a gas chromatograph to detect the presence of pheromones in the column effluent.

9.2.5   Coupled techniques

Mass and infrared spectra provide sufficient information to go a long way towards elucidating the structure of a completely unknown pure compound. Conversely, a mixture of two or more compounds is likely to cause great difficulty in either of these techniques. Chromatography is able to assist in this respect and the combination of chromatography with spectrometric techniques is an obvious goal. The combination of chromatography (an excellent separation method) with MS or FTIR (excellent identification methods) represent the best methods of qualitative analysis. The development of 'coupled' or 'hyphenated' techniques has depended on advances in interfacing, scanning speed of the measuring techniques, and adequate data systems. All three chromatographic techniques, GC, LC and SFC, have been combined with MS and with FTIR. Not all have been equally developed and certainly none has reached the reliability of GC–MS. Details of these techniques are given in chapters on specific techniques. Further developments and combinations can be expected in the future, based on nuclear magnetic resonance, polarography, Raman spectrometry, and fluorescence using laser excitation sources.

9.3   Quantitative Analysis

Quantitative analysis involves the whole process of analysis (Fig. 9.4), from the collection of the sample to the interpretation of the final results; in this Fig., the analytical experiment, includes sampling, sample handling (Chapter 8), chromatographic separation and detection. The basis for all quantitative work is the fact that over the linear response range of the detector, the area underneath a chromatographic peak is directly proportional to the amount of substance giving rise to the peak. This is so, independently of the shape of the peak. However, the response of different compounds will differ depending on their structure and the selectivity of the detector. Thus, after the detector has generated its signal, the chromatographer has three important considerations, namely: data acquisition, that is, collection and storage of the data; data handling, that is, conversion of the detector signal into a number representing an area measurement; and calibration, that is, conversion of that area number to a concentration of analyte in the original sample.

Sources of error are apparent at the various stages of analysis and the entire procedure should be designed to minimize systematic and random error as much as possible. In addition to the problems identified in the introduction to Chapter 8, the very act of measuring the peak size can contribute significantly to the overall error. Interlaboratory comparisons provide a good guide to laboratory and method

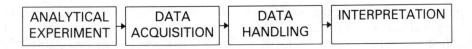

Fig. 9.4.   Schematic outline of the sequence of steps involved in an analysis.

performance. In one typical study [15], two samples were sent to 78 laboratories for analysis by reversed-phase HPLC. Analysis of the results showed that: data from five laboratories were statistically unacceptable; 5% of the 700 results that were reported were in error; the relative standard deviation of results from the acceptable laboratories ranged from 3.1% to 4.6%; and a large plate number does not necessarily produce the best quantitative results. The findings of such studies are sobering and should act as a caution to chromatographers.

Peak area versus peak height measurement

In older literature, the question was frequently asked as to whether peak area or peak height should be used for quantification. This question is still raised occasionally [16] and warrants brief discussion. Fundamentally, peak area is the correct parameter that relates to concentration, but there are problems associated with its measurement in certain situations that make peak height a better estimate of the concentration. However, there is no simple answer to the question, as the accuracy and precision of peak height and area measurements depend on many factors. A 1981 study [15] using LC established that peak height measurements were more reliable, although the committee could not resist stating that 'in the hands of competent chromatographers the choice of peak height vs peak area measurements may be a stand-off'. A later study [17] found that the choice depends on the quality of the chromatographic separation; for good separations, peak areas were preferred, but for poor separations with overlapping peaks, peak heights were better. A newer report [18] concluded that peak area was indeed better although, in fairness, it should be noted that the mixtures concerned were easier to separate and less widely varying in concentration than in the previous studies. With manual methods of measurement, peak area measurements are more likely to be in error than peak height. However, using modern high-performance systems with precise mobile phase flow-rates and computer data handling, area measurements should now be as accurate as height measurements. If the mobile phase flow-rate is not constant, the concentration-type detectors should show greater errors in area measurements, and mass-flow-rate-type detectors should show greater errors in peak height measurements. Halasz and Vogtel [19] have demonstrated that as far as mobile phase flow-rate variations are concerned, it is the average short-term variations that occur while the peak is in the detector which affect quantitative precision. In summary, peak area is the generally preferred measurement especially if there are any changes in chromatographic conditions (e.g., temperature, method of sample introduction) that are likely to change peak height or width (but not area).

9.3.1   Data acquisition and handling

In practically all instrumental analysis methods, the detector converts physicochemical information into an electrical signal which must be conditioned (converted, amplified, scaled) to give an analogue voltage amenable to further processing. The analogue signal is proportional to the instantaneous quantity of analyte eluting through the detector sensing volume. The graphical display of detector output as a function of time is what we have called the chromatogram. It has traditionally been recorded on a potentiometric strip chart recorder, although the increasing demands of qualitative and quantitative analysis have resulted in the widespread use of more sophisticated data-acquisition devices incorporating microprocessor technology.

Manual methods of peak area measurement

The conditioned output signal can be registered by a potentiometric recorder. However, serious errors can arise from a number of sources and the chromatographer must be aware of potential problems. Most significant of these is the dead band of the recorder. This is caused by the fact that a certain minimum input voltage is required before the pen will start moving. In practice the dead band is often higher than the specified value (typically 0.25% of full scale deflection, f.s.d.) simply because the recorder amplifier is not tuned as it should be. Errors caused by the dead band are more significant for smaller peaks. For example, a 1% dead band produces a 2% error in peak area for a full scale peak while the error is 10% for a peak with a height of 20% of f.s.d. Other sources of error frequently encountered with recorders include irregular paper transport and the tolerance of the pen. Despite their shortcomings, including the need for tedious manual processing of the data, strip chart recorders provide excellent visual information on the chromatogram. Moreover, they can provide a continuous output of the detector response.

Manual methods for calculating peak areas all depart from the recorded trace. These methods include planimetry, trapezoidal approximation, triangulation, the product of peak height and width at half-height, and cut and weigh [20]. Most of these are of historical interest only, having been made obsolete by electronic digital integration. Nevertheless, peak height × width at half-height and cut and weigh techniques are still used in some situations (e.g., some teaching laboratories, smaller industrial laboratories) and will be discussed briefly.

The most commonly used method for measuring peak areas manually involves multiplying the actual height times the width at half height. The major factor affecting the precision of this method is the accuracy of measuring peak width, particularly of narrow peaks. It is unreliable for measuring asymmetric and fused peaks. Operational procedures are quite detailed and must be rigidly adhered to in order to minimize errors. This can be appreciated by considering the following extract [21] from an earlier text 'The recorder trace is of appreciable width, and the first point to note therefore is that measurements must take account of this; for example, in measuring the peak width, the measurement should be made from the same side of the two limbs of the peak. Equally, it must be recognized that the marks on a rule are also of finite width, and parallax errors in measuring must be avoided. An interpolation of the baseline, by the drawing of a fine line, must be made for the measurement of peak height, and the rule must be placed parallel to the baseline at exactly the halfway point for measurement of peak width if the method of estimating peak areas by the quantity peak height times peak width at half height is used. Each of these operations has its own error which is unlikely to be less than 0.1 mm.'

Cutting and weighing makes no assumptions about the shape of the peak and can be used to determine the area of asymmetric peaks. The precision of the cut and weigh method depends on the accuracy of the cutting operation and the homogeneity, moisture content and weight of the paper. Better accuracy and precision are usually obtained with a Xerox copy of the chromatogram rather than the actual one on chart paper, apparently because of the greater homogeneity and mass of the Xerox paper. This also preserves the original chromatogram.

Problems with manual methods include the difficulty of defining peak boundaries. Manual measurements of peak area are probably limited to an accuracy of 2%, which puts a rather high limit on the precision attainable. Janak [22] studied the accuracy of several methods useful in the measurement of the area of what were then small peaks (0.25 to 6 cm^2). Some authors have recommended [23] the use of fast recorder chart speeds to increase the peak area of small peaks, thus decreasing the relative error in the measurements. The use of a magnifier with a scale graduated to 0.1 mm is also recommended, particularly for the measurement of peak width. In addition to having limited precision, manual methods are very time consuming. Thus, from an initial investment standpoint, manual methods are cheap, but from an operational standpoint they are quite expensive in terms of time, poor precision and accuracy. The entire December 1967 issue of the *Journal of Gas Chromatography* was devoted to the subject of quantitative analysis and contains valuable discussions relevant to these older manual methods. Detailed theoretical and practical comparisons of the manual methods and errors in these techniques are given in a paper by Ball *et al.* [24]. The problems of accurate measurement are compounded with modern high-performance systems (which produce narrow peaks of very small area) because of the decrease in accuracy of measurements as the peak area decreases. Under these circumstances the use of manual methods cannot be recommended.

Computer-based methods of peak area measurement

Manual methods of integration are tedious, time-consuming and relatively imprecise. Therefore, methods of automating the process have been sought since the advent of instrumental chromatography. Development in this respect has proceeded from mechanical ball-and-disk integrators to digital electronic integrators to computer systems. The microprocessor revolution has now reduced the price of computing integrators and microcomputer systems to the point where it is more cost effective to purchase one or other than to use analysts' valuable time on manual integration. Computing integrators and microcomputers work in essentially the same way and both can be networked to other systems. Computer techniques may be distinguished from other integration methods essentially by the use of computer memory for data storage and this has important consequences as detailed below.

Computerization of chromatographic instruments has progressed to the point that the chromatograph is designed around the computer. The operator interacts through the computer keyboard with the instrument which, in turn, can prompt the operator to enter and check all variables. In an advanced system, the computer performs the following functions: instrument control including sample injection, optimization of separation conditions, data acquisition, data reduction and manipulation, and generation of the final report.

The principles involved in data acquisition and manipulation are treated in this section. Data processing with a microprocessor can be performed 'on the fly' (i.e., in real time) or by storing the complete raw data in memory for analysis at a later time. Using the latter approach, data may be manipulated as many times as desired. In either case, the output data can be displayed directly on a monitor and/or printer. The ability to replot and manipulate data post-run is the major advantage of a computer-based system. Various possibilities for post-run manipulation include the ability to re-plot the chromatogram in order to: (i) increase or decrease the size of peaks and thereby to increase or decrease the sensitivity of the original plot, (ii) alter the time scale in order to expand or condense the chromatographic information, or (iii) alter the parameters used in peak identification and integration.

Certain features of data processing such as threshold, sensing, baseline correction and signal averaging are common to all computational methods. The analogue (continuously variable) voltage signal from chromatographic detectors is unsuitable for computer processing. The conversion of the detector signal to a computer-readable form requires an interface that converts the analogue signal into the appropriate digital form. Digitization implies that the analogue signal is sampled at finite time intervals and rounded to the nearest discrete value, reducing precision to about half the least-significant digit unit. This precision will be unaltered in subsequent manipulation provided only digital data handling is employed.

During data accumulation, a computer reads the detector signal at a certain sampling rate. The resolution of the chromatogram held by the computer (i.e., the ability of the computer accurately to measure narrow peaks such as those found in open tubular column operation) depends on this sampling rate as illustrated in Fig. 9.5. The slow sampling rate in Fig. 9.5(a) results in a chromatogram with distinct steps. The chromatogram in Fig. 9.5(b) was recorded using a much faster data accumulation rate and the step intervals are no longer observable. However, this is

(a)

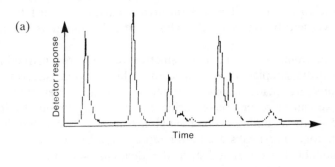

(b)

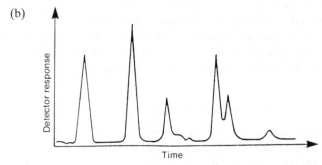

Fig. 9.5. Chromatograms illustrating the effect of sampling rate. Chromatogram (a) was recorded at a sampling rate of two readings per second and (b) at 25 readings per second.

achieved at the cost of increased data storage space. Fortunately, microcomputer systems are now available with memories that can store megabytes of information in the working (random access) memory with additional storage in associated disk systems. A typical sampling rate on a modern system would be 25 times per second, but older instruments were much slower and are therefore not suitable for high-performance systems.

During the course of an isocratic or isothermal analysis the sampling rate could be decreased in order to apply constant peak detection criteria to the broader later eluting peaks. However, modern data systems apply constant sampling frequency and group or bunch the raw data into wider time intervals to produce a smaller number of points. For data bunching it is essential that the data points are summed rather than averaged. In this way, all rounding-off operations are delayed, thus increasing the signal-to-noise ratio. Data bunching saves storage space in the computer memory and improves the ability of the system to recognize the start of a peak.

The first decision to be made from the data is the differentiation between the baseline and a peak. The decision is complicated by the presence of noise and drift on the incoming signal and, although digital smoothing is used to reduce the effects of noise, it cannot be eliminated. In order to recognize the start, maximum and end of each peak, it is necessary to establish a baseline by comparing the slope and rate of change of the slope. This is achieved by comparing the value of each new data point with those of previous points. If the values show a continuing upwards trend that rises above a preset threshold noise level and persists for a predetermined number of points, the program will recognize this as the beginning of a peak. The signal will continue to rise until it passes through a maximum and falls below the threshold noise level and becomes steady. This point is identified as the end of the peak.

The accurate determination of peak parameters requires reconstruction of the original signal from the sampled values collected at discrete time intervals. In other words, assumptions are made about intermediate signal values that were not sampled. The curve can be reconstructed using a curve-fitting routine to fit the data points to a curve or by assuming a polynomial curvature between the data points (Fig. 9.6). Chromatographic peaks are better represented by an asymmetric profile as a result of nonideal behaviour that distorts the ideal Gaussian profile. Models that have been applied to real data include combination Gaussian/triangle/exponential decay, combination Gaussian/hyperbolic/exponential decay, and bi-Gaussian [25]. Based on a number of studies it would appear that the exponentially modified Gaussian model, a convolution of a Gaussian profile with an exponential decay, is the best proposed model to mimic real chromatographic data.

Assigning a peak area—resolved peaks

The value of the peak area is obtained most simply as the sum of the product of the baseline-corrected amplitudes of all number samples between peak beginning and end and the time interval between successive samples. This approach is applied by most integrator and microcomputer systems to assign a value to the peak area. Slow changes in the baseline caused by programming (flow, temperature or mobile phase)

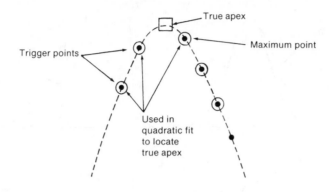

Fig. 9.6. Curve fitting to establish the exact position of the peak maximum.

or similar causes will not affect integration if the slope threshold has been appropriately chosen, as the increase in signal will be insufficient to initiate the start of a peak. Peak identification and integration present few problems with chromatograms containing well-resolved peaks and a stable baseline. This situation is highly desirable for accurate determinations [25] but is not always possible. With poorly resolved peaks, the program must decide what proportion of the total area under the curve to assign to each component.

Assigning a peak area—overlapped peaks

According to Giddings [see ref. 25], a chromatogram must be approximately 90% vacant to provide a 90% probability that a given component of interest will appear as an isolated peak. On this basis, it is not unexpected that the chromatographer will be confronted with chromatograms containing overlapped peaks. Complete quantification in such circumstances is feasible only if the composite peak can be accurately resolved into its constituents. Several methods have been devised for resolving the areas of fused peaks. These are based on either linear separation techniques [26–31] (Fig. 9.7) or mathematical construction of a synthetic chromatogram and use of nonlinear regression analysis to fit this to the composite peak. The most reliable and simple peak area deconvolution methods are the perpendicular drop method and the tangent skim method, even though both are inherently inaccurate. As a result, tangent skim and perpendicular drop are used universally by electronic integrator systems to deconvolute overlapping peaks, although most systems allow manual setting of the integration. However, regardless of the manner in which the peak area is assigned to overlapping peaks there is some degree of arbitrariness and hence inaccuracy. These inherent inaccuracies are much greater than those calculated from ideal Gaussian peak profiles. To date the only viable alternatives are either to accept the inaccuracy or to improve the separation to give baseline resolution of all analytes by altering the chromatographic conditions. In many instances, the latter is either impractical or is not justified because of the increased time and cost.

If there is significant overlap between two peaks, the signal will pass through a

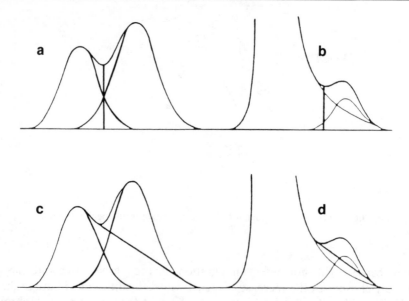

Fig. 9.7. Representation of two methods of peak deconvolution, showing the individual peaks contributing to the overall profile. Perpendicular drop is demonstrated in (a) and (b), and peak skimming in (c) and (d).

minimum value above the threshold level and will then rise again. The system must recognize this minimum as a valley position and use it to define the end of the first peak, and then search for the second peak maximum (Fig. 9.8). After the run is completed, the system will arbitrarily drop a vertical line from each valley point to the baseline and calculate the area (Fig. 9.8). When the resolution of two overlapping tailed peaks of differing magnitude decreases, the perpendicular drop method becomes very inaccurate for the smaller peak; the smaller peak is grossly overestimated whereas the larger peak is underestimated. Under these conditions the tangent skim method is more appropriate (Fig. 9.8). Criteria for deciding which approach to use are not well defined, although a general rule of thumb uses the peak height ratio, and a value of 10:1 has been suggested as the decision point [25]. Very large differences appear to exist among systems from different manufacturers, leading to large differences in peak integration values for the same set of input data [25]. The chromatogram and final report should indicate the method used to assign peak areas with an identifying code. For example, in the case of perpendicular drop, a peak identified as BV indicates that the peak area corresponds to a valley peak from an initial baseline whereas VV indicates that the peak area corresponds to a valley peak from a previous valley.

Report format

At the conclusion of the run, the computer will print out a report typically listing peak heights, areas and retention times, together with any post-run calculations such as relative retention times, relative peak areas and areas as percentages of the total area. Alternatively, if the system has been standardized, concentrations may be

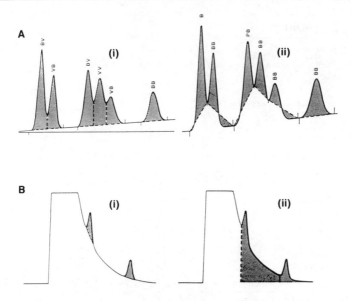

Fig. 9.8. Peak deconvolution techniques. (A) Separation of peaks by the perpendicular drop method showing the (i) correct procedure and (ii) incorrect procedure, in which the baseline has been erroneously assigned because the threshold has been incorrectly set. (B) Separation of peaks by (i) tangent skim method; (ii) erroneous allocation of peak areas using perpendicular drop method.

reported relative to internal or external standard(s). If desired, column parameters such as plate numbers and asymmetry factors may be computed, peak identities allocated and a complete report of the analysis generated (Fig. 9.9).

There is a tendency to uncritically accept the output from a computer system. The speed and sophistication of computers may blind the unwary to their limitations. The precision of microprocessor systems is very high, but this should not be mistaken for accuracy. Computers operate on a series of fixed rules and instructions, using parameters set by the chromatographer or their own default conditions. They are unable to assess the appropriateness of these values for a given chromatogram. Data systems require constant checking by human eye to ensure that the start and end of each peak has been assigned correctly and has not been misled by noise or baseline disturbances, and that any decisions, such as the assignment of the divisions between unresolved peaks, have been made correctly. The trend towards relying solely on the analysed values in the final printout without checking the original chromatogram is to be eschewed. The chromatogram and report should be examined to ensure that the desired approach has been employed.

Integrator parameters

The accuracy and precision of peak integration depends on the correctness of the computer logic decisions. These decisions, however, are determined by parameters that can be varied within fairly wide limits by the operator. Hence, the operator's experience as well as the quality of the chromatogram (e.g., overlapped versus resolved peaks) determine the precision of the final results. An understanding of the

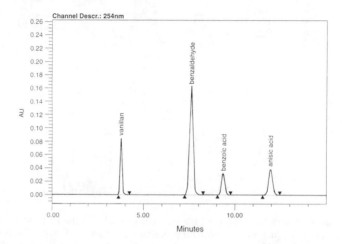

RESULTS

#	Name	Retention Time (min)	Area (uV*sec)	Partition Ratio	Asymmetry @ 4.4	Peak Width 5 Sigma
1	vanillan	3.780	586677	1.70	1.55	5165
2	benzaldehyde	7.565	1990916	4.40	1.26	7379
3	benzoic acid	9.310	414972	5.65	1.17	11235
4	anisic acid	11.913	642439	7.51	1.08	11398

Fig. 9.9 Typical report generated by a computer-based system showing tick marks identifying the start and end of peaks. Reprinted with permission from Waters Chromatography Division of Millipore. Refer also to Fig. 5.14.

integrator parameters and their effect on the integration is therefore essential. These parameters usually include baseline sensitivity, peak width and area sensitivity. Integrator systems have an in-built set of standard or default conditions which may be satisfactory for simple chromatograms but frequently require altering to fit the needs of a particular separation.

Most systems contain a parameter such as peak width which gives the software some indication of the sharpness of peaks at a particular point in the chromatogram. It is this parameter that causes the data to be bunched, producing a smaller number of data points. If the value is too large, sharp peaks will be treated as noise spikes and ignored because insufficient data points will have been tested during the rise of the peak to trigger the start of the integration routine. Conversely, if the value of peak width is too small, broad peaks will be treated as baseline drift and ignored. In addition, noise spikes may be interpreted as signals. The usual indications of an incorrect setting are that the system will miss peaks or will identify only one end of the peak, as recognized from the diagnostic start and stop tick marks which most systems place on the chromatogram. With most systems the value of peak width is upgraded automatically to allow for peak broadening as an isocratic or isothermal run progresses. Alternatively, the value can be upgraded using a timed-event schedule.

The value of baseline sensitivity determines the allowable variation in the baseline signal before a peak is triggered. The threshold value should be set according to the noise level of the signal. A value which is too small will prevent the software from finding a baseline, whereas too large a value will cause parts of some peaks to be treated as baseline. The system can be instructed to ignore peaks below a certain size by setting the peak sensitivity but care must be taken not to exclude legitimate sample components.

9.3.2   Calibration procedures

After the area of individual peaks in the chromatogram has been determined, the next step is to relate these areas to an amount or concentration of the substance. Depending on the analytical problem, the chromatographer may need to establish this for every sample component producing a peak or only for one or a few sample components.

Normalization

Normalization assumes that area per cent is equivalent to concentration per cent. Hence, normalization is achieved by dividing the area of each peak by the total area for all peaks in the chromatogram and multiplying by 100. If a solvent peak is present in the chromatogram, it is common to disregard this peak when totalling the areas and to consider the areas of peaks corresponding to actual sample components only. The method assumes that (i) each sample component gives rise to a peak in the chromatogram and (ii) the detector response is equivalent for all compounds. There are limitations on the use of normalization imposed by both of these assumptions. In relation to the first assumption, there are many instances where a sample component does not produce a peak in the chromatogram because of the separation system (e.g., a non-volatile component which is back-flushed in GC) or detector characteristics (e.g. an alkane and u.v. detector). In this respect, the applicability of normalization is problematic since the absence from an unknown sample of a portion not amenable to chromatography or detection is never certain. Problems relating to the second assumption are easily overcome by using response factors to correct for variations in detector response to individual components. Response factors are determined by preparing and chromatographing standard mixtures of all components of the sample and calculating the individual factors as the ratio of peak area per unit mass. Response factors are less commonly quoted on an equivolume or equimolar basis, but care should be exercised when examining literature data. Usually, these factors are not quoted in absolute terms but rather relative to a given analyte. In subsequent analyses, each peak area is corrected by the appropriate relative response factor before normalization is applied to calculate concentrations.

Normalization is commonly used for analyses that are performed repeatedly on very similar samples, for example, monitoring the composition of a solvent mixture in the paint industry.

External calibration

In this technique standard solutions of pure analyte(s) of varying concentration are prepared and chromatographed. Peak areas are determined from the chromatograms and plotted as a function of the concentration of analyte. The unknown sample is chromatographed and the concentration of analyte determined from the measured peak area and calibration graph. This method requires precise control of the analytical technique, particularly the size of the injected sample which needs to be reproducibly injected from run to run with a high precision. Valve injection provides adequate precision whereas syringe injection is usually inadequate, particularly for syringes that contain sample in the needle. Hence, external calibration is more widely practised in LC than in GC. Reproducibility of injection volume with syringe injection is relatively poor and errors of 10 to 20% are common. Careful control of injection technique can reduce such errors substantially and auto-injectors produce reasonable precision. Under these circumstances, external calibration can be used with syringe injection.

Internal standard

In this method a known amount of a standard substance is added to all samples and standards; hence the name internal standard. The standard chosen for this purpose cannot ever be a component in a sample. Some of the selection criteria for an internal standard were introduced in Section 9.2. The full criteria are:

- The internal standard should resemble the analyte as closely as possible in terms of physical and chemical properties. It must not react with any component of the sample.
- The internal standard must not be a normal constituent of the sample.
- The internal standard should be incorporated into the sample in exactly the same way as the analyte. This ideal can rarely be achieved in practice.
- In general, the analyte and internal standard should elute close together with baseline resolution. There are exceptions to this requirement where the two can be distinguished by the detection system as, for example, with isotopically labelled samples.
- The internal standard and analyte should respond to the detection system in a similar manner and be present in nearly equal concentrations.

If many analytes are to be determined in one sample, several internal standards may be used in order that the preceding criteria are satisfied. Substances used as internal standards include analogues, homologues, isomers, enantiomers and isotopically labelled analogues of the analyte.

After selecting a suitable compound to function as the internal standard, a series of standards is prepared containing a fixed amount of the internal standard plus known but varying amounts of the analyte. The amount of internal standard to be used is selected with respect to the estimated amount of analyte in the sample, such that the heights of internal standard and analyte peaks are about the same. Each standard is chromatographed and the area ratio (area of analyte peak/area of internal standard peak) is calculated and plotted against concentration of analyte. An amount of internal standard sufficient to produce the same concentration as in

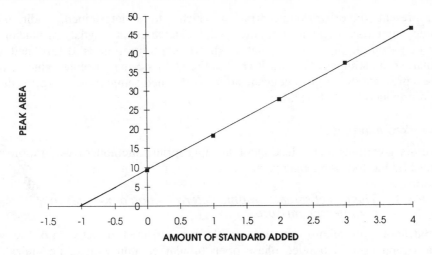

Fig. 9.10. A typical calibration plot illustrating the method of standard addition.

the standards is added to the sample. This addition is performed before any sample manipulation so that losses in the sample-handling procedure can be compensated. This is based on the assumption that the behaviour of analyte and internal standard in any extractions etc. will be similar because of the chemical similarity of the two compounds. The sample with added internal standard is chromatographed, the peak area ratio is determined, and the corresponding concentration is read from the calibration graph. The final concentration of analyte in the sample is reported after allowing for any dilution factors in the sample handling.

The main advantage of the internal standard technique is that it is not necessary to know the volume of the sample injected. However, this advantage has been gained at the expense of adding an extraneous substance to the sample; this operation may be troublesome if for no other reason than the difficulty of finding a suitable substance to function as the internal standard. For example, it is difficult, if not impossible, to find a suitable substance with complex samples containing numerous peaks where there is no free space within the chromatogram.

With both external calibration and internal standardization, concentration may be determined algebraically using a single-point calibration provided that the concentration of standard and sample is within the linear response range.

Standard addition

This technique combines features of both external calibration and internal standardization. It is rarely used and is completely ignored in many texts. Nevertheless, it is a useful technique in some situations as, for example, when working in a nonlinear portion of the detector response. Shatkay [32] has presented a mathematical treatment of the procedure. In this procedure the standard is added to the sample, but the chemical chosen as the standard is the same as the analyte. The additional signal produced by the addition of standard is proportional to the original signal. The original concentration can be calculated either graphically or using an equation. A typical calibration plot is shown in Fig. 9.10. The signal when no standard is added represents the original concentration, which is to be determined. Addition of increasing amounts of standard to the sample increases the signal, producing a straight-line calibration. The original unknown concentration is determined by extrapolating the straight line until it crosses the abscissa; the absolute value of the abscissa represents the original concentration. A detailed summary of the procedure has been provided by Bader [33].

Worked examples

The factors governing the selection of an appropriate method of calibration are illustrated by the following examples.

Case 1.   The gas chromatographic determination of xylenes in a commercial mixture

After filtration, an aliquot of the sample was injected directly into the gas chromatograph using a bonded phase open tubular column with flame ionization detection, producing the results presented in Table 9.3.

For this analysis normalization was an ideal choice as all sample components were eluted and were expected to produce a similar detector response. Using normalization, the composition of the mixture was calculated as in Table 9.4.

The composition of the commercial mixture was reported as 52.7% 1,4-dimethylbenzene, 24.0% 1,3-dimethylbenzene and 23.3% 1,2-dimethylbenzene.

Case 2.   The characterization of a vegetable oil by gas chromatographic determination of fatty acid content

The oil sample was treated with alcoholic potassium hydroxide to hydrolyse triglycerides to glycerol and fatty acids, and to convert the latter to methyl ester derivatives which were then extracted into hexane. The hexane extract was dried over anhydrous sodium sulfate and injected into the gas chromatograph using an

Table 9.3   Analytical data for analysis of xylenes.

Peak	Retention time (min)	Identity	Integrator counts
1	3.86	1,4-dimethylbenzene	24 470
2	4.25	1,3-dimethylbenzene	11 171
3	5.79	1,2-dimethylbenzene	10 816

Table 9.4   Calculation of xylene-isomer concentrations.

Peak	Integrator counts	Percentage composition
1	24 470	(24 470/46 457) × 100% = 52.7%
2	11 171	(11 171/46 457) × 100% = 24.0%
3	10 816	(10 816/46 457) × 100% = 23.3%
Total	46 457	100%

Table 9.5   Peak area data for fatty acids of vegetable oil sample.

Peak	Retention time (min)	Identity	Integrator counts
1	3.83	Lauric acid	10 675
2	4.96	Myristic acid	6788
3	6.22	Palmitic acid	14 574
4	7.61	Stearic acid	26 893
5	8.97	Oleic acid	4691

Table 9.6   Raw data and relative response factors for fatty acid methyl esters.

Peak	Identity	Integrator counts	Relative response factor
1	Lauric acid	15 670	15 670/15 670 = 1.00
2	Myristic acid	16 985	16 985/15 670 = 1.08
3	Palmitic acid	18 211	18 211/15 670 = 1.16
4	Stearic acid	19 483	19 483/15 670 = 1.24
5	Oleic acid	19 364	19 364/15 670 = 1.24

Table 9.7   Fatty acid composition of vegetable oil sample.

Peak	Integrator counts	Compensated area	Percentage composition
1	10 675	10 675/1.00 = 10 675	(10 675/54 995) × 100% = 19.4%
2	6788	6788/1.08 = 6285	(6285/54 995) × 100% = 11.4%
3	14 574	14 574/1.16 = 12 564	(12 564/54 995) × 100% = 22.9%
4	26 893	26 893/1.24 = 21 688	(21 688/54 995) × 100% = 39.4%
5	4691	4691/1.24 = 3783	(3783/54 995) × 100% = 6.9%
Total		54 995	100%

open tubular column and flame ionization detection. The results in Table 9.5 were obtained for the oil sample.

Normalization with response factors was selected for this analysis. A mixture of equal weights of the five fatty acids (as their methyl esters) was prepared and chromatographed. The results, summarized in Table 9.6, were used to calculate relative response factors with lauric acid arbitrarily selected as the standard.

Results for the unknown sample were then calculated from the sample peak area data after allowing for the relative response factors, with the results summarized in Table 9.7.

The fatty acid composition of the vegetable oil was reported as 19.4% lauric acid, 11.4% myristic acid, 22.8% palmitic acid, 39.5% stearic acid and 6.9% oleic acid.

Case 3.   Determination of nicotinic acid in instant coffee

An aqueous extract of instant coffee (1.000 g in 50 ml) was filtered using a 0.45 μm filter and subjected to clean-up on a reversed-phase cartridge column [34]. The eluent was injected by a valve injector (30 μl) onto a reversed-phase column using

Table 9.8 Peak height data for nicotinic acid standards and unknown coffee samples.

Sample	AU
Nicotinic acid standards (mg l^{-1})	
5	0.0109
10	0.0219
15	0.0326
20	0.0437
25	0.0543
30	0.0658
Coffee samples	
1	0.0170
2	0.0209

ion-interaction chromatography for separation with u.v. detection. The results shown in Table 9.8 were obtained for a series of aqueous nicotinic acid standards and two unknown samples of coffee. External calibration was employed because of the excellent precision achieved with a valve injector.

The data for the standards were used to construct a calibration plot, as in Fig. 9.11. From this plot, the nicotinic acid concentration in samples 1 and 2 is 8.0 mg l^{-1} and 9.8 mg l^{-1}, respectively. This corresponds to a nicotinic acid concentration of 40.0 and 49.0 mg per 100 g, dry coffee in samples 1 and 2, respectively.

Case 4.   Determination of bromate in bread dough

A sample of dough (1.00 g) was homogenized with water (100 ml) in a blender [35], filtered through a 0.45 μm filter and diluted with water to yield a final concentration of 1.00 g of dough per 2000 ml water. The final extract was analysed by ion

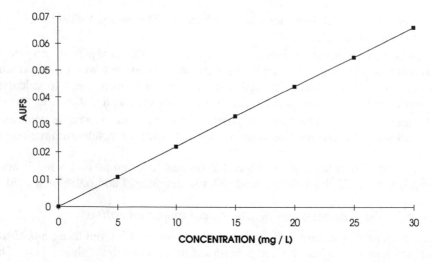

Fig. 9.11.   Calibration plot for the determination of nicotinic acid in instant coffee.

Table 9.9 Detector response data for bromate determination. (S, Siemens.)

Sample	Detector response (mS)
Standard	0.628
Bread dough	0.512

chromatography using a valve injector, anion-exchange column and conductivity detector. An aqueous standard containing 50 mg l^{-1} of bromate was chromatographed using the same set of conditions. Single-point external calibration was chosen for this analysis because of the excellent precision of valve injection and the known linear response range for this analyte (5–120 mg l^{-1}) under the conditions employed. Results of the analysis are shown in Table 9.9.

The concentration of bromate in the bread dough can be calculated using equation 9.2.

$$C_u = A_u \frac{C_{std}}{A_{std}} \qquad\qquad 9.2$$

where C and A represent the concentration and peak area, respectively, and the subscripts denote the unknown sample and standard. Substituting in equation 9.2:

$$C_u = 0.512 \frac{50}{0.628}$$
$$= 40.8 \text{ mg l}^{-1}$$

After allowance is made for the dilution, the bromate concentration in the bread dough is 81.6 mg per gram.

Case 5. The determination of polychlorinated biphenyls in environmental samples

For the purpose of this example, it is convenient to define the terms isomer, homologue and congener. An isomer refers to a compound defined by the numerical arrangement of the chlorine substituent within the moiety of the homologue, e.g. 2,3,3′-trichlorobiphenyl isomer. A homologue represents a group of isomers having a specific number of chlorine atoms, e.g., the trichlorobiphenyl homologue series. The congener means any isomer of any homologue. There is a total of 209 possible polychlorinated biphenyl (PCB) congeners, with significant differences in the toxicity of individual congeners. Analysis of PCBs is complicated by the presence of a large number of congeners in commercial formulations.

A tissue sample was ground with anhydrous sodium sulfate and extracted with hexane. Removal of co-extracted fat was achieved by clean-up using gel permeation chromatography, partitioning between hexane and acetonitrile, and saponification. After a preliminary clean-up, the PCBs were separated from organochlorine pesticides using solid phase extraction on a cartridge column. The final extract was concentrated using a Kuderna–Danish evaporative condenser and injected into the gas chromatograph. A non-polar bonded phase open tubular column and electron capture detection were chosen for the analysis. In all, there were 81 peaks in the chromatogram and, with rigorously controlled conditions, matching of Kovats retention indices provided qualitative identification of 63 peaks.

The response of the ECD is not the same for all PCB congeners so that calibration for the individual congeners is necessary. Multiple internal standards have been used for this purpose, but the selection of suitable internal standards is complicated by the large number of peaks in the chromatogram. For this reason, external calibration is a likely candidate for the determination of individual congeners.

Case 6.   The determination of ethanol in a blood sample

Since ethanol is volatile and thermally stable at typical column temperatures, this determination is conveniently performed by GC. Internal standardization is recommended because of the relatively poor precision of syringe injection in GC, and the simplicity of the chromatogram which facilitates selection of an internal standard. Propan-2-ol was selected as the internal standard because of its similarity to ethanol, its ready availability in a state of high purity, and the fact that it is not a normal constituent of blood. Moreover, an initial chromatogram of the blood sample established that there were no peaks co-eluting with propan-2-ol, and that its retention time was similar to that of ethanol, but with baseline resolution.

A concentration of the internal standard, propan-2-ol was selected to produce a detector response similar to that of the ethanol in the unknown blood, given the anticipated concentration. A series of standard aqueous solutions of ethanol containing this fixed concentration of propan-2-ol was prepared (see Table 9.10). The blood sample (0.10 ml) was treated in the same fashion and samples (1 μl) of each standard and unknown were injected into a gas chromatograph equipped with an FID and a polar bonded phase column at 115°C. The results, summarized in Table 9.10, were used to calculate peak area ratios for standard solutions and the unknown (Table 9.11).

Table 9.10   Peak area data for ethanol determination.

Sample*	Volume ethanol stock (ml; 2.0 mg l^{-1})	Volume propanol stock (ml; 4.0 mg l^{-1})	Ethanol concentration (mg l^{-1})	Integrator counts	
				Ethanol peak	Propanol peak
Standards					
1	1.00	2.00	0.20	1671	8937
2	2.00	2.00	0.40	3387	8843
3	3.00	2.00	0.60	5099	9009
4	4.00	2.00	0.80	6758	8916
5	5.00	2.00	1.00	8456	8882
Blood sample (0.10 ml blood)		2.00	?	6073	8895

*All standards and samples were diluted to a final volume of 10.00 ml.

Table 9.11   Data for calibration curve and unknown blood ethanol sample.

Sample	Ethanol concentration (mg l^{-1})	Area ratio (ethanol peak/propanol peak)
Standard 1	0.20	1671/8937 = 0.187
2	0.40	3387/8843 = 0.383
3	0.60	5099/9009 = 0.566
4	0.80	6758/8916 = 0.758
5	1.00	8456/8882 = 0.952
Blood sample	?	6073/8895 = 0.683

The area ratios were graphed (Fig. 9.12) against the concentrations of the standard ethanol solutions, and the result for the blood sample was read from the graph as 0.721 mg l^{-1}. The concentration of ethanol in the sample of blood after allowance is made for the dilution was 72 mg l^{-1}.

Case 7.   The determination of cocaine

This analysis was conveniently performed by HPLC using a single-point calibration with an internal standard. Although high precision can be obtained with valve injection in HPLC, single-point calibration with an internal standard is rapid and allows for minor changes in response as a result of day-to-day variations in mobile phase composition. The analysis was performed on a fluoroether bonded phase (25 cm × 2.6 mm) using acetonitrile:water (90:10) containing 0.3% phosphoric acid as mobile phase. A u.v. detector at 210 nm was used to monitor column effluent.

An equation for the mass of unknown analyte in a sample can be derived from the

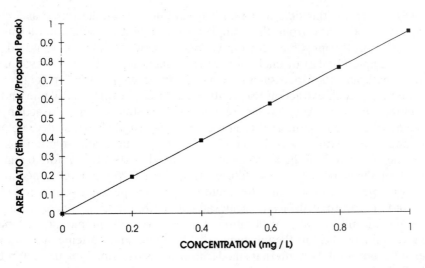

Fig. 9.12.   Calibration graph for the determination of ethanol using propan-2-ol as internal standard.

expression $C_i = f_i A_i$ where f is the detector response factor, C and A are the concentration and area of component i, as equation 9.3

$$C_{a;u} = \frac{C_{a;s}}{C_{is;s}} \cdot \frac{A_{is;s}}{A_{a;s}} \cdot \frac{A_{a;u}}{A_{is;u}} \cdot C_{is;u} \qquad\qquad 9.3$$

In this expression, the subscripts a and is refer to the analyte and internal standard, which are contained in a standard solution (s) or unknown solution (u), respectively.

A standard solution was prepared by diluting 17.6 mg codeine (internal standard) and 16.5 mg cocaine to 100 ml with distilled water. Repeated injections of this solution were made using a valve injector to establish the area of the peaks. Individual peak areas were averaged with the following results: average peak area, cocaine, 936 integrator counts; average peak area, codeine, 1421 integrator counts.

A white powder suspected of containing cocaine was prepared for analysis by weighing a 94.6 mg sample of the powder plus 16.9 mg codeine into a 100 ml volumetric flask. The mixture was diluted to volume and injected into the liquid chromatograph under the same operating conditions as used for the response factor determination. The following peak areas were obtained for the cocaine and codeine peaks in the unknown: cocaine peak area, 924 integrator counts; codeine peak area, 1368 integrator counts. The amount of cocaine present in the sample is calculated from equation 9.3 as:

$$(16.5/17.6) \times (1421/936) \times (924/1368) \times 16.9 = 16.24 \text{ mg}$$

and the concentration of cocaine in the white powder as:

$$(16.24/94.6) \times 100\% = 17.2\%$$

Case 8. The determination of caffeine and theobromine in chocolate products

HPLC was chosen for this determination with an internal standard used as a check on recovery of analytes from the sample. An acetonitrile solution of internal standard (8-chlorotheophylline) corresponding to 5 mg of internal standard was added to a sample (1.00 g) of finely ground chocolate. The solvent was evaporated on a water bath and the dried sample extracted twice with hexane to remove fat. The residue from each extraction was centrifuged at 2000 r.p.m. for 10 min and the supernatant discarded. Any hexane remaining after the second extraction was evaporated on a steam bath. HPLC grade water was added to the dry, defatted residue and the mixture was boiled for 25 min to facilitate solution of the theobromine. A portion of the extract was centrifuged at 2000 r.p.m. for 10 min and an aliquot of the supernatant was filtered through a 0.45 μm filter. The filtered solution was injected directly into the liquid chromatograph using a reversed-phase column and an acetonitrile/acetic acid/water (18:1:81) mobile phase with u.v. detection at 280 nm. Finally, calibration was achieved by preparing a standard solution containing 8-chlorotheophylline, theobromine and caffeine and using a single-point calibration with internal standardization as described for the determination of cocaine.

References

1. Sternberg J.C. (1966). In Giddings J.C., Editor, *Advances in Chromatography, Vol. 2*. Marcel Dekker, New York, p. 205.
2. McWilliam I.G. and Bolton H.C. (1969). *Anal. Chem.*, **41**: 1755 and 1762.
3. McWilliam I.G. and Bolton H.C. (1971). *Anal. Chem.*, **43**: 883.
4. Krejci M. and Dresslre M. (1970). *Chromatogr. Rev.*, **13**: 1.
5. Novak J. (1988). *Quantitative Analysis by Gas Chromatography, Chromatographic Science Library Series, No. 41*, 2nd edn. Marcel Dekker, New York.
6. Halasz I. (1964). *Anal. Chem.*, **36**: 1428.
7. Novak J. (1974). In Giddings J.C., Editor, *Advances in Chromatography, Vol. 11*. Marcel Dekker, New York, 1.
8. Dal Nogare S. and Juvet R.S. (1962). *Gas Liquid Chromatography*. Wiley Interscience, New York, p. 188.
9. Scott R.P.W. (1977). *Liquid Chromatography Detectors*. Elsevier, Amsterdam.
10. Miwa T.K., Micolajczak K.L., Earle F.R. and Wolf I.A. (1960). *Anal. Chem.*, **32**: 1739.
11. Onuska F.I. and Karasek F.W. (1984). *Open Tubular Column Gas Chromatography in Environmental Sciences*. Plenum Press, New York, pp. 153–157.
12. Smith R.M. (1988). *Gas and Liquid Chromatography in Analytical Chemistry*. John Wiley, New York, p. 315.
13. Bierl B.A., Beroza M. and Ashton W.T. (1969). *Mikrochim. Acta*, **3**: 637.
14. Poole C.F. and Schuette S.A. (1984). *Contemporary Practice of Chromatography*. Elsevier, Amsterdam, p. 561.
15. Subcommittee E-19.08 Task Group (1981). *J. Chromatogr. Sci.*, **19**: 338.
16. Kipiniak W. (1981). *J. Chromatogr. Sci.*, **19**: 332.
17. McCoy R.W., Aiken R.L., Pauls R.E., Ziegel E.R., Wolf T., Fritz G.T. and Marmion D.M. (1984). *J. Chromatogr. Sci.*, **22**: 425.
18. Pauls R.E., McCoy R.W., Ziegel E.R., Wolf T., Fritz G.T. and Marmion D.M. (1986). *J. Chromatogr. Sci.*, **24**: 273.
19. Halasz I. and Vogtel P. (1977). *J. Chromatogr.*, **142**: 241.
20. Delaney M.F. (1982). *Analyst*, **107**: 606.
21. Ambrose D. (1971). *Gas Chromatography*. Butterworths, London, 2nd edn., p. 261.
22. Janak J. (1960). *J. Chromatogr.*, **3**: 308.
23. Lichtenfels D.H., Fleck S.A. and Burow F.H. (1955). *Anal. Chem.*, **27**: 1510.
24. Ball D.L., Harris W.E. and Habgood H.W. (1967). *J. Gas Chromatogr.*, **5**: 613.
25. Papas A.N. and Tougas T.P. (1990). *Anal. Chem.*, **62**: 234.
26. Proksch E., Bruneder H. and Granzner V. (1969). *J. Chromatogr. Sci.*, **7**: 473.
27. Novak J., Petrovic K. and Wicar S. (1971). *J. Chromatogr.*, **55**: 221.
28. Kaiser R. and Klier M. (1969). *Chromatographia*, **2**: 559.
29. Westerberg A.W. (1969). *Anal. Chem.*, **41**: 1770.
30. Hock F. (1969). *Chromatographia*, **2**: 334.
31. Anderson A.H., Gibb T.C. and Littlewood A.B. (1970). *Anal. Chem.*, **42**: 434.
32. Shatkay A. (1980). *J. Chromatogr.*, **198**: 7.
33. Bader M. (1980). *J. Chem. Educ.*, **57**: 703.
34. Trugo L.C., Macrae R. and Trugo N.M.F. (1985). *J. Micronutrient Anal.*, **1**: 55.
35. Cox D., Harrison G., Jandik P. and Jones W. (1985). *Food Technol.*, **39**: 41.

Bibliography

Ettre L.S. and McFadden W.H., Editors (1969). *Ancillary Techniques of Gas Chromatography*. Wiley-Interscience, New York.

Kaiser R.E. and Rackstraw A.J. (1983). *Computer Chromatography, Vol. 1*. Huethig, Heidelberg.

Reese C.E. (1980). Chromatographic data acquisition and processing. Part I. Data acquisition. *J. Chromatogr. Sci.*, **18**: 201.

Reese C.E. (1980). Chromatographic data acquisition and processing. Part II. Data manipulation. *J. Chromatogr. Sci.*, **18**: 249.

Qualitative analyses

Budahegyi M.V., Lombosi E.R., Lombosi T.S., Meszaros S.Y., Nyiredy Sz., Tarjan G., Timar I. and Takacs J.M. (1983). Twenty-fifth anniversary of the retention index system in gas liquid chromatography. *J. Chromatogr., Chromatogr. Rev.*, **271**: 213.

Smith R.M. (1987). Retention index scales in liquid chromatography. *Advances in Chromatography, Vol. 26.* Marcel Dekker, New York, pp. 277–320.

Quantitative analyses

Balke S.T. (1984). *Quantitative Column Liquid Chromatography. A Study of Chemometric Methods, Journal of Chromatography Library Series, No. 29.* Elsevier, Amsterdam.

Foley J.P. (1987). Equations for chromatographic peak modeling and calculation of peak area. *Anal. Chem.*, **59**: 1984.

Katz E., Editor (1987). *Quantitative Analysis using Chromatographic Techniques.* John Wiley, Chichester.

Maeder M. (1987). Evolving factor analysis for the resolution of overlapping chromatographic peaks. *Anal. Chem.*, **59**: 527.

Novak J. and Leclercq P.A. (1988). *Quantitative Analysis by Gas Chromatography, 2nd edn., Chromatographic Science Series, Vol. 41.* Marcel Dekker, New York.

Ouchi G.I. (1992). Peak identification and quantitation in chromatography. *LC-GC*, **10**: 524.

Papas A.N. and Tougas T.P. (1990). Accuracy of peak deconvolution algorithms within chromatographic integrators. *Anal. Chem.*, **62**: 234.

Index